WASTE MATERIALS IN CONSTRUCTION

PROCEEDINGS OF THE INTERNATIONAL CONFERENCE ON ENVIRONMENTAL IMPLICATIONS OF CONSTRUCTION WITH WASTE MATERIALS, MAASTRICHT, THE NETHERLANDS, 10 – 14 NOVEMBER 1991

WASCON '91

ENVIRONMENTAL
IMPLICATIONS OF CONSTRUCTION
WITH WASTE MATERIALS

Studies in Environmental Science 48

WASTE MATERIALS IN CONSTRUCTION

PROCEEDINGS OF THE INTERNATIONAL CONFERENCE ON ENVIRONMENTAL IMPLICATIONS OF CONSTRUCTION WITH WASTE MATERIALS, MAASTRICHT, THE NETHERLANDS, 10 – 14 NOVEMBER 1991

Edited by

J.J.J.M. Goumans

Netherlands Agency for Energy and the Environment (NOVEM)
P.O. Box 8242, 3503 RE Utrecht, The Netherlands

H.A. van der Sloot

Netherlands Energy Research Foundation (ECN)
P.O. Box 1, 1755 ZG Petten, The Netherlands

Th. G. Aalbers

National Institute of Public Health and Environmental Protection (RIVM)
P.O. Box 1, 3720 BA Bilthoven, The Netherlands

ELSEVIER
Amsterdam — London — New York — Tokyo 1991

ELSEVIER SCIENCE PUBLISHERS B.V.
Molenwerf 1
P.O. Box 211, 1000 AE Amsterdam, The Netherlands

Distributors for the United States and Canada:

ELSEVIER SCIENCE PUBLISHING COMPANY INC.
655, Avenue of the Americas
New York, NY 10010, USA

ISBN 0-444-89089-0

This book is printed on acid-free paper.

Printed in The Netherlands

Studies in Environmentel Science
Other volumes in this series

1 **Atmospheric Pollution 1978** edited by M.M. Benarie
2 **Alr Pollution Reference Measurement Methods and Systems** edited by T. Schneider, H.W. de Koning and L.J. Brasser
3 **Biogeochemical Cycling of Mineral-Forming Elements** edited by P.A. Trudinger and D.J. Swaine
4 **Potential Industrial Carcinogens and Mutagens** by L. Fishbein
5 **Industrial Waste Management** by S.E. Jørgensen
6 **Trade and Environment: A Theoretical Equiry** by H. Siebert, J. Eichberger, R. Gronych and R. Pethig
7 **Field Worker Exposure during Pesticide Application** edited by W.F. Tordoir and E.A.H. van Heemstra-Lequin
8 **Atmospheric Pollution1980** edited by M.M. Benarie
9 **Energetics and Technology of Biological Elimination of Wastes** edited by G. Milazzo
10 **Bioengineering,Thermal Physiology and Comfort** edited by K. Cena and J.A. Clark
11 **Atmospheric Chemistry. Fundamental Aspects** by E. Mészáros
12 **Water Supply and Health** edited by H. van Lelyveld and B.C.J. Zoeteman
13 **Man under Vibration. Suffering and Protection** edited by G. Bianchi, K.V. Frolov and A. Oledzki
14 **Principles of Environmental Science and Technology** by S.E. Jørgensen and I. Johnsen
15 **Disposal of Radioactive Wastes** by Z. Dlouhý
16 **Mankind and Energy** edited by A. Blanc-Lapierre
17 **Quality of Groundwater** edited by W. van Duijvenbooden, P. Glasbergen and H. van Lelyveld
18 **Education and Safe Handling In Pesticide Application** edited by E.A.H. van Heemstra-Lequin and W.F. Tordoir
19 **Physicochemical Methods for Water and Wastewater Treatment** edited by L Pawlowski
20 **Atmospheric Pollution 1982** edited by M.M. Benarie
21 **Air Pollution by Nitrogen Oxides** edited by T. Schneider and L. Grant
22 **Environmental Radioanalysis** by H.A. Das, A. Faanhof and H.A. van der Sloot
23 **Chemistry for Protection of the Environment** edited by L. Pawlowski, A.J. Yerdier and W.J. Lacy
24 **Determination and Assessment of Pesticide Exposure** edited by M. Siewierski
25 **The Biosphere: Problems and Solutions** edited by T.N. Veziroğlu
26 **Chemical Events in the Atmosphere and their Impact on the Environment** edited by G.B. Marini-Bettòlo
27 **Fluoride Research 1985** edited by H. Tsunoda and Ming-Ho Yu
28 **Algal Biofouling** edited by L.V. Evans and K.D. Hoagland
29 **Chemistry for Protection of the Environment 1985** edited by L. Pawlowski, G. Alaerts and W.J. Lacy
30 **Acidification and its Policy Implications** edited by T. Schneider
31 **Teratogens: Chemicals which Cause Birth Defects** edited by V. Kolb Meyers
32 **Pesticide Chemistry** by G. Matolcsy, M. Nádasy and Y. Andriska
33 **Principles of Environmental Science and Technology (second revised edition)** by S.E. Jørgensen and I. Johnsen
34 **Chemistry for Protection of the Environment 1987** edited by L. Pawlowski, E. Mentasti, C. Sarzanini and W.J. Lacy

35 **Atmospheric Ozone Research and its Policy Implications** edited by T. Schneider, S.D. Lee, G.J.R. Wolters and L.D. Grant
36 **Valuation Methods and Policy Making in Environmental Economics** edited by H. Folmer and E. van Ierland
37 **Asbestos in the Natural Environment** by H.Schreier
38 **How to Conquer Air Pollution. A Japanese Experience** edited by H. Nishimura
39 **Aquatic Bioenvironmental Studies: The Hanford Experience, 1944–1984** by C.D. Becker
40 **Radon in the Environment** by M. Wilkening
41 **Evaluation of Environmental Data for Regulatory and Impact Assessment** by S. Ramamoorthy and E. Baddaloo
42 **Environmental Biotechnology** edited by A. Blazej and V. Prívarová
43 **Applled Isotope Hydrogeology** by F.J. Pearson, Jr., W. Balderer, H.H. Loosli, B.E. Lehmann, A. Matter, Tj. Peters, H. Schmassmann and A. Gautschi
44 **Highway Pollution** edited by R.S. Hamilton and R.M. Harrison
45 **Freight Transport and the Environment** edited by M. Kroon, R. Smit and J.van Ham
46 **Acidification of Research in The Netherlands** edited by G.J. Heij and T. Schneider
47 **Handbook of Radioactive Contamination and Decontamination** by J. Severa and J Bár

FOREWORD

The organizing and scientific committees of the international conference WASCON '91 herewith present you the proceedings of this conference, which will be held from November 10th till November 14th, 1991 in Maastricht the Netherlands.
Due to the Gulf War the conference was postponed from March 1991 till November of this year. We are aware of the fact that this postponement has caused extra work for many persons, but the committees decided that it was not possible to organize an international conference under those circumstances.
We are now looking forward to a conference with over 75 oral presentations and 25 poster presentations from 20 different countries, and a technical exhibition. We hope that this conference will be of interest to all participants, together with an audience from all over the world, and will contribute to the solution of the great environmental problems concerning waste materials, of which application in construction is one of the main environmentally acceptable possibilities.

SCOPE OF THE CONFERENCE

Many western countries are facing the problem of a growing burden of waste materials, accompanied by a shortage of primary materials. Serious problems with cleaning-up old landfills and pollution of the groundwater are currently making disposal of waste very difficult in many countries. The need for a new trend in environmental protection policy is clearly spelt out in the report "Our Common Future" issued by the UN Brundlandt committee. The protection of soil and water, the limitation of waste production and the re-use of waste materials are key items in a concept the committee termed **"Sustainable Development"**.
With respect to waste materials, extensive research has been carried out to find a market for these materials, e.g. powder coals fly ash in concrete and incinerator slag in road construction. Beneficial use of products derived from waste materials can in fact contribute to sustainable development. However, the market for such waste-derived products mostly involves their re-use as construction materials, implying close contact with the soil. If not properly managed, this may result in pollution of the soil, or even of the groundwater, due to the uncontrolled release of contaminants.
In order to predict and control potential contamination, laboratory leaching tests have been developed in several countries, e.g. the USA, Canada, Germany and the Netherlands. The knowledge gained from this research can be used to control or eliminate possible contamination. One problem is the fact that the various tests being used are not comparable.
The initiative for this conference was generated by the observed need to achieve consensus on methodologies for assessing the environmental behavior of waste residues and the consequences of using them as building materials and to establish criteria and

standards to ensure environmentally safe re-use.
The second part of the conference addresses the state of the art in technical solutions and procedures to use waste materials for the production of construction materials such as concrete, calcium silicate bricks, artificial gravel and other products. Various contributions regarding environmental policy and legislation complete the conference.
The organizing committee hopes that this conference will contribute to the solution of the environmental problems concerning the re-use of waste materials and that a sustainable development in building practice will be one of the results.

SCIENTIFIC COOPERATION

In cooperation with the WASCON conference a working group has been formed with the aim of forming an international body for scientific cooperation. The goal of scientific cooperation being exchange of knowledge and results of research on an international level in order to provide solutions to the environmental problems of waste materials.
We hope that this initiative will indeed lead to international cooperation and are looking forward to participate in this organization.

ACKNOWLEDGEMENT

Organizing an international conference means a lot of work for many persons, therefore we wish to express our thanks to the following persons and organizations:

The members of the Honorary Committee and the Scientific Committee.

The participants in the organization: The National Institute of Public Health and Environmental Protection, The Netherlands Energy Research Foundation, Environment Canada, The United States Environmental Protection Agency, The Netherlands Ministry of Housing, Physical Planning and the Environment and The Netherlands Agency for Energy and the Environment.

Van Namen and Westerlaken Congress Organization Services, De Boer and Van Teylingen Public Relations and the staff of Elsevier Science Publishers.

All authors, participants of the conference and all others who have contributed to WASCON '91.

On behalf of the Organizing Committee,
Utrecht, The Netherlands, August 19th 1991,
dr. J.J.J.M. Goumans
Novem bv

CONTENTS

Section 1:
Policy and Legislation

Waste Materials in Construction.
J.J.J.R. Goumans, H.A. van der Sloot and Th.G. Aalbers (Editors)

POLLUTION PREVENTION--U.S. ENVIRONMENTAL POLICY

Alfred W. Lindsey[1] and Beverly J. Campbell[2]

[1]*U.S. Environmental Protection Agency, Washington, DC*
[2]*Technical Resources, Inc., Rockville, MD*

During the past two decades, the U.S. Environmental Protection Agency (EPA) has made considerable progress in improving environmental quality, but these efforts have focused largely on treating and controlling pollutants that have already been generated. EPA's "end-of-the-pipe" approaches have achieved significant reductions in the discharge of pollutants. Many of our streams that were formerly dead, now support sport fishing--the Potomac River in Washington, D.C. is an example. We accomplished this by installing secondary sewage treatment plants and industrial "end-of-pipe" controls. We have also achieved substantial success in air pollution control. Cities like Pittsburgh and London no longer gasp under a blanket of soot and smog; and Southern California's air pollution problems would be significantly worse without catalytic converters. Hazardous waste management has become a science and an industry--we do not just dump our waste and ignore the problem. We now require treatment to reduce the toxicity of the waste before it is landfilled. These are just a few of the advances that can be attributed to "end-of-pipe" controls forced by regulation.

The U.S. currently spends nearly $115 billion each year (about 2 percent of the U.S. gross national product) on environmental protection. These expenditures have largely been for end-of-pipe controls and the amount has been increasing. Despite the increasing expenditures for pollution control, many environmental problems remain and complex new problems have arisen that pose serious environmental and health risks. Among these are:

- Volatile organic chemicals (VOCs), hazardous air pollutants (HAPs), and tropospheric ozone problems in our cities
- Acid rain
- Continued ubiquitous spread of toxics throughout the environment and into our food chain
- Where to put or how to manage ever increasing volumes of waste
- Continued cleanup of the sins of the past--chemically contaminated dump sites.

We will extend use of "end-of-pipe" controls through incentives and regulations. We will put scrubbers and other stack controls on coal-fired power plants. We will learn better how to contain and clean up oil spills, and we will treat more municipal solid waste in incinerators or perhaps other more innovative techniques.

However, "end-of-pipe" controls are not the key to future progress. Our current problems are not as simple as they were in the "old days" of 10 to 30 years ago. As world population continues to grow and assuming that our standard of living and that of the rest of the world continues to improve, we have a situation of increasing stress on the world's resources. One of those finite resources is the capacity of the environment to assimilate the residuals of human activity--the residues, effluents, wastes, and emissions--the pollutants. Given the finite ability of the environment to absorb this increasing load, we must remove more and more of the objectionable components in our air emissions, effluents, and wastes--just to stay even. However, it is an axiom of "end-of-pipe" control engineering that removing increasingly higher percentages of pollutants from a waste, emission, or effluent increases the cost by orders of magnitude.

Additionally, many of the new problems we face do not lend themselves readily to end-of-pipe controls. Some new problems are generated by many small diverse sources--e.g., the myriad of combustion sources spewing carbon dioxide that are contributing to the global warming problem. Others are not point sources, but can best be described as area sources--e.g., pesticides or fertilizer impacts on our ground and surface waters. It is hard to visualize how sources such as these can be controlled by an "end-of-pipe" strategy.

The old tools in the environmental toolkit are no longer adequate to address these problems. We need creative new strategies for reducing environmental risk. Pollution prevention is the answer, and it holds the key to future gains in environmental protection. If we are to preserve the quality of our environment for future generations, we must adopt a prevention strategy for environmental protection. President George Bush has openly endorsed EPA's new strategy:

> *"Environmental programs that focus on the end of the pipe or the top of the stack, on cleaning up after the damage is done, are no longer adequate. We need new policies, technologies, and processes that prevent or minimize pollution -- that stop it from being created in the first place."*

What is Pollution Prevention?

Pollution prevention is, very simply, any activity undertaken to reduce or eliminate the generation of pollutants or wastes or to reduce their toxicity at the source. It involves the use of processes, practices, or products that reduce or eliminate the generation of pollutants and wastes, or that protect natural resources through conservation or more efficient utilization. The application of pollution prevention techniques varies depending on the economic sector in which they are used. For example, there are three basic approaches to preventing pollution in the manufacturing sector:

- ***Changing the inputs to processes to reduce reliance on toxic raw materials***. A manufacturer may substitute non-toxic for toxic feedstocks in making a product.
- ***Changing processes to reduce the amount and toxicity of waste generated***. The production process may be altered to reduce the volume of materials released to the environment and/or the toxicity of these materials; in addition to avoiding waste management costs, these changes often improve efficiency by reducing raw material losses and conserving water. Process changes may include equipment modifications or less expensive maintenance and housekeeping measures, as well as in-process, closed loop recycling that returns waste materials directly to production as raw materials.
- ***Changing outputs to reduce reliance on toxic or environmentally harmful products***. The manufacturers or users of products may switch to non-toxic (or less toxic) or less polluting substitutes.

In the above examples we have focused on reduction and elimination of toxics, but the pollutant could be any objectionable characteristic or component. In the agricultural sector, pollution may be prevented by developing and adopting low input sustainable agricultural practices that eliminate the wasteful use of inputs, such as water, fertilizers, and pesticides. In addition, soil conservation and land management practices that prevent sediment erosion and the runoff of pesticides and fertilizers also prevent pollution. In the energy and transportation sectors, pollution from energy production can be prevented by increasing efficiency to reduce the generation of pollutants associated with extraction, refining, and use of fuels; and by increasing reliance on clean, renewable energy sources or alternative, less polluting fuels.

While recycling, reuse, and reclamation are not included in the Agency's definition of pollution prevention, EPA recognizes the important role they play in reducing the amount of waste generated that requires subsequent treatment and disposal. EPA has recommended various voluntary activities in the November 1989 report *The Solid WAste Dilemma: An Agenda for Action* and more recently in *The Environmental Consumer's Handbook* that can be undertaken by federal, state, and local governments, as well as industry and private citizens to reduce the amount of waste generated and increase the amount of municipal solid waste recycling.

Benefits of Pollution Prevention

Pollution prevention not only offers an approach to reducing the risks associated with most of the serious environmental problems facing the U.S., it also makes good economic sense. There are benefits of pollution prevention, as well as incentives for prevention, that affect many sectors of society. The benefits can be significant, and can serve to encourage voluntary action to implement pollution prevention approaches in both the public and private sectors. Pollution prevention is often in the self-interest of manufacturing enterprises, since it has potential to save raw material costs (including energy), reduce present and future waste management costs, minimize liability, and earn public goodwill. Major corporations like 3M, Monsanto, and DuPont are committed to pollution prevention as a cost effective means of sustaining environmental health and economic growth. Each of these corporations has pledged to achieve at least an 85 percent reduction in the amount of toxic chemicals that they release into the environment. They believe that such a pollution prevention strategy will save them money in the long run by increasing operating efficiency, as well as limiting future liability.

It has been 16 years since the Minnesota Mining and Manufacturing Corporation (3M) established the landmark Pollution Prevention Pays (3P) program to make pollution prevention a way of life throughout the corporation--from the boardroom to the laboratory to the manufacturing plant. The tools of prevention utilized over the years have ranged from high-tech innovation to simple housekeeping. The result has been twofold--substantial elimination of pollution and significant cost savings. By 1987, worldwide annual releases of air, water, sludge, and solid waste pollutants (hazardous and nonhazardous) from 3M facilities had been reduced by nearly 450,000 tons,with about 95 percent of the reduction coming from U.S. operations. The process and product changes made to achieve these reductions have yielded cumulative worldwide savings of $420 million.

Similar to 3P, Chevron initiated its Save Money and Reduce Toxics (SMART) program in 1987. During the first year of SMART, hazardous waste disposal dropped 44 percent, from 135,000 to 76,000 tons, saving the company $3.8 million. This reduction was, in part, achieved by substituting non-hazardous drilling mud additives for compounds that were considered hazardous. Chevron has set a goal of a 65 percent across-the-board reduction by 1992.

A New Philosophy

These are just a few of the hundreds of pollution prevention "success stories" published by industry, that support EPA's belief that pollution prevention can benefit both the environment and the economy. The old philosophy--that a healthy economy and a healthy environment are fundamentally at odds--is no longer valid. The new philosophy embraced by EPA and industry in the U.S. is that environmental protection is not a luxury bought at the expense of economic health; rather, it is a prerequisite for a healthy economy and sustainable growth.

Pollution prevention is EPA's preferred approach for protecting human health and the environment. Whenever possible, environmental protection efforts first should be aimed at eliminating or minimizing wastes or pollutants at the source, or as close to the source as possible.

This does not mean that all wastes from every production process will be eliminated. Rather, it offers a more cost-effective means of minimizing the generation of waste. Another way to look at prevention (or source reduction) is as the first step in a hierarchy of options for reducing the risks to human health and the environment from pollution. The next step in such a hierarchy would be the responsible recycling or reuse of any wastes that cannot be reduced at the source. When recycling is conducted in an environmentally sound manner, it shares many of the same advantages as prevention, such as conserving energy and other resources, and reducing reliance on raw materials and the need for "end-of-pipe" treatment or containment of wastes. Any wastes that cannot feasibly be prevented, recycled, or reused should be treated in accordance with environmental standards that are designed to reduce both the hazard and volume of waste streams. Finally, any residues remaining from the treatment of waste should be disposed of safely, to minimize their potential for adverse impacts on public health and the environment.

What is the U.S. Policy on Pollution Prevention?

The American public has become increasingly aware of the potential health and environmental risks associated with pollution. The information reporting requirements of Title III of the Superfund Amendments and Reauthorization Act (SARA) have made the public more aware of the massive amounts of pollution that are released by industry each year. Congress, recognizing the public's growing concern and the importance of preventing further decay of the environment, drafted legislation that established a national pollution prevention policy in the U.S.

Pollution Prevention Act of 1990

On October 27, 1990, Congress passed the Pollution Prevention Act. Enactment of this legislation strengthened and accelerated efforts to promote pollution prevention throughout the nation. The Act declares the national policy of the United States to be that pollution should be prevented or reduced at the source whenever feasible; and it establishes source reduction as the first priority in the pollution prevention hierarchy, followed by recycling, treatment, and proper disposal. Source reduction, as defined in the bill, is any "practice which reduces the amount of any hazardous substance, pollutant, or contaminant entering any waste stream or otherwise released into the environment prior to recycling, treatment, or disposal; and reduces the hazards to public health and the environment associated with the release of such substances, pollutants, or contaminants."

The Pollution Prevention Act of 1990 requires EPA to establish a separate office specifically to carry out its functions under this Act, and to develop and implement a strategy to promote pollution prevention within the public and private sectors. As part of the strategy, the EPA Administrator will:

- Coordinate pollution prevention activities in each Agency office and promote similar practices in other federal agencies and industry.
- Establish standard methods for measuring source reduction; and develop, test, and disseminate model source reduction auditing procedures.
- Establish a training program on pollution prevention opportunities.
- Make recommendations to Congress to eliminate barriers to pollution prevention and identify opportunities to use federal procurement to encourage source reduction.
- Establish a source reduction clearinghouse containing information on management, technical, and operational approaches to source reduction; and develop improved methods for providing public access to data collected under federal environmental statutes.
- Establish an advisory panel of technical experts to advise the EPA Administrator on ways to improve the collection and dissemination of data.
- Provide grants to states for programs to promote pollution prevention by local businesses.
- Identify research needs relating to pollution prevention and set priorities for research to target the most promising opportunities for source reduction.

The Act also requires facilities reporting under the Toxic Release Inventory (TRI) provisions of Section 313 of the Superfund Act amendments to provide information on pollution prevention and recycling activities with each annual filing, and the information is made available to the public.

Pollution Prevention at EPA

In a 1988 report to EPA entitled, *Future Risk: Research Strategies for the 1990s*, the Agency's Science Advisory Board (SAB) recommended that prevention or reduction of environmental risks should be the long-term goal for the Agency. The report advised EPA to shift the focus of its environmental protection efforts from "end-of-pipe" treatment to preventing the generation of pollution. The SAB defined a hierarchy for risk reduction to help in setting priorities and in achieving the Agency's overall goal of protecting human health and the environment. This hierarchy--endorsed by Congress as national policy in the Pollution Prevention Act of 1990--clearly indicates that pollution prevention should consistently be the first option for reducing risks. In addition, the SAB recommended that EPA plan, implement, and sustain a long-term research program to support the new source reduction strategy.

Shortly after he took office in early 1989, EPA Administrator William Reilly asked the SAB to review EPA's 1987 report on relative environmental risk, *Unfinished Business: A Comparative*

Assessment of Environmental Problems, to evaluate its findings and develop strategic options for reducing risk. The results of the SAB's review were published in *Reducing Risk: Setting Priorities and Strategies for Environmental Protection.* This report recommended that EPA target its environmental protection efforts on the basis of opportunities for the greatest risk reduction. It also recommended that EPA emphasize pollution prevention as the preferred option for reducing risks:

> *"A fundamental restructuring of the way the Agency approaches risk reduction is in order; the Agency's primary focus should be to prevent the creation of risks, as opposed to trying to control such risks once created."*

The SAB cited seven reasons for focusing on pollution prevention to reduce risk:

- For some environmental problems, such as stratospheric ozone depletion and global climate change, pollution prevention is the only solution.
- Pollution prevention is often the most effective solution. For instance, in the case of lead, asbestos, PCBs, and certain pesticides, the most effective solution has been to ban their use.
- There can be a tremendous cost benefit for pollution prevention in terms of avoiding costs of control, cleanup, and liability; and in terms of decreasing costs by increasing efficiency and productivity.
- Pollution prevention is the key to sustainable development. In many areas the U.S. is approaching or even exceeding the capacity of the environment to absorb pollutants. It is clear that economic and industrial strategies for the future that minimize pollution and the consumption of resources are more likely to be sustainable.
- Pollution prevention often prevents the solution to one environmental problem from re-emerging as another kind of environmental problem in another medium, sometime in the future or in another place.
- Pollution prevention can help improve international relations in two ways--first, it can help developing countries avoid the environmental problems that we had in the U.S. by moving directly to low polluting, low waste technology; second, because of the worldwide impact on the U.S. generation of pollution and consumption of resources.
- Pollution prevention protects the natural resources on the planet for future generations by reducing the amount of destruction caused by excessive pollution and slowing the depletion of resources.

From its study of 13 priority environmental problems identified in *Unfinished Business*, the SAB noted a substantial number of strategy options which involve pollution prevention approaches. Upon reviewing these options, the SAB distinguished several cross-cutting themes:

- EPA's pollution prevention program should be directed broadly to address products and many productive sectors, not just industrial production processes. EPA should promote pollution prevention in all sectors, from manufacturing to agriculture to construction.
- EPA and other federal agencies should go beyond problem-by-problem pollution prevention to focus on comprehensive multi-problem solutions, such as toxics use reduction, energy

efficiency and conservation; and on altering specific production technologies for products which contribute to multiple problems, such as the automobile.

- Federal agencies should identify and eliminate standards, subsidies, activities, or approvals that promote polluting or damaging activities or technologies, and instead promote non-polluting activities, technologies, and products, through incentives, research, technical assistance, procurement, and other means.

- EPA should actively work with representatives of many interests to promote better understanding of pollution prevention. Collaborative research, education, and technology development and transfer with industry, state agencies, organized labor, and public interest groups should be considered.

- Community right-to-know and other related programs should be given special attention and possibly expanded. These possibilities include having more producers and users of toxic chemicals and pesticides report publicly on such production and usage.

- In the long run, economic incentives and disincentives need to promote pollution prevention. Energy policy should encourage conservation, tax policy should encourage recycling and reuse, etc.

EPA's Pollution Prevention Strategy

Even before the Pollution Prevention Act was passed, EPA published a proposed policy statement that established pollution prevention as the Agency's preferred approach for protecting human health and the environment. EPA has already begun to incorporate this policy into the decision-making processes--integrating pollution prevention policy into how the Agency conducts business. Since issuing the policy statement, EPA has established the Pollution Prevention Office (PPO) which is charged with promoting an environmental ethic founded on the prevention of pollution both within and outside the Agency. PPO is the focal point for the U.S. EPA's pollution prevention activities and was a major impetus behind the establishment of an Agencywide pollution prevention program. This office was responsible for coordinating the Agency's strategic planning efforts for the program and preparing an integrated, Agencywide, cross-media strategy for pollution prevention.

EPA recognized that a clear and coordinated federal strategy for pollution prevention was needed both to remove obstacles to preventing pollution and to foster preventive initiatives in the future. EPA believes that its environmental protection goals will be best served in the long run by a pollution prevention strategy that proposes roles for industry, agriculture, the energy and transportation sectors, government, the American public, and the international community. The Pollution Prevention Office has recently drafted such a strategy. The *U.S. Environmental Protection Agency Pollution Prevention Strategy*, published in January 1991, presents the Agency's blueprint for a comprehensive strategy designed to serve two purposes--(1) to provide guidance and direction for efforts to incorporate pollution prevention within EPA's existing regulatory and non-regulatory programs, and (2) to set forth a program that will achieve specific objectives in pollution prevention within a reasonable timeframe.

The first objective reflects EPA's belief that for pollution prevention to succeed, it must be a central part of the Agency's primary mission of protecting human health and the environment. The goal is to incorporate prevention into every aspect of the Agency's operations in program and regional offices. To address the second objective, the *Pollution Prevention Strategy* includes a plan for targeting high risk chemicals that offer opportunities for prevention. This industrial toxics project (ITP), referred to as the 33/50 project, targets 17 specific chemicals that are reported on the Toxics Release Inventory (TRI). Over a billion pounds of these chemicals are released into the environment each year. This project involves the development of focused prevention strategies for each of the 17 chemicals and sets a voluntary goal of reducing total environmental releases of these chemicals by 33 percent by the end of 1992, and at least 50 percent by the end of 1995. EPA has developed this list of targeted chemicals based upon five criteria--(1) high levels of emissions, (2) technical or economic opportunities for pollution prevention, (3) potential for health and ecological risk, (4) potential for multiple exposures or cross-media contamination, and (5) limitations of treatment technologies. The list of targeted chemicals is presented in Exhibit 1. Most of the 17 pollutants targeted by the *Pollution Prevention Strategy* are slated for even greater regulatory controls under the Clean Air Act Amendments of 1990, but the controls would not go into effect until 1995.

EPA is seeking voluntary, measurable commitments from major industrial sources of these contaminants to reduce environmental releases through prevention. Beginning in early 1991, EPA sent letters to 600 corporate polluters asking them to help achieve the Agency's goal of substantially reducing releases of the 17 chemicals over the next four years. EPA asked these companies to make commitments to the project and to develop prevention plans to carry them out. EPA will rely on data from the Toxics Release Inventory (TRI) to track reductions in releases of targeted contaminants from industrial facilities, and will develop more appropriate indicators for sources not covered by the TRI.

The industrial toxics project is only the first step. EPA recognizes that there are abundant opportunities to promote pollution prevention in other sectors, such as agriculture, energy, transportation, municipal water and wastewater, and EPA is working with other federal agencies to develop specific strategies for these sectors.

What Research is Needed to Support EPA's Pollution Prevention Policy?

Research is the primary vehicle for enhancing our pollution prevention knowledge base. It is needed to provide the scientific and technical knowledge necessary to implement pollution prevention initiatives on a cross-media, cross-program basis. In the September 1988 report *Future Risk: Research Strategies for the 1990s*, the Science Advisory Board (SAB) recommended that EPA plan,

EXHIBIT 1

INDUSTRIAL TOXICS 33/50 PROGRAM TARGET CHEMICALS

Benzene

Cadmium and Cadmium Compounds

Carbon Tetrachloride

Chloroform (Trichloromethane)

Chromium and Chromium Compounds

Cyanide Compounds and Hydrogen Cyanide

Lead and Lead Compounds

Mercury and Mercury Compounds

Methylene Chloride (Dichloromethane)

Methyl Ethyl Ketone

Methyl Isobutyl Ketone

Nickel and Nickel Compounds

Tetrachloroethylene (Perchloroethylene)

Toluene

1,1,1-Trichloroethane (Methyl Chloroform)

Trichloroethylene

Xylenes (All Xylenes)

implement, and sustain a long-term research program to support the Agency's new philosophy of preventing the generation of pollution. The SAB noted:

> *"Just as EPA's regulatory role will change as it incorporates this broader approach to environmental protection, its R&D role will change as well. EPA must conduct research that will support materials substitution, industrial process changes, and recycling technologies, because it is unlikely that any individual community or small business will have the incentive or resources to do it."*

Over the past five years, EPA has attempted to redirect the nation's pollution control strategy toward prevention by adopting a waste management hierarchy that placed priority on pollution prevention. In 1987, the Agency initiated a waste minimization research program that focused on encouraging the development and demonstration of processes and techniques that result in a reduction or prevention of hazardous pollutants. In response to the SAB's recommendation, EPA significantly expanded the waste minimization research program to include both hazardous and non-hazardous wastes. The expanded research program also adopted a multimedia approach to pollution prevention. EPA's overall plan for expanding the Agency's pollution prevention research program was described in a report to Congress published in March 1990.

Pollution Prevention Research Plan: Report to Congress

The *Pollution Prevention Research Plan: Report to Congress* is a multi-year plan that addresses the critical research elements needed to support an Agencywide multimedia pollution prevention initiative. This plan described a comprehensive program that includes both technological and non-technological research to address a broad range of pollution prevention issues. Preparation of the *Pollution Prevention Research Plan* was the first step in developing the research component of EPA's pollution prevention initiative.

The report to Congress was founded on the premise that pollution prevention should be a guiding principal for all environmental protection efforts and general human activities. We have learned that it can be enormously costly to clean up and dispose of pollutants after they have been generated. "End-of-pipe" controls and waste disposal should be the last line of defense, rather than the front line. Preventing pollution at the source offers great environmental and health benefits, and is almost certain to be the most economical approach in the long run.

The report to Congress identified six fundamental goals for the pollution prevention research program:

- Stimulate the development and use of products that result in reduced pollution--research is needed on methods for conducting product assessments and identifying pollution prevention opportunities, development and use of less polluting products, and the impacts of products on the environment at each stage of their life cycle.

- **Stimulate the development and implementation of technologies and processes that result in reduced pollution**--research is needed to identify and evaluate those aspects of production, use, maintenance, repair, and disposal processes that generate pollutants and waste. Research is also needed to assess pollution prevention opportunities, to develop less polluting processes, and to transfer these techniques to other industries.

- **Expand the reusability and recyclability of wastes and products and the demand for recycled materials**--research is needed on ways to improve the reusability and recyclability of wastes and products and to increase the capacity and demand for recycled materials in production processes.

- **Identify and promote the implementation of effective socioeconomic and institutional approaches to pollution prevention**--research is needed to understand the socioeconomic and institutional factors that motivate behavior and foster changes in behavior, as they relate to incentives for adopting pollution prevention techniques; and the impact of these factors on the effectiveness of pollution prevention programs.

- **Establish a program of research that will anticipate and address future environmental problems and pollution prevention opportunities**--research is needed to assist EPA in anticipating and responding to emerging environmental issues and in evaluating new technologies that may significantly alter the status of pollution prevention programs in the future.

- **Conduct a vigorous technology transfer and technical assistance program that facilitates pollution prevention strategies and technologies**--it is imperative that the results of research investigations conducted under this program or by industry and academia are communicated expeditiously to appropriate audiences.

Each of these six goals corresponds to a research area that ORD needed to address in its comprehensive pollution prevention research program. These areas are displayed in Exhibit 2.

The report to Congress formed the foundation of EPA's pollution prevention research efforts, but it did not delineate specific themes for future research efforts nor did it define the projects to be undertaken. The report to Congress provided representative examples of the types of research projects that EPA expected to conduct, but an implementation strategy was necessary to clearly delineate themes for future research efforts and projects that could be conducted to achieve the goals and objectives of the program.

Pollution Prevention Research Strategic Plan

In August 1990, EPA drafted the *Pollution Prevention Research Strategic Plan*, which provides the blueprint for the pollution prevention research program by focusing the Agency's research efforts on high priority environmental problems and the pollution prevention research projects that address these problems. By identifying and selecting priority environmental problems on which to focus the strategic plan, EPA is optimizing the use of limited resources and increasing the potential for significant impact in reducing the risks associated with these priority problems.

EXHIBIT 2

POLLUTION PREVENTION RESEARCH PROGRAM AREAS

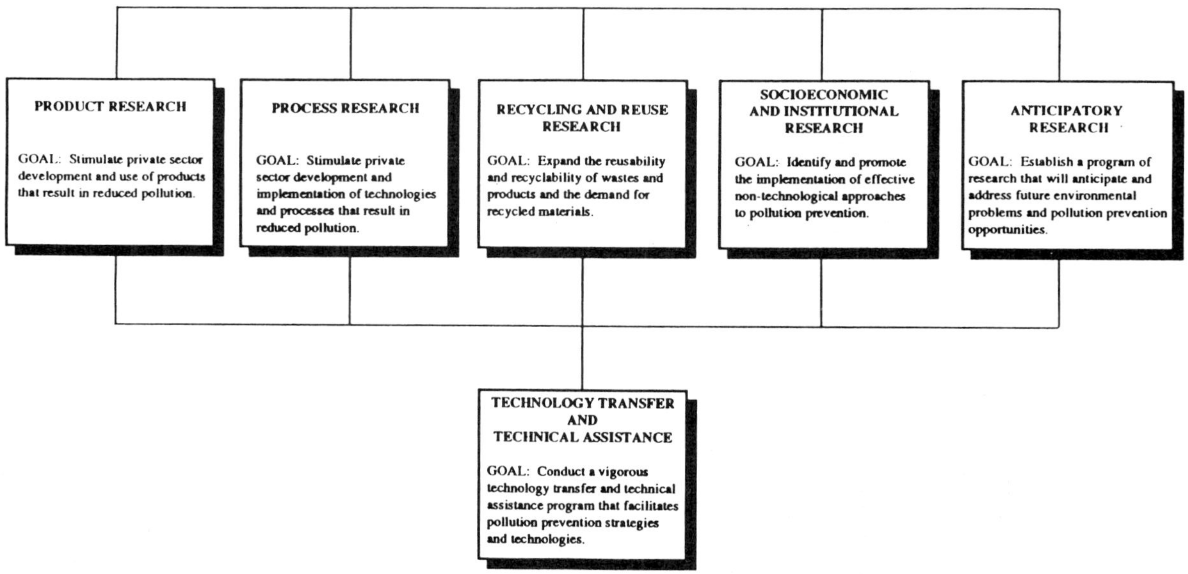

In addition to providing a focus for future research efforts, the research strategy enables the Agency to investigate a variety of tools that could potentially impact more than one environmental problem, and priority can be given to the research projects that impact multiple problems. For example, the manufacturing and use of paints containing toxic solvents can contribute to multiple environmental problems--criteria and toxic air pollutants, indoor air pollution, nonpoint source pollution, hazardous waste, municipal solid waste, and worker/consumer exposure. Therefore, a project investigating alternate formulations eliminating the toxic solvents could beneficially impact all of these environmental problems. Through the pollution prevention research program and other Agency efforts, EPA is attempting to establish pollution prevention as a cornerstone of national environmental protection strategies, to communicate the message to all members of the environmental protection community, and to provide assistance in implementing pollution prevention programs. The Agency recognizes the significant role that pollution prevention can play in preserving and protecting human health and the environment since it is applicable to a broad array of environmental problems and can be implemented through a variety of approaches and tools. Preparation of the *Pollution Prevention Research Strategic Plan*, as well as the EPA-wide *Pollution Prevention Strategy*, are important steps in institutionalizing pollution prevention at EPA and throughout the nation.

The *Pollution Prevention Research Strategic Plan* identifies 10 high priority environmental problems that will be used to focus ORD's pollution prevention research efforts over the next five years. It describes and rates the potential for various research approaches that may be employed to meet the research needs associated with the 10 high priority problems. Exhibit 3 graphically depicts the 10 priority problem areas selected for EPA's focus and displays the potential for each of six pollution prevention research approaches to effectively address each problem.

EPA selected the 10 priority problems based on the following criteria, each of which require subjective evaluation since no real data exist to allow quantification:

- ***Risk to human health and the environment***--those problems that pose the greatest risk relative to other environmental problems when taking into account the risk of cancer, chronic non-cancer health effects, reproductive, developmental, and neurotoxic risks, and the potential for toxic and non-toxic ecological damage; as well as the risk for multiple exposures.

- ***Amenability to pollution prevention solution and limitations of other solutions***--the potential contribution of a pollution prevention approach to solution or elimination of the environmental problem, particularly when pollution controls, treatment, and disposal options are limited or relatively ineffective in reducing the associated risks.

- ***Probable benefits and costs of reducing risks***--the expected benefits (economical and environmental) of preventing the generation of sources contributing to the environmental problem outweigh the costs associated with implementing a pollution prevention approach.

EXHIBIT 3
PRIORITIZATION OF POLLUTION PREVENTION APPROACHES FOR TARGETED ENVIRONMENTAL PROBLEMS

APPROACH / PROBLEM	Product Research	Process Research	Recycling/Reuse Research	Socioeconomic Research	Anticipatory Research	Technology Transfer/Technical Assistance
Indoor Air Pollutants	HIGH	LOW	LOW	HIGH	MEDIUM	MEDIUM
Criteria Air Pollutants	HIGH	HIGH	MEDIUM	LOW	MEDIUM	LOW
Ozone Depleting Substances	HIGH	HIGH	MEDIUM	LOW	LOW	MEDIUM
Greenhouse Gases/GC Change	HIGH	HIGH	LOW	MEDIUM	MEDIUM	LOW
Toxic Air Pollutants	HIGH	HIGH	LOW	LOW	MEDIUM	MEDIUM
Pesticides Application	HIGH	HIGH	LOW	MEDIUM	LOW	MEDIUM
Nonpoint Source Water Discharges	MEDIUM	LOW	MEDIUM	HIGH	LOW	HIGH
Hazardous Waste	HIGH	HIGH	MEDIUM	LOW	LOW	MEDIUM
Consumer Products	HIGH	LOW	HIGH	MEDIUM	LOW	MEDIUM
Municipal Solid Waste	HIGH	LOW	HIGH	HIGH	LOW	MEDIUM

- ***Degree to which the problem is addressed and funded by programs other than pollution prevention***--environmental problems which are already being effectively addressed through other programs should not be priority targets of the pollution prevention program.

The *Pollution Prevention Research Strategic Plan* identifies pollution prevention approaches to address each of the 10 priority problems, as well as the research needs associated with these approaches. In identifying potential pollution prevention approaches and research needs, the work group members that developed the document considered the nature and controllability of the risks associated with the priority problem areas. For example, risks associated with individual lifestyle choices may be more effectively reduced through market incentives and risk communication than through conventional regulatory approaches. Greater emphasis was placed on those approaches which concurrently address multiple risks; for example, pollution prevention initiatives to reduce fossil fuel use in the energy sector would help to address human health risks posed by criteria and toxic air pollution from fossil energy powerplants, and ecological risks posed by the threat of global climate change resulting from the emission of greenhouse gases.

To evaluate which research approaches hold the most promise for addressing each problem (see Exhibit 3) the work group members considered the following criteria:

- ***Contribution of pollution prevention research in reducing the risks***--the potential contribution of the research project in preventing, reducing, or eliminating the risks associated with the environmental problem.

- ***Value added by EPA research***--the necessity of EPA conducting the research because of information needs that others are not addressing, and the importance of this research in implementing pollution prevention approaches to the problem.

- ***Impact on multiple environmental problems***--the contribution of the research results to better understanding of and capability to implement pollution prevention approaches that address multiple priority environmental problems.

- ***Cost effectiveness***--the cost of the research relative to the absolute amount of expected environmental improvements.

EPA will use these ratings to determine, for example, what kind of research to sponsor to address the problem of pesticides application--technology transfer or product research. EPA will periodically review the research priorities and update them as new problem areas emerge and as new information leads to revised evaluations of the risks associated with existing and new problem areas.

The comprehensive pollution prevention research program described in the strategy focuses on developing specific prevention strategies for individual contaminants, clusters of contaminants, or sources targeted in the industrial toxics project. The strategy also focuses on prevention approaches to address problems outside of the manufacturing sector and on research designed to understand and overcome social, economic, and institutional obstacles to pollution prevention. The pollution prevention research program is intended to promote a fundamental change--one that will make prevention an integral part of public programs and private activities.

The research program outlined in the *Pollution Prevention Research Strategic Plan* is an important component of EPA's Pollution Prevention Strategy. The Agency has already committed substantial resources toward pollution prevention research--EPA is currently conducting or funding dozens of research projects with a cumulative projected cost of over $10 million. The *Pollution Prevention Research Strategic Plan* focuses the Agency's research efforts on priority problems and the pollution prevention approaches to address these problems. The research strategy is the culmination of efforts to define projects, set priorities, and implement a cooperative research program designed to further the adoption of pollution prevention approaches within both the public and private sectors.

Can Pollution Prevention Make a Difference?

While Congress and the federal government can make some progress by incorporating prevention into statutes and regulations, each sector of society must become a partner in this endeavor to achieve the full promise that pollution prevention offers. Only with widespread participation can we sustain economic growth without inviting ecological disaster. As members of a global society, all nations must begin to integrate pollution prevention into the way we design, manufacture, regulate, buy, consume, and dispose. The investment that we make today in preventing the problems of the future is an invaluable part of the legacy that we leave to the children, grandchildren, and great grandchildren of the world.

Waste Materials in Construction.
J.J.J.R. Goumans, H.A. van der Sloot and Th.G. Aalbers (Editors)

SYSTEMATIC LEACHING BEHAVIOUR OF TRACE ELEMENTS FROM CONSTRUCTION MATERIALS AND WASTE MATERIALS.

H.A van der Sloot

Netherlands Energy Research Foundation (ECN),
P.O Box 1, 1755 ZG Petten, The Netherlands.

SUMMARY

This paper focuses on systematic trends observed in the characterization of contaminant leaching from solid wastes including stabilized wastes, sintered and cement-based products containing waste materials. The improved experimental methods available to evaluate these products are briefly reviewed. The interpretation of test results in terms of systematic trends is shown to be essential for judging the environmental acceptability of waste disposal or utilization activities.

1 INTRODUCTION

Many assessments of the environmental consequences of waste disposal and utilization activities focus on compliance with existing regulations. Unfortunately, the correlation between reglatory test results and actual field conditions on the short and the long term is very poor. A variety of experimental methods is available [1,2] some of which lead to a better simulation of the actual behaviour than the single batch extractions presently in use in the regulatory framework [3,4,5,6]. These new mathods allow the identification of systematic trends and the elucidation of mechanisms controlling the release of potentially harmful constituents. The resulting improved understanding of the fundamental aspects of leaching leads to better founded decisions and to more rational regulations, especially when long term effects are the main concern.

2 TESTING METHODS

2.1 Availability test

In recent years, it has been shown conclusively that the total concentration of contaminants in a waste material is not correlated with release to the environment [2,7]. The chemical form of contaminants in the matrix (speciation) and the distribution over different solid phases in the material (fractionation) largely dictates the availability for leaching and the potential for release through external influences. For example, elements tied up in silicate phases or poorly soluble mineral phases are only released after complete destruction/dissolution of the matrix. Under environmental conditions, it is not very likely that these phases will be severly attacked on the long-term. Therefore, the "availability for leaching" is more relevant for environmental assessment purposes

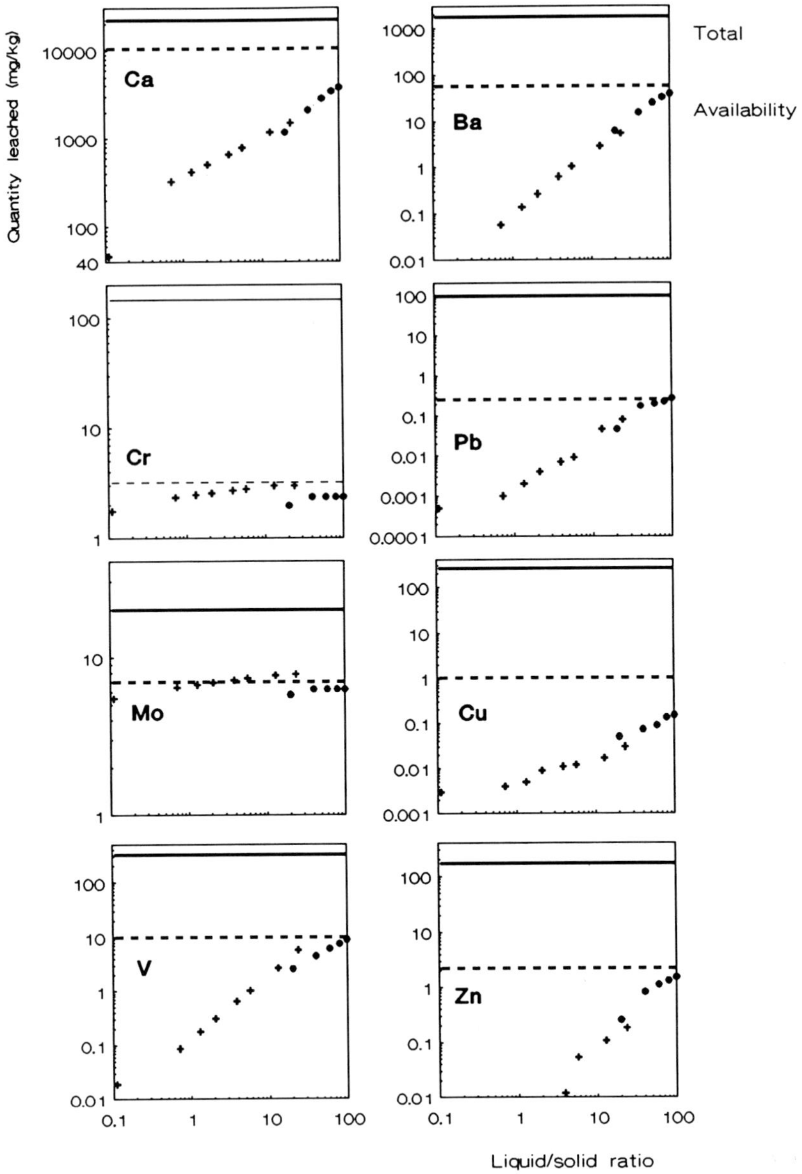

Fig 1.Release data for Ca, Ba, Cr, Pb, Mo, Cu, V and Zn from coal fly ash as a function of the liquid to solid ratio.

than the total concentration.

The availability test is based on the extraction of fine grained material at a controlled pH of 4 using a liquid to solid ratio of 100 [9]. The high dilution of LS 100 minimizes solubility limitations. which are apparent at lower LS ratios even at low pH. A controlled pH of 4 is applied as a lower limit of pH in natural environments (sandy soil, peat soil). Lower pH values do occur in practice (e.g. acid drainage in sulfide bearing deposits), but these environments are not likely chosen for disposal. The material is tested in fine grained form (95 % < 125 μm) to avoid sensitivity to solid phase diffusion. At the specified grain size, constituents with effective diffusion coefficients of 10^{-14} m^2/s are completely leached within the duration of the test. Recently, control of pH at pH =7 is applied prior to pH 4 extraction to optimize the leachability of oxyanionic species. By recording the acid consumption, the acid neutralization capacity of the material can be estimated simultaneously.

2.2 Leaching tests

To estimate actual release from waste or waste products, the release by percolation or by diffusion is assessed using tests in which the material is allowed to dictate the leachate composition. Leaching test methods can be divided into methods appropriate for powder and granular materials (d < 40 mm) and methods more suitable for stabilized waste monolith (d > 40 mm).

2.2.1 Powder- and granular materials

Leaching protocols for granular materials are described in the Dutch standard NVN 2508 [9]. Three quantities are distuiguished which are related to the degree to which contaminants are tied up in the material:

- Total concentration (expressed in mg/kg).
- Availability (expressed in mg/kg).
- Actual leaching as a function of liquid to solid ratio, which is translated into a time scale through the infiltration rate [7,10].

In figure 1 the results of such tests for leaching of Ca, Na, Mo, Pb, V and Ba from a neutral coal fly ash are presented. The results show that the availability data reflect a worst case scenario.

The results of laboratory column studies have been compared with large scale column experiments within the Mammoet project [7]. The results are in good agreement. The availability test defines an upper limit in the release of contaminants up to liquid to solid ratios of 100. This observation is consistent for a wide range of materials [7]. It strongly supports the use of the availability test for a first evaluation of materials.

2.2.2 Monolithic materials

Leaching tests for monolithic materials are described in the Dutch protocol draft NVN5432 [11].A distinction similar to that for granular materials can be made among total concentration, availability and actual leaching as a function of time. For monolithic products the third quantity is determined by a tank leaching test.

In the tank leaching test, a test specimen is immersed in water. At regular intervals, the contact solution is renewed and analyzed to determine the release of components from the specimen. The physical retardation caused by the pore structure of the product can be calculated from the release of an inert component (e.g. Na). By comparing the release of other components to that of the inert component, a chemical retention factor can be calculated describing the degree to which the potentially mobile fraction of a

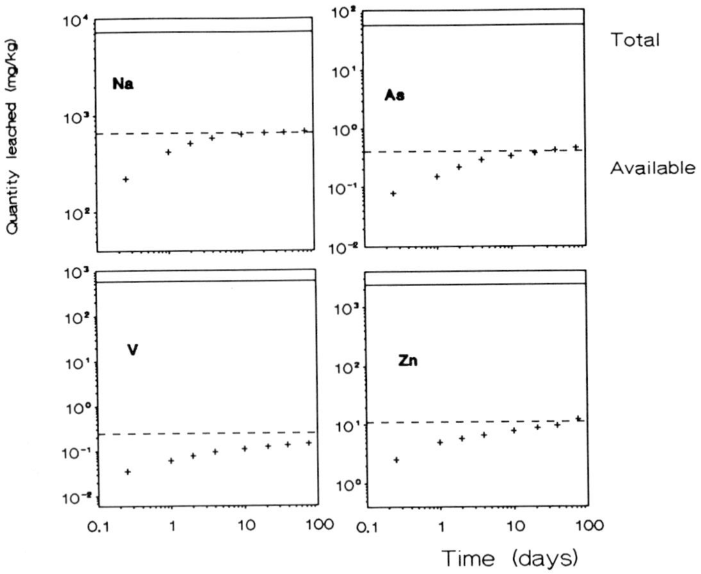

Fig 2. Release data of Na, As, V and Zn from coal ash aggregate as a function of the leaching time showing depletion.

component is retained by the chemical environment in the pores of the product. The effective diffusion coefficient derived from these experiments can be used to estimate long term release from different waste form sizes once diffusion control is established as the main release mechanism [12,13].

The upper limit defined by the availability test is not reached within the time frame of most tank leaching tests, because the test are designed to avoid substantial depletion of any constituent. Nevertheless, a few materials have been leached to such an extent that elements do approach the limit indicated by the availability test. Figure 2 shows leaching data from a granular material with particles of 1 cm diameter. Depletion is evident within the duration of the leaching test [14]. The asymptotic approach to the availability limit indicates the relevance of availability data to monolith leaching.

2.3 Regulatory test methods

Single batch extraction tests, such as the EP tox [8], the Toxicity Characteristic Leaching Procedure [6], the French [4], the Swiss [5] and the German extraction procedures [3], give an indication of the amount of each element which is leachable under specific experimental conditions. They do not in themselves allow an extrapolation to long term effects nor do they provide information on leaching mechanisms.

When the results of the more extensive tests are compared with single extraction tests, such as are currently applied in the regulatory framework, the tests can be grouped into those assessing a potential leachability (EP, TCLP, Availability test), those indicating a release at some point in the future dictated by the waste (DIN 38414 S4, X-31-210) and those assessing the release as a function of time (NVN2508, NVN

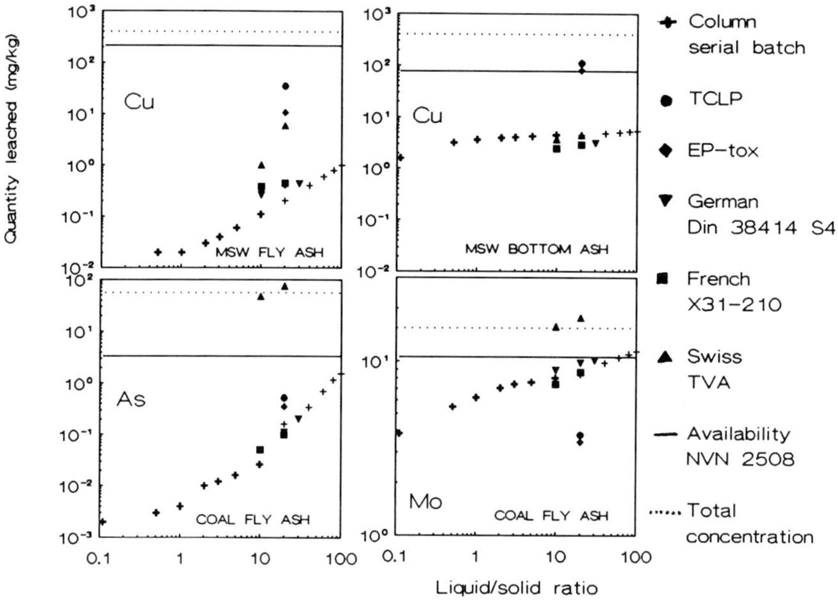

Fig 3. Comparison of regulatory leaching test data for Cu from MSW bottom ash and from fly ash and for As and Mo from coal fly ash in terms of the quantity leached as a function of liquid/solid ratio.

5432). The concentrations measured in the single extraction tests are by no means comparable to actually occuring concentrations in the field. Expressing the test results in quantity leached (mg/kg), which can be converted easily to a mass flux, is more appropriate for assessing environmental impact.

In figure 3, a comparison of different test methods is shown for three elements in a MSW incinerator fly ash [16]. The results for oxyanions (e.g. Mo, As, Se) obtained with the EP and TCLP are systematically lower than those obtained with the availability test. The German and French tests yield comparable results to those obtained in the column test, when expressed in mg/kg leached. However, if the final pH or redox condition is markedly different, due to rapid leaching of pH or redox controlling species in the column test, the German and French procedure may underestimate leachability.

2.4 Concentration profiles

Concentration profile analysis of soil exposed to waste or of the waste product itself yields valuable information on processes occuring under actual conditions. A concentration profile is obtained by slicing a soil or waste sample into sections and determining the concetration ofreactive elements in each slice. The results give direct information about leaching rates and mechanisms in diffusion dominated systems. For example, profiling of a stabilized waste block retrieved from the ocean revealed a sealing mechanism which prevents uptake of seasalts and release of soluble contaminants [17].

In studying leaching behaviour of waste and waste products, the mutual interaction of waste and soil (water, air) should not be overlooked. The release obtained from laboratory leaching tests can be substantially modified by both mobilization and precipitation reactions occurring at the waste/soil (water, air) interface [18,19]. These phenomena will have significant effects on the net release to the environment. They are discussed in more detail in another paper in this proceedings [20]. Concetration profiling is an essential tool in identifying and quantifying their effects.

3 MECHANISMS CONTROLLING LEACHING BEHAVIOUR

3.1 Chemical mechanisms

The role of major element chemistry in controlling parameters that affect the release of contaminants, such as pH, redox and formation of soluble complexes, is not well addressed in the field of waste research. The influence of these parameters on release either resulting from the properties of the waste or influenced by the surroundings of the waste, may result in order of magnitude changes in estimated release.

3.1.1 pH dependence

Among the parameters controlling release, pH has been studied most extensively. The leaching behaviour of contaminants as a function of pH is very systematic. This point is best illustrated by considering a number of different waste materials.

Coal fly ash – A general feature of the leaching test results for coal ash is the minimum metal leachability in the pH range 7 to 10. For the oxyanions, such as MoO_4^-, AsO_4^{3-}, SbO_3^{3-}, SeO_3^{2-}, VO_4^{3-}, a maximum leachability is observed in this pH range. Major elements also show systematic leaching behaviour, for example Al and Si feature minimum leachability from coal fly ash at neutral pH, a maximum around 10 – 11, and a minimum at pH values over 11.5 [21].

Figure 4 shows the leachability of Zn, As, Mo and Al from pulverized coal ash as a function of pH. The acidic, neutral and alkaline nature of the ashes (50) is largely dictated by their CaO content and results in a pH range from 4 to 12.5 [21]. To date the behaviour of the oxyanions at high pH is not satisfactorily explained. With time acidic ashes will increase in pH, whereas alkaline ashes will be neutralized (self-neutralization/ carbonation). Thus the active working range in the field on the longer term will be limited to pH 7 – 10.

Municipal solid waste incinerator residues – The sensitivity of municipal solid waste incinerator residues to pH variations has been studied earlier [22,23]. In spite of the heterogeneous nature of incinerator residues, systematic trends are observed in the leachability – pH relations. Figure 5 shows the leachability of Cd, Pb, Cu and Zn in relation to pH. The data obtained by modelling the leachability using Minteqa2 is indicated in the individual element plots. The agreement between modelling and actual data is good for Zn, reasonable for Pb, Cd and Cu. Chloride complexation is largely responsible for the discrepancy of the Cd data. In the case of Cu, the leachability in the range of pH 8 – 11 is probably caused by an organically complexed form of Cu [22].

Refuse derived fuel ash – The ash obtained from the combustion of refuse derived fuel (RDF) was tested as a function of pH following the same procedure as for coal ash and MSW residues. Figure 6 shows the results for Pb, Cd, Cu and Zn. The modelling data obtained with Minteqa2 are given for comparison. The modelling of Cd and Zn is in good agreement with the measured leachate concentrations. Apparently, Pb is present in

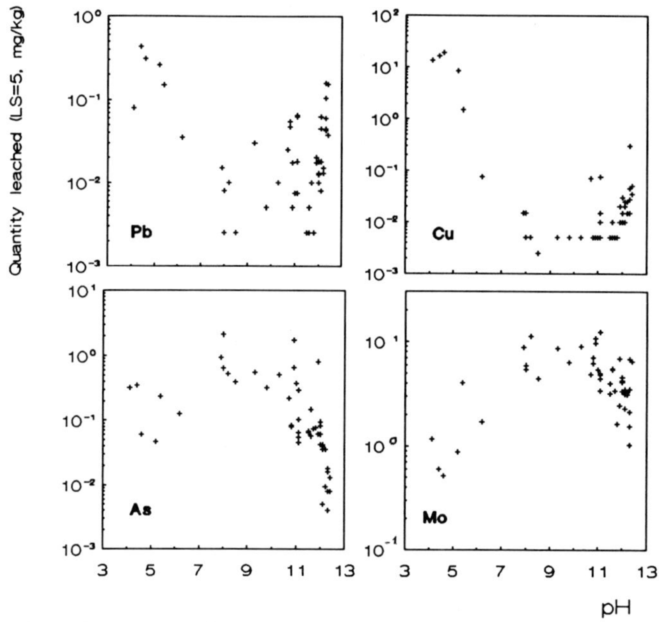

Fig 4. Leaching behaviour of coal fly ash as a function of pH (n=50).

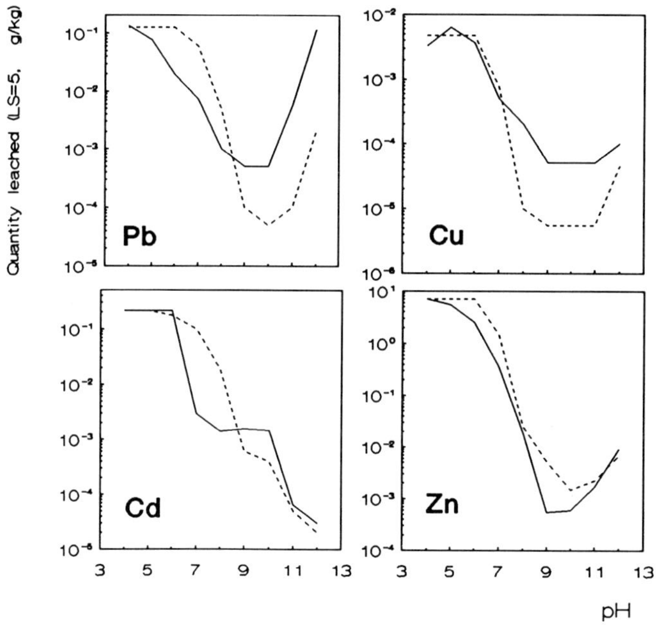

Fig 5. Leaching behaviour of municipal solid waste incinerator fly ash as a function of pH.

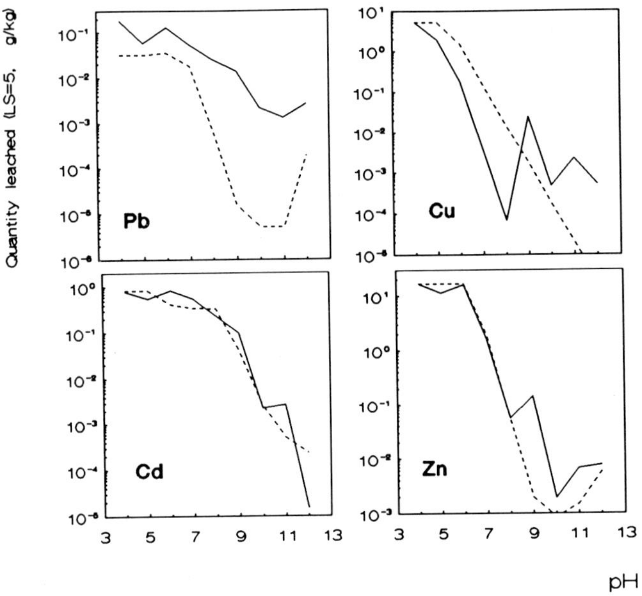

Fig 6. Leaching behaviour of refuse derived fuel as a function of pH.

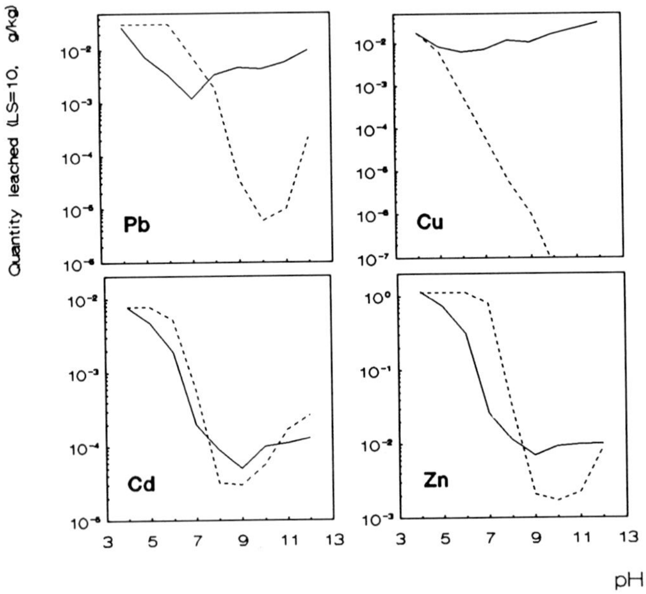

Fig 7. Leaching behaviour of shredder waste as a function of pH.

an as yet unidentified chemical form. The leaching of Cu from RDF ash is very strong. In the pH range 8 - 11 the results are scattered. The RDF matrix is characterized by an extremely high chloride content, which explains the considerable shift in Cd solubility in comparison with coal ash and MSW incinerator fly ash.

Shredder waste - The waste obtained from car shredding contains a fairly high loss on ignition (e.g. fibres, plastics). This material was leached at pH values ranging from 4 to 12. At high pH a strong brown coloration of the extracts was noted. The results of the leaching experiments are shown in figure 7. The Cd data agree well with the modelling. For Pb, Zn and Cu the modelling fits poorly on the measured data. The brown color observed in the extracts is due to dissolved organic carbon (DOC). Since metals are known to be complexed by these substances, the discrepancy may be explained by the presence of DOC-metal complexes. Copper, in particular, is prone to compexation with DOC. Addition of active carbon to remove DOC resulted in substantial reductions in dissolved Cu and Pb.

3.1.2 Redox conditions

The oxidation-reduction state of a waste and its surroundings has a significant effect on the leaching of some contaminants. Under reducing conditions, metal leachability will drop significantly, while the leachability of Ba, Mn, Fe and sometimes As may increase substantially in comparison with oxidized conditions.

Redox conditions can be controlled by the waste material itself. Industrial slags, for example, can create strongly reducing conditions through sulfide leaching. Redox conditions can also be controlled by the environment into which the waste is placed. Biologically active environments can create reducing conditions, whereas contact with air or surface water leads to oxidizing conditions.

The leaching behaviour of materials under reducing conditions is not well addressed by the current test procedures, which are carried out in open contact with the air. To assess the environmental impact of reducing materials, an important question arises in estimating whether the material will remain reducing or become oxidized. A leach test performed without precautions may lead to higher leach rates than are likely to occur in the field. On the other hand, components leaching under reducing conditions may not show up in the standard test and lead to problems in the ultimate application.

Steel slag - Figure 8 shows the leaching behaviour of V for steel slag from a blast-furnace. Under reducing conditions dictated by the steelslag, the leachability of V is controlled by the formation of quadrivalent vanadium, which has an higher affinity for the solid phase than pentavalent vanadium. The availability test, carried out under standard conditions leads to a low estimate of V leaching potential for fully oxidized conditions. In another paper at this conference, the translation of the leaching behaviour of steel slag from lab to field (coastal protection) is addressed [24].

Coal fly ash in a reducing environment - In cases where waste will be exposed to externally controlled reducing conditions, testing under reducing conditions may be required. For this purpose, a leachate based on the sulfur system is being developed [25]. Leaching experiments with coal fly ash using this reducing leachate are presented in figure 8. The metal leachability is clearly decreased, whereas Mo leachability is hardly affected.

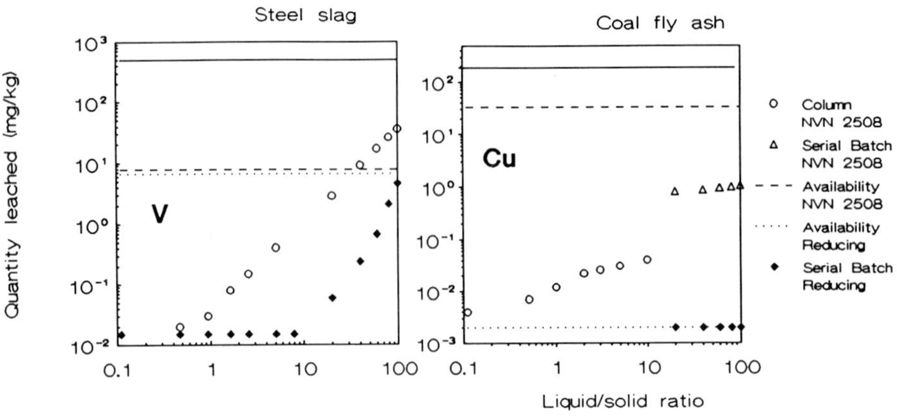

Fig 8. Leaching data for V from steel slag and for Cu from coal fly ash under oxidized and reducing conditions.

3.2 Physical mechanisms

Physical mechanisms controlling contaminant release depend strongly on the geometry of the waste form. In particular, a distinction must be made between granular and monolithic materials.

3.2.1 Granular materials

In most environments, the relatively high permeability of granular wastes means that water percolation will dominate leaching. Strongly soluble components will be washed out of the system within 2 – 3 pore volumes. Components controlled by lower solubilities will leach at a consistent rate leading to a continuous increase in a release–time (LS) plot. Sometimes a component is only leached after wash–out or substantial reduction of another component. In this case, the leachability suddenly increases at some point in time. An example of this behaviour is the leaching of As from coal ash, which is controlled by the interaction with Ca [27].

3.2.2 Monolithic materials

Physical mechanisms controlling the leaching of contaminants from waste forms can also be determined from leaching test data. Mechanistic understanding allows a better prediction of release on the long term and, consequently, a better control over undesired adverse effects at some time in the future. The following mechanisms of release have been identified [12]:
- Surface dissolution. An example is the release of Ca from stabilized gypsum waste [25].
- Initial wash–off. Slag type materials often exhibit initial surface wash–off, as salts coating the surface are leached immediately upon contact with water. After subtraction of the initial wash–off peak, the diffusion controlled release from the matrix can often be observed [12].
- Diffusion control. Diffusion controlled release is apparent for a large number of

components in a wide range of products [7,14,13].

Several studies have shown that physical retardation (tortuosity) may vary over orders of magnitude depending on the nature of the material [7,12,14]. In table I order of magnitude estimates are given for a variety of (waste) products. In spite of the high porosity, the physical retardation factor for light weight concrete is relatively high. This is caused by the relatively high proportion of unconnected porosity. The high physical retardation factor for bituminous concrete is caused by the hydrophobic nature of the material which causes a certain resistance to wetting and, consequently, to transport of contaminants.

Table I. Physical retardation in (waste) products

Material	Physical retardation factor
Unconsolidated granular waste	2.5
Stabilized coal fly ash	10 - 30
Stabilized incinerator slag	40
Lime stone	70 - 100
Light weight concrete	220
Concrete	340
Fly ash concrete	400 - 900
Bituminous concrete	2000 - 10000

3.3 Combined chemical and physical mechanisms

In figure 9, the respective contributions of free mobility, tortuosity, chemical retention and mineral incorporation to the overall mobility (m^2/s) of components in a stabilized product are presented. The relatively high mobility of salts and some anionic species is apparent.

The sensitivity of the release of one element (zinc) to several of the above mentioned parameters is combined in figure 10. It shows a simulated release pattern for Zn from coal ash at three different LS ratios, in which the most important release controlling factors for zinc are included. Relative to the normal release, the effects of increased concentrations of Cl and DOC are indicated. The effect of complexation with chloride is only relevant in the initial phase of leaching as chloride is released from the material much faster than zinc. If the waste contains biodegradable matter, the degradation products may not be important in the very beginning, but increase in importance as time progresses to decrease again once the degradable matter gets depleted. Under reducing conditions the release of zinc is strongly decreased by the formation of zinc sulfides (cf. fig 10). If however at the long-term a reducing material is oxidized, an increase in the release can be expected, since sulfide precipitation is reversible. The combined information shows that a single point in this 3-dimensional space is completely useless for an assessment of environmental impact. By defining the actual field conditions in terms of pH range, redox conditions, complexants and liquid to solid ratio, the accuracy of predictions for release of a specific component can be improved substantially. The release curves for points A and B in figure 10 coincide. Due to the difference in the dominant chemical species involved in both cases, the release may be the same in terms of quantity, but the net effect in the surrounding will be markedly different as a

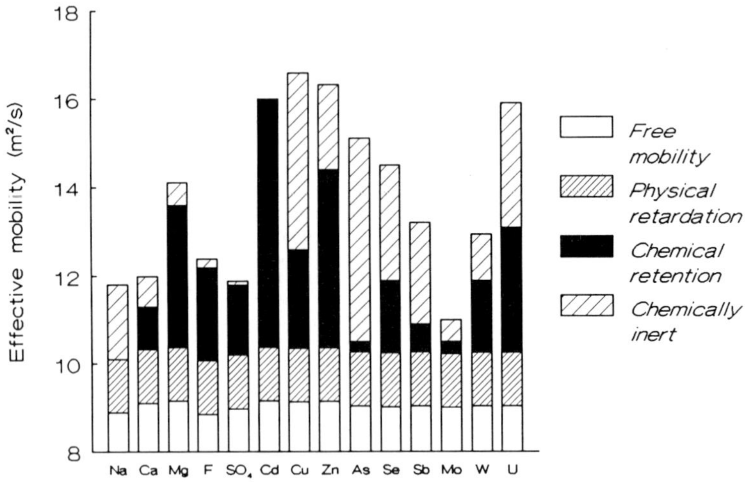

Fig 9. Contribution of free mobility, tortuosity, chemical retention and mineral incorporation to the overall mobility in waste products.

zinc chloride complex behaves differently than an organic-complexed zinc species. This aspect of speciation is important for assessing the net impact of a waste on its surroundings, but it is at present not covered in any existing regulation.

4 SYSTEMATIC TRENDS WITHIN WASTE CATEGORIES

Within a waste or waste product category, the behaviour of individual contaminants or groups of contaminants has so far been proven to be quite systematic. Large variations in major element chemistry may cause differences in behaviour, but these can be largely explained by identifying the factor controlling the variability.

4.1 Coal fly ash

The leaching results obtained for coal fly ash prove to be very consistent (figure 4) indicating that after a detailed characterization sufficient knowledge is accumulated to use simplified methods for quality verification. In addition, the number of elements relevant for an assessment can also be reduced substantially.

In a leaching study of 50 coal fly ashes from different sources, the variability in leaching behaviour is largely explained by the behaviour of components as a function of pH [21]. The chemical phases controlling the release of trace contaminants still needs to be quantified to allow more accurate modelling. Relevant elements for coal ash leaching in the short term are Mo, B and sulfate.

4.2 Municipal solid waste incinerator residues

The results of leaching tests on MSW incinerator residues are more consistent than the chemical composition would lead one to expect. Table II shows leaching data from different MSW incinerators [28].

Table II. Leaching data from MSW incinerator residues (mg/kg).

Element	Bottom ash[28]		Bottom ash[7]	Fly ash[7]
	Average	Std	Range	Range
As	.044	.006	0.2	0.1
Cd	.003	.001	< 0.005	0.05
Cr	.31	.06	0.2	9
Cu	8.4	5.5	5 - 8	0.2
Mo	1.2	1.2	0.3 - 1.1	7
Ni	.12	.08	< 0.2	< 0.2
Pb	.85	.7	1.4 - 2.6	120
Zn	.7	.3	< 1	20

In MSWI bottom ash, Cu and Mo are elements of concern in terms of quantities released. From MSWI fly ash, the release of Pb, Mo and Cr is of importance, although in this case the final storage pH is crucial. For any assessment of environmental impact, these data have to be interpreted in terms of the conditions encountered in the site of application or storage [2].

4.3 Cement-based products

The leaching behaviour of cement-based products is largely dictated by the high alkalinity of the matrix. This condition dictates a fairly narrow range in leachability of major elements, trace metals and oxyanions. Figure 11 shows the retention factor in cement-bound products as a function of pH. At very high pH, leachability of metals tends to increase, whereas that of oxyanions decreases at extremely high pH [13,21].

4.4 Bituminous products

The leachability from bituminous materials is largely dictated by the hydrophobic nature of the materials, which leads to a limited uptake of water in the product and consequently a low release of all constituent from the interior of the product [14].

4.5 Sintered products

The leaching of sintered products, which have been exposed to temperatures higher than 1000 °C, is characterized by a relatively high leachabilty of arsenic and molybdenum. Metals, such as Pb, Cu and Zn, are largely incorporated in the silicate matrix. Major elements Ca and Na are also largely incorporated in the silicate matrix [29].

4.6 Industrial slags

The leachability of industrial slags, such as blast-furnace slag, steel slag and phosphate slag, is characterized by the occurence of surface wash-off effects and by the reducing properties due to the presence of sulfides [30]. This aspect dictates a sulphur speciation in the contact solution which strongly influences the solubility of many

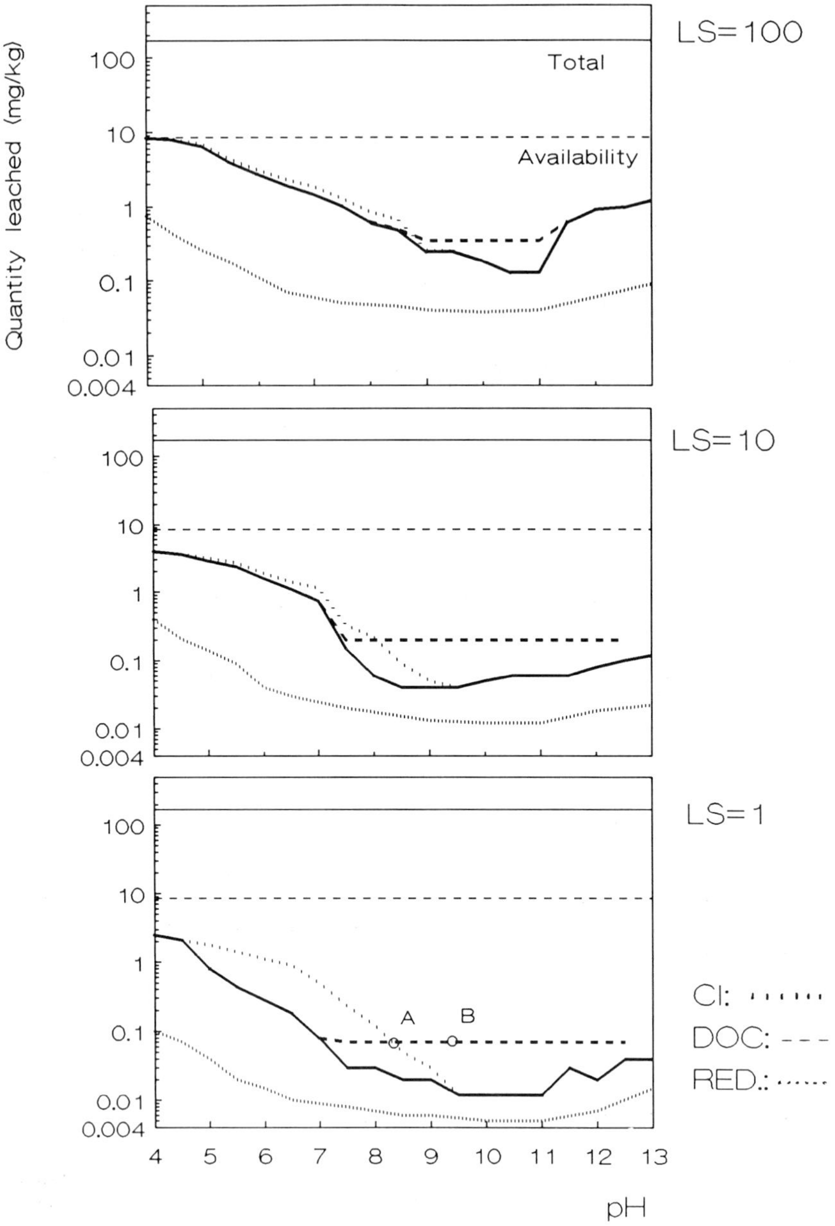

Fig 10. Evaluation of long term release of Zn from coal fly ash as a function of pH and liquid/solid ratio (time) under influence of increased Cl– or DOC–concentration or under reducing conditions.

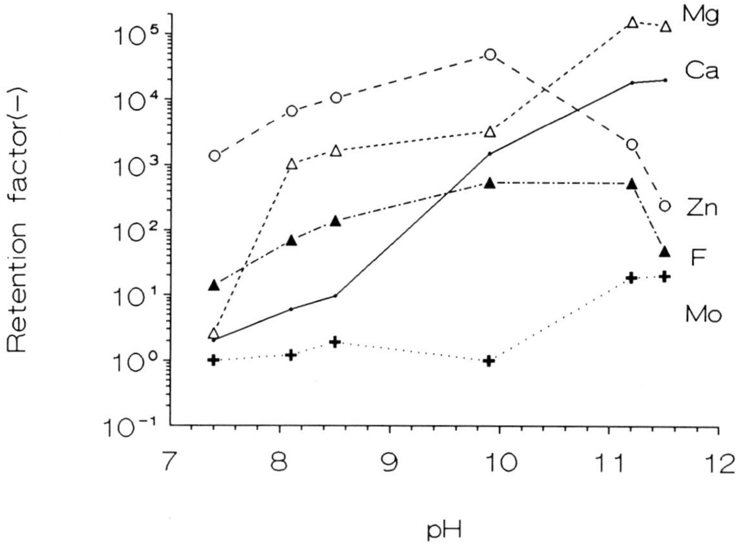

Fig 11. Retention factors for Ca, Mg, Zn, F and Mo from cement-stabilized products as a function of pH.

major and trace elements. Steel slag and phosphate slag are characterized by V and F release, respectively [7].

5 DISCUSSION

5.1 Application of waste products

In the assessment of different applications of secondary materials, similarities in factors such as pH, redox condition, exposure CO_2, temperature, amount and mode of water contact need to be taken into account. For example, the use of stabilized waste as coastal protection may involve variable redox conditions, a relatively constant pH, continual contact with water at a constant temperature [24]. Other applications requiring a similar characterization include:

- road base
- embankments
- construction above groundwater
- construction in contact with groundwater

Relatively few secondary materials can be utilized in all of these options. Certain materials may exhibit properties excluding application in a particular area. In the Netherlands, the utilization of MSW incinerator bottom ash is restricted to applications at least 0.5 meter above groundwater level. Conversely, blast-furnace slag, which exhibits reducing properties, should be placed under the same restrictions. Consideration of leaching systematics along with application systematics can lead to more rationally based regulations.

5.2 Leaching database and certification

Since collecting data by the detailed testing protocols described above is expensive and time consuming, it is important to insure that maximum information is gained. A well organized database is a powerful tool to extract useful information from raw data. The leaching database discussed in another paper at this conference [31] compiles results from leach tests and diffusion tube studies of many wastes and waste products. Effective use of this tool allows systematic leaching behaviour to be identified. Knowledge of these systematics can ultimately reduce costs associated with environmental testing by directing emphasis to the most important parameters and critical components.

Once the leaching behaviour of a waste materials has been adequately characterized, simpler procedures can be applied to reach the same level of confidence. These short tests may well be used in the certification of those waste materials that show consistent behaviour [12,31].

5.3 Optimization of technical measures

The systematic trends observed in the laboratory can suggest improvements that would result in a more environmentally acceptable product. Both physical and chemical changes are possible and the physical and chemical retardation factors measured in the laboratory provide a key to assessing the environmental quality of the product.

Physical – Options to reduce release of contaminants from waste by physical means include increasing the proportion of unconnected porosity and turning the matrix hydrophobic. An effective method of encapsulation is to render the surface layer of a monolith more impervious, thus sealing the bulk of the stabilized waste from the surroundings [17]. All of these methods can be characterized by their effect on physical retardation (tortuosity).

Chemical – control of chemical retardation can result in order of magnitude changes in leachability, but its use is now limited by poor understanding of chemical fundamentals. Factors such as pH, redox and complexing agents can be modified to affect the release properties of a mix design. However, improving the retention for one component may result in a substantial increase in the leach rate of another. The influences of chemical additions on the physical properties of waste products should also not be overlooked.

5.4 Evaluation of long term effects

To arrive at a conclusion on the acceptability of a material in a given situation requires that the magnitude of the contribution of different factors to the contaminant release be known. Starting from the major variables, liquid solid ratio, pH and redoxpotential, a release pattern can be identified for individual contaminants. Other variables such as complexation, which can modify this general pattern, should be identified for each material. Modeling the chemical influences on release with chemical speciation models, such as Minteqa2 is very useful to assess the magnitude and sensitivity range for individual variables.

It is important to realize that it will be very hard to create a 1:1 relation between lab tests and field data for all constituents of interest. In the translation, a number of factors need to be taken into account:

- temperature
- mode of contact with water (permanent, intermittent, superficial).
- channeling effects
- pH changes over time
- redox changes over time

- contact with the air (O_2 and CO_2)
- waste/soil interactions

For a proper assessment of environmental impact, it will be necessary to model a number of field situations of utilization and disposal of waste materials to identify factors between lab tests and field data to improve and maintain the balance between predicting tools and actual observations in the field. The interaction between waste and soil should be addressed in these evaluations, as a substantial reduction or increase in release can be observed as a result of direct contact between different media at waste/soil interfaces.

6 CONCLUSIONS

Systematic behaviour of waste materials and products has been established at a limited scale to date. Release mechanisms and factors controlling release have been identified and to some extent quantified. Systematic behaviour can be identified with respect to individual contaminants or groups of contaminants. Systematic trends are also noted in the behaviour of a specific waste material or group of waste materials of similar origin. Finally, specific applications can be classified and each category treated in a common way to assess the environmental impact.

A full characterization of a waste category involves the determination of primary release mechanisms, relevant components and the pH and redox sensitivity of specific components. Once reproducible results are obtained and the controlling factors have been identified, the testing program can be substantially reduced. A leaching database will be an essential tool in identifying systematic trends and in achieving optimal use of existing information.

For utilization of waste materials in construction, more control over contaminant release is needed. Prediction of the environmental impact of a waste material utilization requires a fundamental understanding of leaching systematics and knowledge of the conditions to which the material will be exposed during and after its useful life. By defining the environmental conditions more precisely, the range in the predicted long term release of contaminants can be narrowed down substantially in comparison with the pretentious results of most regulatory tests.

The combination of the geotechnical and geohydrological aspects of applications of waste materials in construction and the leaching characteristics of these materials is not sufficiently addressed in assessing and controlling the potential long term impact from the beneficial use of industrial residues.

Acknowledgement – The financial support from the Netherlands Agency for Energy and Environment (NOVEM), the Ministry of Waterworks (RWS–RIZA), the Ministry of Public Housing, Physical Planning and Environment and the Ministry of Economic Affairs is gratefully acknowledged.

7 REFERENCES

1. Compendium of waste leaching tests. Environment Canada Report EPS3/HA/7, 1990.
2. H.A. van der Sloot. Waste Management and Research, 8, 1990, 215–228.
3. DIN 38414 S4: Determination of leachability (S4). Institüt für Normung, Berlin, 1984.

4. Déchets: Essai de Lixiviation X 31-210, 1988. (AFNOR), Paris.
5. Bericht zum Entwurf für eine technische Verordenung über Abfälle (TVA), 1988.
6. TCLP. Federal Register, Vol No 261, March 29, 1990 (final version).
7. C.W Versluijs, I.H Anthonissen and E.A Valentijn. Integrale evaluatie van Mammoet '85. Report 738504008. RIVM, June 1990.
8. EP Tox. Appendix II, Federal Register, Vol 45(98),1980,33127. Washington D.C.
9. NVN 2508 Determination of leaching characteristics of inorganic components from granular (wastes) materials. Dutch Standardization Institute NNI. Revision Dec. 1990
10. O. Hjelmar. Leachate from incinerator ash disposal sites. Int. Workshop on Municipal Waste Incineration. Montreal, Canada Okt 1987.
11. NVN5432 Draft. Determination of the maximum leachable quantity and the emission of potentially hazardous components from construction materials and stabilized waste products of mainly inorganic character.1990.
12. G.J. de Groot and H.A. van der Sloot. Proc. Int. Symp. on Stabilization/solidification of Hazardous, radioactive and mixed wastes, May 1990, Williamsburg, VA.
13. H.A. van der Sloot, G.J de Groot and J. Wijkstra. 1987. In: Environmental aspects of stabilization and solidification of hazardous and radioactive wastes, ASTM STP 1033, P.L. Côté and T.M. Gilliam, Eds., ASTM, Philadelphia, 1989, pp 125-149.
14. G.J de Groot, H.A van der Sloot, P.Bonouvrie and J. Wijkstra. Karakterisering van het uitlooggedrag van intakte produkten. Mammoet deelrapport 09. March 1990.
15. R. Smith. Proceedings Second Int. Symp. on Stabilization/ solidification of Hazardous, Radioactive and Mixed Wastes, May 29-June 1,1990, Williamsburg, VA.
16. H.A van der Sloot, D. Hoede and P. Bonouvrie. Comparison of regulatory leaching test procedures (in preparation 1991).
17. D. Hockley and H.A. van der Sloot, Long-term processes in a stabilized waste block exposed to seawater. 1990. Accepted ES&T 1991.
18. H.A. van der Sloot, O. Hjelmar and G.J. de Groot. In: Flue gas and fly ash, Eds. Sens, P.F. and Wilkinson, J.K., Elsevier applied science, London, 1989.
19. H.A van der Sloot and P.L.Côté. Environm. Technol. Lett., 10, 969 - 976, 1989.
20. D. Hockley and H.A avn der Sloot. Modelling of interactions at waste-soil interfaces. These proceedings.
21. G.J. de Groot, H.A. van der Sloot and J. Wijkstra. 1987. In: Environmental aspects of stabilization and solidification of hazardous and radioactive wastes, ASTM STP 1033, P.L. Côté and T.M. Gilliam, Eds., ASTM, Philadelphia, 1989, pp 170-183.
22. J.V. Di Pietro, M.R. Collins, M. Guay and T.T. Eighmy. Leachability of Municipal Solid Waste Incinerator Residues. Int. Conf. on Municipal Waste Combustion, Vol. 1 , Hollywood, Florida U.S.A, April 1989.
23. H.A van der Sloot, G.J. de Groot, J. Wijkstra and P. Leenders. Ibid.
24. R. Comans, H.A van der Sloot, D. Hoede and P. Bonouvrie. Chemical processes at a redox/pH interface arising from the use of steel slag in the aquatic environment. These proceedings.
25. ECN Unpublished results (1990).
26. MINTEQA2. Metal Speciation Equilibrium Model.1988 U.S. EPA, Georgia, U.S.A.
27. R.G. Robins. The aqueous chemistry of arsenic in relation to hydrometallurgical processes. Proc. 15th Int Hydrometallugical Meeting. Vancouver, August 1985.
28. Kwaliteits controle AVI slakken '88-89. Tauw infra consult. Report 51467.15, 1989.
29. R. Gerritsen. Personal communication. 1990.
30. F.P. Richardson. In: Physical chemistry of Melts in Metallurgy. Vol 2, Chap. 8: Slags and mattes. Academic Press, 1974, 291-327.
31. G.J. de Groot. Netherlands leaching database: A useful tool for product quality control, environmental certification and evaluation of leach test results. These proceedings.

Waste Materials in Construction.
J.J.J.R. Goumans, H.A. van der Sloot and Th.G. Aalbers (Editors)

WASTE POLICY RELATED TO THE NATIONAL ENVIRONMENTAL POLICY PLAN

Gerard Delsman

MINISTRY OF HOUSING, PHYSICAL PLANNIGN AND THE ENVIRONMENT, P.O.BOX 450, 2260MB, LEIDSCHENDAM, THE NETHERLANDS

The problems with waste are big. In particular in a country like the Netherlands, with its high population density, shortage of space, and, due to geological and hydrological circumstances, vulnerable to environmental pollution, it is almost a matter of survival to overcome the problems. Therefore it is no wonder that waste-policy has top-priority within the environmental policy. It also is no wonder that technology development is an item which is very good looked at, in order to supply instruments needed to seek environmentally sound solutions.
Before I continue to discuss all this, I like to start with an overview of the Netherlands' environmental policy, the place of waste-policy in general and after that, the role technology can, or even has to play in this policy. And last but not least I shall tell something about the use of waste as construction material.

1. Environmental policy

As most people will know, the starting point of the Netherlands' environmental policy is the so-called sustainable development. This conception, which derives from the report of the Brundtland Commission, contains a development which provides for the needs of the present generation, without endangering the possibilities for the future generations to provide for their needs too.
This sounds complicated, but in plain English this means, that we will have to save some of the natural resources of our earth, and that we will have to leave the earth clean.
In the National Environmental Policy Plan (NEPP), published in 1988, this philosophy has been translated into workable elements, which are needed to reach sustainable development. These elements are:

1. integral chain management, aimed at the closing of as many material cycles from raw materials to waste materials as

possible, through which the development of emissions and waste flows will be limited;

2. energy-extensification, aimed at the reducing of the use of fossil fuels by saving energy and a more intensive use of sustainable energy sources like the sun and wind;
3. improvement of quality, through which especially durable consumption goods can be used longer and will cause less environmental problems being waste.

It should be clear that the first and the third element are the most important with respect to waste policy. The importance of these elements is lying in the fact that practically everything we do in the environmental policy can be traced back to these elements, with the exception of recuperation of environmental damages like soil clean up and noise abatement measures.

Besides these elements, we also handle another classification in eight central themes. For the waste policy the most important themes are Disposal and Squandering. But relations exist also with other themes. One example is the theme Climatic Change. I will, however, not go further into that matter, but I will concentrate on Disposal. Before that I will give you a brief summary of what we are talking about in the Netherlands.

2. Waste in the Netherlands

In my country there is an annual production of waste of approximately 55 million tonnes excluded dredging sludge, which adds to this figure some 60 million tonnes per annum.

Looking at the contribution of the different targets groups of our environmental policy, the following picture can be shown:

- agriculture 23.4%, of which most is manure surplus
- traffic 1.7%
- chemical industry 6.3%
- refineries 0.2%
- power plants 1%
- consumers 8.8%
- building industry 12.8%
- the remainder of which most is generated by the not mentioned industries 45.8%.

In 1986 35% of all waste, excluded dredging sludge and manure surplus, was re-used or otherwise useful applicated. 10% was incinerated and 55% dumped. With respect to dumping, over 15 million tonnes per annum is dumped, which means the use of 75

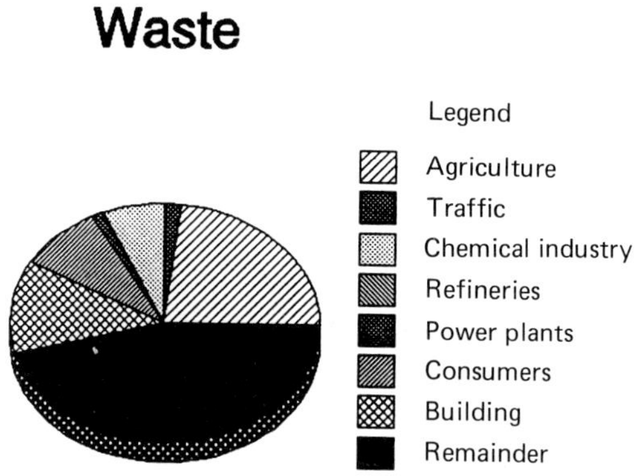

hectares with an average height of 15 metres. You can take for granted that this all gives big problems in a country like the Netherlands, due to its high population density, land scarcity and geological and hydrological circumstances.

3. Problem analysis

As I said, the disposal of waste has almost always been a problem. In the past it used to be a public health problem, because all waste was just dumped without thinking of the consequences. Nowadays, we identify a wide range of problems, that will all have to be solved. I will name a few, without being really exhausting:

- the space that dumping grounds need, while space is very scarce in the Netherlands;
- the quality of waste has evidently changed in the past decades, it contains much more chemical or badly biodegradable materials;
- the quantity of waste is still increasing, by which many bulky things are found, especially in domestic waste, while the prognoses say that the quantity of waste is still growing. One simple example: in 1986 all Dutch people received per capita 5 kilos of advertising pamphlets in their mailboxes, in total 70,000 tons. The prognoses is that this quantity will be tripled by the year 2000, 220,000 tons, about 15 kilos on a per capita basis;

- waste incineration is chosen as a solution to the volume problem, but I hardly need to tell that burning itself creates another problem: air pollution. Especially dioxines are a problem, which is, however, solvable. Moreover, burning gives residues, which cannot be just dumped either;
- at this moment there is a lack of processing capacity, which can, besides that, hardly be solved in the short term. This has to do with two matters:
  1. in the past the lack of capacity was disguised by the export of waste, something we very much object to;
  2. if choices were made for processing installations, than the so-called NIMBY-syndrome comes around: "Not In My BackYard". Perhaps this will bring us the situation that the government is urged to commission locations on the basis of the Physical Planning Act.

In short, the problems can be understood with the words quantity and quality. The questions then, of course will be how to solve the problems. This brings me to the waste policy itself.

4. Waste policy

The waste material policy aims itself on full control of waste flows, from the creation of waste until the final destination.
In this respect a distinction can be made between:

- the prevention of creating waste, recycling and useful application of what then can better be called rest materials; in short this is called prevention;
- the improvement of the disposal structures and where necessary the development of new structures in order to have a so-called "leak-proof" disposal by the year 2000.

The first is a matter of policy recorded in the note Re-use and Prevention of Waste. In this note the aims are a drastic decrease of the quantity of waste to be dumped, from 55% now to 10% by the year 2000, an increase of the incineration of waste and a very large increase of re-use and useful application. Besides this, 10% in quantity less waste should be produced by then. This will be attained by determining 29 priority waste streams. In consultation with the industry prevention and re-use programmes are developed. These priority waste streams are divided in big

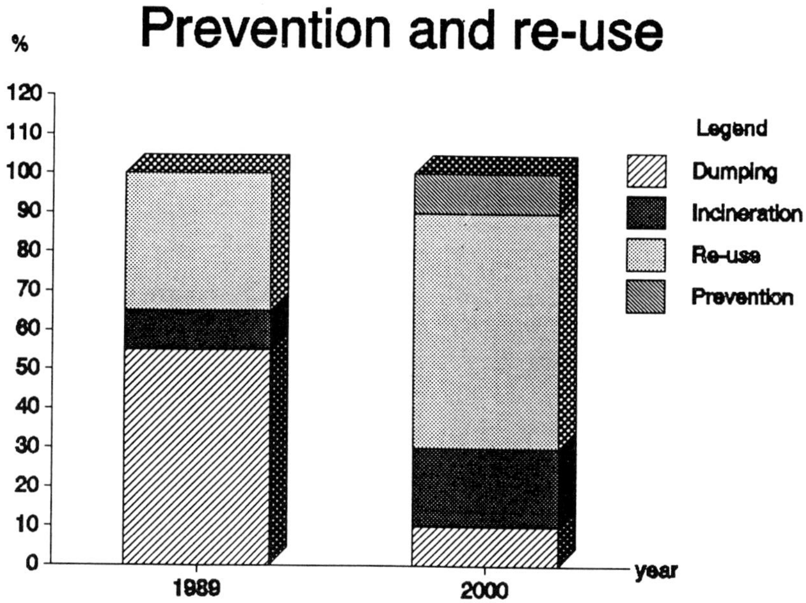

streams, i.e. over 100.000 tonnes per annum (jarosite, manure surplus, building and demolition waste, dredging sludge, slags from the incineration of domestic and industrial waste, sewage purification sludge, synthetic material waste, packing waste, oxi-lime sludge, phosphoric acid gypsum, contaminated soil and cargo remainders, wash-water, chemicals) and little streams, i.e. less than 100.000 tonnes per annum (batteries, fly ashes from the incineration of domestic and industrial waste, halogenated hydrocarbons, painting waste, shredder waste, exhausted oil, mordant baths of thermic galvanization plants).

Most of the prevention programmes have already been developed, some of them are in the stage of implementing.

In this respect waste policy can be seen as a chain consisting of four links in order of preference: prevention, re-use, incineration and dumping.

In order to achieve these goals, in particular reduction of waste, a policy has been developed consisting of three main features:

- the responsibility of producers for goods being waste should be strengthened.

- the introduction of a duty for producers to take back their products being waste linked to a re-process regulation.
- volume measures.

The latter refers in particular to exceptionally growing waste streams, which will or cannot be re-used and to environmentally hazardous waste.

The second point, improvement of the disposal structure, is a programme that is in development at this moment.

In the NEPP strategic choices have been made with respect to waste policy. Two of these choices are in particular important:

1. Technical and logistic improvement of the disposal structure in order to limit the risks of the disposal of all in the Netherlands generated waste streams to an acceptable level before the year 2000;
2. Preparation and stimulation of structural measures with respect to re-use and prevention of waste materials.

It should be clear that both choices require the use of existing technology and the development of new technology.
From these strategies a number of activities have been derived, two of which I will mention:

* screening waste streams to determine high priority waste, and
* research of new systems and improvement of existing disposal systems.

Before I go on talking about technology, I will give a rough drawing of the matters one should watch while developing waste programmes. These aspects itselves indicate how complicated the waste removal is, when at first it seemed so simple.

5. Important aspects

The aspects are the following:

- Integration with other environmental policy fields . As I said in my introduction, disposal of waste has to do with almost all central environmental themes. Then it is logical, that with creating the policy, all the other fields should be watched;

- make priorities in waste flows; especially because there are so many other problems, it is not very efficient to put one's energy in waste flows which are only a small problem;
- chain management, aimed at creating as little unnecessary looses of raw materials, and with that of waste, as possible.
- improvement of the quality of waste on the one side to make re-use easier, on the other side to make sure that waste is causing as little environmental damage as possible while dumping or incinerating;
- professionalize the waste disposal system; nowadays the complaint is often that the structure is not transparent; that there are too many provisions for waste that is relatively simple to process, but that there is a lack of more progressive techniques and too many collectors of chemical waste with special licenses. In short, the ones who have to get rid of waste, do not always know to whom they can give this waste;
- in principle disposal of waste in the country where this waste was created, with the exception of some small special waste flows, which can better be processed in cooperation with neighbouring countries;
- and last but not least, international cooperation and especially harmonization of laws and regulations at EC-level.

6. Waste programmes

After this expose, no one will wonder that a few programmes are being developed aiming especially at these specific matters. I shall not bore you with all the programmes, but I will aim especially at the subject technology. This programme, called T-2000 (T is standing for technology), is still in development.
In general it can be said that developing technology for so-called end-of-pipe solutions and for the promotion of recycling and useful application has been a major aim for a considerable time. Even great progress has been made in the last ten years. Recycling itself has increased, while there have been developments in the improvement of techniques for sampling, measurement and analysis, and the setting of standards. An effect-specific approach does not usually coincide with an

integral approach to environmental problems, nor with integral chain management.
Therefore much attention is now being paid to source specific measures and improving the quality of raw materials. Source oriented measures lead to more economical use of raw materials, at the same time reducing the need to clean materials and dispose of waste. Developing new technology is time consuming. Yet we have reason to be optimistic. But to accelerate the development, we use the programme T-2000.
The goals of this programme are as follows:

* stimulation of innovations in the field of technology with respect to waste disposal;
* stimulation of the use of technologies in order to prevent the creation of waste;
* stimulation of the use of technologies in order to improve the quality of waste;
* making available more basic technologies for waste disposal;
* taking care of the availability of processes and techniques for appropriate processing of waste;
* striving after diminishing the risks of the dispersal of environmentally hazardous waste.

These goals can be translated into activities which give shape to the technology policy.
The activities are clustered according to the earlier mentioned chain of the waste policy:

Prevention:

* stimulating that designers of production processes take into account the integration of the environmental aspects;
* reduction of secondary waste flows by:
  - innovations in the field of purification processes;
  - integral consideration of environmental interests with respect to the different environmental compartments (air, soil, water).

Useful application:

* stimulation of the development of new markets for secondary goods;
* technology development aiming at less dumping, relatively less incineration and more useful application of waste.

"Leakproof" disposal:

* stimulation of broader employment of existing technology for the disposal of waste;

* stimulation of broader employment of existing techniques for the quality improvement of waste;
* development of new technology for the disposal of waste and for quality improvement of waste;
* developing of technologies for the treatment of extracts and remnant-materials, which get off at waste disposal;
* improvement of existing disposal technologies.

Doing so, you have to look carefully to every wastestream separately, but also to wastestreams in connection to other wastestreams in order to avoid double work. Let me give you an example. The wastestreams zinc and cadmium can be seen apart from each other. But since cadmium is a residue of zinc production it is better to tackle these problems together. In other words, to avoid cadmium, you have to get to grips to zinc production.

And last but not least, it has to be taken into account that the factor time is playing an important role in policy development. Assuming that environmental policy is anticipating, the question arises how decisions on infrastructure and production investments can be attuned to the prior conditions of sustainable development. A common and broadly supported image of sustainable development is to be considered most important, if anticipation of industry is more desirable than the "actual rules". Next to the discounted cash flow, the discounted environmental burden should be an appreciation criterion with respect to the weight of sacrifices now against profit in the future. In other words: even if technology development seems not to be profitable in the short term from an economic point of view, it can be worth while to go on developing that particular technology looking at the future environmental profit. Moreover, in the long run these investments can even be profitable in economic terms.

7. Negative factors

Formulating targets and setting up programmes are one thing. To get things done another, because there are a few negative factors which hamper the smooth introduction of technologies. Without being exhausting, I mention a few:

- if dumping is too cheap, incineration and re-use of waste will be bogged down;
- also if incineration is relatively too cheap, re-use and prevention will not come off the ground;

- it is hardly possible to force industry to develop technologies if there is no incentive;
- secondary goods give sometimes environmental problems, for instance fly ash and slags used in construction materials can lixiviate and pollute soil and groundwater. Therefore secondary goods in general have a bad name and are difficult to sell;
- New materials are being made, for instance engineering thermo plastics, which consists of various compounds. This also hampers recycling, because on the one hand these complex materials are hard to recycle, on the other hand the variety of materials make the quantities relatively small, so recycling cannot be done economically.

So it is not enough to stimulate the use of existing technology and the development of new technology to get to grips with the waste problem. More need to be done in terms of creating instruments and prior conditions.

8. Instruments

This brings me to the development of instruments. The easiest way to get things done seems to be giving a wagonload of money to the R&D departments of industry. Apart from the budget deficit of the government, we don't think it is the right way. Of course there have to be some financial incentive, but that should at least be temporarily.

So we decided that the main instrument to get technology development aimed at sustainable development, is setting striding standards. It means that in the course of time more severe standards will set on products and production processes, which at the time those are set, cannot be meet. This sounds strange, but it proofed to be working. For instance, the State of California acted the same way years ago with respect to standards on exhausting gas of cars. The result was a technology development which nobody had expected.

In the NEPP the following line of strategy to pursue has been formulated:

"Challenging and progressive standards for products and production processes, to be developed jointly with industry as much as possible. The standard setting should be aimed at emission reductions, possibilities for integral lifecycle management, energy extensification and quality promotion."

It is expected that doing so encourages technology development. One Netherlands' example shows that it must be possible. In 1989 the Incineration Guideline has been published, in which high standards with respect to emissions into the air. All new incinerators will now be build according to the newest available technology in order to meet the standards. The old-ones, which cannot be modified, will be closed by 1993.

Next to setting standards, other activities are required in order to encourage technology development. Most important in this respect are:

* stimulation of research and development programmes with respect to environmental technology and environmentally sound products;
* stimulation of displays and implementation of clean technology;
* stimulation of know how transfer of nationally and internationally available applications of clean technology;

Of course industry and research institutes have to participate in this respect.
It should be clear that technology development itself is not an aim, but just a means like legislation and financial regulations. A mixture of the means will be chosen, aimed at effective and efficient solution to the waste problem.

9. Construction

The next subject I want to talk about as promised, is construction and the use of secondary materials. In the Netherlands approximately 100 million tonnes of new raw materials per annum are used in construction. Apart from this quantity, approximately 10 million tonnes of secondary materials are used.
One of the main problems is that reserves of some primary raw materials that can still be won ore mined economically, will run out between 15 and 50 years time. Therefore there is an urgent need for re-use of construction and demolition waste. But also this need exists because of the still growing amount of waste.
Therefore a programme has been developed, sustainable construction, which inter alia aims at a doubling of the use of secondary materials in order to save primary materials. Therefore it is needed that the possibilities of separating waste demolishing a building are taking into account while designing

and building it. Also the use of materials which cannot be re-used have to be diminished or even stopped.
A problem, which I already mentioned, is the fact that some secondary materials are polluted, so using these can cause pollution of soil of groundwater. I mentioned earlier the possibility of lixiviation of fly ashes and slags. Consequently ways have to be found in order to prevent these consequences. There are some immobilisation techniques available, but as far as we know, they are not good enough in terms of durability. So other techniques have to be developed.
Apart from this specific problem, research is being done in general into how the use of secondary materials can be increased and the reserves of primary materials can be saved. Research is also being done into the improvement of the quality of raw materials by removing pollutants in order to increase the possibility of their re-use being waste.

10. Conclusion

I have tried to explain the way the Netherlands' government is acting with respect to the environmental problems in general and to waste in particular. In all humility I dare say, that we are well en route in tackling the problems. Still we are not at the end.
I want to stress here that international cooperation is also required if environmental targets are to be achieved in the Netherlands and elsewhere. This applies with even greater force if a source oriented approach is chosen, as in fact the Netherlands did.
Moreover, the Netherlands is too small to develop and adapt technologies on its own, also because the necessary standards to get things from the ground, only can be established and enforced internationally.

Waste Materials in Construction.
J.J.J.R. Goumans, H.A. van der Sloot and Th.G. Aalbers (Editors)

MANAGEMENT OF WASTES RESULTING FROM BUILDING ACTIVITIES IN THE FEDERAL REPUBLIC OF GERMANY

J. KUEHN

Department WA II 3 - Avoiding and Re-use/Recycling of Noxious Wastes, Federal Minister for the Environment, Nature Conservation and Nuclear Safety, P.O. Box 12 06 29, D-5300 Bonn 2 (Germany)

SUMMARY

This paper gives a short overview about the approaches to noxious and non-noxious wastes resulting from construction work. There are two ways the political will is expressed, by objectives and by statutory ordinance.

1. INTRODUCTION

The building trade of the Federal Republic of Germany within the boundaries prior to October 3rd, 1990 had, by estimate of the Statistisches Bundesamt, to manage following amounts of materials:

	Mio. t,	thereof re-used Mio. t,	%
structures	22,6	3,7	16
site waste	≈ 10,0	0	0
soils	167,9	53,3	32
roads (bed & coat)	20,4	11,2	55

Those materials are problematic in two ways:

1. The large amount of material disposed yet more and more reduces the disposal capacity which is limited anyway.
2. The co-disposed portion of noxious waste leads to a potential hazard to ground and surface waters.

The rather small portion of re-used material leads, beside the claim of limited disposal capacity, to a ruinous exploitation of the resources. These include not only raw materials, but also accompaniments like destruction of the landscape by the winning of sand and gravel, interference to water economy, changes to the scenery by winning of minerals in quarries etc.

One of the reasons why the recycling-quote is that low is due to discrimination of re-used material in regulations and standards. They require - without technical substantation - un-used, new materials.

Another reason for the rightnow low, not to say unsatisfactory recycling-quotes, is the portion of noxious admixtures especially to waste from old structures. These are not only substances hazardous to health or environment, but in the same way admixtures which hamper or prevent re-use.

There are a lot of noxious admixtures in wastes resulting from building activities because

a) noxious admixtures are not recognised as noxious or they are underestimated in possible effects,

b) they are concidered negligible by quantity.

2. THE SCOPE FOR THE LEGISLATOR

On grounds of the last two mentioned topics the legislator is forced to react. Article 14 of the Federal Waste Act authorizes the Federal Government to release statutory ordinances.

In the approaches to statutory ordinances there are some other possibilities to move the concerned parties to conformable actions.

Aimes:
Reducing the endangerings to waters and soil
Preservation of the mineral resources

Obstructions:
Discriminating regulations and standards
Noxious admixtures to potentialy recyclable materials

Solution:
Removal of discriminating regulations and standards
Reducing the noxious load in potentialy recyclable materials

Measures:
Statutory ordinance bases upon art. 14 Waste Act with the obligation of separate disposal (keep, collect and manage separately; prohibition of mixing; prohibition of deposal of re-usable material)

Self-commitment of the industry

The first opportunity is a voluntary engagement of the industry. So done in the accepting of returned discharged batteries.

Objectives specified by the Federal Government

A bit more compulsory but still based upon agreement are the objectives to be reached within an adequate period of time for avoiding, reducing or re-using/recycling waste arising from certain products. This instrument is provided in the Waste Act to reduce the quantities of waste and to point out the political aimes to the concerned parties. If the objectives are not reached the legislator is authorized to interfere with statutory ordinances to the problem of waste quantities. Such objectives were provided in concern of quantities of waste from building activities.

Statutory ordinances

By article 14 of the Waste Act the Federal Government is authorized to provide statutory ordinances,

a) to avoid or reduce noxious substances in waste or to ensure their environmentally compatible management (Art. 14, Para 1),

b) to the extent this is required for avoiding or reducing the quantities of wastes produced or for environmentally compatible management, especially to the extent this is not possible by specifying objectives (Art. 14, Para 2).

Two conditions must be fulfilled prior to publication:

a) there must have been a hearing of the parties concerned,

b) the consent of the Bundesrat is necessary.

Remark:

Concerned Parties:	Commerce, industry, trade, consumers, agencies for environmental protection (Non-Governmental Organisations)
Bundesrat:	Parliment of the Federal States (Länder)

3. MEASURES CONCERNING MANAGEMENT OF RESIDUES AND WASTES FROM BUILDING ACTIVITIES

Since materials resulting from pulling down buildings, residues leftover on construction sites, removed soils and the break of old roads are a problem of quantity als well as in treating as waste, there are objectives for avoidance, reducing or re-use/recycling as well as an ordinance concerning the management of noxious wastes from building activities.

3.1 STATUTORY ORDINANCE ABOUT THE MANAGEMENT OF WASTES FROM BUILDING ACTIVITIES

The ordinance concerns all materials resulting from building activities and so they are not re-used/recycled have to be treated as wastes.

It is adressing als those, who perform building activities which are subject to authorization; that includes both builders and contractors and parties who give the building orders.

In general the contents of the ordinance are:

1. **The obligation of separate disposal**; that means noxious waste from constructions has to be kept, collected and treated separately from other residues from constructions, removed soils and removed road materials.

2. **The prohibition of mixing**; that means noxious wastes from building activities must not be mixed with other residues thereof, removed soils or removed road materials.

3. **Prohibition of disposal** for noxious wastes from building activities on disposal sites or other areas provided for the disposal of harmless residues from building activities, removed soils or removed road materials.

4. **Obligation of documentation**; the once whose work delivers wastes are defined as producers for the purpose of the ordinance, so they are required to produce documentation on type, quantity and managemant of such wastes and to keep record books about.

The term **noxious wastes** includes those materials which by improper handling or storage cause probable health or environment endangering effects.

In expansion of this definition for the purpose of the ordinance it also includes those materials which hamper or prevent re-use/ recycling.

With the separate disposal of noxious materials the amount of potentially re-usable/recyclable residues from constructions is increased. The re-use/recycling is therefore subject to objectives.

3.2 OBJECTIVES

The objectives are expression of the political intention; therefore they are not actionable. Nevertheless there is a remarkable constraint, because the legislator in case of missing the objectives will use his authorization to provide ordinances in order to realize his aimes.

The **Objectives of the Federal Government for Avoiding, Reducing or Re-use/Recycling of Residues from Constructions, Waste from Construction Sites, Removed Soils and Removed Road Materials** shall promote, that

1. producers of construction materials and products in the design of new materials and products consider the needs for avoiding wastes, material re-use/recycling and environmentally compatible disposal;

2. the rise of waste from constructions, construction sites, removed soils and removed road materials will be minimized by appropriate measures on construction sites;

3. materials hampering or preventing re-use/recycling are collected, kept and treated separately;

4. re-use/recycling have priority over disposal;

5. re-useable/recyclable portions are not mixed or disposed together with non-re-useable/non-recyclable portions;

6. noxious wastes from constructions be treated separately.

The objectives are defined as rates of re-use/recycling staggered in time:

	Rise <1> Mio. t/a	Re-use <1> Mio. t/a	%	1992 entire Germany %	1993 %	1994 %	1995 %
structures	22,6	3,7	16	30	40	50	60
site wastes	10,0	-	-	10	20	30	40
roads (bed&coat)	20,4	11,2	55	60	70	80	9
soils	167,9	53,3	32	100			

<1> data from prior to oct 3rd, 1991

Removed soils are separated from the other materials because they are not to be disposed in future. If there is no immediate use for it, it has to be disposed temporaryly and to be managed via "soil exchanges".

With respect to the new contries these objectives are very ambitious if one considers the enormous need for building activiities on the one hand and the partially problematic substances used on the other hand.

ART. 14 MARKING/LABELLING, SEPARATE DISPOSAL, MANDATORY RETURN OF CERTAIN GOODS, OBLIGATION TO ACCEPT RETURNED GOODS

(1) To avoid or reduce noxious substances in waste or to ensure their environmentally compatible management, the Federal Government is herewith authorized to provide by statutory ordinance, after hearing the parties concerned and with the consent of the Bundesrat, that

1. certain products, due to the content of a noxious substance in the waste expected to arise from their intended use, shall only be put into circulation if they are provided with an appropriate marking/labelling which points out in particular the necessity of return to the manufacturer, distributor or specified third parties, in order to ensure the required special type of waste management (obligation of marking/labelling);

2. waste with a particularly high content of noxious substances, in appropriate re-use/recycling or other disposal routes of which require special treatment, shall be kept, collected, transported and treated separately from other wastes and that corresponding records and documentation shall be submitted (obligation of separate disposal);

3. distributors of certain products shall be obliged only to put them into circulation if they offer the possibility of return or if they place a deposit on the product (obligation of accepting returned goods, mandatory deposit);

4. certain products shall only be put into circulation if they are either used in a certain form and for certain uses, guaranteeing appropriate management of the resulting waste, or not at all if the release of noxious substances during their management cannot be avoided or only be prevented at disproportionately high expenditure.

(2) To avoid or reduce the quantities of waste produced and to promote re-use and recycling, the Federal Government, after hearing the parties concerned, shall specify objectives to be reached within an adequate period of time for avoiding, reducing or re-using/recycling waste arising from certain products. It shall publish these objectives in the Bundesanzeiger[1]. To the extend this is required for avoiding or reducing the quantities of wastes produced or for environmentally compatible management, especially to the extent this is not possible by specifying objectives pursuant to the first sentence of this Para, the Federal Government, after hearing the parties concerned, may provide by statutory ordinance, with the content of the Bundesrat, that certain products, especially packings and containers,

1. shall be marked/labeled in a specific manner;

2. shall only be put into circulation in a certain form which shows considerable advantages for waste management, especially in a form that makes it possible to use it more than once or which facilitates re-use/recycling;

3. shall be taken back by the manufacturer, distributor or third parties acting in their behalf to ensure environmentally sound re-use, recycling or other management and that return must also be ensured by appropriate reception an deposit systems;

4. after use, shall be delivered by the owner in a certain manner, especially separate from other wastes, to facilitate their re-use/recycling or other environmentally compatible management as waste;

5. shall only be put into circulation for certain purposes.

[1]Official Gazette of the Federal Republic of Germany

Waste Materials in Construction.
J.J.J.R. Goumans, H.A. van der Sloot and Th.G. Aalbers (Editors)

THE U. S. EPA PROGRAM FOR EVALUATION OF TREATMENT AND UTILIZATION TECHNOLOGIES FOR MUNICIPAL WASTE COMBUSTION RESIDUES

C. C. Wiles[1], D. S. Kosson[2], T. Holmes[3]

[1]Risk Reduction Engineering Laboratory, United States Environmental Protection Agency, Cincinnati, Ohio 45268, U.S.A.

[2]David S. Kosson, Rutgers, The State University of New Jersey, Piscataway, New Jersey, U.S.A.

[3]Teresa Holmes, United States Army Corps of Engineers, Waterways Experiment Station, Vicksburg, Mississippi, U.S.A.

SUMMARY

Vendors of solidification/stabilization (S/S) and other technologies are cooperating with the United States Environmental Protection Agency's (U.S. EPA) Office of Research and Development (ORD), Risk Reduction Engineering Laboratory (RREL) to demonstrate and evaluate the performance of the technologies to treat residues from the combustion of municipal solid waste (MSW). Solidification/Stabilization is being emphasized in the current program. This technology may enhance the environmental performance of the residues when disposed in the land, when used as road bed aggregate, as building blocks, and in the marine environment as reefs or shore erosion control barriers.

The program includes four S/S process types: cement, silicate, cement kiln dust and a phosphate based process. Residue types being evaluated are fly ash, bottom ash and combined residues. An array of chemical leaching tests and physical tests are being conducted to characterize the untreated and treated residues. This paper discusses program design and general observations based on available results. The S/S evaluation program is the first phase of ORD's Municipal Solid Waste Innovative Technology Evaluation (MITE) program; a program to demonstrate and evaluate technologies for managing municipal solid waste.

The U.S. EPA is also supporting research to address the scientific and other issues associated with utilizing MSW Combustion residues. This paper discusses these issues and research directions.

INTRODUCTION

During the past several years there has been a significant concern about the management of the residues from the combustion of municipal solid waste. In the United States, much of this concern is based on the fact that when the residues are subjected to the Extraction Procedure for Toxicity (EP tox) and the Toxicity Characteristics Leaching Procedure (TCLP) concentrations of lead and cadmium in the leachate will sometimes exceed those levels defined as hazardous by these tests. This occurs more often for the fly ash, less for the combined fly ash and bottom ash, and least often for the bottom ash alone. Because of this, a controversy exists as to whether or not the residues should be considered and regulated as a hazardous waste or exempted because they originated from burning municipal solid waste. Currently they are excluded for a 2 year period based on provisions of the Clean Air Act. Several states, however, are requiring that these residues be disposed into landfills with designs and operating procedures as, or more, stringent than those for hazardous waste. Municipal Waste Combustion (MWC) ash characteristics are extremely variable as is the leachate from these ashes. Ranges of metal concentrations observed in bottom and fly ashes from many sources are presented in Table 1 (1). Detailed descriptions of the chemical and physical characteristics of MWC residues are available (2,3,4,5).

TABLE 1

Ranges of Total and Leachable Metals in United States MSW Combustor Ash as Determined by Researchers(1)

Compound	Bottom Ash mg/kg	Bottom Ash Leachate mg/l	Fly Ash mg/kg	Fly Ash mg/l
Pb	31 - 36,600	0.02 - 34	2.0 - 26,000	0.019 - 53.35
Cd	0.81 - 100	0.018 - 3.94	5 - 2,210	0.025 - 100
As	0.8 - 50	ND(0.001) - 0.122	4.8 - 750	ND(0.001 - 0.858)
Cr	13 - 1,500	ND(0.007) - 0.46	21 - 1,900	0.006 - 0.135
Ba	47 - 2000	0.27 - 6.3	88-9000	0.67 - 22.8
Ni	ND(1.5) - 12,910	0.241 - 2.03	ND(1.5) - 3,600	0.09 - 2.90
Cu	40 - 10,700	0.039 - 1.19	187 - 2,300	0.033 - 10.6

ND = Not Detectable; () = Detection Limit

Because of the growing concern about the residues and anticipating the need for appropriate treatment techniques, the United States Environmental Protection Agency (U.S. EPA) designed and implemented a program to evaluate the use of solidification//stabilization technologies for treating the residues. Originally known as the U.S. EPA MWC Ash Solidification/

Stabilization Evaluation Program, it is now the Municipal Innovative Technology Evaluation program (MITE). This paper summarizes the design of the program and observations based on the results available at this writing. The paper also discusses ORD's other MWC ash research and issues associated with utilization of the ashes.

1. THE MITE PROGRAM

The MITE program is a U.S. EPA Research program designed to conduct demonstrations of technologies for managing municipal solid waste. The objective is to encourage development and use of innovative technology for municipal solid waste management. It is a cooperative program in which the technology developer and/or vendor pays the cost of conducting the demonstration. The U.S. EPA pays the cost of testing and evaluation, including analytical cost and will report the results of the evaluations in an unbiased manner. This provides a means for assisting municipalities and others to better evaluate and select technologies more appropriate for their given situation.

The current program is demonstrating and evaluating alternatives for the treatment and utilization of residues from the combustion of municipal waste. While it is uncertain if treatment will be required prior to disposal, it is likely that treatment will be necessary for many utilization options. S/S technology was selected for initial evaluations based upon experience and knowledge of the technology for treating hazardous waste and experimental studies on solidifying municipal waste combustion (MWC)residues(6). The program objective is to provide a credible data base on the effectiveness of S/S technology for treating the residues. S/S, in general terms, is a technology where one uses additives or processes to transform a waste into a more manageable form or less toxic form by physically and/or chemically immobilizing the waste constituents. Most commonly used additives include combinations of hydraulic cements, lime, pozzolans, gypsum, silicates and similar materials. Other types of binders, such as epoxies, polyesters, asphalts, etc. have also been used, but not routinely. More detailed descriptions of S/S technology are available(7).

Preliminary design of this program was completed by the U.S. EPA. To assure that results are scientifically credible, a panel of international experts was assembled to provide technical oversight to the program. This Technical Advisory Panel (TAP) consisting of experts from academia, industry, state and federal governments, and environmental groups assisted in developing the final design for the physical, chemical, and analytical tests conducted on the treated and untreated MSW residues.

1.1 Mite Program Organization and Design

The organization, design and implementation procedures for the MWC Ash S/S program have been reported (8,9). The program included a comprehensive list of chemical, physical, and analytical tests conducted on the untreated and treated residues. This testing included unconfined compressive strengths before and after immersion, freeze/thaw, wet/dry, several leaching procedures (e.g., Toxicity Characteristics Leaching Procedure [TCLP], distilled water leach test, etc.), and numerous analytical determinations for metals, suspended solids, dissolved solids, and similar analysis required to fully characterize the untreated residues, the treated residues, and the extracts.

1.1.1 Ash Types Tested

Residues tested were limited to that collected from a modern state-of-art waste to energy facility (i.e., high burn out, lime scrubber with fabric filter, etc.). One reason for limiting the number of residues was that the prime objective was to evaluate solidification/stabilization for treating the residues, rather than determine how characteristics of different residues may affect the performance of the technology. The program included four different S/S process types plus one control. Another reason for limiting the residues tested was that the analytical cost for the program is the major U.S. EPA expense. For each additional residue added these costs must be duplicated. Adding more residues would have reduced the number of processes which could be evaluated to an unacceptable number. The program is also evaluating testing protocols that can be used to evaluate selected S/S processes on different residues if required.

Fly ash (including the scrubber residue), bottom ash, and combined ash were tested. The MWC facility has the following process sequence: (i) primary combustor with vibratory grates, (ii) secondary combustion chamber, (iii) boiler and economizer (iv) dry scrubber with lime, and (v) particulate recovery using baghouses (fabric filters). Bottom ash sampled was quenched after exiting from the combustion grates. Fly ash sampled was mixed residuals from the scrubber and baghouses. The fly ash was screened to pass a 0.5 inch square mesh. The bottom ash and combined ash were screened to pass a 2 inch square mesh at the MWC facility. Materials not passing through the 2 inch mesh were rejected. Each ash type was dried to less than 10% moisture, crushed and screened to pass a 0.5 inch mesh (nominally 3/8 inch after clogging), and homogenized.

1.1.2 Selection of Processes for Evaluation

Process selection was competitive based upon evaluation of proposals submitted by parties interested in participating. A formal Request For

Participation (RFP) was issued which provided information required to respond. Evaluation criteria were developed to make final selections.

Twenty-one responses were evaluated. The responses included 11 S/S processes, 6 vitrification processes and 4 miscellaneous processes. The S/S process proposals were judged to be superior. In order not to select similar S/S process types (e.g., two cement based) the best proposal was selected out of the different types available. The vitrification process proposals were generally incomplete and failed to address some major issues. This, in conjunction with the potential high quantities of residues required for most of these processes, resulted in the decision not to select one for evaluation. Alternatives for evaluating vitrification processes are being pursued. Proposals in the miscellaneous category were not acceptable.

1.1.2.1 <u>Description of Process Types Selected</u>

Process types selected in the program are cement based, silicate based, cement kiln dust and phosphate based. The control was a non-vendor cement process performed by research project personnel.

A brief description of each process selected follows:

- Cement Based Process - This process involves the addition of polymeric adsorbents to a slurry of MWC ash prior to the addition of portland cement. The final product is soil-like rather than monolithic.
- Silicate based process - The process immobilizes heavy metals through reactions involving complex silicates. This patented process uses soluble silicates as an additive with cement to promote reactions with the polyvalent metal present to produce insoluble metal compounds, gel structures, and promote hydrolysis, hydration and neutralization reactions. The final product is clay-like material.
- CKD process - This patented process mixes MWC ashes with quality controlled waste pozzolans and water. Good quality control of reagents is required because they are secondary materials. Therefore, the pozzolanic characteristics critical to the process are subject to change. The finished product is similar to moist soil, but hardens to a concrete-like mass within several days.
- Phosphate process - This patented process uses soluble phosphate to convert lead and cadmium to insoluble forms. The process mixes fly ash with lime; then this is mixed with bottom ash and treated with water soluble phosphate. The process does not alter the physical state of the ash.

1.2 Demonstrations

1.2.1 Process Used to Conduct Demonstrations

Characteristics and samples of ashes were furnished to the vendors to pretest their process prior to the demonstration. Vendors supplied any specialized equipment or ingredients required. Each agreed to observation by U.S. EPA selected observers. The demonstrations were conducted at the U.S. Army's Waterways Experimental Station (WES), Vicksburg, Mississippi and observed by U.S. EPA designated staff.

During the process demonstration, each vendor treated three replicate batches for each ash type. A total of between 50 and 100 gallons of each ash type was treated for each process. Process additives were provided by vendors to U.S. EPA for analysis and archiving.

1.2.2 Scale

The processes were demonstrated at bench scale. Reasons for this include the technologies being tested, the large amount of resources required for full scale demonstrations and the desire to include as many different processes as possible within available resources. The program plan was to conduct a full scale field demonstration of a selected process if deemed necessary. Because of the nature of S/S technologies, U.S. EPA and the TAP believed that bench scale demonstrations were adequate to prove if the technology is an effective treatment for MWC residues. Sufficient experience is available for conducting the engineering and design required for scaling to a specific situation. Furthermore, the bench scale permitted much more detailed testing and thus more exploration of the basic mechanisms involved in the process. This in turn will assist in the determination of expected long-term behavior. A drawback with this scale however, is the difficulty in sampling and potentially wide variability associated with bottom ashes.

1.2.3 Status

The S/S process demonstrations have been completed. The very large volume of data generated is still being compiled, organized and interpreted. The final report is expected by the end of December 1991.

1.2.4 Future MITE Demonstrations

Future MITE demonstration candidates have been solicited by notice in the Commerce Business Daily, through appropriate MSW trade organizations, interested developers and similar means. Emphasis for these demonstrations is on processes for recovering marketable products from the MSW stream. Additional industry and state cooperative evaluations of MWC ash treatment and/or utilization processes are being pursued under separate programs.

2. RESULTS FROM THE MWC ASH S/S EVALUATIONS

Complete results are not available at this writing. Data from the monolith leach test is still being generated and modelling efforts are underway to determine diffusivity constants for the various treated constituents. This information is required to determine rates of release for the various species of concern.

Examples of the results being generated are available in other papers and details will appear in the final report for the program currently in 2preparation. (9, 11, 12) Results generated from the program will permit many comparisons of the different leach tests, efficacy of S/S treatment processes, relative amounts of species release and similar observations.

One objective of conducting the leach testing and physical tests was to evaluate the effectiveness of the various processes to retain physical durability and the metals of concern when exposed to different stresses. One potentially valuable observation is how well the physical structure can be expected to withstand degradation under exposure to wet conditions (e.g., marine environment, road base, construction blocks, etc). If one assumes that physical durability will improve the capability of the treated form to resist leaching, then strength before and after immersion will provide insight about this characteristic.

Compared to the other processes, the cement control process showed somewhat better effectiveness in retaining physical strength after immersion. Also of interest is the general trend that the APC residues (as compared to bottom and combined) appeared to be more difficult to treat as measured by lower strengths. While this confirms observations from other researchers, one must note that UCS measurements results often have wide ranges.

In comparing strengths of solidified waste forms, one must note that there is little scientific evidence that directly relates increased strength with decreased release rates of pollutants of environmental concern. Also increased strength may not be important in the case of placement in a landfill. In such cases strength concerns deal with sufficient load bearing capacity necessary to support equipment and landfill covers, etc. In some cases, these may be 12 to 15 psi or lower. These relatively low strengths often can be easily achieved through routine compaction. Additionally many MWC residues contain sufficient pozzolanic properties which when combined with the excess lime from wet scrubbers will result in some hardening of the ashes without additional additives (13).

Strengths greater than those generated in these tests would be required for potential uses such as shore erosion control and some construction applications. Higher strengths have been routinely achieved (14).

In reviewing the leaching data, one readily determines that the quantities of the indicated metals in the extracts of the TCLP and the distilled water leach test did not exceed levels defined as hazardous in the U.S.A. by the regulatory TCLP test. The untreated bottom and combined residue did not exceed these levels. The value of the data is still apparent, however, in permitting comparison of processes and different leach tests.

Based on the data, one can note the difference between the total concentration value obtained for several species using the U.S. EPA's recommended SW846 acid digesting procedure as compared to neutron activation analysis (NAA). In many cases the SW846 recovered only a fraction of that determined by NAA.

Based upon these and similar results one can determine the relative aggressiveness of the various tests to the treated and untreated residues. In combination with information on buffering capacities of the various treated materials, equilibrium pH, rates of release and similar information one will be better able to predict long term behavior expected from the materials when exposed to different disposal and utilization schemes.

An important component for evaluating the processes was to determine the effectiveness of treatment in retaining species of interest. In order to do this and make comparison on an equal basis, (e.g., amount of dry ash treated) calculations were made to account for the effects of diluting the untreated residues with the process additives. Based on these calculations, the processes were generally not effective in treating the ash to retain the constituents indicated.

Data also indicates that the processes were generally effective in treating the combined ash when measured by the TCLP with the exception of Barium. This data, in conjunction with other results in the program indicate that the vendor treatment processes were designed to pass requirements of the TCLP. Of considerable interest to note is the amount of total dissolved solids released from the treated combined residues as determined by the distilled water leach test. The results indicate that significant amounts of the material could be expected to dissolve or erode over time if exposed to wet conditions. If not properly designed for, this would have significant impact on potential utilization options.

2.1 <u>General Observations</u>

Based on the data currently available from evaluating the treatment processes listed in the program, several general observations are possible. One must note, however, that data to determine rates of release are not yet available. These observations are summarized as follows:

- Only 1 process attenuated most of the heavy metals.

- Three processes caused minimum attenuation of metals (except for apparent attenuation caused by pH effects).
- One process did attenuate some of the heavy metals after considering pH and dilution effects.
- For Pb, data indicates similar behavior for treated as untreated (i.e., significantly increased release at ph < 5). In some cases Pb became more mobile after treatment compared to before treatment.
- For Cd, similar behavior as Pb was observed but the threshold pH is around 6.
- Results support preliminary conclusion that the releases are pH controlled and the treatments evaluated did not alter species except for the phosphate based process.
- At least 2 processes evaluated caused an increased TDS release for the bottom ash.
- Based on selected tests, approximately 60% of the ash will dissolve over time (in some cases treated more than untreated).
- Results indicate that S/S processes can be designed which will effectively treat the residues. In the case of the processes tested, some could be combined which would be significantly more effective than any of the ones evaluated would be alone.
- Observations of buffering capacities can play an important role in estimating treatment process behavior over time.

3. THE U.S. EPA MWC ASH RESEARCH PROGRAM

In addition to the ash S/S evaluation program, U.S. EPA is conducting or planning to support several research programs.

3.1 Effects of Municipal Waste Combustor Ash Leachate on Clay Liners, Flexible Membrane Liners, and a Geosynthetic

Landfill lining components are being tested to determine if they will maintain their integrity and designed performance when exposed to MWC ash leachate. Natural lining components are being tested in accordance with EPA Test Method 9100; a triaxial hydraulic conductivity test where a baseline value is obtained with water, then the permeant is changed to leachate. Variations in performance are measured directly. Synthetic materials are being tested using EPA Test Method 9090 where baseline physical and polymer properties are measured as the material is manufactured, then coupons are immersed in leachate under elevated temperatures and removed after 30, 60, and 120 days. Observations are made to determine if the physical and polymer properties have changed.

3.2 The Nature of Lead, Cadmium, and other Elements in Incineration Residues and their Stabilized Products

This project is investigating raw and stabilized incinerator residues to test how the chemical nature and binding state of metals affect their leachability. Three residues--two from municipal waste combustors and one from a hazardous waste incinerator--and three S/S formulations will be tested. Sophisticated surface analysis techniques will be applied to characterize metals in these inhomogeneous, poorly-crystalline materials. The resulting knowledge of metal speciation, enhanced by output from geochemical models, will be used to interpret the laboratory leaching behavior of raw and stabilized residues. Project results should enhance our understanding of how common binders achieve stabilization, suggest other binders based on residue chemistry, and improve estimation of long-term treatment performance.

3.3 Investigation of Mobility of Dioxins and Furans from Stabilized Incineration Residue

This project was conducted to evaluate the fate of dioxins and furans - found in incinerator ash - when the ash is solidified and utilized to build artificial reefs and embankments. Blocks of solidified ash and cement blocks were exposed to sea water for extended time. The blocks are retrieved to monitor the physical, chemical and biological activity at several intervals of exposure. Results show that the blocks retain their structural integrity after prolonged exposure. No evidence was found to indicate the transport of dioxins and furans from the blocks. Marine organism growing in the blocks contained no detectable quantities of chemicals of concern.

3.4 Effects of Materials Separation on the Characteristics of MSW Combustion on Residues

This is a planned cooperative program involving several organizations. The objective is to determine the fractions of the municipal waste streams that contain the trace metals which most adversely affect combustion ash quality.

3.5 MWC Ash Utilization Criteria

This effort is concentrating on generating data to support development of technical criteria for the safe utilization of MWC residues. The program includes the compilation and assessment of existing criteria applied to MWC residue utilization, support of residue utilization demonstrations to monitor environmental affects, and the identification and conduct of identified special research needed to develop technical criteria for utilization. Effects of institutional and public attitudes toward waste utilization will also be considered.

4. ISSUES ASSOCIATED WITH MWC ASH UTILIZATION IN THE UNITED STATES

4.1 Utilization Options

There are several potential options for using MWC residues in a beneficial manner. Examples are as in aggregate, construction blocks, erosions control, artificial reefs, landfill cover, etc. While there has been significant interest in using the residues very little actual utilization has occurred in the United States.

4.2 MWC Ash Utilization Issues

There are several issues which are impeding the utilization of MWC residues in the United States. These include:

4.2.1 Environmental Consequences and Human Health Concerns

Environmental concerns focus on the heavy metals in the ashes, their form and their ultimate fate when the ashes are applied to different uses (e.g., roadbed, building blocks, etc.)

4.2.2 Long-term Performance and Prediction of Performance

The long-term performance issue concerns the ability to accurately measure and predict the environmental behavior of the ashes over extended periods for different utilization options.

4.2.3 Liability Issues

Potential liability for future problems and the uncertain regulatory situations in the United States has been a deterrent to utilization.

4.2.4 Federal Guidance

The lack of guidance on MWC ash management in general, and utilization specifically, is a deterrent to utilization. In many cases, this has impeded the initiation of field demonstrations much needed to quantify the benefits, risks, and other factors associated with ash utilization. This creates uncertainty for the industry, the users, and the public.

4.2.5 Criteria for Utilization

There is a need for technically sound criteria for utilizing MWC ashes. Physical properties and characteristics must meet or exceed specifications of the materials the ashes will replace. These criteria are known or can be ascertained. Criteria are needed, however, to guarantee that adverse environmental affects will not result.

4.2.6 Markets

Markets for the ash will be limited. In cases where virgin materials (e.g., gravel) are plentiful, these markets may not exist. In such cases, alternative markets may need to be developed and/or incentives (e.g., regulatory, economic) may be required to assist the ash utilization.

5. CONCLUSIONS

This paper presented preliminary observations from evaluating the effectiveness of 5 S/S processes to treat MWC residues. Although final conclusions must wait until all analyses are complete, preliminary findings indicate that the processes tested generally did not change the species of metals in the MWC. Attenuation of metals observed was attributed to pH and dilution effects. One process, however, did appear to transform some of the heavy metals to less soluble forms.

ORD, EPA is supporting research on MWC residue treatment and utilization options. This research is currently focusing on the development of criteria for the safe utilization, investigating the sources and affects of metals in MSW on the combustion residues, and the behavior of selected metals in the untreated ashes and stabilized ashes.

Several issues were identified which are impeding MWC ash utilization in the United States. Research and demonstrations are required to assist in resolving these issues.

References

1. Wiles, C. C. "Characterization and Leachability of Raw and Solidified U.S.A. Municipal Solid Waste Combustion Residues" ISWA 86 Proceedings of the 5th International Solid Waste Conference, Copenhagen, Denmark. September 1988.

2. U.S. EPA (Environmental Protection Agency) <u>Characterization of MWC Ashes and Leachates from MSW Landfills, Monofills and Co-Disposal Sites</u>. EPA 530-SW-87-028A, Office of Solid Waste. October 1987.

3. U.S. EPA (Environmental Protection Agency) <u>Addendum to Characterization of MWC Ashes and Leachates from MSW Landfills, Monofills and Co-Disposal Sites</u>, Office of Solid Waste, June 1988.

4. J. L. Ontiveros, T. L. Clapp and D. S. Kosson. "Physical Properties and Chemical Species Distributions Within Municipal Waste Combustor Ashes." In <u>Environmental Progress</u>, Vol. 8, No. 3, pp 200-206, August 1989.

5. H. A. van der Sloot, et. al. "Leaching Characteristics of Incinerator Residues and Potential for Modification of Leaching." In Proceedings of the International Conference on Municipal Waste Combustion, Vol. 1, p 2B-1, April 1989.

6. D. R. Jackson, "Evaluation of Solidified Residue from Municipal Solid Waste Combustors." U.S. Environmental Protection Agency, EPA/600/S2-89/018, February 1990.

7. Wiles, Carlton C., "A Review of Solidification/Stabilization Technology." Journal of Hazardous Materials, 14:5-21, 1987.

8. Wiles, C.C., David S. Kosson and Teresa T. Holmes, in proceedings, "The United States Environmental Protection Agency Municipal Waste Combustion Residue Solidification/Stabilization Evaluation Program", First United States Conference on Municipal Solid Waste Management, U.S. Environmental Protection Agency, Washington, D.C., June 13-16, 1990.

9. Wiles, Carlton C., "The U.S. EPA Program for Evaluation of Treatment and Utilization Technologies for Municipal Waste Combustion Residues", in proceedings, The Second International Specialty Conference for Municipal Waste Combustion, Tampa, Florida, U.S.A., April 15-19,1991

10. Federal Register, 40CFR Part 261 et. al. "Hazardous Waste Management System; Identification and Listing of Hazardous Waste; Toxicity Characteristics Revisions; Final Rule, Environmental Protection Agency, March 29, 1990.

11. Kosson, David S. et al, "A Comparison of Solidification/Stabilization Processes for Treatment of Municipal Waste Combustor Residues, Part II - Leaching Properties", in proceedings, The Second International Specialty Conference on Municipal Waste Combustion, Tampa, Florida, U.S.A., April 15-19, 1991

12. Holmes, T., David Kosson, and Carlton Wiles, "A Comparison of Five Solidification/Stabilization Processes for Treatment of Municipal Waste Combustion Residues, Part I - Physical Testing", in proceedings, The Second International Specialty Conference on Municipal Waste Combustion, Tampa, Florida, U.S.A., April 15-19, 1991

13. R. W. Goodwin, Ph.D., P.E., "Utilization Applications of Resource Recovery Residue" Proceedings. First U.S. Conference on Municipal Solid Waste Management, Washington, D.C. pp. 898 - 915. June 13-16, 1990.

14. F.J. Roethel, V.T. Breslin, "Interactions of Stabilized Incineration Residue with the Marine Environment". Proceedings of the First International Conference on Municipal Solid Waste Combustor Ash Utilization. October 13-14, 1988, Philadelphia, PA. (Eds. T. Eighmy and W. Chesner).

Waste Materials in Construction.
J.J.J.R. Goumans, H.A. van der Sloot and Th.G. Aalbers (Editors)

THE USE OF WASTE MATERIALS IN CIVIL ENGINEERING

AVI slag can replace gravel in concrete production

D. Stoelhorst, Concrete Association (Betonvereniging), P.O. Box 411,
2800 AK Gouda, The Netherlands

Research has shown that gravel can be replaced as an aggregate for concrete by slag from refuse processing installations. This applies to unreinforced concrete poured on site and factory-made unreinforced concrete units. The time is ripe for more applications in practice.

The availability of natural aggregates from the province of Limburg, the Netherlands, is seriously at risk due to the decision-making of the regional authorities. The province has already declared that it is prepared to do its best to grant further concessions of up to 70 million tonnes, in addition to the concessions already in operation and being negotiated. That permits for 70 million tonnes will in fact be granted up to the year 2000 or 2010 is, however, not yet certain. In addition, devlivery on short term will certainly cause problems, owing to the delay in granting concessions for the Stovel area.

There are a number of solutions to these problems. The supply of coarse aggregates from abroad will have to be considerably increased. However, the use of secondary raw materials will also have to be much more widely disseminated. In many places in the Netherlands, research is being carried out into the possibilities of making real use of these materials. One of the materials involved is AVI slag, a slag which is released when burning domestic refuse in refuse incinerators.

Research has been carried out by the CUR into the possible use of AVI slag as an aggregate in concrete, and by the CROW for use as an foundation material. A few small-scale experiments have been carried out using AVI slag as a filter for asphalt.

In this article, attention is paid to the research carried out by the CUR and CROW. Intron and TNO/IBBC have been closely involved in the CUR research.

AVI slag

Intron have carried out a study of the literature to take stock of existing knowledge on the use of AVI slag as an aggregate in concrete and to ascertain the gaps in this knowledge.

The conclusion is that AVI slag concrete is of lower quality that a comparable concrete made with river sand and gravel.

Moreover, the researchers are of the opinion that concrete made with large quantities of AVI slag cannot be considered for use in reinforced concrete, due to the high chloride content of the slag. Only "processed" (crushed, screened and iron-free) AVI slag can be considered for use as an aggregate for concrete.

To replace river sand and/or gravel, it is necessary to compensate with a higher cement content, the use of adjuvants, particularly to reduce excessively high water/cement ratios and a greater structural thickness (in road surfaces, for example). These factors must in turn be compensated for by the low cost of processed AVI slag.

The possibilities of using AVI slag will probably remain limited to low-value unreinforced structures. Therefore, it was investigated whether this field of use is sufficiently large to justify further research. It appeared that a large market can be anticipated, specifically in this group of possible applications, and that further research was, therefore, in order.

Guidelines for these applications would be able to promote use considerably. These must be established on the basis of further research into the influence of variation in quality of AVI slag on the variation in quality of the AVI slag concrete.

In the literature study, it was further recommended that it be investigated whether certain negative aspects can be reduced with the aid of adjuvants. One can think, for example, of the use of (super)plasticisers, setting accelerators, water-repellent agents and corrosion inhibitors.

Alternatives were mentioned; the use of a less sensitive, cheaper binder, such as, for example, sulphur, and the manufacture of a synthetic aggregate by sintering or remelting the AVI slag. With regard to the environmental aspects in the use of AVI slag in concrete, it was concluded that there was too little information available to justify a full statement.

Quality of slag concrete

The influence of the variation in quality of AVI slag on the variation in quality of the concrete made with it was looked into by Intron. For this purpose, thirteen samples were taken from each of four refuse processing plants, those at Amsterdam, Dordrecht, The Hague and Rotterdam.

These samples were characterised by main and secondary components, particle shape of the main components, gas formation in an alkaline environment (due to metallic zinc and aluminium), the effect on the setting of the cement (due to organic components and zinc salts), sulphate content, free CaO and MgO due to the occurrence of destructive expansion, chloride content due to the risk of corrosion of the reinforcement and components which cause spots.

From six of the thirteen samples per plant, concrete mixtures were made on the following basis:

- 340 kg/m^3 of Class B blast furnace cement;
- content of AVI slag (dry) in the total aggregates mix (topped up with river sand and gravel) is 75 % w/w;
- particle size distribution within the AC range of the particle group of 0-16 mm according to VBT 1986;
- age of AVI slag when preparing the concrete is six weeks; temperature of the AVI slag is the 40°D;
- slump about 70 mm.

The slump, air content and apparent volume weight of the fresh concrete mortars are determined. The cube compressive strength and the apparent volume weight of the hardened concrete are determined after 3, 7, 28 and 91 days. When AVI slag is stored outdoors (screened from rain), its properties for concrete can alter. This may be due to conversion of the organic fraction or of zinc salts or free CaO and MgO. To investigate this, a repeat test was carried out on two samples per plant which had been "aged" for a year in the manner described above.

Properties

The samples taken in April and May 1988 had a somewhat lower content of moisture and stone and a somewhat higher content of cinders and sulphate that the samples from the second half of 1987. The other investigated properties of the AVI slag were comparable in value for both periods of sampling.

Aging of the AVI slag in the open air for a year resulted in an increase (average 50 %) in the time between initial and final setting (the setting period). It was further found that the aging resulted in the CaO (and Ca(OH)2) being almost completely converted into CaCO. A number of conclusions can be drawn from the results of the characteristics investigation. AVI slag mainly consists of glass, cinders and stone-like material (together approximately 85 %) with additional ceramics, metals and miscellaneous substances (each about 5 %). The composition varies both between the plants and at different times for each plant. In all plants examined, it was found that the composition varied greatly for each fraction. The content of glass, cinders, ceramics, metals and miscellaneous substances in the fraction larger than 4 mm is generally higher than in the 0.5 to 4 mm fraction. On the other hand, the content of stone-like material is significantly higher in the 0.5 to 4 mm fraction. Although the particle size distribution of the AVI slag varies between plants and periods, the range of variation in virtually all cases is found to be within the AC range of the 0-16 mm particle size group. The content of sulphate, free CaO and free MgO in the AVI slag is less than 1 % (w/w), which if uniformly present does not represent any risk of destructive expansion. The chloride content is sufficiently high for there to be a potential risk of reinforcement corrosion when concrete has a high content of AVI slag. Use of AVI slag in cement-bound products can in a number of cases result in considerable retardation of initial setting of the cement. It is also of importance that, when AVI slag is used, the time between initial and final setting generally increases, which results in increased vulnerability in after-treatment. Outdoor storage of AVI slag for a year does not result in any improvement in the investigated properties compared with slag which has been aged for only six weeks.

Concrete research

With a quantity of AVI slag of about 1000 kg/m^3 of concrete and a cement content of about 340 kg/m^3, concrete can be made with a cube compressive strength of which the mean varies per investigated plant between 17 N/mm^2 and 27 N/mm^2 after 28 days. Concrete made with AVI slag from the plants in Amsterdam and The Hague shows on average a higher compressive strength than that made with AVI slag from Rotterdam or Dordrecht.

In a number of cases, the development of compressive strength of concrete made with AVI slag from the plants in Rotterdam and Dordrecht was found to be initially slow, which can be attributed to retardant components in the AVI slag concerned. In one case, the retardation was extreme; after 28 days the compressive strength was still 0 N/mm^2 and after 91 days only 13 N/mm^2. This means that when AVI slag is used an appropriate preliminary must take place.

The factors of content of glass, stone-like material, ceramics, metals and fraction larger than 8 mm appear in a number of cases to have a beneficial effect on the compressive strength of the AVI slag concrete.

The compressive strength of AVI slag concrete appears in a number of cases to be adversely affected by the content of AVI slag, as well as by the content of cinders and miscellaneous substances, the loss on ignition and the fraction larger than 63 mm of the AVI slag. Outdoor storage of AVI slag for a year does not result in any improvement in the investigated properties of concrete made with it compared with concrete made with AVI slag which has been aged for only 6 weeks.

Mean cube compressive strength as a function of time, per refuse incineration plant.

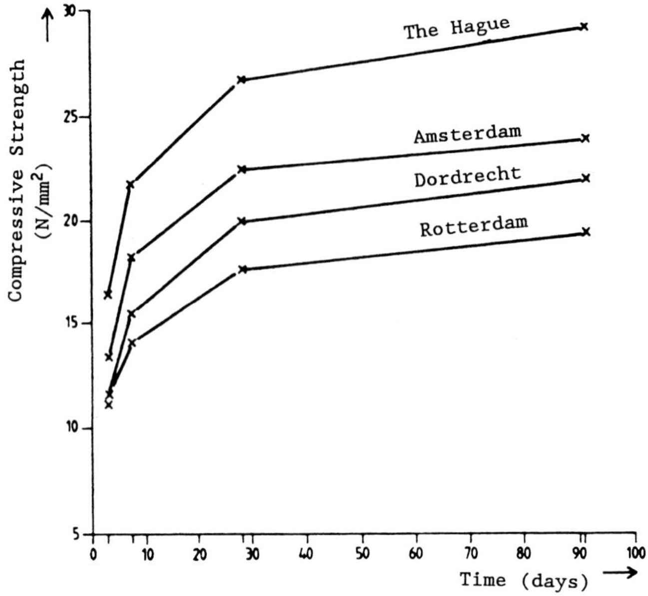

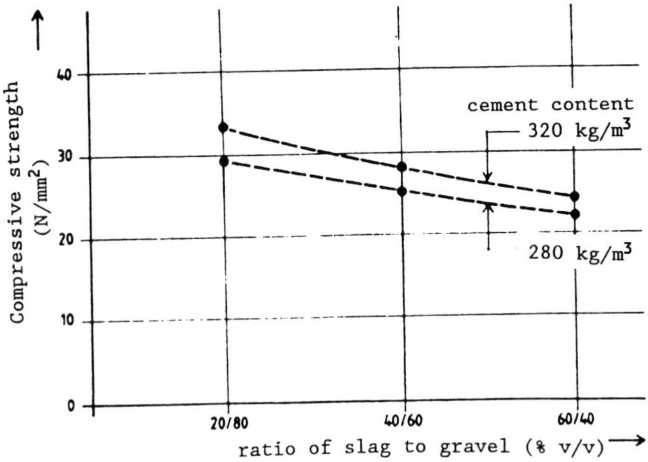

Cube compressive strength after 28 days as a function of the slag/gravel ratio in the 2 to 32 mm fraction aggregate.

The highest possible slag/gravel ratio in the 2 to 32 mm fraction was 40 : 60, when the cement content was 320 kg/m^3 of concrete, Class A blast furnace cement.

B20 concrete

An investigation was made to ascertain with what composition a plastic concrete mortar can be made, consistency range 3 and strength class B20, using AVI slag (0 to 33 mm) from Rotterdam. This slag contained so much fine material that it was decided first to remove the fraction larger than 2 mm (48.2% (w/w)). In its place, sand was added. In the 2 to 32 fraction, various ratios of AVI slag to gravel were tested. For B20, a mean cube compressive strength of 29 to 30 N/mm^2 is required with a standard deviation of 4 N/mm^2.

With this composition, the concrete properties were determined which are asked for in part G of VB 1974/1984 in order to get an idea of the deformation properties of concrete in a given strength class. For comparison sake, some properties of gravel concrete with the same cube compressive strength were determined. It was also investigated whether any improvement could be obtained with a superplasticiser. The water content was repeatedly adjusted in such a way that consistency range 3 was reached (slump 100 to 150 mm). A few observations can be made here. To achieve the same strength class, (B20) about 15% less cement was needed in the gravel concrete. In many cases, the environmental class is decisive for the composition of the concrete. For environmental class 2 (damp) the water/cement factor may not be higher than 0.55.

The degree of resistance to frost was also investigated in a test in which heat absorption took place from one side. In such a test, spalling can take place from the surface. There was no evidence of this. Minor loss of sand from the cement skin occurred only with the reference gravel concrete. This was not abnormal with the given water/cement factor of 0.61. Tests to determine creeps and long-term compressive strength are still being carried out, and if the requirement for the long-term compressive strength is met, then concrete with AVI slag is technically acceptable for the use proposed here.

In view of these satisfactory results, supplementary tests were carried out in which the complete AVI slag material of 0 to 32 mm was used. Concrete of consistency class 3 and strength class B20 was again aimed at. The slag content (0 to 32 mm) was made higher than in the previous tests. Therefore, more cement was used, which was moreover more rapid in hardening (Class B). At the same time, a superplasticiser was also mixed in. It was found from this that strength class B20 could be reached with a 0 to 32 AVI slag content of 1146 kg/m^3 and 102 kg of cement per m^3.

This was also possible with a 0 to 32 AVI slag content of 759 kg/m^3 and 361 kg cement per m^3. In both cases, environmental class 2 was just met.

Building bricks

In NEN 7027 "Building blocks and bricks of concrete", requirements are set for concrete units intended for use in masonry walls. The most important requirements are the strength classes and the maximum shrinkage permitted. Special aggregates such as foam lava, sintered clay, broken bricks, cinders and slag may be used, provided they are durable and contain no deleterious components.

The strength class and shrinkage were determined for four different compositions of concrete with AVI slag. Two of them were coarse concrete, by leaving out the fine aggregate material (0 to 2 mm), with a low cement content. In the other two compositions the concrete had a dense structure.

As in the test for B20 concrete, consistency range 3, the fine material fraction (0 to 2 mm) was first removed from the slag (48.2% (w/w)). Moreover, in this test, the fraction larger than 16 mm was also removed (5.0% (w/w)) because the dimensions of the concrete bricks made (212 x 101 x 75 mm) were relatively small. For the concrete with a dense structure, sand (0.2 mm) was added. In the 2 to 16 mm fraction, various contents of AVI slag were used with regard to gravel. The water content of the concrete mortar was repeatedly adjusted in such a way that the adequate "green strength" was obtained for immediate removal from the mould (consistency range 1, compacting factor 1.26).

The results call for comment. Coarse concrete bricks with 100% AVI slag (2 to 16 mm) as aggregate in a quantity of 1526 kg/mm^3 complied with strength class 10. The mean compressive strength was 17.5 N/mm^2. The mean shrinkage, which may reach a maximum of 0.60 per mille, was 0.55 per mille. Coarse concrete bricks with 50% gravel met strength class 20. The shrinkage of these was 0.50 per mille. In both series, about 193 kg of class B blast furnace cement per m^3 was used. Bricks with a dense structure and with 50% AVI slag (603 kg/m^3) and 50% gravel as coarse aggregate and sand as a fine aggregate easily met strength class B30. The mean compressive strength of these was 55.9 N/mm^2 and the mean shrinkage 0.35 per mille. With a proportion of 20% AVI slag (252 kg/m^3) compared with 80% gravel, a mean compressive strength of 72.1 N/mm^2 was reached. This also satisfied strength class B30. In these two series, about 285 kg/m^3 of class B blast furnace cement was used.

Bricks were stored both indoors and in the open air, and were inspected from time to time for any occurrence of pop-outs or cracks. So far, none have been found. In the units stored in the open air, bits of iron (wire, nails) were found to be rusting at the surface.

The measured compressive strengths were higher than expected. It can also be noted that the concrete was very well compacted by vibrating with a high-pressure ram. Moreover, class B cement was also used. In addition, in the compression test on blocks with a height-to- width ratio of about 0.75, compressive strengths were obtained which were about 15% higher than would be the case for standard test cubes.

The results were an inducement to also carry out supplementary tests in this case. In these, 0-16 mm AVI slag was used and more cement incorporated.

Economy

Apart from the technological considerations, it is important to ascertain whether the use of AVI slag in concrete is economically justifiable and if so, in what areas its use offers the most possibilities. Various factors are of importance in considering these possibilities: the achievable strength, the presence of reinforcement, the cost price of AVI slag and the percentage replacement by AVI slag.

From the investigation carried out, it appears that the strength class of concrete in which 75% (w/w) of the sand and gravel is replaced by AVI slag is B15 to B20 at the most. If a smaller part of the sand and/or gravel is replaced by AVI slag, a higher strength is obtainable. Experience has especially been obtained in this respect in the production of various types of concrete unit (building bricks).

The use of AVI slag in reinforced concrete is not as yet acceptable. The chloride content of AVI slag is so high that its use represents a potential risk of corrosion of the reinforcement.

The cost price of AVI slag has a decisive influence on the feasibility of using it. The price level is determined by the possibility of selling the material elsewhere. The economic possibilities are thus largely dependent on the strategic estimation of the manufacturers. In view of the quantities available both now and in the future and the market position of the material, it may be expected that the cost price will remain sufficiently low.

The quantity of the material which will be used depends , as has been previously indicated, on the desired strength of the concrete. In the economic survey, it is important to consider whether complete replacement is aimed at and thus lower strength with a relatively limited market, or partial replacement with higher strength, so that a larger market is possible.

If the above factors are considered in turn, then two areas of use appear to be available. One is a low concrete strength for unreinforced use of concrete poured in situ. The size of this area is relatively large and complete replacement makes a large sales market possible. The other is a relatively high concrete strength in the production of unreinforced concrete units. Production takes place under strictly specified circumstances, so that it is possible to control the process well and high concrete strengths are feasible.

On the basis of cost calculations in which it is assumed that the AVI slag aggregate will not be made available "for a song", it is economically feasible to use the material.

Practical application

The results of the research that has been carried out until now show that the use of AVI slag as a replacement for gravel is technically justified. The practical feasibility of using it depends on the readiness, certainly at the start, to adjust to specific requirements. The developments which present themselves with the certification of AVI slag can be assessed as positive, and give a better idea of the quality of the product. The time is ripe for its use in practice. At present, a working group will endeavour to carry out practical applications as demonstration projects. The economic feasibility has been examined and it would appear that the use of slag is possible in two fields. A low-value one, in unreinforced concrete poured in situ, where complete replacement of the gravel fraction can be achieved. A higher-value one is used in unreinforced factory made concrete units. Here only part of the gravel fraction is replaced.

In the near future, the working group will prepare directives for use and accompany them with practical applications as demonstration projects.

Literature

CUR report 87-1
Van der Wegen, G.J.L. and O. Kliphuis
1988. Research into the variation in quality of AVI slag in relation to the variation in quality of AVI concrete. Intron Report No. 88094.

VEABRIN 1988. Quality control of AVI slag
'87-'88, NOH research. TAUW/ZWL Report No. 305/JJS/avd

1988 Waste is no longer refuse,
refuse incineration slags. CROW publication.

The average composition of the concrete and the researched properties of the fresh concrete mortar and hardened concrete per refuse processing plant.

	Values per refuse incineration plant [1]							
	Amsterdam		Dordrecht		The Hague		Rotterdam	
	x1)	s	x	s	x	s	x	s
Composition								
Cement, b-f B (kg/m^3)	331	28	351	32	353	24	345	11
Water (total) (kg/m^3)	320	40	367	38	335	39	356	27
AVI slag (dry)(kg/m3)	956	26	908	39	1003	33	906	61
Sand and gravel (dry) (kg/m3)	303	37	313	35	324	35	304	37
Mortar properties								
Slump (mm)	78	24	65	8	66	13	65	9
Volume weight (kg/m^3)	1915	23	1961	15	2015	30	1921	36
Air content (%)	8.7	1.0	4.4	0.5	5.3	0.7	5.9	1.3
Cube compressive strength[2]								
after 3 days (N/mm^2)	13.3	7.1(11.5)				16.5	7.6(10.9)	
7 days (N/mm^2)	18.1	12.0(15.4)				21.6	13.0(14.9)	
28 days (N/mm^2)	22.4	16.6(19.9)				26.7	17.5	
91 days (N/mm^2)	23.8	20.6(22.0)				29.2	19.4	

[1] Mean values for 6 samples per plant

[2] All observations have been processed in the tables. In the case of the figures in brackets, mixtures which are very slow to develop strength have been ommited.

Results of research on characteristics of samples of AVI slags, taken in 1987 at the refuse processing plants in Amsterdam, Dordrecht, The Hague and Rotterdam.

Characteristics	Values per refuse incineration plant							
	Amsterdam		Dordrecht		The Hague		Rotterdam	
properties and composition	x[1]	s	x	s	x	s	x	s
Moisture content (%w/w)	23	8	26	5	21	4	25	4
Particle size (%w/w)								
> 8 mm	16	6	33	6	34	7	24	4
4-8 mm	20	4	14	1	15	2	17	1
0.5-4 mm	36	4	28	3	27	3	34	4
< 0.5 mm	28	4	25	2	24	5	25	3
components in > 0.5 mm fraction (%w/w)								
Cinders	27	9	29	8	38	10	32	7
Stone	34	7	26	7	21	8	30	5
Glass	27	8	27	5	21	5	20	4
Ceramics	5	3	7	2	9	3	5	2
Metals	3	2	4	3	7	2	6	3
Miscellaneous	4	2	6	2	4	3	7	3
Setting times (min)[2]								
Initial	239	37	300	44	224	16	279	21
Final	320	36	369	40	300	22	351	30
Gas formation (ml/10g)	116	45	54	34	119	36	72	53
CaO (%w/w)	0.3		0.2		0.2		0.2	
MgO (%w/w)	0.7		0.6		0.5		0.6	
Sulphate (%w/w)	0.36	0.05	0.19	0.06	0.34	0.10	0.41	0.13
Chloride (%w/w)	0.17	0.04	0.14	0.04	0.17	0.06	0.17	0.06
Spot index	20		20		20		20	

[1] x - mean value, s - standard deviation

[2] Values for an extract of AVI slag in saturated limewater; as a reference, saturated limewater was taken with initial setting at 200 minutes and final setting at 265 minutes.

Composition and average concrete properties of B20 concrete, consistency range 3, with 0 to 32 mm AVI slag and Class B blast furnace cement.

Slag concrete	Measured values (mean of 3 batches)			
Slag:gravel ratio in 0-32 mm aggregate (V/V)	80:20	80:20	50:50	50:50
Composition				
Cement, blast-fur. B (kg/m^3)	402	377	361	333
Water[1] (kg/m^3)	219	215	196	188
Water/cement ratio	0.55	0.57	0.54	0.56
0-32 mm AVI slag incl. 9.7% (w/w) moisture (kg/m^3)	1146	1168	759	769
0-32 mm sand/gravel (dry) (kg/m^3)	337	343	894	915
Superplasticiser in relation to cement (%(w/w))	0.7	0.7	0.7	0.7
Mortar properties				
Slump (mm)	140	120	110	110
Shaking (mm)	410	400	380	470
Volume weight (kg/m^3)	2102	2102	2212	2210
Air content (%)	2.9	2.8	1.4	1.7
Properties				
Cube compressive strength				
after 7 days (N/mm^2)	24.5	24.1	24.7	23.9
after 28 days	29.0	27.8	30.8	28.3
Volume weight (kg/m^3)	2082	2083	2202	2196

[1] Excluding absorbed water in the AVI slag.

Waste Materials in Construction.
J.J.J.R. Goumans, H.A. van der Sloot and Th.G. Aalbers (Editors)

MANAGEMENT OF RESIDUES FROM COAL UTILISATION: AN OVERVIEW OF FBC AND IGCC BY-PRODUCTS

L.B. Clarke * and I.M. Smith *

* IEA Coal Research, Gemini House, 10-18 Putney Hill, London, SW15 6AA, UK

SUMMARY

The chemical and physical characteristics of residues produced by FBC and IGCC power generation are reviewed. Legislation, leaching tests, and disposal practices in different countries are discussed. By-products from FBC and IGCC power generation may be utilised in agriculture, building and structural materials, pollution control, and materials recovery.

The successful management of FBC and IGCC residues requires detailed understanding of the nature and quantity of the waste products, knowledge of the legislative constraints that control the use and disposal of waste products and their leachates, and optimal disposal and utilisation methods in order to minimise environmental impact. Whilst appropriate uses for FBC by-products have been demonstrated, the variability of some residues is a major handicap to commercialisation. IGCC residues are similar to bottom ashes discharged from conventional combustors and should therefore be easier to use in applications traditionally associated with coal-use residues.

1. INTRODUCTION

New coal technologies for power generation, such as fluidised-bed combustion (FBC) and integrated gasification combined cycle (IGCC), may provide superior environmental performance compared with more conventional pulverised coal firing (PCF) power plants, but produce new and different types of residues.

In this paper FBC refers to atmospheric FBC (AFBC) using limestone sorbents for sulphur control, which can be sub-divided into bubbling FBC (BFBC) and circulating FBC (CFBC) types. Residues from pressurised FBC (PFBC) are not considered here. IGCC systems may be divided into two broad groups in terms of the solid residues produced:

those using slagging or non-slagging gasifiers.

This paper provides an overview of various aspects of the management of FBC and IGCC residues. It is based on more comprehensive reviews of FBC and IGCC residues are provided in reports published by IEA Coal Research (1, 2).

2. CHEMICAL AND PHYSICAL CHARACTERISTICS

FBC solid residues consist of coal, ash, unburned carbon, and unreacted sorbent and desulphurisation products. In BFBC the residues are collected from up to three locations: the bed-offtake, from the cyclone if material is not recycled, and from particulate control devices. In the case of CFBC all residues collected by the cyclone are recycled. Coal ash does not fuse at the relatively low combustion temperature of around 850°C used in AFBC, and particles generally retain the shape produced by the grinding process (3, 4). FBC residues typically contain illite (up to 50%), together with quartz, hematite and magnetite (5). Additional crystalline phases appear in the residues due to limestone or dolomite added as sorbent: lime (CaO), calcium sulphate anhydrite ($CaSO_4$), and calcium carbonate ($CaCO_3$) (6). The use of different coals and sorbents affects the composition and mineralogy of the residues.

IGCC residues consist of ash and/or slag particles, unburned carbon, fluxes which may be added to the coal, and in some processes sorbent and desulphurisation products. Particulates carried over from the gasifier in the raw product gas are captured in downstream gas clean-up systems and may be discharged as filter cake or recycled to the gasifier. In slagging IGCC systems the mineral matter in the coal is converted into molten slag, usually at temperatures above 1500°C. It is collected and solidified, and usually discharged as a black granular frit. The slag has the appearance of coarse sand, and is composed of spherical or cylindrical particles, or broken angular fragments of these shapes (7, 8). It is fairly homogeneous on a macroscopic scale, but in detail shows small disseminated mineral grains within a glassy silicious matrix, internal vesicles, and open pores and fractures. IGCC slags are physically similar to glassy bottom ashes from conventional combustors, and blast furnace slags. Most non-slagging gasifiers use fluidised-bed systems which operate below the ash fusion temperature. Residues are discharged as ash or partly bound agglomerates, composed of a mixture of crystalline and glassy materials, which may be partially melted. They are typically pale grey to black in colour (depending on the carbon content) and vary in consistency from fine powder to agglomerate. Some more refractory minerals from the coal may pass through the gasifier without completely melting. Several fluidised-bed gasifiers have been operated with in-bed desulphurisation using limestone sorbent (2). The addition of limestone produces additional phases in the residues, such as lime, calcium sulphide (CaS), calcium sulphate anhydrite, and calcium carbonate. CaS must be oxidised to $CaSO_4$ before discharge to prevent formation of H_2S gas. These residues are more similar to AFBC residues.

3. LEACHATES

Although coal-use residues are not usually considered as toxic wastes, residues from new coal utilisation technologies may require special consideration before disposal is permitted. FBC residues and IGCC residues with sorbents may present special problems due to their high alkalinity. Differing discharge criteria, water quality limits, and waste disposal guidelines are used in different countries. Legislation is typically implemented through standard leaching tests. Testing procedures vary between countries, but are frequently based on US EPA standard methods. Table 1 summarises the maximum permitted trace element leachate levels for the FRG, the EC, and the USA. Details of the legislation affecting FBC and IGCC and the leaching tests involved are reviewed elsewhere (1, 2).

In other countries, such as the Netherlands, current regulations based on a single extraction test are considered to be inadequate for evaluating the range of conditions observed in practice. New approaches are being developed in order to more accurately assess leaching from coal-use residues, and the waste-utilisation materials which contain them. This new approach attempts to appraise the release of potentially hazardous elements from the residues or materials with time. Distinctions may be made between different leaching mechanisms, such as dissolution, surface wash-off and matrix diffusion (9). Field tests based on lysimeter experiments or borehole data are more appropriate than laboratory conditions for assessing the long term leaching effects from disposal sites for coal-use residues (10).

Table 1

Selected maximum concentrations of trace elements and other compounds in leachates from waste, in drinking, irrigation, and surface waters (mg/l) (2)

Substance/ Property	A RCRA	B PDWS	C SDWS	D IWS	E EC-SW	F EC-HL	G EC-IL	H FRG-1	I FRG-2	J FRG-3
Ag	5	0.05	-	-	-	-	-	0.01	0.1	0.5
As	5	0.05	-	10	0.1	1	0.1	0.04	0.1	1
B	-	-	-	2	-	-	-	1	1	10
Ba	100	1	-	-	1	-	-	0.5	1	5
Be	-	-	-	0.5	-	-	-	0.004	0.005	0.05
Cd	1	0.01	-	0.05	-	0.5	*	0.04	0.5	2
Co	-	-	-	5	-	-	-	0.05	0.5	2
Cr (+6)	5	0.05	-	1	-	0.5	*	0.05	1	10
Cu	-	-	1	5	-	10	*	0.1	1	10
Fe	-	-	-	20	-	-	-	1	2	-
Hg	0.2	0.002	-	-	1	0.1	*	0.001	0.005	0.05
Mn	-	-	0.05	10	-	-	-	0.05	1	-
Mo	-	-	-	0.05	-	-	-	-	-	-
Ni	-	-	-	2	-	2	*	0.05	1	10
Pb	5	0.05	-	10	0.05	2	*	0.04	0.5	2
Sb	-	-	-	-	-	-	-	0.05	0.1	1
Se	1	0.01	-	-	-	-	-	0.01	0.05	0.5
Sn	-	-	-	-	-	-	-	0.5	0.5	10
Tl	-	-	-	-	-	-	-	0.01	0.1	2
V	-	-	-	-	-	-	-	0.05	0.2	2
Zn	-	-	5	-	5	10	*	0.1	5	10
fluoride (F^-)	-	2.4	-	-	-	50	5	1.5	5	20
chloride (Cl^-)	-	-	250	-	-	6000	500	200	-	-
cyanide (CN^-)	-	-	-	-	-	-	-	0.05	0.5	20
nitrate (as N)	-	10	-	-	-	-	-	11.3	22.6	-
nitrite (as N)	-	-	-	-	0.05	-	-	0.03	0.3	-
phosphate (as P)	-	-	-	-	-	30	3	1.6	3.3	-
sulphate (SO_4^{2-})	-	-	250	-	-	500	200	240	-	-
sulphide (S^{2-})	-	-	0.05	-	-	-	-	0.08	4.1	-
pH	2-12.5	-	6.5-8.5	-	5.5-9	4-13	4-13	5.5-10	5.5-12	5.5-12

* total for these seven metals should not exceed 5 mg/l

A - US Resource Conservation and Recovery Act, 1976 & 1984, TCLP limit values
B - US Primary Drinking Water standards
C - US Secondary Drinking Water standards
D - US Irrigation Water standards
E - EC limiting values for surface water
F - EC proposed guidelines for leachates from hazardous wastes disposed to landfill
G - EC proposed guidelines for leachates from inert wastes disposed to landfill
H - FRG draft regulations for disposal categories, Class 1: Natural wastes/harmless materials
I - FRG draft regulations for disposal categories, Class 2: Waste causing minimal groundwater changes
J - FRG draft regulations for disposal categories, Class 3: Municipal and similar wastes

The leachates from AFBC residues are alkaline, often with pH >12. Long term tests with acidic leaching media suggest that alkaline leaching would occur for a long time in disposal sites. Leachates from batch tests, which indicate "worst case" leaching, carried out on different FBC wastes (BFBC, CFBC, and PFBC) and PCF+FGD are similar in terms of the trace element content and pH of leachates, despite the differences in composition (11). Most BFBC and CFBC residues satisfy present US Resource Conservation and Recovery Act (RCRA) requirements and other regulatory limits for classification as non-hazardous wastes.

Leaching tests using RCRA guidelines have been carried out on a variety of glassy IGCC slags (2), and all have been classified as non-hazardous solid wastes. The majority of leaching values are much lower than the required limit value (see Table 1). Their leaching behaviour is similar to that of bottom ash from conventional dry bottom boilers (12). The bulk of IGCC residues are composed of slag, and under current regulations it is not anticipated that their disposal would present problems.

Leachates from particulates discharged as filter cake from gas clean-up systems have higher contents of some trace element and heavy metals (13, 14), although there is little published information. In one study (14) the trace element contents of leachates from filter cakes were typically an order of magnitude higher than the corresponding slags, and certain elements (Ni, Sb, Se, Zn) were high enough to require more specialised disposal or treatment. Trace element contents of coals vary considerably and so the leachates from various filter cake residues need to be examined on a case by case basis.

4. DISPOSAL

Potential problems arise with the disposal of AFBC residues where sorbent has been used and with IGCC residues from fluidised-bed systems with in-bed desulphurisation. This is due to the high contents of CaO and $CaSO_4$ in these residues and the alkalinity of any leachate. Dust problems occur during the handling of dry residues with a high CaO content. Trucks must be covered and sealed to prevent hydration of residues, and care must be taken to avoid detrimental health effects. Water is added and the wastes are compacted to decrease volumes and permeability, thus reducing the amount of leachate formed. Measures can be taken during the design of landfill sites to minimise potential pollution. Appropriate site selection may permit adequate dilution of leachates, or dykes and impermeable clay or synthetic liners may be required to reduce the amount of pollutants reaching surrounding waters. Neutralisation of alkaline leachates is possible using direct addition of acid, aeration, or recarbonation (15). IGCC slags are relatively inert and require no special processing before transportation and disposal.

Land resources for disposal sites are becoming more expensive and there are increasingly stringent controls concerning the possible contamination of ground waters from dumps. In many countries there is a greater emphasis on utilisation rather than disposal of coal-use residues.

5. UTILISATION

A variety of utilisation options have been demonstrated for FBC and IGCC residues (Figure 1), but few of these have reached commercial status. Building and structural uses appear to provide the most important sector for future use.

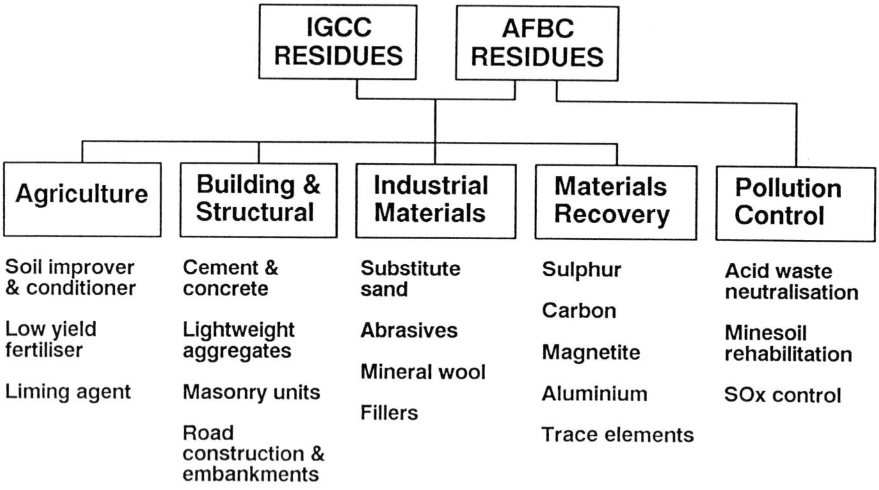

Figure 1 Utilisation of FBC and IGCC residues

Because FBC residues are not glassy pozzolans they cannot be used in the same way as PCF ash in cements and concrete. However, as a result of the high CaO and $CaSO_4$ content of FBC residues; cements, mortars, and concretes made with these residues have unique cementitious properties, including the slow development of good strength characteristics. The direct use of FBC residues in Portland cement production, rather than as a substitute for it, has been demonstrated in the USA (16), but this use is not recommended in other countries, such as the Netherlands and Sweden, because of the high sulphate content and variability of the residues.

Following size grading or crushing, glassy IGCC slags are suitable for use as a substitute for the natural aluminosilicate materials in the manufacture of Portland cement clinker (2). Some slags also possess pozzolanic properties allowing them to be used as a partial replacement for Portland cement in mortars and cements. Slag can be used to substitute directly (25-50%) for the sand fraction in cement (17).

Examples of the use of FBC residues in road construction as a soil stabiliser or filler in roadbase and asphalt have been demonstrated in several countries (1). Although the performance of the residues is good in these applications there is still concern about long term performance, especially under freeze-thaw conditions. IGCC slag could be used in civil engineering projects, in a similar way to PCF ashes, for example in embankments, structural fills, and backfills.

Synthetic aggregates using FBC residues have been produced in several countries. Production methods tend to use mixes of PCF ash and FBC residue, which are either pelletised or briquetted and crushed prior to use as a substitute for road gravel or in concrete. IGCC slags could be processed in a similar way. The fusibility of IGCC slag also makes it suitable for the manufacture of synthetic lightweight aggregate for use in concrete and other applications (16), and tests have been carried out on various slags to assess their performance and potential markets.

Masonry units from BFBC residues have been produced in the FRG and USA. Mixes of BFBC residues with PCF ash appear to give satisfactory performance as hollow non-load bearing concrete masonry blocks (1). However, in most cases FBC residues need to be improved and made more consistent in order to find uses in high quality construction materials. Once IGCC residues are manufactured into lightweight aggregate they can be utilised in lightweight pre-cast products such as roofing tiles and masonry blocks.

Miscellaneous applications of FBC and IGCC residues in building materials include abrasives, bricks, roofing shingles, tiles, mineral wool, and specialist ceramics. These uses are reviewed elsewhere, together with more esoteric applications and utilisation strategies (1, 2).

6. CONCLUSIONS

The management of residues from FBC could pose a major obstacle to the widespread introduction of the technology on a larger scale. This is because greater quantities of a different kind of residue are produced by FBC using sorbent for the control of sulphur emissions than by PCF power plants fitted with FGD. Residues from IGCC are more similar to bottom ashes from conventional combustors and appear to present less of a problem in terms of disposal or utilisation.

Legislation concerning disposal of residues, leachates from landfill sites, and water quality is fragmented and varies from country to country. Variability of residues means that leachate trace element contents and properties such as pH must be examined on a case by case basis. FBC residues may require special handling and disposal techniques because of their high lime contents.

Utilisation of coal-use residues is becoming more important. Bonded applications control leaching of trace elements most efficiently. Building and structural uses of residues appear to provide the greatest potential for future use. As the quantities of residues increase, more varied and novel uses may become more important. Continuous uses, especially masonry blocks and tiles, may provide a long term outlet for these residues, but must be balanced with discontinuous applications, such as road construction, in order to maximise residue use and minimise environmental impact.

REFERENCES

1. I.M. Smith, "Management of AFBC residues", IEACR/21, London, UK, IEA Coal Research, 83pp (Feb 1990)
2. L.B. Clarke, "Management of by-products from IGCC power generation", IEACR/38, London, UK, IEA Coal Research, 73pp (May 1991)
3. K. Kautz, Fernwärme International, 15 (4), 237-240 (Jul-Aug 1986) (In German)
4. K. Kautz, VDI Berichte, 601, 319-334 (1986) (In German)
5. VDEW/VGB Joint Committee - Residues and Waste, VGB Kraftwerkstechnick, 68 (11), 1049-1057 (Nov 1988)

6. Coal Research Establishment, "Disposal and utilisation of ash residues", Final report, ECSC project no.7220-ED/803, Cheltenham, UK, Coal Research Establishment, British Coal, 91pp (Sep 1986)
7. D.M. Deason and V. Choudhry, "Potential uses for the slag from the Cool Water demonstration plant", EPRI-AP-5048, Palo Alto, CA, USA, Electric Power Research Institute (EPRI), vp (Feb 1987)
8. B.H. Thompson and H.E. Vierrath, "The BGL gasifier - experience and application", In: Proc. 6th Annual International Pittsburgh Coal Conference (25-29 Sep 1989), Pittsburgh Coal Conf. MEMS, Greensburg, PA, USA, vol 1, 530-538
9. H. van der Sloot, G.J. de Groot, and J. Wijkstra, "Leaching characteristics of construction materials and stabilisation products containing waste materials", In: Environmental aspects of stabilisation and solidification of hazardous and radioactive wastes, ASTM STP 1033, P.L. Côté and T.M. Gilliam (eds.), ASTM, Philadelphia, PA, USA, 125-149 (1989)
10. J. Frigge, VGB Kraftwerkstechnick, 68 (2), 143-150 (Feb 1988) (In German)
11. C. Nilsson, "Restprodukter från förbränning i fluidiserande bädd - egenskaper vid deponering och åter-användering", Bränsleteknik 276, Malmö, Sweden, Sydkraft AB, 125pp (Aug 1987) (In Swedish)
12. N. Bolt, W.F. van den Broeke, G.D. Enoch and J.B. Lefers, "ICGCC: Slag utilisation, hot gas clean-up and waste water treatment research", Presented at: Coal and Power Technology '90, Amsterdam, Netherlands (21-23 May 1990)
13. K. Hufen, "Origin and properties of slag and fly ash obtained in the PRENFLO process", Presented at: IEA Expert Meeting on the use of Coal Gasification Slag, Arnhem, Netherlands (10-11 May 1990)
14. G. Baumgärtel, "Disposal categories and utilisation of slag and fly ash obtained in the PRENFLO process", Presented at: IEA Expert Meeting on the use of Coal Gasification Slag, Arnhem, Netherlands (10-11 May 1990)
15. Dearborn Environmental Consulting Services, "Prediction of wastewater characteristics from alkaline combustion wastes", EPS 3/PG/11, Ontario, Canada, Environment Canada, 104pp (Mar 1988)
16. O.E. Manz, B.A. Collings, J.S. Perri, and D.M. Golden, "Utilisation of advanced SO_2 control by-products: laboratory test results", EPRI-CS-5362, Palo Alto, CA, USA, Electric Power Research Institute (EPRI), 8/1-8/19 (Oct 1987)
17. V. Choudhry, "Evaluation and testing of coal gasification slag from the Cool Water facility", Praxis Engineers, Inc., CA, report for Southern California Edison Co., Ref. No. C1826902, 1-45 (March 1987)

Waste Materials in Construction.
J.J.J.R. Goumans, H.A. van der Sloot and Th.G. Aalbers (Editors)

APPLICATIONS OF WASTE MATERIALS AT INFRASTRUCTURAL WORKS

a vision from the Ministry of Transport and Public Works

by Winden, R. van[1]; Zwan, J. Th. van der[1]; and Zeilmaker, J.[2]

ABSTRACT
This contribution is meant to give an overview on the use of residual products. Furthermore it goes into the aspects of a technical, administrative and legal nature involved in the renewed use of these materials. Finally, it will pay attention to the market-mechanisms and the manner in which the administration can influence them.

1. INTRODUCTION

Rijkswaterstaat is a division of the Ministry of Transport and Public Works. This Ministry is the primary responsible agent for the excavation policy in Holland [1]. Rijkswaterstaat, as an operational division of the Ministry, is accountable for the actual realisation of this policy. The extraction of primary surface minerals meets more and more opposition as a result of the negative aspects. The excavation policy strongly advocates the application of alternative materials, including waste materials and industrial residuals. The policy of the national administration regarding waste materials in general offers an additional strong stimulation towards minimizing the dumping of waste materials and, wherever possible, an efficient use of residuals.

Rijkswaterstaat, apart from its operational task with regard to development and implication of general policies, is also charged with the management of the national infrastructure and with the protection of land from water. This aspect makes Rijkswaterstaat a major commissioner in the sector of hydraulic engineering, road construction and earth works. Roughly 25% of all contracts in this sector is commissioned by Rijkswaterstaat. This situation enables Rijkswaterstaat to fulfill an obvious rôle as pioneer and example for regional and local authorities. From this point of view, Rijkswaterstaat aims to further strongly the ecologically justifiable application of alternative materials.

This paper aims to enlighten the use of alternative materials in the Rijkswaterstaat operations concerning the sector of hydraulic engineering, road-building and earth works.

---

[1]Regional office South Holland of the Ministry of Transport and Public Works.

[2]Road and Hydraulic engineering division of the Min. of Transport and Public Works

2. HISTORY

The application of waste materials and industrial by-products in road-building is not a specifically recent development. The Romans already used brickrubble and slag from forges in imepuzzolane mixtures for road-building. History does not tell us whether this was caused, as nowadays, by scarcity-problems or area-planning.

It was the problem of scarcity which stimulated the employment of alternative materials in recent years. The yearly extraction of large quantities of surface minerals more and more met with problems and objections. The objections concerned mainly the loss of agricultural areas, the disturbance of areal structures and the loss or affectation of areas which represent great values from the point of view of nature or landscape. In Holland this concerns a modest series of surface minerals, being sand, gravel, limestone and clay. Some minerals are available, geologically, in confined parts of the country (figure 1). Furthermore, the extraction of these minerals takes place in areas that are highly valued for their landscape. Social resistance against extraction increased continually during the last decades. The administration was confronted more and more with the problem how to balance the pro's and cons without the availability of adequate legal steering instruments for integral policymaking. For a better balanced policy the talks were structured, both between ministries and with the provincial autorities concerned. Administrative responsabilities were delegated to the provincial authorities in provincial areas, and to the national administration in statewaters, to issue licenses to delve surface minerals. In these gremia it was considered useful to search for applicational possibilities of alternative materials in the supply of basic materials [2] This was in correlation with two other developments.

Primarily this concerns the aim to efficiently use waste materials, one of the aims of the actual waste material act. Secondarily this consideration was in accordance with the aim, re-instating charcoal to produce energy, to apply residuals and wastes occurring from the incineration of coal, in an efficient and effective manner. The administration researched and stimulated research to possible applications of alternative materials. In many ways, the administration created financial and organisational conditions to enable and effect research. The industry participates in various projects.

All this goes to show that already in the seventies, social developments strongly stimulated attention to application of alternative materials in the building industry. Very early the road-building industry came into focus, obvious, because this industry uses large amounts of building materials.

Rijkswaterstaat researched many possible applications of alternative materials. And, as it is habitual in this country, Rijkswaterstaat discussed and negociated with the industry on equal footing, in order to promote research and to translate its results into technical standards, acceptable to both parties.

During the last 15 years many applications have become operational, specifically in the road-

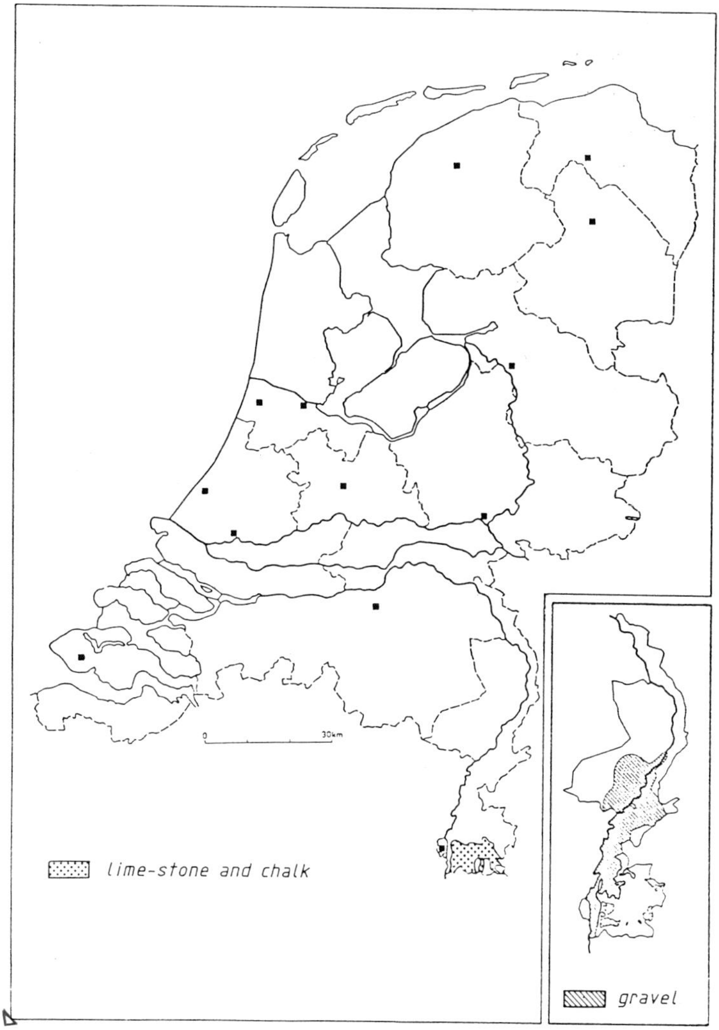

Figure 1. Map of some raw materials in the Netherlands

building industry. In paragraph 4 of this paper more details are given. Meanwhile there are still two large-scale research-programmes aimed at the application of alternative materials in hydraulic engineering and in concrete constructions.

During recent years the targets have shown deepening, resulting in the aim to make top-quality use of residuals and waste. Top-quality to be understood in the way of replacing scarce basic materials, in Holland specifically gravel.

An other important point that influences strongly both technical aspects and aspects of policy is the creation of an integral policy concerning building materials and environment. Messrs Eikelboom and Delsman will detail this point further. In this paper it may suffice to mention that integration is necessary because targets of one policy may conflict with aims in a different field of policy. E.g. the protection of the soil could oppose the application of alternative materials.

3. USE OF WASTE MATERIALS

3.1 Civil-engineering aspects

Of old, road-building is a strongly empirical industry. Much knowledge is achieved through feed-back from experience. Through taking of measurements and behaviourial models we are able to predict the behaviour of mixtures and constructions. This enables us also to use a quality-control-system to judge materials, mixtures and works in progress. The demands and conditions etc. that are used on the basis of experiences however are per definition only valid for materials and constructions we are used to apply. The introduction of new materials, as is the case when waste materials and residuals are applied, leads to an assessment of demands, conditions and testingprocedures. Often new testingprocedures and different conditions are required. It is obvious that traditional materials are only subject to demands concerning properties that can be influenced. Other, equally important requirements, are guaranteed by the nature of the materials concerned. It is a very time-consuming exercise to formulate requirements based on empirical experiences. A further disadvantage is that the insight in the actual behaviour of any material in a construction in not enhanced, so that it is hardly possible to achieve an optimalisation for the application of materials. As an escape, the choice was made to characterize both these new and old materials in such a way that it becomes possible to calculate roadconstructions and embankments along existing materialmodels.

In order to decide the measure of embankment or the thickness of a roadbase it is necessary to define stability and compressability of the embanking or basing material, apart from information concerning the subsoil.

Compressiontests are executed to define compressability of a material, stability is decided upon by means of a tri-axial test. A correllation exists between grainsize-distribution of a material or mixture and stability. An optimal stability is achievable when the cavitypercentage is minimal and the contents of grain of relatively large size is maximal. The "Fuller"-curve is the theoretical solution of this optimalisation-problem. This implies certain requirements applied to grainsize-distribution in a material or mixture in order to use the material or mixture to its optimum.

Next to insight in stability and compressability we need knowledge on the subject of resistance. This means the aspects of mechanical, physical, biological and chemical resistance. Should this be below standards, the percentage of fine grains increases and the stability due to overfilling decreases. Besides, the increase of the fine-grain-percentage causes an increase of the sensibility for frost and moisture, which has its consequences for the overall stability. As a result materials and mixtures in Holland are subject to a requirement in the field of steadiness.
When objects are demolished, rubble and rubbish is produced. It is hardly ever possible to separate the various elements like wood, brickrubble and glass completely. As a result these recycling materials, as offered on the market, like granulates of brick and/or concrete, will always be slightly polluted. Since these pollution influence the mechanical behaviour of these products, a maximum percentage for polluting elements has been defined.

Conclusive: Civil engineering will express requirements and conditions concerning materials and mixtures to be used now or in the future in embankments and/or roadbasing, in the fields of:

1. spreading of grain sizes
2. compressability
3. steadiness and
4. composition.

These requirements and their testing procedures are given in a standard. [3]

3.2 Environmental aspects

Often environmental aspects stand in the way of the application of wastes. Uncontrolled "use" of waste and residual materials in the past resulted in clean-up activities nowadays.
In essence there is one objection to the use of wastes, environmentally speaking: the materials may emit polluting matter and thus endanger the environment, more specifically soil, groundwater and surface water. That aspect leads to three questions:

1. How great is this risk?
2. How can this risk be influenced or contained?
3. Which level of risk is acceptable?

Answers to these questions will result in standards as well as to an appropriate system of protective measures to comply with these standards.

The extent of risk

This concerns the unwanted consequences of a certain activity, related to the chance that these consequences occur. The chance of unwanted emmission of polluting matter to the environment can be estimated by means of research, both in laboratory conditions and in the actual situation, e.g. to the effect of lixiviation (see the paper of Mank and Brulot). The consequences of these emissions can vary greatly. They might result in endangering certain water-transporting soil-layers (drinking water supply), pollution of surface water or surface pollution with effects on the ecosystem.

Influencing risks

The need to influence risks follows from the chances that polluted matter is emitted as well as from the effect of such an emission. The risk that a certain emission exceeds a predefined limiting value can be influenced through technical measures: isolation and control. Regular control will give timely indications concerning the effect of the measures.
In other papers these technical measures will be discussed. The effect of them, though, can be influenced by defining limiting values and/or an approach varying per district.

Acceptability of risks

Balancing risks as to which level of risk can be accepted, leads to standards. For a number of material directives have been made. This leads to the conclusion that the environmental risk of certain applications and materials is not considered acceptable without providing protective measures. One roadconstruction, made of incineration-slag is actually constructed in total containment and undergoes research.

The general process of defining standards is now underway. It is to result in a system of standards for the use of materials in or on the soil (Building materials decree). Mr Eikelboom will enhance this subject, so in this paper a few remarks may suffice.
When standards are decided upon, many interests play their part, often opposing parts: the interest of a clean soil demands a strict standard; a less strict standard is required to obtain economic use of primary materials and an effective removal of wastes. The administration has not only a legislative task, but is also clearly a producer in this aspect.

Where protection of the soil is concerned, defining standards is not an easy task. Parties must know what actually is an "unwanted emission", in other words when the extent of the emission causes undesired consequences for the public health and/or for the ecological system. That goes for hundreds of materials with each their own characteristics, dozens of environmentally dangerous substances and numerous different soil- and ecosystemtypes. Much knowledge is still to be acquired.

Well known problems are retardation as to the actual occurrence of effects, and the coupling of effects and causes.
Primarily responsible for the defining of standards is the Ministry of Public Health, Area Planning and Environment. But other Ministries including the Ministry of Transport and Public Works have their responsability too when the process leads to legislation. Since the Ministry of Transport and Public Works is responsable for the excavation policy, it aims wherever possible to maximal use of residuals and wastes, in order to contain the actual extraction of primary surface minerals. Furthermore it values highly that applicational possibilities, resulting from major research investments by the industry (sometimes heavily subsidized) remain practically feasible also after implementation of the Building Materials Decree.

3.3. Administrative aspects

Licensing

In Holland the "waste material act" regulates the application of waste materials. More specifically the flow of wastes. The act delegated power to the Provincial Administration. This administration therefore decides whether an application of "waste" is subject to a license and then under what conditions the license is issued.
The annex to the "Decree concerning works" explains that the waste material act is not applicable when suitable waste materials are applied directly in "road-constructions, building constructions and building materials". In these exceptions waste materials are regarded as raw materials or auxiliary materials. Yet they must comply with technical standards and environmental criteria. Thus a license is not required, but a license may become a necessary document if third parties object to the proposed application.
If the appropriate authority agreed that the application is permissible with only an "announcement" however, this does not imply that a license will be issued automatically should it be required.

Ownership

In those cases that no specific agreement is made, the legal owner of the soil becomes automatically owner of the material that is applied. That implies that when possibly polluting waste is applied, the environmental consequences are his responsability. When the life-cycle of the object is terminated, the owner is obliged to remove, recycle or cleanse the outcoming waste materials. This is, as a result of the uncertainty concerning costs and risks of possibly polluting agents, an extreme one-way-situation. Rijkswaterstaat judges that the use of residuals and waste should be budgettary neutral, compared to sand and gravel.
For the time being the view is that extra expenses in construction, exploitation and maintenance should be covered by the producer of these materials, from the theory "Who pollutes will have to

pay up". For this purpose the risks of application of such materials must be estimated, as well as the financial consequences in case of environmental damages. These risks may be covered through the insurance of objects or through the establishment of funds by contributions from producers. Right now, Rijkswaterstaat prefers separate funds and researches the conditions under which such a fund could operate.

TABLE 1

Use of raw materials in different sub-markets.

Sub-markets	use in million tons	change in volume since '87/'88	
	'87/'88	1995	2000
embankment & supplementations	50 (combined)	+	?
dyke building			++?
(un)bound road foundations		+	-
asphalt	8	0	0
shore protection	3	0	+?
concrete mixture	15	+/-	+/-
concrete products	13	+/-	+/-
ceramics	5	+?	+?
cement	5,5	+/-	+/-
calcium silicate	4	+	+
gypsum using industry	0,6	+	+
Total	± 100	+	++?

- reduction + increase
0 unchanged ++ strong increase
? great uncertainty +/- opposite developments may be expected.

A corporate body

A foundation might be a legal means to separate funds for this specific purpose. The foundation can be charged with control and exploitation of possibly polluting materials. In order to balance correctly the interests of all involved parties, the foundation should be boarded by representatives of the appropriate authorities, the producers and the national administration as owner of the soil. Producers of dumpable materials are to pay for the right to dump materials, in order to enable the foundation to cover the environmental risks. Part of the income of the foundation is then used to "insure" the risks, whereas an other part may be used to research (or to finance research) with a view to the reduction of the production of waste materials.
If this enables the foundation to accept efficiently the responsability an important obstacle to the application of residuals and waste in this sector of industrial activity can be reduced.

4. MATERIALS AND QUANTITIES

4.1 Quantities of primary elements

An insight is needed of the total flow of primary elements that are used for the Dutch building industry in bulk, in order to judge the potential use of alternative materials. During the past years much research had been done in this field, and models have been developed for long-term-prognoses. Table 1 shows a prognosis of the sales of materials in market segments. An annual quantity of 100.000.000 tons of materials in bulk is the outcome, with a slight tendency of growth towards the end of this century.

4.2 Potential alternatives

The production of alternative materials is well known. In this paper a number of large flows, of at least 100.000 tons annually, will be discussed. Table 2 shows these flows. The table demonstrates that, although these flows are fairly large, the quantity in total replaces only a modest part of the total flow of primary materials, apart from the flow of dredged material.
The application of alternative materials achieved a high level as it is. During 1989 the future market development of the use of alternative materials was studied, and the past period was subjected to an evaluation. [4] This study demonstrated that during the past years recycling of alternative materials reached a fairly high level. (10.000.000 tons annually) (figure 2). Remarkable is that high percentages of application were reached in relatively small flows (< 1.000.000 tons annually) [5].
The bulkflows will be subject of the next paragraphs, including the policy of Rijkswaterstaat in this field.

TABLE 2

Expected production of secundary bulk raw materials.

secundary raw materials	Production in Holland (x million ton dry material)		
	supply	estimation at 1989	
	1988/89	2000	2010
■ blast furnace slag	1,3	1,3	0
■ steel slag	0,4	0,4	0
■ phosphorus slag	0,7	0,7	0
■ chemical gypsum			
- phosphorus gypsum	2,4	2,4	?
- "Ro"gypsum (1)	0,2	0,2	?
■ construction and demolition waste	3,0	1,0	0
■ construction and demolition rubble	5,0	8,0	+
■ asphalt rubble	0,9	0,9	0
■ dredging spoils	(2)	60 (3)	0
■ fly ash (1)	0,8	1,0	+
■ waste incinerated slag	0,7	2,0	++

(1) out of fume from a cole fired power plant.
(2) no information available
(3) 60 million cubic meters accountable dredging waste can be processed into about 25 million tons of usable spoile.
0 unchanged
+(+) (strong) increase
-(-) (strong) reduction
? great uncertainty

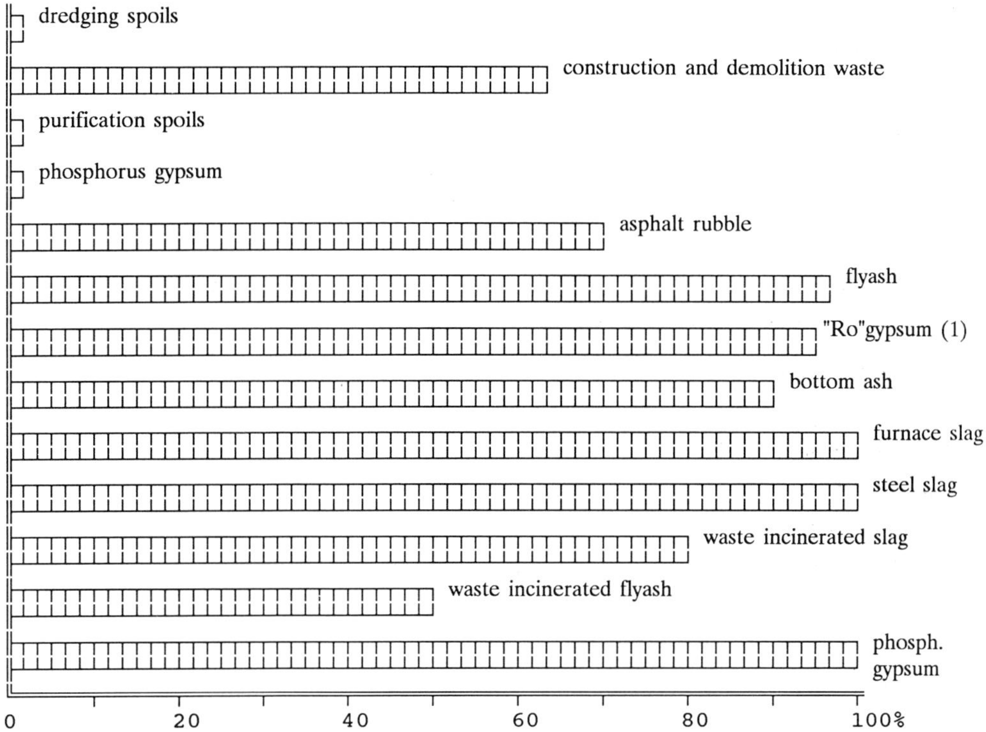

(1) out of fume from a cole fired power plant.

Figure 2. Percentage reused secundary raw materials

Granulated brick and concrete rubble

Annually app. 12 million tons of rubble result from demolition an reconstruction of buildings. Roughly 9 million tons of this quantity is of a stony nature and sorted. During 1989 some 5 million tons were reduced to granulates by brick crushers and similar installations. Nearly all of this is used in roadbuilding as base material. The constructive value of such road foundations has been researched, which led to the conclusion that this material can be marketed economic, i.e. competitively. This is a direct consequence of the policy of the administration to check dumping materials through a sharp increase of dumping rates. (Rates exceeding $50/ton are not exceptional)

The application as base material is of a relatively low value; a higher value could be achieved when applied as a replacement for gravel in concrete. Various research projects have resulted in conditions under which this material can be applied in an acceptable way. The conditions resulted in standards laid down in regulations [6 & 7].

Rijkswaterstaat now dictates the use of concrete granules in new concrete constructions. Wherever possible, gravel is to be replaced by granulated concrete between 20% and 100% in engineering works of Rijkswaterstaat. The economic conditions do not yet lead to an application in conformity to the market, this requires a further increase of dumping rates and gravelprices.

Granulates of bituminous materials

Recycling bituminous materials has its roots in the oilcrisis of the seventies. In cooperation with two important combinations of contractors two recycling techniques were developed [8]. In the early years Rijkswaterstaat promoted recycling strongly, dictating the use of recycled bituminous material in its specifications. Further technological developments manifested themselves since then. We mention parallelbarrels and mixing barrels that are introduced on the Dutch market. At this moment practically 90% of the present equipment is able to recycle bituminous material. In Holland the application of recycled bituminous material is accepted in all layers of bitumen. It goes without saying that for the various applications different conditions are laid down, based on the desired functional properties. Apart from warm recycling, granulated bituminous material is applied in road basing material, in the form of a stabilizing layer with sand and cement. An estimate of 1.5 million tons annually results from the reconstruction of roads. Of this quantity over one third is recycled warm, one third is used in stabilizing layers and the remaining is used unbound as material for the pavement of yards and similar applications.
Although warm recycling is nearly in conformity with the market situation, Rijkswaterstaat will specify warm recycling if the local market situation gives cause.

Incineration slag

An important aim of the national administration is the reduction of the quantity of domestic waste. A policy aimed at the cources of waste is seen as a way to reduce the continously growing mountain of waste. Nevertheless, a further growth is expected. The quantity of waste to be dumped is also reduced by incineration. Actually some 700.000 tons of incineration slag per annum is the result. In the year 2000 this quantity will show an increase to roughly 2 million tons of slag. The diversity of domestic waste and its uncontrollable character give cause to environmental objections to this material. Potentially various applications are possbile, such as application in concrete and foundations bound by cement. In each case the environmental aspects must be considered.
Rijkswaterstaat stated that its social responsability motivates cooperation towards the solution of a social problem. However, considering the character of this material, Rijkswaterstaat does not advocate diffuse spreading. Therefore it is preferred to use incineration slag only in large-scale embankments, applications to be realised under existing rules and standards, the so-called IBC-criteria (Isolation, control and cross-checking).
Isolation means to prevent the admission of water that might cause lixiviation. The isolating conditions must be controlled and the effectiveness should be checked periodically. A hard

condition is the periodical inspection of the isolation as well as examination of soil and groundwater in relation to emitted components. These requirements are basic for the application of all materials that might be harmful.

For the next decade an inventory is made of the projects of Rijkswaterstaat that might be suitable for this application, in order to determine a percentage of the market. It is agreed between national and provincial administrations that the latter will make a similar effort.

Fly ash

For its electricity Holland depends mainly on coal as a source of energy. A residual product, when coal is burned, is fly ash. Annually some 700.000 tons of fly ash is produced. A large part is sold to the cementindustry, both in Holland and in surrounding countries. Another part is sold as filler in bitumen and to be added to concrete. A further 120.000 tons is sintered into artificial gravel. There is practically no residual quantity, the whole quantity is use profitably. Modifications of the coalburning process as a result of stricter standards for the emission of gases into the air will result in a changed quality of the filler. This may result in a surplus quantity because a part of the market, specifically in cement, might decrease. This led to an investigation of the possible application of fly ash as an embankment material. Then, of course, the same standards and conditions that are reported concerning incineration slag will be applicable. (Chapter 3.3)

Blast furnace slag and steel slag

This slag is a by-product of blast furnace operations. The own production of blast funace slag is app. 1.8 million tons; steel slag amounts to app. 450.000 tons. During 50 years blast furnace slag has been used in the cement-industry. In a granulated form, owing to its hydraulic character, an other market has been found in the road-building industry.

Steel slag is from old applied as armour layer in hydraulic engineering, owing to its high density. Large quantities were used at the well-known Oosterschelde project.

Since that market was reduced, new markets had to be found. When combined with blast furnace slag, it can be sold as base material for road-building. The possibility of use in concrete and bitumen is now being investigated. Steel slag poses a problem: its steadiness is unsure since it contains free lime.

Phosphor slag

When phosphorus is disclosed from ore, phosphor slag forms a residual product. Annually this amounts to 70.000 tons. Nowadays three quarters of this quantity is sold as road-base material in combination with blast furnace slag. The emission of fluoride and the incresed radiaton-level limit the applicational possibilities. Radiaton will prohibit application in concrete. Application in bitumen seems very well possible.

5. INFLUENCING THE MARKET

Economics teach us that the price of goods is influenced by its availability or scarcity. Increasing or decreasing production as well as affecting the demand are means to affect the price. This goes also for the sand and gravel trade.

The substantial quantities that are imported however reduce the direct influence of the national administration on the prices of these materials by means of their excavation policy. All the same, the dumping rates or dumping limitations can have a steering effect.

For residuals the general principles of economics are valid. As there is no demand, they have no commercial value. Furthermore, since the producer of these materials must pay for transport, storage etc., their value is negative. These expenses will therefore be incorporated in the costing price of the other products.

For any material, and likewise for residuals, a demand can be created. Then a difference should be made between products that **are** and that **are not** hazardous towards environment.

For the latter the administration should create a demand, by reduction of excavation concessions and by imposing dumping limitations or even prohibitions. The policy should be aimed to stabilize offer and demand at a reasonable pricelevel, in order to stimulate producers of residual products to modify them in such a way that application as a replacement for sand or gravel becomes a real possibility.

For environmentally hazardous materials all this is much more complicated. Here the administration should guarantee that the dumping rates, as a negative value, are sufficiently high to finance environmental measures now or in the future.

A further advantage of dumping fees is the tendency of producers to reduce the production and/or to cleanse them in order to enable selling them in the commercial market.

None the less, the administration must aim to realise functional dumping sites, where possible, and based on environmental considerations of all pro's and cons.

The administration must also try to prevent that environmentally hazardous materials are "diluted away".

One example of a functional dumping site is the embankment of a traject of highway 15, where incineration slag is used in stead of sand. The size of dumping sites like this enables a financially acceptable way to take environmental measures.

It is necessary though, in order to guarantee justifiable applications for this category of materials, to reach reliable agreements between producers, issuers of licenses (administration) and owners.

6. POLICY TARGETS

Rijkswaterstaat has two separate functions, defined by their responsabilities: legislator and commissioning contractparty. The legislative function in combination with the responsability of Rijkswaterstaat for the excavation policy in this country, leads to the aim to recycle all waste

materials that can be used optimally, in the most valuable way. The scarcity of gravel leads to an accent on the application of alternative materials to replace gravel. Next, Rijkswaterstaat aims to increase the quantity of alternative materials to be applied.
This function is fulfilled by creating conditions and specifications, of course often cooperating with other authorities involved, and the industry concerned. Technical and environmental research and investigations help to specify engineering conditions. The actual applications depend strongly on a consistency in the policy and in the effort of the industry. Obviously, industry is not prepared to invest until an application promises sufficient return of investment.
The administration is able to, and actually does, influence the economics of this process through increasing dumping rates, announcing dumping prohibitions, and introducing levies.
For the effectivity of these measures a good insight in market developments is essential.
Furthermore a very important target is the reduction of obstructions. A recent investigation to the application of alternative materials in the building industry (...) demonstrated a large number of obstructions to a justifiable application of alternative materials. They are of various natures: legislative and legal obstrustions (liability and legislation), technical obstructions and financial aspects.
Rijkswaterstaat aims within its own responsibilities to streamline and to uniform standards and legislating in this area.

The other function is contractparty commissioning contracts to the industry, as Rijkswaterstaat is manager of an important part of the national infrastructure.
Rijkswaterstaat commissions contracts for as much as 25% of the total value of contracts in hydraulic engineering, roadbuilding and earth works. This is the area where 98% of all recycling takes place. Rijkswaterstaat aims to maximalize recycling in all its projects. This requires the creation of conditions under which an increased application is possible. In this policy though the principle goes that the application of alternative materials should not exceed the expense of the application of primary minerals. But in the initial stages higher expenses can well be acceptable. It is even acceptable to specify certain applications to stimulate them. Exemplary is e.g. the specified recycling of bituminous material in the eighties, and tthe specified use of granulated concrete rubble in engineering works.
Yet, when a market is created, one should operate very carefully. After all, to stimulate the use of marketable waste may imply a decreasing tendency of waste-producers to improve the environmental quality of their product. Sufficient stimuli for quality-improvement must remain built into this policy.

Rijkswaterstaat judges, as explained before, that **all** extra expenses concerning the use of alternative materials have to be passed on. Both extra expenses in the building process and extra expenses of management and after-care. An important item is how to quantify these extra expenses and how to provide coverage. One of the options is to delegate this to a private corporation.

It is difficult to specify quantitative targets. A sufficiently reliable insight in flows of material might help, but is not yet available. Friction between offer and demand, both overall and in marketsegments, should be avoided. If demand is too high, the effects will be undesirable, such as the import of waste materials with unknown environmental consequences.

Another aspect to be influenced by Rijkswaterstaat as a contracting party is the prevention of future waste problems in the recycling phase by using materials now. A general rule is that materials, used in projects of Rijkswaterstaat today, must be recycleable in a later stage. That implies that beforehand the later possibilities of recycling have to be known. This aspect may well lead to investigations beforehand.

7. LITERATURE

1 "Gegrond ontgronden", Landelijke beleidsnota voor de oppervlaktedelfstoffenvoorziening voor de lange termijn. Ministerie van Verkeer en Waterstaat, 's-Gravenhage, 1987.

2 De Jong, B., Grondstofvoorziening voor de bouw - verandering troef -, Preadviezen voor het Nederlandswegencongres, blz 55 - 75, 's-Gravenhage, 1990.

3 Standaard RAW-bepalingen, Stichting Centrum voor Regelgeving en Onderzoek in de Grond-, Water- en Wegenbouw en de Verkeerstechniek, Ede, 1990.

4 Hulst, J.G.A., Stralen, van J.M. en Ruiten, van L.H.A.M., Evaluatie en actualisatie kwantitatieve inventarisatie gebruik van secundaire grondstoffen, Distributiecentrum DOP, Den Haag, 1990.

5 Meijer, G.B. en Zwan, v.d. J.Th., Toepassing alternatieve materialen in de bouw, Preadviezen voor het Nederlandswegencongres, blz 55 - 75, 's-Gravenhage, 1990.

6 CUR-VB Aanbeveling 4, Betonpuingranulaat als toeslagmateriaal voor beton, CUR Civieltechnisch centrum uitvoering research en regelgeving, Gouda, 1984

7 CUR-VB Aanbeveling 4, Betonpuingranulaat als toeslagmateriaal voor beton, CUR Civieltechnisch centrum Uitvoering, Research en regelgeving, Gouda, 1984

8 Zwan, J.Th. van der en Hopman, P.C., Hot mix recycling of asphalt concrete,an evaluation of ten years experiance in the Netherlands, Roads and traffic 2000, Volume I

Section 2:
Methodology of Environmental Impact Assessment

Waste Materials in Construction.
J.J.J.R. Goumans, H.A. van der Sloot and Th.G. Aalbers (Editors)

A COMPARISON OF FIVE SOLIDIFICATION/STABILIZATION PROCESSES FOR TREATMENT OF MUNICIPAL WASTE COMBUSTION RESIDUES - PHYSICAL TESTING

Teresa T. Holmes[1], D.S. Kosson[2], C.C. Wiles[3]

[1]United States Army Corp of Engineers, United States Army Engineer Waterways Experiment Station, Vicksburg, MS 39180-6199, USA

[2]Rutgers, The State University of New Jersey, Piscataway, NJ 08855-0909, USA

[3]Carlton Wiles, Risk Reduction Engineering Laboratory, United States Environmental Protection Agency, Cincinnati, OH 45224, USA

SUMMARY

This paper presents the results of physical testing included in the United States Environmental Protection Agency (USEPA) program to evaluate the use of solidification/stabilization (S/S) technologies for treating municipal waste combustion (MWC) residues. These evaluations are part of the Municipal Innovative Technology Evaluation (MITE) Program sponsored by the USEPA, Risk Reduction Engineering Laboratory (RREL), Cincinnati, Ohio. The MITE program is demonstrating and evaluating technologies for managing municipal solid waste. The physical properties of the treated MWC residues are important for determining utilization applications. Consequently, considerable emphasis was placed on the structural properties and long-term durability during exposure to varied environmental conditions. Five solidification/ stabilization (S/S) processes were demonstrated for the treatment of bottom ash, air pollution control (APC) residue, and combined ash collected from a state-of-the-art MWC facility. For each process, side-by-side comparisons were made of physical tests results from (1) unconfined compressive strength (UCS), (2) UCS after immersion, (3) wet/dry weathering, (4) freeze/thaw weathering, and (5) permeability determinations. In general, the portland cement only process produced the most durable test specimens. The APC residue test specimens were the least durable test specimens.

1. INTRODUCTION

The proper management of municipal waste combustion (MWC) residues has become an important waste management issue. MWC processes concentrate potentially toxic metals, originally present in municipal solid wastes, in the MWC residues. Studies have shown that the residues may leach metals, primarily cadmium and lead, at levels exceeding the concentrations specified in the U.S. Environmental Protection Agency's (USEPA) Toxicity Characteristic Leaching Procedure (TCLP) for classifying wastes as hazardous.

Solidification/stabilization (S/S) studies have shown the potential for successful treatment of contaminated materials to control the release of toxic constituents. A S/S process involves mixing a contaminated material with a binder material, to enhance the physical and chemical properties of the material and to chemically bind any free liquid (USEPA 1986a). Typically, the binder is a cement, pozzolan, or a thermoplastic with or without proprietary additives. Comprehensive general discussions of S/S processes are given in Malone and Jones (1979); Malone, Jones, and Larson (1980); and USEPA (1986b).

To address environmental concerns, the USEPA designed a program to evaluate the use of S/S technologies for treating MWC residues through extensive testing of physical, chemical and leaching properties of untreated and treated MWC residues. These evaluations are part of the Municipal Innovative Technology Evaluation (MITE) Program sponsored by the USEPA, Risk Reduction Engineering Laboratory (RREL), Cincinnati, Ohio. A detailed description of the background and overall design for this program is provided by Wiles, et al (1991), and Wiles (1991). An overview of results of the chemical leaching tests are provided by Kosson et al (1991). An overview of results of the physical testing are presented in this paper. Side-by-side comparisons are presented for untreated and treated residues. The applicability of these processes for utilization of MWC residues is discussed. Complete program details, results, and conclusions will be provided in the project report to be completed during the fall of 1991.

2. PROGRAM DESIGN AND APPROACH

2.1 Residue Collection and Preparation

The MWC residues were collected from a state-of-the-art waste-to-energy facility. The residue types collected were bottom ash, air pollution control (APC) residue, and combined ash (APC residue mixed with bottom ash). The MWC facility uses the following operation process sequence: (1) primary combustor with vibratory grates, (2) secondary combustion chamber, (3) boiler and economizer, (4) dry scrubber with lime, and (5) particulate recovery using baghouses (fabric filters). Bulk grab samples were obtained and air dried to less than 10% moisture, crushed, and screened to pass a 0.5-inch square mesh screen. Each residue was homogenized prior to testing.

2.2 Vendor Processes Evaluated

Four commercial vendors were selected to demonstrate their treatment process on each type of the MWC residues. A portland cement only process was applied to each residue type for use as a control. Each process was applied in triplicate to allow verification of resulting data. The processes evaluated in the program are listed as follows with a brief description of the process:

Process 1 - Cement based with polymeric additives;
Process 2 - Addition of soluble silicates and portland cement;
Process 3 - Addition of quality controlled waste pozzolans;
Process 4 - Addition of water soluble phosphate; and,
WES Control - Addition of Type 1 portland cement only.

2.3 Test Specimen Preparation

One hundred pound test specimen batches were prepared by each participant using pre-determined proprietary formulations. Untreated residue test specimens were prepared by mixing residue with water to the optimum moisture content. The products resulting from the demonstrated S/S processes were collected following the treatment demonstration and prepared for testing by (1) vibratory compaction or compaction into molds of varying sizes using the Modified Proctor compaction effort, and (2) spread into a flat container and allowed to cure without compaction. The residues were cured in an environmental chamber at 20°C and 98 percent relative humidity until testing. The compacted test specimens were extruded from the molds when they acquired enough strength to be free standing (e.g., could withstand slight fingertip pressure without deformation). After 28 days of curing, if the treated residue was unconsolidated the residue that cured in the flat container was used for testing. Otherwise the monoliths prepared by compaction into the molds were used for testing. All the treatment processes yielded test specimens classified as monolithic except the APC residue treated by Process 4. Table 1 lists the physical tests performed on the untreated and treated residues, the objective of each test, and designates each test as one applied to a monolithic product, unconsolidated product, or both.

TABLE 1

Physical Tests Conducted On Treated And Untreated Residue

Test	Objective	Monolithic (M), Unconsolidated (U), or Both
Moisture Content	General data	B
Loss on Ignition	Residual/Organic Matter and hydrated water	B
Modified Proctor Density	Optimum moisture content	U
Bulk Density	Volume changes	B
Particle Size Distribution	Range of particle sizes	U
Cone Penetrometer	Curing rate and hardness	B
Pozzolanic Activity	Self-cementing potential	U
Porosity/Surface Area	Potential for liquid-solids contact and diffusion effects	B
Permeability	Assist in determining contaminant release mechanisms	M
Unconfined Compressive Strength	Load bearing capacity	M
UCS after Immersion	Effects of immersion on test specimen durability	M
Freeze/Thaw	Physical weathering effects	M
Wet/Dry	Physical weathering effects	M

2.4 Physical Testing

Each of the test protocols carried out are briefly summarized and referenced in the subsequent paragraphs.

2.4.1 Unconfined Compressive Strength (UCS). The UCS was determined using American Society of Testing and Materials (ASTM) Method C109-80. Laterally unconfined test specimens were axially loaded until failure, using a compression machine. UCS readings were obtained after 7, 14, 21, and 28 days of curing. The UCS is indicative of the load bearing capacity of the material.

2.4.2 UCS after Immersion. UCS after immersion was determined using ASTM Method C109-80. Two test specimens that cured for 28 days were completely submerged in a dilute lime solution (0.10 g lime/L distilled water). This environment mimics the natural pore water of cured cement when wetted. After 24 hours of immersion, one test specimen was removed and a UCS determination made. The remaining test specimen was removed after 28 days of immersion and a UCS determination made. The UCS after immersion test assesses the effects of exposure to wet conditions on the durability of the solidified waste form, an important consideration for utilization options.

2.4.3 Wet/Dry Weathering Test. The Wet/Dry test was conducted according to ASTM Method D4843. The test is a cyclic weathering test in which test specimens are subjected to 12 successive cycles of being submerged in water for twenty-four hours, followed by drying in a nitrogen purged oven for twenty-four hours. Wet/dry weathering control test specimens are subjected to 12 successive cycles of being submerged in water for twenty-four hours, followed by placement in a chamber maintained at 20° C and 98% relative humidity for twenty-four hours. Based on weight loss of the test specimen throughout the series of cycles, determinations are made concerning the ability of the test specimens to maintain physical integrity.

2.4.4 Freeze/Thaw Weathering Test. The Freeze/Thaw test was adapted from ASTM Methods C666-80 and D560-57. The Freeze/Thaw test is a cyclic weathering test that subjects test specimens to 12 successive cycles of being submerged in water for twenty-four hours followed by freezing at 20° C for twenty-four hours. Test control specimens were carried out in conjunction with the wet/dry weathering tests. Based on weight loss of the test specimen throughout the series of freeze/thaw cycles, determinations are made concerning the ability of the test specimens to maintain physical integrity.

2.4.5 Permeability. The permeability were conducted according to the draft ASTM Method for Measurement of Hydraulic Conductivity of Saturated Porous Materials Using Flexible-Wall Parameters. The permeability measurements were made by placing a cylindrical test specimen surrounded by a thin flexible rubber membrane in a triaxial cell. Lateral pressure was applied to minimize the flow of water between the sample wall and the membrane and water flow with a pressure of 230 kPa was forced through the test specimen. The amount of water flowing through the test specimen was recorded periodically and hydraulic conductivity calculations were made. Calculations made from the hydraulic conductivity were used to determine the flow rate of liquids through the matrix and to estimate the rate of contaminant leaching by convective transport (US Army Corp of Engineers "Laboratory Soils Testing" manual EM 1110-1-1906, Appendix VII).

3. RESULTS

The results from the UCS, the UCS after immersion, the wet/dry weathering tests, the freeze/thaw weathering tests, and the permeability determinations are discussed separately. A side-by-side comparison of UCS and permeability determinations for each S/S process is made. The UCS and UCS after immersion test results are compared to determine the effect of immersion on the durability of the test specimens. The wet/dry and freeze/thaw weathering tests results are evaluated to determine durability of the test specimens during long-term cyclic weathering conditions.

3.1 UCS

Figures 1a, b, and c present the UCS's of the test specimens following 7, 14, 21, and 28 days of curing. Figure 1a compares the UCS-cure time curves of the untreated and treated bottom ash. The untreated bottom ash test specimens acquired negligible strength at the onset and showed no increase as cure time progressed. Bottom ash treated with the portland cement only process (WES Control) exhibited the highest strength formation. Bottom ash treated using Process 1 acquired slightly less strength. The remaining process test specimens acquired only twenty percent of the strength that the portland cement only process test specimens acquired. All treated bottom ash samples asymptotically approached a maximum strength with time indicating little potential for further strength development.

Figure 1b compares UCS cure-time curves of the untreated and treated APC residue. Process 4 is not shown because the treated residue was unconsolidated. The untreated APC residue test specimens initially acquired low strength which decreased as cure time progressed. The APC residue treated with the portland cement only process (WES Control) exhibited the highest strength formation. The UCS of APC residue treated using Processes 1, 2, and 3 were not significantly different from each other and were one-third of the UCS of the WES control test specimens. The UCS for Processes 1, 2, 3, and the untreated test specimens asymptotically approached a maximum strength with time indicating little potential for further strength development. Conversely, the UCS-cure time curve for the portland cement only process (WES control) test specimens did not asymptotically approach a maximum strength by the 28th day, indicating a potential for further strength development.

The UCS of the untreated and treated combined ash is presented in Figure 1c. Untreated combined ash initially acquired low strength and showed no strength increase with cure time. The portland cement only process (WES

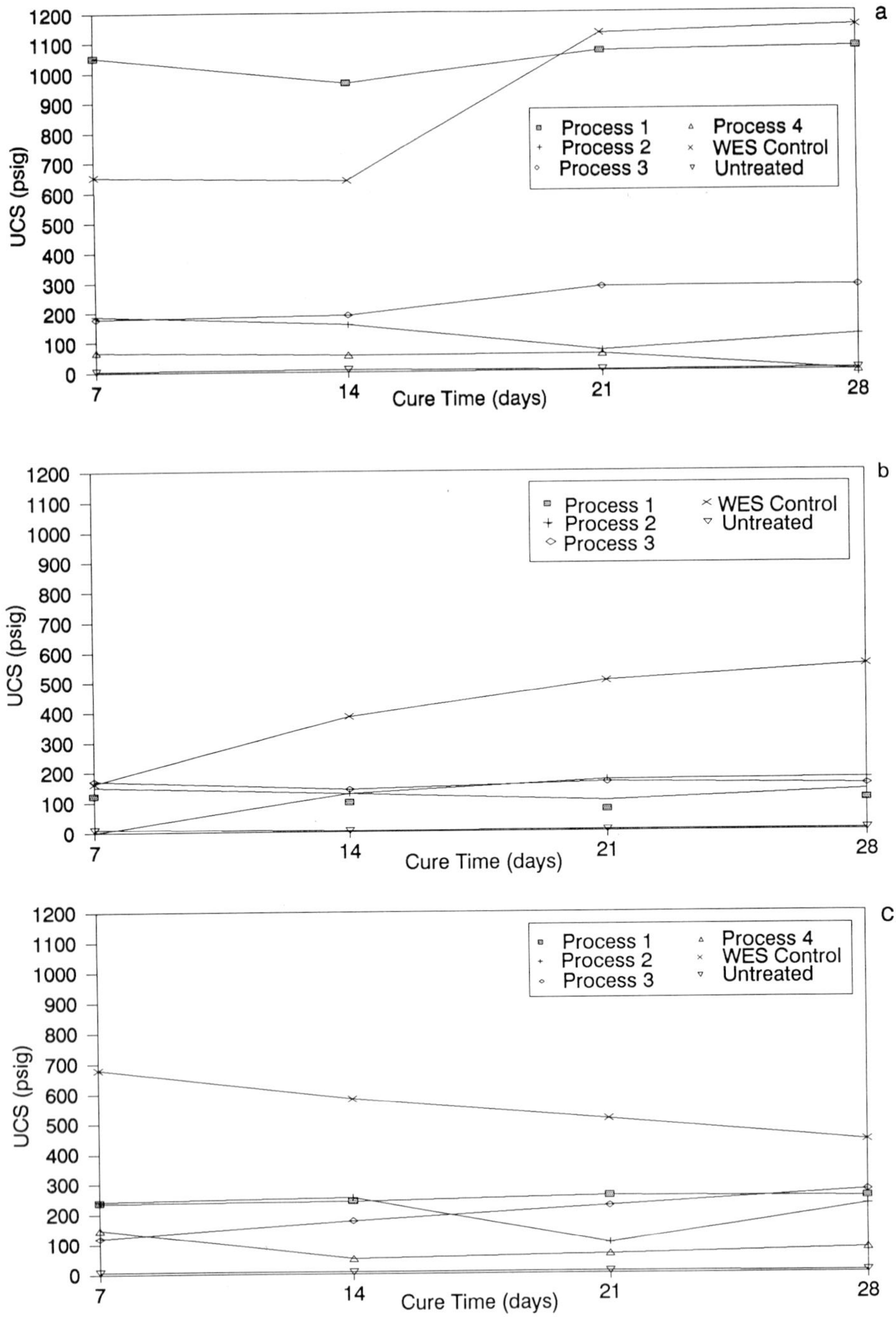

Fig. 1. Unconfined Compressive Strength. (a) Bottoms Ash. (b) APC Residue. (c) Combined Ash.

control) resulted in the highest strength formation, but the strength decreased by approximately 250 psig as cure time progressed. Process 1, 2, and 3 acquired a 28-day UCS of approximately fifty percent of that acquired by the portland cement only process. Process 4 test specimens acquired only about fifteen percent of the strength of the portland cement only process. The Process 1, 2, 3, 4, and untreated test specimens UCS asymptotically approached a maximum strength with cure time, indicating little potential for further strength development. The UCS of combined ash treated using portland cement only decreased with time indicating that strength loss would continue. This decrease may be attributed to increased dryness of the treated material as available water was depleted during the setting reaction. The addition of larger proportions of water in the S/S process should be investigated to provide improved strength formation.

In summary, the untreated residue test specimens developed little if any strength. The portland cement only process exhibited the highest strength formation for all three ash types. Bottom ash treated with portland cement only and Process 1 developed three times the strength of the other treatment processes. The APC residue and the combined ash samples treated using portland cement only developed over twice the strength of these residues treated using the other treatment processes. There was not a significant difference between the strengths of the other process test specimens. The APC residue treated with portland cement only exhibited the potential for further strength development. Combined ash treated with portland cement only decreased in strength with cure time.

3.2 UCS AND UCS After Immersion

Figures 2a, b, and c, compare UCS at 7, 14, 21, and 28 days with UCS after 1 day and 28 days of immersion. All of the untreated residue test specimens deteriorated from a free standing monolith to an unconsolidated form during immersion and therefore are not discussed. All three ash types when treated using Process 3 also deteriorated from a free standing monolith to an unconsolidated form during 28 days of immersion.

Figure 2a compares the UCS and UCS after immersion for the treated bottom ash. The portland cement only process (WES control), Process 2 and Process 4 resulted in increased UCS as the immersion period increased. Process 1 resulted in decreased UCS by approximately fifty percent as the immersion period progressed from 1 to 28 days.

Figure 2b compares the UCS and UCS after immersion for the treated APC residue. Process 2 and Process 3 test specimens deteriorated to an unconsolidated form. The APC residue treated using Process 4 initially was unconsolidated and therefore not tested. The UCS for the portland cement only process and Process 1 increased as the immersion period increased.

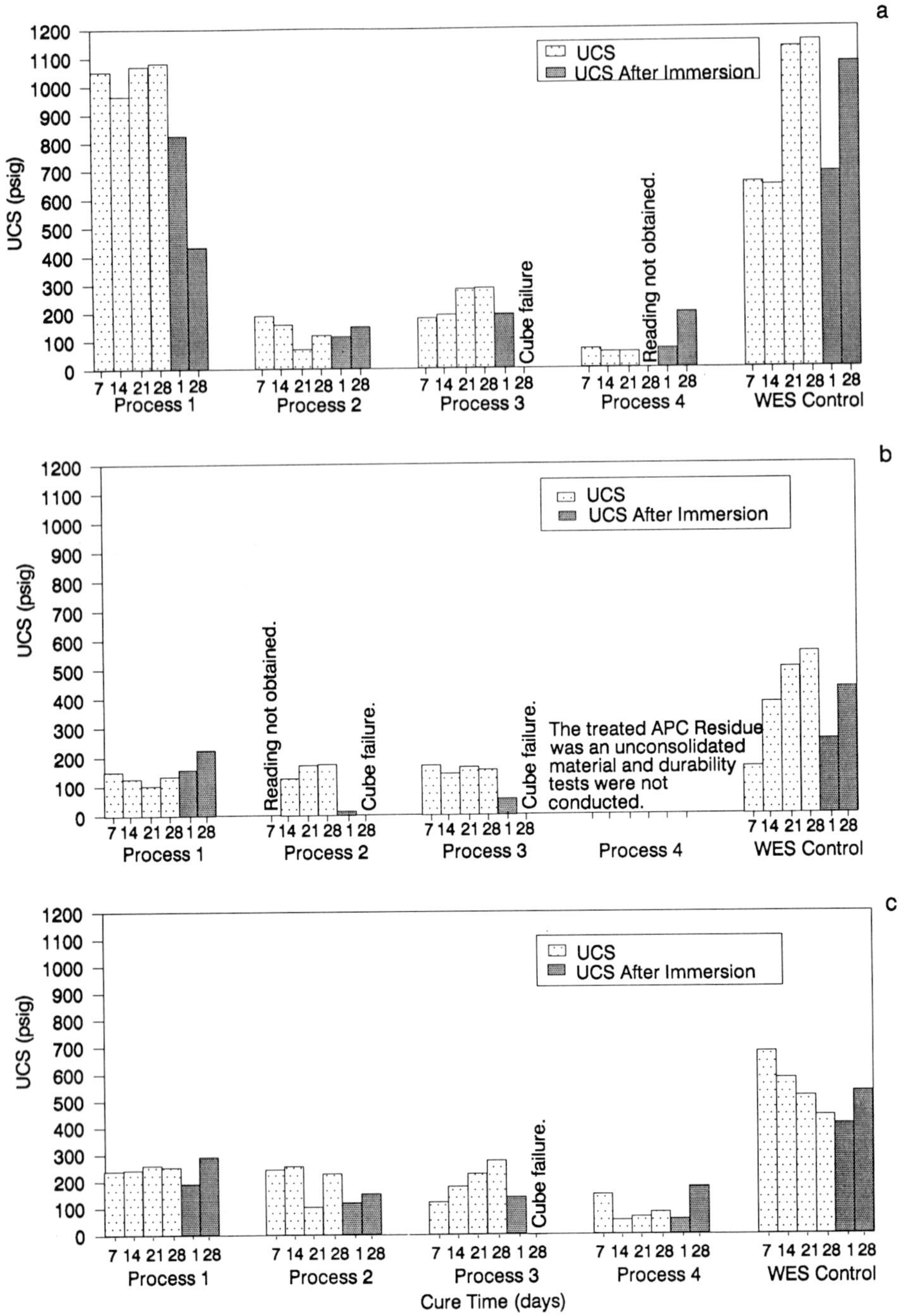

Fig. 2. UCS and UCS After Immersion. (a) Bottoms Ash. (b) APC Residue. (c) Combined Ash.

Figure 2c compare UCS with UCS after immersion for the treated combined ash. Except for Process 3, the UCS for the treated combined ash increased with increasing immersion periods.

3.3 Permeability

Table 2 lists the average permeability of each treated ash type. There is no data listed for APC residue treated using Process 4 because the resultant product was unconsolidated. The three permeability test specimens for the bottom ash and the combined ash treated using Process 2 fractured during curing and therefore no permeability determinations were made. In summary, the permeabilities varied between 1E-04 to 1E-06 cm/s for all ash types treated and processes applied. There was no trend in the data from process-to-process or between ash types.

TABLE 2.

Permeability of Monolithic Test Specimens

Process	Bottom Ash Permeability (cc/s)	APC Residue Permeability (cm/s)	Combined Ash Permeability (cm/s)
WES Control	1.62E-05	1.59E-06	1.22E-04
Process 1	6.49E-05	2.92E-05	3.67E-05
Process 2	ND	4.33E-06	ND
Process 3	2.62E-04	4.07E-05	4.41E-04
Process 4	3.79E-05	ND	5.59E-04

ND - Permeability not determined.

3.4 Wet/Dry and Freeze/Thaw Weathering Tests

Figures 3a, b and c present the wet/dry and freeze/thaw weathering test results are presented as the cumulative weight percent eroded at the conclusion of twelve cycles. The same test control was used for both weathering tests and these results also are presented. The WES control (portland cement only process) should not be confused with the wet/dry and the freeze/thaw weathering test control (cycled without freezing or drying).

Weathering test results for treated bottom ash are presented in Figure 3a. The freeze/thaw weathering test had the most adverse effect on the test specimens for all processes except Process 2. The test control specimens for each process, except Process 3, had less than 10 percent erosion. The portland cement only process (WES control) and Process 1 test specimens sustained both weathering tests with less than 30 percent erosion. Between 45 and 90 percent of the test specimen mass eroded from bottom ash treated using Processes 2, 3, and 4.

Weathering test results for treated APC residue are presented in Figure 3b. (Process 4 is not represented because the treated APC residue was

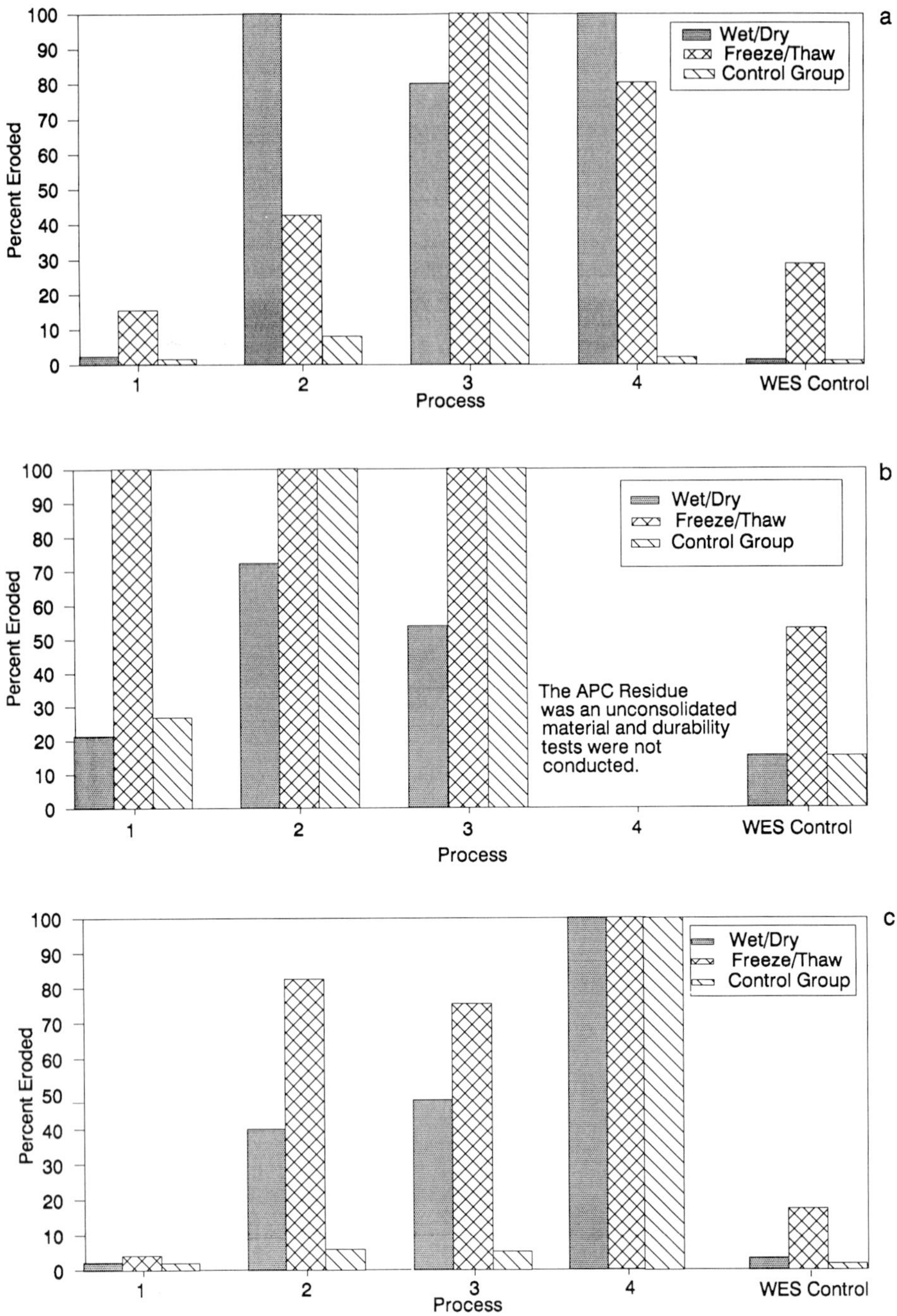

Fig. 3. Cumulative Percent Eroded. (a) Bottoms Ash. (b) APC Residue. (c) Combined Ash.

unconsolidated.) The portland cement (WES control) and Process 1 test specimens sustained the wet/dry weathering test with less than 25 percent erosion. Process 2 and 3 test specimens eroded between 45 and 75 percent for the wet/dry weathering test. One hundred percent of the test specimen eroded for Process 1, 2, and 3 when subjected to the freeze/thaw weathering test.

Weathering tests results for combined ash are presented in Figure 3c. The freeze/thaw weathering test had the most adverse effect on the test specimens for all processes. The Process 4 test specimens eroded by 100 percent for the wet/dry, freeze/thaw, and control test specimen. The control test specimens for the remaining processes eroded by less than 10 percent. The portland cement only process (WES control) and Process 1 test specimens sustained both weathering tests with less than 20 percent erosion. The Process 2 and Process 3 test specimens eroded from 30 to 85 percent for the wet/dry and the freeze/thaw.

4. CONCLUSIONS

The following conclusions about the structural properties and long term durability of MWC residues treated by solidification/stabilization are based on the physical property data obtained in this study. These conclusions do not imply performance characteristics of these processes with respect to leaching or other resultant properties.

1. Use of portland cement only resulted in the development of unconfined compressive strengths greater than or equal to all of the processes with proprietary additives. This indicates that the additives tested did not enhance the strength of the treated residues.

2. The treated APC residues exhibited the poorest performance in all of the durability tests, including UCS after immersion, wet/dry cycling and freeze/thaw cycling. Thus, APC residues treated with the processes tested have the least potential for utilization in applications requiring structural durability.

3. The portland cement only process and Process 1 produced the most durable treated bottom ash and combined ash products. This conclusion is based on the results of UCS after immersion, wet/dry and freeze/thaw testing.

4. The test specimens with the highest UCS were the most durable during cyclic weathering tests. Thus, UCS may be used as a preliminary indicator of durability.

5. The UCS after immersion test with a 28 day immersion period is useful for assessment of structural durability in exposed utilization applications. Processes for which products disintegrated or resulted in decreasing strength may not satisfy structural requirements in these applications. Processes resulting in stable or increasing strengths should be evaluated further.

6. The freeze/thaw weathering test was the most aggressive of the durability tests applied in this study.

7. Permeability was not correlated with the strength or durability of the test specimens.

The tests described and the resulting information presented herein, unless otherwise noted, were obtained from research conducted by the U.S. Army Engineer Waterways Experiment Station, Vicksburg, MS and Rutgers, The State University of New Jersey. Collection of data used in this study was sponsored by the U.S. Environmental Protection Agency.

This paper has been reviewed in accordance with the U.S. Environmental Protection Agency's peer and administrative review policies and approved for presentation and publication.

REFERENCES

1. Kosson, D. S., van der Sloot, Hans, Holmes, T. T. A Comparison of Solidification/Stabilization Processes for Treatment of Municipal Waste Combustion Residues, Part II - Leaching Properties, In: Second International Conference on Municipal Waste Combustion, Tampa, Florida, April 1991.
2. Malone, P. G., and Jones, L. W. 1979. "Survey of Solidification/Stabilization and Technology for Hazardous Industrial Wastes," EPA-600/2-79-056, U.S. Environmental Protection Agency, Cincinnati, Ohio.
3. Malone, P. G., Jones, L. W., and Larson, R. J. 1980. "Guide to the Disposal of Chemically Stabilized and Solidified Waste," SW-872, Office of Water and Waste Management, U.S. Environmental Protection Agency, Washington, DC.
4. U.S. Environmental Protection Agency. 1986a (Nov). Best Demonstrated Available Technology (BDAT) Background Document for F001-F005 Spent Solvents, EPA-1530-SW-B6-056, Vol I, Office of Solid Waste, Washington, DC.
5. U.S. Environmental Protection Agency. 1986b (7 Nov). Federal Register, Vol 51, No. 142, Office of Solid Waste, Washington, DC.
6. Wiles, C.C. The U.S. Environmental Protection Agency, Municipal Waste Combustion Residue Solidification/Stabilization Program, Proceedings of the Seventeenth Annual Hazardous Waste Research Symposium, USEPA, Cincinnati, OH, April 1991.
7. Wiles, C.C. The U.S. Environmental Protection Agency Program for Evaluation of Treatment and Utilization Technologies for Municipal Waste Combustion Residues, In: Second International Conference on Municipal Waste Combustion, Tampa, Florida, April 1991.

Waste Materials in Construction.
J.J.J.R. Goumans, H.A. van der Sloot and Th.G. Aalbers (Editors)

LEACHING PROPERTIES OF UNTREATED AND TREATED RESIDUES TESTED IN THE USEPA PROGRAM FOR EVALUATION OF TREATMENT AND UTILIZATION TECHNOLOGIES FOR MUNICIPAL WASTE COMBUSTOR RESIDUES

D.S. KOSSON[1], H. VAN DER SLOOT[2], T. HOLMES[3] and C. WILES[4]

[1] Rutgers University, Dept. of Chem. & Biochem. Engineering
P.O. Box 909, Piscataway, NJ 08855-0909, USA

[2] Netherlands Energy Research Foundation, Westerduinweg 3
P.O. Box 1, Petten, The Netherlands 17 55 ZG

[3] U.S. Army Corps of Engineers/WES, CE WES-EE-S
3909 Halls Ferry Rd, Vicksburg, MS 39180-6199, USA

[4] US Environmental Protection Agency, 5955 Center Hill Ave
Cincinnati, OH 45224, USA

SUMMARY

This paper will present the results of leaching tests carried out on the untreated and treated residues incorporated in the USEPA program evaluating technologies for treating and utilizing municipal waste combustor residues. Assessment and prediction of contaminant releases during residue utilization is essential for the determination of environmental acceptability. To date, this program has evaluated and compared several stabilization/solidification and vitrification techniques for treating bottom ash, fly ash with scrubber residue, and combined ash. All MWC residues employed in this program were obtained from a state-of-the-art MWC facility incorporating mass burn combustion, energy recovery, semi-dry scrubbers and fabric filters. Leaching protocols included the USEPA regulatory leaching test (TCLP) as well as leaching tests designed to evaluate constituent release in varied pH environments and long-term releases. In addition to the TCLP, leaching tests included were serial distilled water leach test, acid neutralization capacity, total availability leach test, and monolith leach test.

1. INTRODUCTION

Increasing reliance on municipal waste combustion (MWC) for disposal of solid waste has focused concern on management of MWC residues. MWC residues represent approximately ten percent by volume and 25 percent by mass of the solid waste combusted and are comprised primarily of bottom ash and air pollution control (APC) residues. Bottom ash is generally a combination of partially or completely combusted waste that is discharged from the primary combustion grates and materials that pass through these grates. APC residues are comprised of acid gas scrubber residues and baghouse dust. APC residues typically are 25 percent of the total MWC residue stream. In the United States, bottom ash and APC residues most often are mixed during generation to produce what is referred to as "combined ash."

Significant issues for the management of MWC residues include:

- Should bottom ash and APC residues be managed as separate or combined waste streams?
- Should MWC residues be treated prior to landfill disposal?
- Can a significant fraction of MWC residues be beneficially utilized?

A key consideration in resolving these issues is the release of contaminants from MWC residues to the environment and the effectiveness of treatment and utilization techniques to minimize contaminant release. Leaching has been identified as the most important contaminant release mechanism from MWC residues to the environment. To address these issues and others, USEPA initiated the Municipal (Waste) Innovative Technology Evaluation Program (MITE). The initial activity of this program has been to carry out a laboratory testing and evaluation program of vendor processes for the treatment and utilization of MWC residues. Processes selected for evaluation during phase one of this program are:

Process 1 - Solidification/stabilization (S/S) with portland cement and a polymeric additive;
Process 2 - S/S with portland cement and soluble silicates;
Process 3 - S/S with quality controlled waste pozzolans;
Process 4 - Reaction with soluble phosphate; and,
WES Control - S/S with portland cement only.

Subsequent phases of this program will include testing of at least two vitrification processes.

Technology evaluation under this program includes extensive testing of the physical, chemical and leaching properties of untreated and treated MWC residues. Physical testing includes measurement of fundamental properties such as density, moisture content, permeability, unconfined compressive strength, etc. and materials durability under environmental cycling. Chemical testing includes measurement of principal and trace constituents of concern in the untreated and treated residues. A detailed description of the background and overall design for this program is provided by Wiles [1]. Details and results of physical testing are provided by Holmes et al [2]. This paper presents an overview of significant results from the leaching tests carried out under this program. Processes 1, 2, 3, 4 and the WES Control are compared. Complete results of testing will be presented in a forthcoming project report.

2. LEACHING TESTS SELECTED

Leaching tests included in this program were selected to provide a broad understanding of contaminant release under a variety of potential environmental conditions. The basic leaching properties sought to be evaluated and the leaching tests chosen to estimate these properties were:

2.1 Regulatory Benchmark

TCLP was selected to be carried out as a regulatory benchmark and to allow a comparison with a broad database of results obtained from testing of other materials. This test is carried out on a sample crushed to less than 9.5 mm. Extraction is carried out at a 20:1 liquid to solid ratio using dilute acetic acid. The extraction solution is either buffered or unbuffered depending on the alkalinity of the material to be tested. Only a fixed quantity of acid is added for the extraction, and therefore the final pH of the extract is widely variable. Thus, metals concentrations observed in the extract reflect the pH dependent solubility constraints of the specific element. Previously, considerable controversy existed because MWC residues sometimes failed the predecessor to TCLP, the EP Toxicity test, for lead and cadmium.

2.2 Maximum Release

The Availability Leach Test was selected to assess the maximum amount of specific elements or species which could be released under an assumed "worst case" environmental scenario. This test was developed by the Standardization Committee for Leaching of Combustion Residues [3]. The test is carried out on a sample crushed and size reduced to less than 300 um. Two serial extractions are carried out, each at a 100:1 liquid to solid ratio, using distilled water. The pH is controlled to pH 7 during the first extraction and pH 4 during the second extraction, using an automatic pH controller which delivers dilute nitric acid. Thus, the final extraction pH is controlled, not the amount of acid used. The first and second extracts are combined for analysis. The very large liquid to solid ratio insures that the contaminant release is not constrained by its solubility at the final pH and the amount of contaminant extracted is the maximum amount which would be available at that pH. This test generally extracts all species which are not tightly bound in a mineral or glassy matrix. The test does not provide information on the rate of contaminant release.

2.3 Release During Prolonged Exposure to Precipitation/Infiltration

The Distilled Water Leach test (DWLT) was selected to assess the amount of specific elements or species which could be released under continued exposure to precipitation or nominally clean water percolation. Synthetic acid rain solutions were not selected as the extractant because the limited acidity of these extractants would have minimal impact on the extraction of untreated or treated MWC residues due to the residues' very high natural alkalinity. The test is carried out on a sample crushed to less than 2.0 mm. Four serial extractions were carried out, each at a 10:1 liquid to solid ratio using distilled water as the extractant. No acid was added and no pH control was used. Thus, the natural buffering capacity of the material controlled the final extract pH, which was typically between pH 10 and 12 for the materials tested. The first and second extracts were combined for analysis, as were the third and fourth extracts. This test indicates the amount of contaminant release over prolonged exposure and limited information on the rate of contaminant release. Results from this test were reported on a mass of species released per mass of treated or untreated residue extracted (e.g. mg/kg ash or mg/kg product).

2.4 pH Dependent Solubility of Metals

The Acid Neutralization Capacity (ANC) test was selected to assess the solubility of specific metals over a broad pH range. The test was carried out on a sample crushed and size reduced to less than 300 mm. Eleven separate extractions were carried out using separate size reduced subsamples at a liquid to solid ratio of 5:1. The low liquid to solid ratio results in the extraction being solubility constrained. Each extraction received a different amount of dilute nitric acid, varying from 0 to 12 meq/g dry waste, resulting in a broad range of final pHs. A titration curve was obtained for each material tested. Metals solubility as a function of equilibrium pH also is obtained from this procedure. Results are reported as a titration curve (meq/g product) and on a concentration basis for metals (mg/l or mg/l).

2.5 Contaminant Release Rate

The Monolith Leach Test was selected to assess the release rate of metals and species from untreated and treated MWC residues. The test was carried out using 4 cm dia. by 4 cm cylindrical, monolithic samples. Treated residues were either vibrated or compacted using modified proctor compactive effort into PVC plastic molds immediately after being treated. Samples were cured at 98% relative humidity and 20C for 28 days prior to testing. Untreated bottom ash and combined ash monoliths were prepared by compaction at optimum moisture content using modified proctor compaction effort and cured as above prior to testing. Monolithic samples are extracted by contacting with distilled water for up to 64 days. Contacting water is replaced at 1, 2, 4, 8, 16, 32 and 64 days and is analyzed for metals and other species. Modeling of the release data in conjunction with the results of the availability leach test was used to determine effective diffusion coefficients, tortuosity and chemical retardation factors for estimating long term species release rates. This leach test is a modified version ANSI 16.1[4].

All leach tests were carried out on each of three replicate process demonstrations with single replication. Extracts were analyzed for an extensive list of metals and anions. In addition, DWLT extracts were analyzed for total dissolved solids (TDS), total organic carbon (TOC) and chemical oxygen demand (COD). Contaminant release results from the TCLP, Availability and DWLT leach tests were back-calculated to mass released per mass of ash initially treated on a dry weight basis (e.g., mg/kg ash dry solid (ds)). This calculation corrects for variations in moisture content and dilution during processing.

3. RESIDUE SAMPLING AND PREPROCESSING

MWC residue used in this study was collected from a state-of-the-art mass burn facility. The MWC facility has the following process sequence: (i) primary combustor with vibratory grates, (ii) secondary combustion chamber, (iii) boiler and economizer, (iv) wet/dry scrubber (spray drier) with lime, and (v) particulate recovery using baghouses (fabric filters). Bottom ash, APC residue, and combined ash were sampled in bulk (5-10 tons of each residue type) during two days of typical facility operation. Bottom ash sampled was quenched after exiting from the primary combustor. APC residue was mixed residuals from the acid gas scrubber and the baghouses. Combined ash was the mixed bottom ash and APC residue as normally managed by the facility. The APC residue was screened to pass a 0.5 inch square mesh. The bottom ash and combined ash were screened to pass a 2 inch square mesh at the MWC facility. Materials not passing through the 2 inch mesh were rejected. After shipment to the Army Corps of Engineers, Waterways Experiment Station (WES), bottom ash and combined ash were air dried to less than 10% moisture, crushed and screened to pass a 0.5 inch mesh (nominally 3/8 inch after clogging), and homogenized. APC residue was less than 5% moisture as collected and therefore was only homogenized after shipment to WES. Residue characterization is presented by Kosson, et.al [5].

4. SAMPLE PREPARATION FOR CHEMICAL AND LEACHING TESTING

Samples of untreated residues used for leaching tests were derived from random grab samples of the preprocessed residue as described previously. Vendor processes were carried out in triplicate on random grab samples of untreated residues to produce treated residue samples for analysis and testing. All treated samples were molded (compacted using modified proctor compaction effort) and cured at 20C and 98% relative humidity for 28 days prior to testing.

Following curing and prior to chemical and leaching testing, samples required further size reduction. Approximately 3 kg of initial sample was crushed using either a mortar and pestle or a hammer to less than 9.5 mm. Subsamples were removed after this step for TCLP and moisture testing. The next step was further manual crushing to reduce the particle size to less than 2.0 mm. The amount of reject (uncrushable material) at this step was specified to be less than 15% of the initial sample mass. Following 2.0 mm screening, subsamples were removed for the DWLT and moisture determination. The subsequent step employed a mechanical parallel ceramic plate grinder to further reduce particle size to less than 50 mesh (300 um). Greater than 65% of the initial sample mass was required to pass the 50 mesh screen. After passing this screen, subsamples were removed for metals analysis, anions analysis, availability leach test, acid neutralization capacity and moisture. Moisture analysis (to constant weight at 105 C) was carried out after each crushing step to facilitate correction of analytical results to a dry weight basis. Some samples require partial drying during intermediate particle size reduction steps to permit screening of material. When this was necessary, samples were dried at 60C.

5. LEACHING TEST RESULTS

Leaching test results provided in this paper are intended to be an overview of significant findings and are preliminary. A direct comparison of the results of the DWLT, TCLP and availability leach tests for the WES Control process as applied to combined ash is provided by Wiles [1].

All untreated and treated bottom ash and combined ash samples passed the TCLP extract concentration criteria. However, the untreated APC residue failed the TCLP criteria for lead and mercury.

Release results for lead from the distilled water leach test are presented in Table 1. The results presented are the cumulative release for all four serial extractions. Lead was chosen as the example because its amphoteric behavior can result in increased solubilities in alkali solutions. Most of the final pHs for the DWLT were between 11 and 12.5. Lead behavior for bottom ash and combined ash was very similar for each process. The amount of lead released was low, but all treatment processes except Process 4 resulted in increased release as compared to the untreated residue. This effect was most pronounced for Process 3 and was most likely the result of additional alkali added in the form of a process additive. Process 4 resulted in lead release similar to the untreated residue for bottom and combined ash. Lead behavior for the untreated and treated APC residue was significantly different from that of the bottom ash and combined ash. Untreated APC residue lead release was over 1000 times greater than that of the other two residues and represented approximately 30% of the total lead present in the residue. Processes 1, 2 and the WES Control resulted in significant reduction in the fraction of lead released to approximately 20-35% of that released from the untreated residue. Process 3 resulted in no significant reduction in lead release under these test conditions. Process 4 resulted in a reduce in lead release to less than 1% of that release from untreated APC residue.

Table 1. Comparison of lead released for the distilled water leach test (mg released/kg ash, dry solid).

	Bottom Ash	APC Residue	Combined Ash
Untreated	0.5	1079	0.2
Process 1	1.6	369	0.9
Process 2	2.9	272	2.9
Process 3	10.6	1030	10.6
Process 4	0.6	10	0.3
WES Control	2.0	185	1.2

Table 2. Comparison of total dissolved solids released for the distilled water leach test (g released/kg ash, dry solid), and in parenthesis, the weight % of the material released.

	Bottom Ash	APC Residue	Combined Ash
Untreated	58 (6%)	289 (29%)	60 (6%)
Process 1	53 (4%)	640 (32%)	54 (4%)
Process 2	187 (12%)	565 (26%)	208 (13%)
Process 3	126 (7%)	578 (24%)	144 (8%)
Process 4	47 (4%)	194 (15%)	56 (5%)
WES Control	59 (5%)	671 (30%)	79 (6%)

Table 2 presents the results of the DWLT for the release of TDS. TDS provides a useful estimate of the total amount of salts released from the material tested. Release of TDS from untreated and treated combined ash were slightly greater than the release observed for untreated and treated bottom ash. Differences in release between untreated and treated residues also were slight with the exception of Processes 2 and 3. Process 2 and 3 resulted in over twice the release of TDS as compared to the untreated residue. Note that for all treated and untreated bottom ash and combined ash, the release of TDS resulted in a release of between 4 and 8% by weight of the initial material. TDS release from untreated and treated APC residue was 5 to 10 times greater than the release from untreated and treated bottom ash and combined ash. TDS release from the treated APC residues, except for Process 4, was approximately twice the amount released from the untreated residue, per mass of residue treated. Note that for all treated and untreated APC residue, the release of TDS resulted in a release of between 24% and 32% by weight of the initial material. This most likely is the reason for the poor durability of treated APC residues as observed by Holmes, et al [2]. Process 4 resulted in a substantial reduction in TDS release. DWLT release results indicate that release of salts may be a much greater concern in the management of these materials than release of potentially toxic metals.

Metals release results from the availability leach test for untreated and treated bottom ash, APC residue and combined ash are presented in Tables 3, 4 and 5, respectively. For treatment of the bottom ash (Table 3), Process 1 resulted in decreased release for the principal metals (aluminum, calcium, potassium, silicon and sodium). Process 2 and the WES Control resulted in no significant change for release of the principal species as compared to the untreated residue, while Process 3 resulted in either no significant change or an increase in release. Release of all of the trace metals of concern (arsenic, barium, cadmium, chromium, copper, lead and zinc) was reduced by Process 1, while only cadmium, copper and lead release were reduced by Process 2. The WES Control resulted in reduced release only of cadmium and copper.

Release of the principal metals from treated APC residue and combined ash (Tables 4 and 5) generally indicated no reduction or a significant increase in release as compared to the untreated residues. Release of the trace metals of concern generally show no reduction or a significant increase in release. For treatment of APC residue, Process 1 generally resulted in no significant increase in release as compared to the untreated residue. It is also interesting to note that mass of cadmium, chromium, copper and lead released from treated APC residue for the poorer performing processes was approximately the same mass as the total amount present.

In summary, the availability leach test results indicate that reductions in species release observed as a consequence of the S/S treatments are predominantly the result of pH effects (increased matrix alkalinity) or physical retardation. The lack of difference between treated and untreated implies that no long term bonding in resistant phases

Table 3. Comparison of metals release for leaching tests on untreated and treated MWC residues (mg released/kg ash treated).

LEACH TEST: Availability **ASH TYPE:** Bottom Ash

					WES		Total	Total
	Process 1	**Process 2**	**Process 3**	**Process 4**	**Control**	**Untreated**	**(SW-846) (1)**	**(NAA) (2)**
Aluminum	2,400	5,100	5,800	410	4,600	5,600	3,200	5,200
Antimony	10 U	20	29	10	16 A	29 B	NA	320
Arsenic	3 B	4 B	16 A	1 A	4	10 A	16	NA
Barium	41	100 A	99	110	130	140	550	NA
Cadmium	8	10	16	17	18	28	36	35
Calcium	60,000	120,000	204,000	88,000	104,000	73,000	NA	110,000
Chromium	4	10	9 A	1 U	8	13	200	780
Copper	77	270	200	230	150	360	2,100	1,500
Iron	320	520	360	350	180	2,700	NA	76,000
Lead	130	150	310	54	330	420	1,600	NA
Magnesium	2,100	2,500	6,800	3,600	4,500	3,600	NA	NA
Manganese	49	190	250	150	140	220	NA	1,100
Nickel	11 A	20	26	32	21	49	430	NA
Potassium	2,030	4,600	20,200	2,200	4,500	3,900	NA	9,500
Silicon	3,200	9,500	8,900	3,900	9,100	7,400	NA	120,000
Silver	2 U	2 U	3 U	1 U	2 U	2 U	4	18
Sodium	3,500	47,000	74,000	3,300	4,500	5,600	NA	20,000
Tin	14 U	18 U	23 U	3 U	14 U	14 B	250	NA
Titanium	5 B	4 U	5 U	0 U	7 A	90 B	NA	7,000
Zinc	670	7,100	1,600	1,800	2,100	2,800	4,800	6,800

U=undetected, A=U(1 of 3 replicates), B=U(2 of 3 replicates), NA= not analyzed

(Undetected values are reported as the detection limit)

Notes: (1) via USEPA Test Methods for Evaluation of Solid Waste, SW-846, 3rd Revision.

(2) via Neutron Activation Analysis.

Table 4. Comparison of species release for the availability leaching test on untreated and treated residues (mg released/kg ash, dry solid).

ASH TYPE: APC Residue

					WES		Total
	Process 1	**Process 2**	**Process 3**	**Process 4**	**Control**	**Untreated**	**(SW-846)**
Aluminum	4,100	7,900	8,400	150	18,000	2,800	14,000
Antimony	78	170	120	120	330	110	NA
Arsenic	9	24 A	19	61	31	14	43
Barium	150	220	180	46	370	89	370
Cadmium	220	170	220	120	150	130	200
Calcium	300,000	280,000	470,000	250,000	570,000	190,000	NA
Chromium	10	20	17	1 U	37	6	41
Copper	160	280	390	130	520	190	360
Iron	37	1,300	120	20	1,800	29	NA
Lead	1,200	2,100	2,300	4	3,700	980	3,000
Magnesium	8,300	6,000	15,000	7,400	15,000	4,100	NA
Manganese	77	150	190	94	310	62	NA
Nickel	11	13	16	5	25	4	26
Potassium	22,000	17,000	41,000	17,000	26,000	13,000	NA
Silicon	9,900	18,000	16,000	7,600	43,000	4,100	NA
Silver	3 U	3 U	4 B	1 U	5 U	2 A	44
Sodium	24,000	38,000	25,000	19,000	26,000	14,000	NA
Tin	22 U	41 B	29 U	3 U	49 B	11 U	603
Titanium	5 U	190 B	8 B	1 U	200	4 B	NA
Zinc	6,700	11,000	12,000	6,700	16,000	7,700	3,000

U=undetected, A=U(1 of 3 replicates), B=U(2 of 3 replicates), NA= not analyzed

(Undetected values are reported as the detection limit)

Table 5. Comparison of species release for the availability leaching test on untreated and treated residues (mg released/kg ash, dry solid).

ASH TYPE: Combined Ash

	Process 1	Process 2	Process 3	Process 4	WES Control	Untreated	Total (SW-846)
Aluminum	6,300	15,000	12,000	340	5,500	4,000	32,000
Antimony	13 B	18 B	34 A	21	12 B	8 U	NA
Arsenic	5	18	11 A	9	5	6	16
Barium	130	250	150	140	200	130	550
Cadmium	32	20	20	26	27	27	36
Calcium	160,000	160,000	190,000	110,000	130,000	79,000	NA
Chromium	5	16	19	1 U	6	3	200
Copper	240	400	390	240	360	380	2,100
Iron	190	3,700	4,500	690	200	670	NA
Lead	260	490	1,400	46	370	500	1,600
Magnesium	5,200	4,400	7,700	4,000	4,900	4,700	NA
Manganese	240	470	580	310	390	680	NA
Nickel	14	28	22	15	18	19	430
Potassium	6,800	7,600	20,000	4,800	6,200	5,800	NA
Silicon	7,000	21,000	17,000	4,900	9,400	4,300	NA
Silver	2 U	2 U	3 U	1 U	2 U	2 B	4
Sodium	7,000	51,000	7,400	5,200	5,700	5,800	NA
Tin	16 U	18 U	24 U	2 U	16 U	11 U	240
Titanium	3 U	237 B	186 A	0 U	3 U	2 U	NA
Zinc	1,700	2,500	2,000	2,200	2,300	2,900	4,800

U=undetected, A=U(1 of 3 replicates), B=U(2 of 3 replicates), NA= not analyzed

(Undetected values are reported as the detection limit)

has been achieved. This does not preclude significant reduction in contaminant release rates or extents as a consequence of these two effects. The primary exception to this statement would be Process 1 as applied to bottom ash only.

Figures 1 and 2 present examples of the ANC leach test results for Untreated APC residue and APC residue treated, using the WES Control process, respectively.

In general, the most significant effect of all the treatment processes on bottom ash, combined ash, and APC residue was a modification of the alkalinity of the material. This result was reflected in the pH titration curves (Figures 1a and 2a). No significant changes between untreated and treated residues were observed with any, except two, of the metals solubility curves as a function of pH. The two exceptions observed were the suppression of the amphoteric behavior of lead for APC residue treated by the WES Control process and Process 4 (results not presented). A decrease in solubility of lead was observed between pH 11 and 12 for both processes. The effect was substantially more pronounced for Process 4. This was the only indication of any metals respeciation for any of the processes.

Table 6 provides examples of the results of diffusion modelling based on the monolith leach test results. Calculated effective diffusion coefficients are presented for sodium, chloride and lead for treated bottom ash, APC residue and combined ash. Release rates for non-interacting ions such as sodium and chloride were two to five orders of magnitude more rapid than for a chemically interacting metal such as lead. Reduction in lead diffusion rates relative to sodium and chloride was most likely the result of interstitial pH effects. Diffusion rates for lead migration in treated APC residue was approximately one hundred times greater than those estimated for bottom ash or combined ash. Also note that the greatest diffusion rates were observed for treated materials that performed poorly in the other leaching tests and the physical testing described by Holmes et al [2] (e.g., Processes 2 and 3). Figure 3 presents cumulative release curves as a function of time for different effective diffusion coefficients. A 10 cm cube was chosen as the assumed geometry for illustration purposes.

6. CONCLUSIONS

This paper presented the basis for selection of a series of leaching tests designed to provide fundamental information on the leaching properties of s/s treated MWC residues. Preliminary findings indicate that untreated and S/S treated MWC bottom ash and combined ash pass TCLP criteria. Treated APC residues also pass TCLP criteria. Release of lead at alkaline pH may be significantly reduced by some S/S processes based on the distilled water leach test. Metals release, based on the availability leach test, was reduced significantly by S/S treatment for bottom ash but not for APC residue and combined ash, except for Process 4. The S/S treatments evaluated had limited effectiveness when applied to APC residues because of high releases of salts, which may account for up to 32% by weight of the treated residue.

Figure 1. Acid Neutralization Capacity results for untreated APC residues (Fig 1a. pH titration curve; 1b. Cadmium and chromium solubility; 1c. Copper and lead solubility; 1d. Zinc solubility)

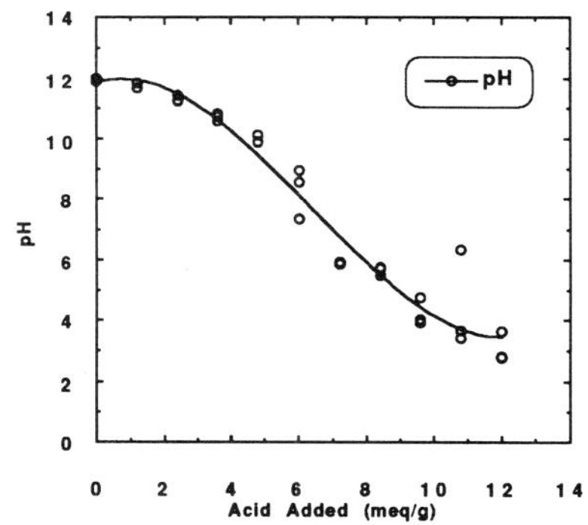

Fig 1a

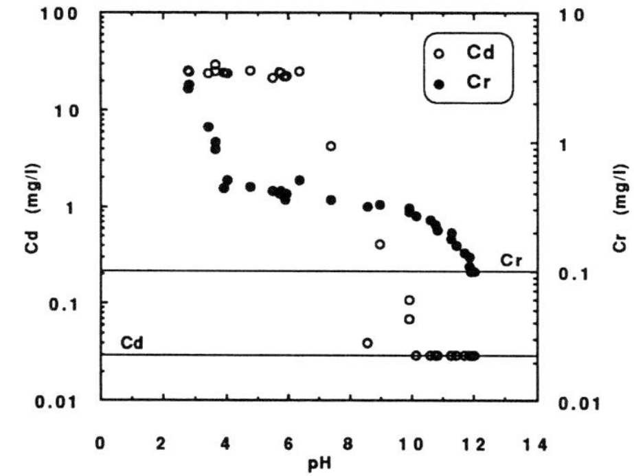

Fig 1b

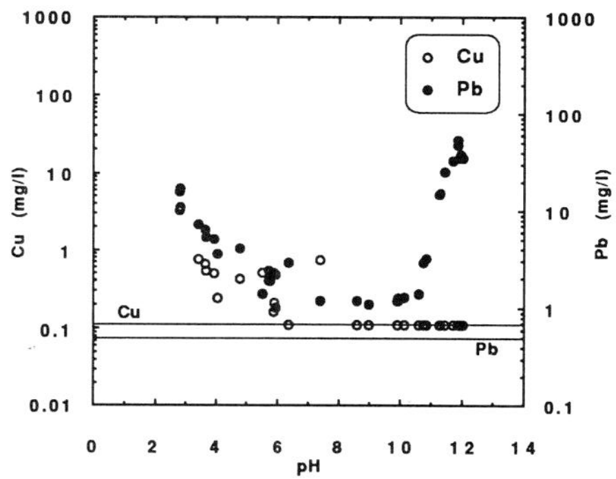

Fig 1c

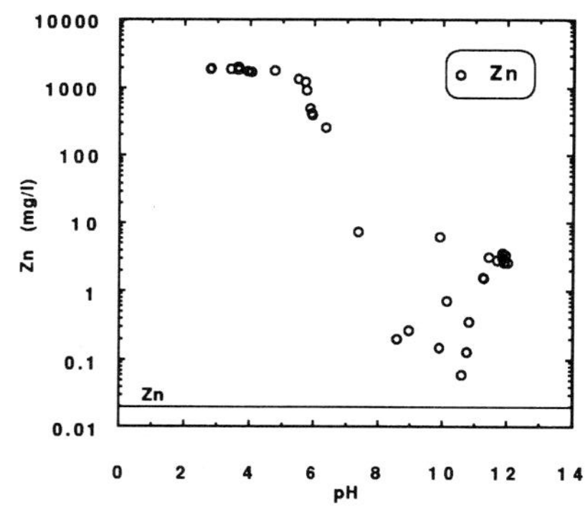

Fig 1d

Figure 2. Acid Neutralization Capacity Results for APC residue treated by the WES Control process (Fig 2a. pH titration curve; 2b. Cadmium and chromium solubility; 2c. Copper and lead solubility; 2d. Zinc solubility)

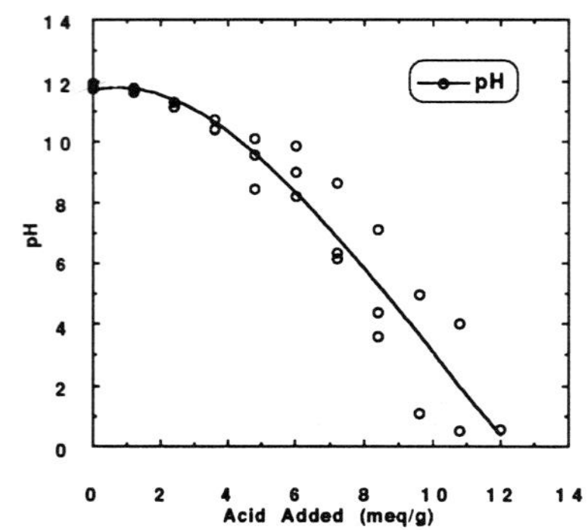

Fig 2a

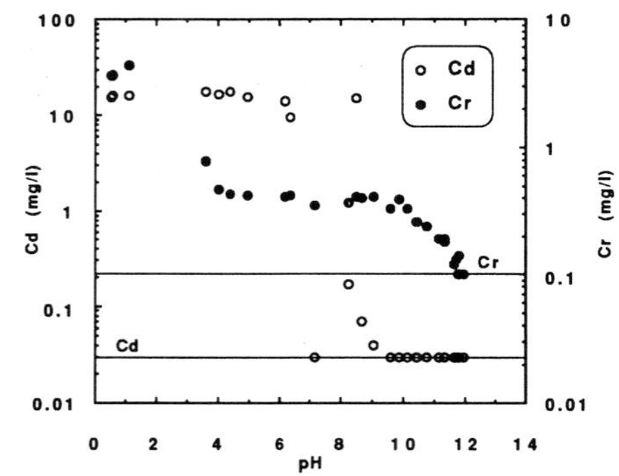

Fig 2b

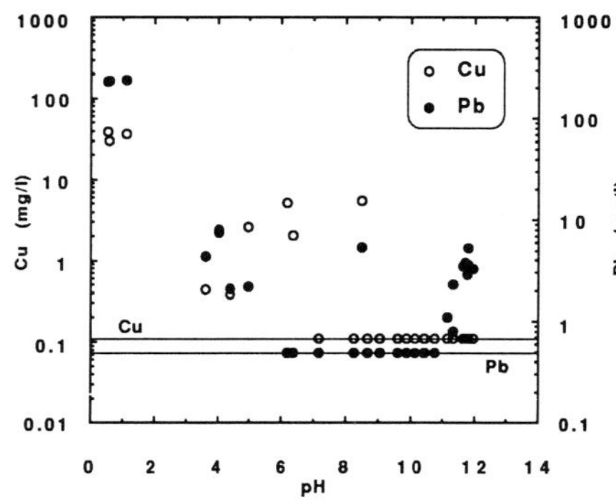

Fig 2c

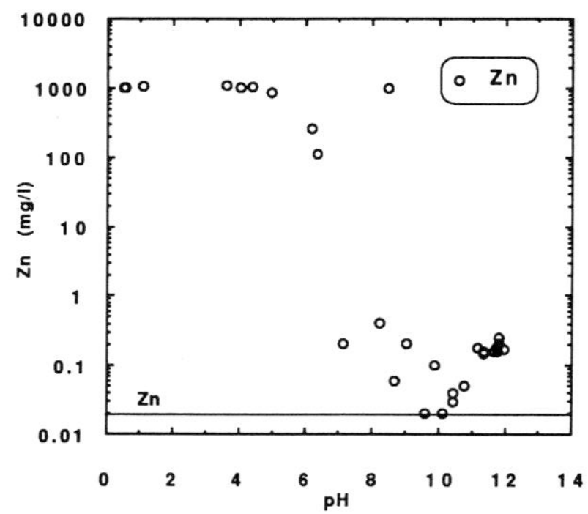

Fig 2d

Table 6. Effective diffusion coefficients (pD_e)[1] for release of sodium, chloride and lead for treated MWC Residues

	Bottom Ash			APC Residue			Combined Ash		
Process	Na	Cl	Pb	Na	Cl	Pb	Na	Cl	Pb
1	9.3	9.8	14.3	9.5	9.8	11.9	9.6	9.9	15.4
2	9.4	9.3	13.1	8.5	8.7	12.3	9.6	9.6	14.8
3	8.9	9.0	13.0	8.5	8.6	10.7	9.0	9.0	15.5
4	9.9	10.2	14.3	NA[2]	NA	NA	10.4	10.4	13.9
WES	9.1	10.0	15.8	8.9	9.1	13.7	9.2	10.2	15.2

1 Effect diffusion coefficients (D_e) are reported as pD_e, where $pDe = -\log D_e$. Units for D_e are (m^2/S)

2 NA = Not analyzed; Process 4 resulted in a granular treated product for APC residue.

Figure 3. Predicted cumulative release of contaminants based on effective diffusion coefficients (pDe) from a 10 cm cube.

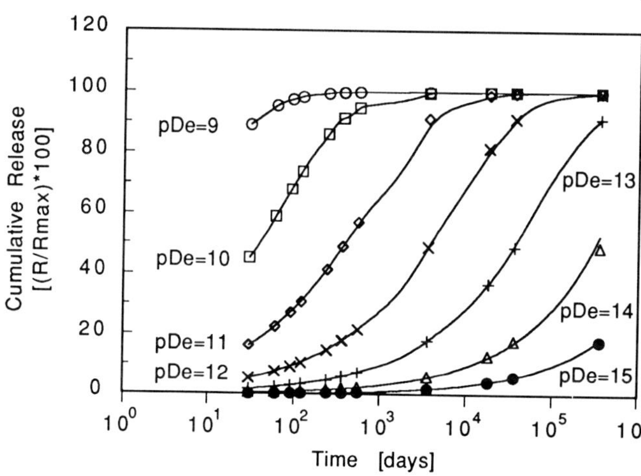

Release of salts also appears to be responsible for loss of sample integrity. Estimation of release rates based on monolith leaching tests also indicate that release of salts may be a much more significant concern than release of potentially toxic metals.

REFERENCES

1 Wiles, C. C. The Unites States Environmental Protection Agency Municipal Waste Combustion Residue Solidification/Stabilization Program, In: Second International Conference on Municipal Waste Combustion, Tampa, Florida , April 1991.

2 Holmes, T. T., Kosson, D.S., Wiles, C.C. A Comparison of Five Solidification/Stabilization Processes for Treatment of Municipal Waste Combustor Residues, Physical Testing. In: Proceedings of WASCON'91, Maastricht, The Netherlands, November 1991.

3 van der Sloot, H.A., Piepers, O., and Kok, A. A Standard Leaching Test for Combustion Residues. Technical Report Bureau Energy Research Projects BEOP-31. 1984.

4 van der Sloot, H.A., de Groot, G.J. and Wijkstra, J. Leaching Characteristics of Construction Materials and Stabilization Products Containing Waste Materials. In: Environmental Aspects of Stabilization and Solidification of Hazardous and Radioactive Wastes. ASTM STP 1033, P.L. Cote and T. M. Gilliam, Eds., American Society for Testing and Materials, pp. 125-149. Philadelphia, 1989.

5 Kosson, D.S., van der Sloot, H., Holmes, T.S., Wiles, C. A Comparison of Five Solidification/Stabilization Processes for Treatment of Municipal Waste Combustor Residues, Part II: Leaching Properties. In: Proceedings of Municipal Waste Combustion , Tampa, FL, April 1991.

Waste Materials in Construction.
J.J.J.R. Goumans, H.A. van der Sloot and Th.G. Aalbers (Editors)

LEACHING POTENTIAL OF MUNICIPAL WASTE INCINERATOR BOTTOM ASH AS A FUNCTION OF PARTICLE SIZE DISTRIBUTION

J.A. STEGEMANN[1] and J. SCHNEIDER[2]

[1]Wastewater Technology Centre, P.O. Box 5050, Burlington, Ontario, Canada, L7R 4A6

[2]Laboratorium für Isotopentechnik, Kernforschungszentrum Karlsruhe, Postfach 3640, 7500 Karlsruhe 1, Germany

1. INTRODUCTION

Although the incineration of municipal solid waste (MSW) results in a mass reduction of approximately 70%, the amount of residue remaining to be disposed of after incineration is substantial. In 1989, approximately 3 million tonnes of incinerator residues were produced in the Federal Republic of Germany (FRG) (1).

More than 90% (by mass) of incinerator residues consist of bottom ash, the slag-like material which is dumped from the grate after combustion. At the present time, approximately half of the bottom ash generated in the FRG is used in road construction (2). This practice diverts a significant volume from landfill, and results in conservation of natural aggregate. Investigation of the suitability of bottom ash for this purpose has centred mainly around the required structural and mechanical characteristics. There is a scarcity of information about the environmental acceptability of such a practice. Consequently, the Laboratory for Isotope Technology (KfK/LIT), as part of a larger effort in cooperation with the University of Karlsruhe, has undertaken to examine the characteristics of bottom ash which will affect its short and long-term behaviour in the environment, particularly in a utilization scenario. The primary characteristics of concern are content and leachability of heavy metals and salts.

The work reported here concerns the chemical characteristics of different physical fractions of bottom ash from the Göppingen regional energy-from-waste facility (3). This facility incinerates approximately 200 000 t/a of municipal solid waste containing a maximum of 10% sewage sludge. The incineration temperature is approximately 850°C. After quenching, the bottom ash is dressed by sieving and magnetic separation of metals. Particles smaller than 32 mm are retained for utilization. Boiler ash, which is compositionally more similar to the more hazardous fly ash, is now collected separately at the Göppingen facility, but the grate siftings are still collected together with the bottom ash.

2. OBJECTIVES

The specific objective of the work reported in this paper was to examine the composition and leachability of the different particle size fractions found in MSW bottom ash. The work was intended to investigate possibilities for improving the suitability of MSW bottom ash for use in road construction.

3. APPROACH

3.1 Bottom Ash Testing

A bottom ash sample of approximately 1 tonne was separated from the main pile of bottom ash at the outlet from the preparation plant using a bulldozer. Chemical analysis and leachability testing were performed for a subsample of a few kilograms.

Particle-size characterization was performed by sieving the dried sample into eleven fractions. Total contaminant concentrations were measured for four particle size fractions (<0.4 mm, 0.4 to 2 mm, 2 to 8 mm, and >8 mm). Leachability of the four bottom ash fractions was examined using components (see Table 1) of a set of test methods assembled at Environment Canada's Wastewater Technology Centre (WTC) for the evaluation of the suitability of solidified wastes for utilization or disposal (4,5). Bottom ash is a highly alkaline alumino-silicate based material which is similar in many ways to cement-based solidified wastes; the major difference is its physical structure, which is particulate rather than monolithic. Since any contaminants in bottom ash must be chemically immobilized in order for the material to be environmentally acceptable for utilization, the test methods applied to the bottom ash fractions in this study focused on chemical containment of contaminants.

TABLE 1
Test Methods Applied to Particle Size Fractions of Bottom Ash

Test Method	Property of Interest
Chemical analysis of solids	Total metal and anion concentrations Residual unburnt carbon
Sequential chemical extraction	Fraction of metals available for leaching
Acid neutralization capacity	Response to acid additions
Equilibrium extraction	Initial leachate concentrations
German standard leachability test	Leachate concentration (Regulatory test)
Swiss leachate test	Leachate concentration (Regulatory test)

First, the distribution of total concentrations of heavy metal contaminants and anions in the different particle sizes was investigated. Total carbon and loss on ignition measurements for each fraction were used as an indicator of the completeness of combustion.

A sequential chemical extraction, measurement of acid neutralization capacity, and an equilibrium extraction were conducted on samples which had been dried and finely ground (<0.2 mm) to ensure that chemical rather than physical factors determined the concentrations measured in the leachates.

The intent of the sequential chemical extraction (6) was to measure the total amount of contaminant available for leaching over an infinite time, with an infinite supply of leachant. In a disposal or utilization scenario, the chemical environment initially provided by the bottom ash will be gradually modified by the chemical characteristics of the surrounding environment. With time, the alkalinity of the bottom ash could be neutralized by infiltrating acidic rain or groundwater, and oxidation of residual carbon to carbonic acid. Such phenomena, would, in turn, increase the solubility of heavy metal contaminants over the long term.

In the sequential chemical extraction, a sample of finely ground bottom ash was contacted with a series of five chemical leachants, each of which was more aggressive than the previous one (see Table 2). The chemical environment provided in each of the successive extractions may be interpreted as that in a "worst case scenario". It has been suggested that the amount of contaminant contained in fractions A and B represents a conservative estimate of the fraction of the contaminant which is ultimately available for leaching in a segregated landfill, while the amount contained in fractions C might be available under the more severe reducing conditions of codisposal with municipal solid waste, and fractions D and E remain unavailable for leaching (7). This test has also been used in the past to examine contaminant partitioning in MSW incinerator fly ashes (8).

TABLE 2
Sequential Chemical Extraction Test

Fraction	Leachant	Release subject to changes in
A	0.25 M $CaCl$ 0.75 M $LiCl$ 60% CH_3OH	Ionic Composition
B	1 M CH_3COONa in CH_3COOH (pH=5)	pH
C	1 M $NH_2OH.HCl$ in 25% CH_3COOH	Eh
D	0.02 M HNO_3 in 30% H_2O_2 1.2 M CH_3COONH_4 in 20% HNO_3	Eh
E	30% H_2O_2, HF, HCl	Eh

The rate at which the total amount available for leaching measured in the sequential chemical extraction is actually released into the environment is affected by the conditions of disposal/utilization, and by the bottom ash's resistance to chemical changes. Most important is its ability to neutralize acid additions, which was measured by extraction of eleven subsamples of ground waste, each with an increasing amount of nitric acid, for 48 hours. The pH's of the extracts were measured and plotted as a titration curve.

In the equilibrium extraction, the finely ground sample was mixed with distilled water at a liquid-to-solid ratio of 4 to 1 for a period of 7 days. Under these conditions, the bottom ash established the chemical environment in the leachate, and steady-state concentrations in the leachate could be measured. The contaminant concentrations measured in this test represent the concentrations in an undiluted leachate at time zero, and are indicative of the solubility of the contaminants in the bottom ash fraction in the short term.

The German standard leachability test (9), and the Swiss leachate test (10) were performed as representative regulatory tests. They are practical procedures designed for estimating the leaching hazard associated with a waste in a short amount of time. Both tests are batch tests with distilled water at a liquid-to-solid ratio of 10:1. In the German test, the waste is allowed to establish the chemical environment in the extract, which is separated for analysis after 24 hours of tumbling. In the Swiss test, the waste is mixed with distilled water by sparging with CO_2. The dissolution of CO_2 in the water has the effect of lowering the pH, providing a more aggressive leaching environment. The Swiss protocol requires two successive 24 hour extractions of the waste material; the average of the concentrations measured in the two extracts, adjusted for the soluble fraction of the waste, is compared with regulatory limits.

Although both tests specify the use of intact (i.e., unground) waste material, finely ground material from each of the particle size fractions was used in this study to provide comparable surface areas in contact with leachant, and ensure comparability of the resulting leachate concentrations.

In principle, the German leach test is intended to measure initial leachate concentrations (similar to the equilibrium extraction), while the Swiss test is designed to measure amounts available for leaching over a longer time period (although the results of the Swiss test are expressed as leachate concentrations, rather than as fractions leached). However, steady-state concentrations might not occur in either test because of the short extraction period (particularly if unground material were used). The resulting concentrations may be used for comparison with regulatory guidelines, but are not suitable for rigorous scientific evaluation of the potential for leaching.

3.2 Chemical Analysis

Metal extracts of solid samples were prepared by acid digestion. Leachates and extracts were analyzed for heavy metals by Total X-ray Fluorescence Analysis (TXFA). This technique measures a number of elements simultaneously, but only chromium, copper, lead, nickel and zinc measurements are reported here. These were the only toxic contaminants found to be present in appreciable quantities in the solids. Chloride and sulfate were measured in the leachates and extracts by ion chromatography. The solids were analyzed for total carbon using a cerimetric method. Extraction and analysis of all samples were carried out at least in duplicate.

4. RESULTS AND DISCUSSION

The particle size distribution for the Göppingen bottom ash examined is shown in Figure 1. It can be seen that 12% of the total mass of ash was smaller than 0.4 mm, 13% was between 0.4 and 2 mm in size; 23% was between 2 and 8 mm in size, and 52% was larger than 8 mm. Contaminant distribution and mobility were examined for these four particle size fractions.

Figures 2(a) and (b) show the distribution of contaminants in these four particle sizes. A clear tendency for the contaminant concentration to decrease with increasing particle size may be observed for zinc, chloride, and sulphate. This tendency may also be observed to a lesser degree for chromium, nickel, and copper. Of the metals analysed for, lead is the only one for which this trend is not apparent. The percentage of unburnt carbon remaining in the sample also decreased with increasing particle size. This observation was confirmed by the loss on ignition measurements (i.e., 1.44%, 2.04%, 3.42%, and 8.3%, from largest to smallest particle size). Only the smallest particle size fraction exceeded the loss on ignition limits of 5% proposed in guidelines for the use of bottom ash in road construction (11). Overall, 36% of the metals, and 42% of the anions and carbon were present in the two smallest particle size fractions, which constituted only one quarter of the mass of the bottom ash.

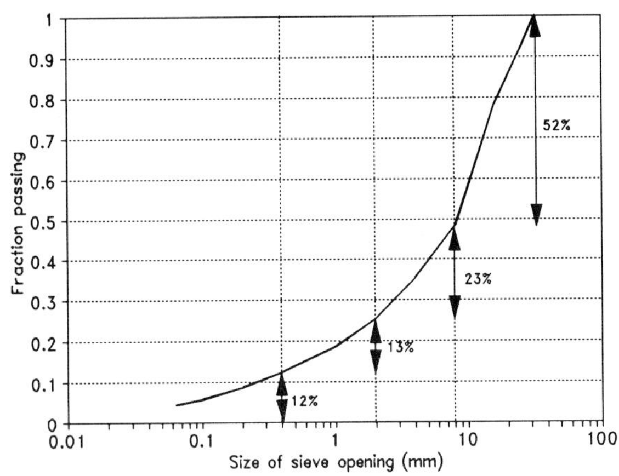

Figure 1 Particle size distribution for the Göppingen bottom ash sample

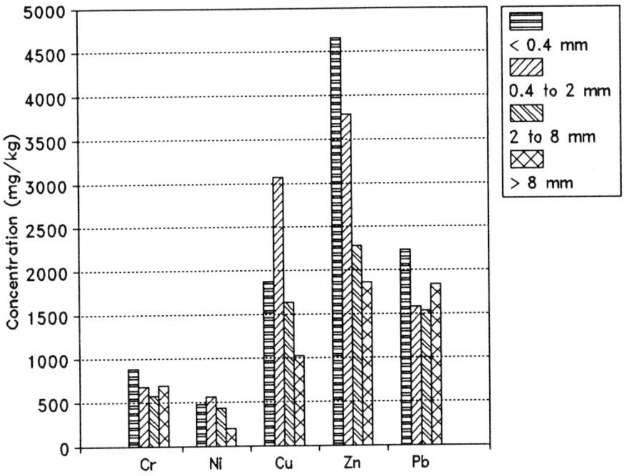

Figure 2(a) Total metal concentrations in four particle size fractions

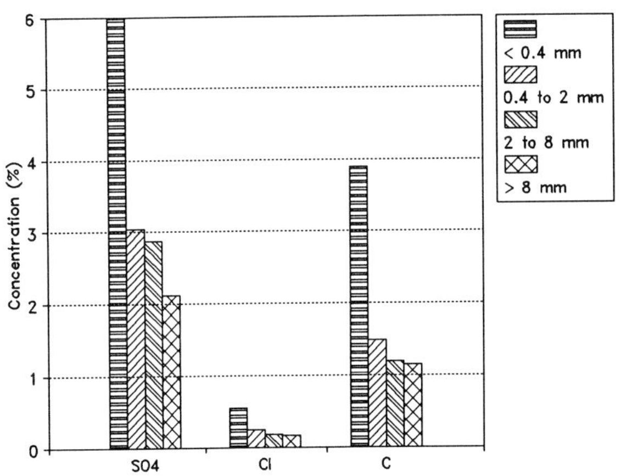

Figure 2(b) Total anion and carbon concentrations in four particle size fractions

Although the data is not presented here, a clear increase in metal, anion, and carbon concentrations as a function of decreasing particle size was also observed in the three other bottom ashes studied recently at the KfK/LIT (12). Not surprisingly, the effect was most pronounced in bottom ashes containing boiler ash.

Results from the sequential chemical extractions of the particle size fractions show that, not only was the total amount of contaminant in smaller bottom ash particles greater, this amount was also more leachable, as shown in Figures 3(a) and (b). This trend is apparent whether the amount ultimately available for leaching is assumed to be the sum of fractions A and B (shown), or A, B and C (not shown).

Measurement of the response of the different bottom ash fractions to additions of nitric acid, presented graphically in Figure 4, showed that the smallest particle size fraction appeared to have a greater acid neutralization capacity at higher acid additions, than the other three particle size fractions.

"Initial leachate" concentrations measured in the equilibrium extraction also indicated that leachability of the smaller bottom ash particle sizes was greater for most contaminants. For amphoteric metals, this may be a function of the increase in leachate pH, which was observed for decreasing particle size. The equilibrium extraction data are not presented in this document, as they are very similar to the results of the German standard leachability test, as would be expected for two distilled water extractions.

The results of the German standard leachability test are shown in Figures 5(a) and (b). Because of dilution by the higher liquid-to-solid ratio of the German standard leachability test, anion concentrations were higher in the equilibrium extraction. Metal concentrations were also slightly higher, which may be attributable to the longer leaching period of the equilibrium extraction.

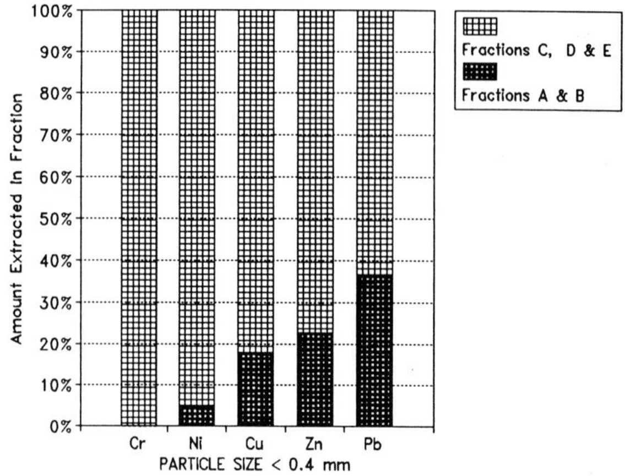

Figure 3(a) Sequential chemical extraction results for particles smaller than 0.4 mm

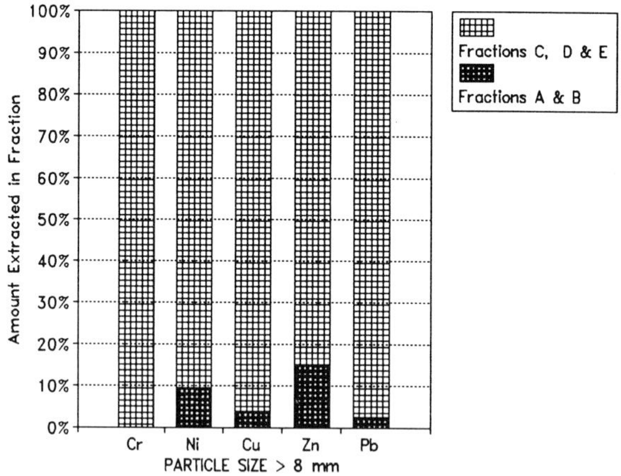

Figure 3(b) Sequential chemical extraction results for particles larger than 8 mm

In the Swiss leachate test, the averaged concentration for the two successive extracts was distinctly higher for copper and zinc in the smallest particle size fraction (see Figure 6 (a) and (b)). In fact, the concentrations of copper and zinc measured for this particle size fraction were above the regulatory limits, while the concentrations measured for the larger particle size fractions were well below the regulatory limits. As would be expected, concentrations of the highly soluble anions were again higher in the leachates from the smallest particle size fraction (this data has not been shown). The higher buffering capacity of the smaller particle size fraction apparently resulted in a slightly higher leachate pH.

Data gathered in two studies by other researchers has also shown that the total contaminant concentration increases, as the particle size decreases (13,14). A study at the Swiss EAWAG showed peak contaminant concentrations in the particle size from 1 to 4 mm, and confirmed increasing leachability of heavy metals with decreasing particle size. An increase in pH as a function of decreasing particle size was also observed (14).

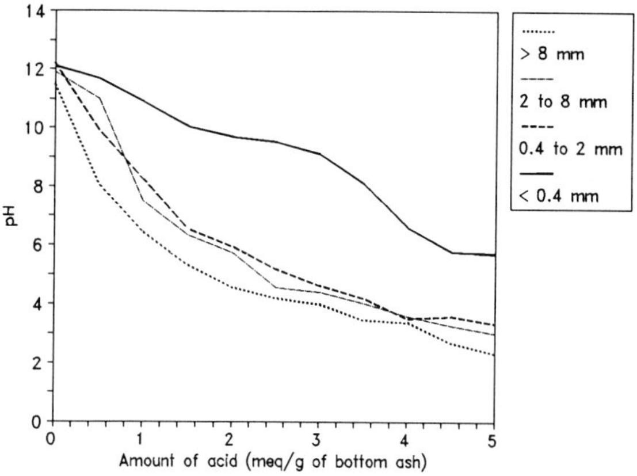

Figure 4 Acid neutralization capacities of bottom ash particle size fractions

An explanation for the concentration of heavy metals in the smaller particle size fractions has not been verified, but it is postulated that inciner-ation of materials which contain relatively high amounts of heavy metals (e.g., paints and inks on paper and packaging, and additives in plastics and rubber) may contribute a major portion of the fines in the combustion residue. In addition, other work has shown that metal concentrations in the fine grate siftings, which have not been exposed to the full required residence time in the incinerator, may be higher, since they have not had an opportunity to volatilize and concentrate in the fly ash (15). Disposal of the grate siftings with

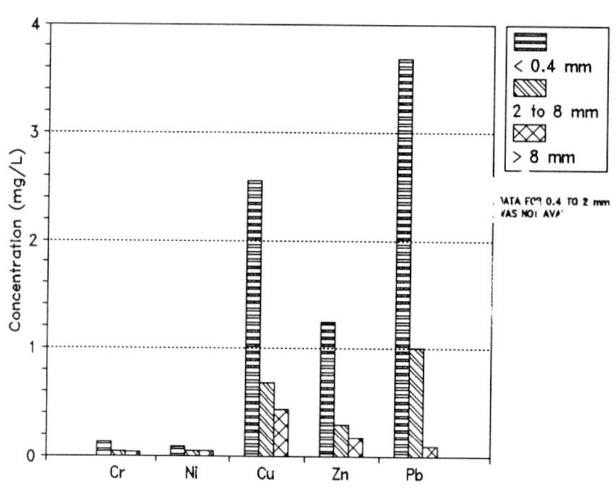

Figure 5(a) Metals concentrations leached from bottom ash fractions in the German standard leachability test

the remaining bottom ash may also lead to fractions of uncombusted material that are between five and six times higher for the smallest particle size fraction, than for the largest particle size fraction.

5. CONCLUSIONS

In summary, this study indicates that more than a third of the metals, and more than 40% of the anions and unburnt carbon were present in the two smallest particle size fractions, which constitute only one quarter of the bottom ash. The leachability of these smaller particles also appeared to be greater than that of larger particles.

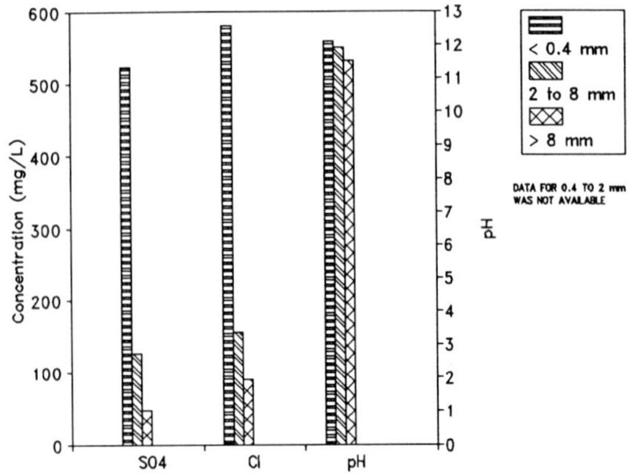

Figure 5(b) pH and anion concentrations leached from bottom ash fractions in the German standard leachability test

Since constructive use of bottom ash is clearly preferable to land disposal, the implications of these findings upon bottom ash utilizability should be carefully considered. While it appears that removal of the smaller bottom ash particle sizes would result in a higher quality product for utilization, characterization studies of incinerator input and output streams may yield explanations and solutions for these findings.

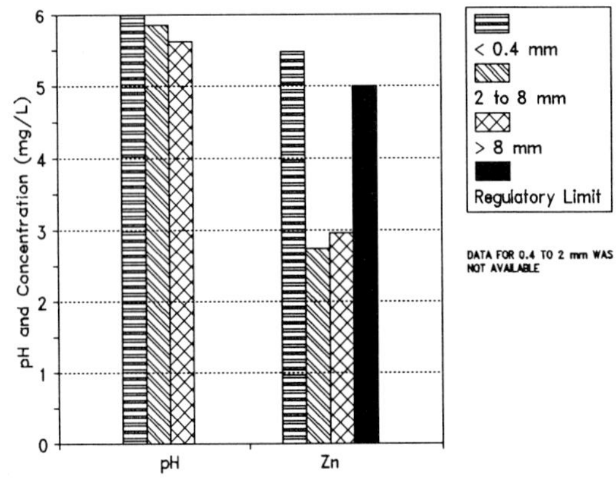

Figure 6(a) pH and zinc concentrations leached from bottom ash fractions in the Swiss leachate test

REFERENCES

(1) Hiersche, E.U. and Wörner, T., Verwendung von Müllverbrennungsaschen im Straßenbau, VDI Berichte Nr. 753, 1989.

(2) Mühlenweg, U., and Brasser, Th., "Reststoffe bei der Hausmüllverbrennung", Müll und Abfall, 2, 1990.

(3) Landratsamt Göppingen, "Das Müllheizkraftwerk Göppingen", June, 1985.

(4) Stegemann, J.A. and Cote, P.L., "Investigation of Test Methods for Solidified Waste Evaluation", Environment Canada, Ottawa, January, 1991.

(5) Wastewater Technology Centre, "A Proposed Protocol for Evaluation of Solidified Wastes", document in press, Environment Canada, Ottawa, 1991.

(6) Tessier, A., Campbell, P.G.C., and Bisson, M. , "Sequential Extraction Procedure for the Speciation of Particulate Trace Metals", Anal. Chem. 51: 844-851; 1979.

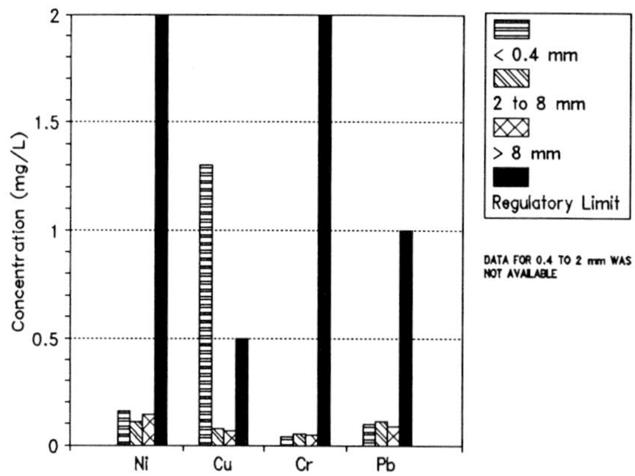

Figure 6(b) Metals concentrations leached from bottom ash fractions in the Swiss leachate test

(7) Kramer, J.R., Gleed, J., Brassard, P., and Collins, P.V., "Comparison of Various Leachate Extraction Procedures for the Characterisation of Inorganics in Wastes", Proceedings of the Ontario Ministry of the Environment Technology Transfer Conference, MOE, Toronto, November, 1988.

(8) Sawell, S.E., Bridle, T.R. and Constable, T.W., "Heavy Metal Leachability from Solid Waste Incinerator Ashes", Waste Management and Research 6: 227-238; 1988.

(9) DIN 38414 Teil 4: Deutsches Einheitsverfahren zur Wasser-, Abwasser- und Schlammuntersuchung, Schlamm und Sedimente (Gruppe S), Bestimmung der Eluierbarkeit mit Wasser (S4), Benthe-Vertrieb GmbH, Berlin und Köln, 1989.

(10) "Bericht zum Entwurf für eine technische Verordnung über Abfälle (TVA)", Swiss Federal Department of the Interior, 1988.

(11) Merkblatt über die Verwendung von industriellen Nebenprodukten im Straßenbau, Teil: Müllverbrennungsasche (MV-Asche), Forschungsgesellschaft für Straßen- und Verkehrswesen Arbeitsgruppe "Mineralstoffe im Straßenbau", Köln, 1986.

(12) Stegemann, J.A. and Schneider, J., "Composition and Leachability of Municipal Waste Incinerator Bottom Ash as a Function of Particle Size Distribution", document in preparation, Kernforschungszentrum Karlsruhe, 1991.

(13) Gavasci, R., Mangialardi, T, and Sirini, P., "Slag and Fly Ash from MSW Incineration Plants - Characterization and Reuse", Presented at the 6th International Conference on Solid Waste Management and Secondary Materials, Philadelphia, PA, USA, December 4-6, 1990.

(14) Grabner, E., Hirt, R., Petermann, R., and Braun, R., "Müllschlacke - Eigenschaften, Deponieverhalten, Verwertung", Schweizerische Vereinigung für Gewässerschutz und Lufthygiene, Rieker + Amman AG, Glattbrugg, Zürich, 1979.

(15) Schneider, J., "Bestimmung der elementaren Müllzusammensetzung durch Analytik der Müllverbrennungsrückstände", Recycling International. VI., ed. K.J. Kozmiensky, EF-Verlag, Berlin, 1986.

Waste Materials in Construction.
J.J.J.R. Goumans, H.A. van der Sloot and Th.G. Aalbers (Editors)

IMPROVEMENT OF FLUE GAS CLEANING CONCEPTS IN MSWI AND UTILIZATION OF BY-PRODUCTS

Y. Volkman*, J. Vehlow, H. Vogg

Karlsruhe Nuclear Research Centre, P.O. Box 3640, D-7500 Karlsruhe (FRG)
* Guest from Nuclear Research Centre Negev (Israel)

SUMMARY

To minimize the amount of residues from air pollution control systems in waste incineration a conception is presented which links the filter ash inertizing and metal recovering capability of the 3R Process to the electrochemical recovery of chlorine and the reuse of sulfates. This process is furthermore characterized by considerable savings of water and neutralizing chemicals.

1 INTRODUCTION

Residues from flue gas cleaning in municipal solid waste incineration (MSWI) are classified as toxic wastes. The filter and boiler ashes (approx. 20 kg/ton of waste) contain leachable heavy metal as well as toxic organic compounds and in wet systems the flue gas scrubbing solutions (FGSS) comprise hydrochloric acid (approx. 5 kg/ton), chloride salts, mercury and sulfates (approx. 4 kg/ton). The safe final disposal of such residues causes problems.

Recently there have been made some efforts to treat and recover parts of these materials. The 3R Process, developed in the years 1984 to 1989 [1, 2], inertizes filter ashes by acid leaching and thermal treatment. The process offers the chance to recycle Cd, Zn, and - to a certain extent - other metals. An additional advantage is the total recovery of mercury. The chloride emission into the waste water however is increased by the chloride inventory of the filter ash.

Other proposals have been made with respect to the disposal of soluble salts. In one German MSWI NaCl is separated for reuse in the chlorine-alkali-electrolysis. Other authors suggest the recovery of hydrochloric acid from the FGSS by distillation [3, 4]. Electrolysis also could be applied to chlorine recovery [5]. The reuse of sulfates by producing gypsum is common practice in air pollution control (APC) systems of power stations and has been proposed for waste incineration too [3, 6]. All those processes, however, do not affect the filter ash quality.

In the following a concept is proposed which integrates both alternatives in one process.

2 PROCESS PRINCIPLES

Such concept has to include fly ash treatment and a most extensive utilization of flue gas cleaning products. The proposed combined process is distinguished by closed circulations of scrubbing solutions and has dry outlets only. It comprises fly ash treatment and metal recycling by the 3R Process, partial chloride recovery and gypsum production. The water is evaporated and reused in the process so that only small amounts are needed for make up. The consumption of neutralizing reagents is minimized by the full utilization of the filter ash basicity. Figure 1 gives a flow-scheme of the concept of such a combined process.

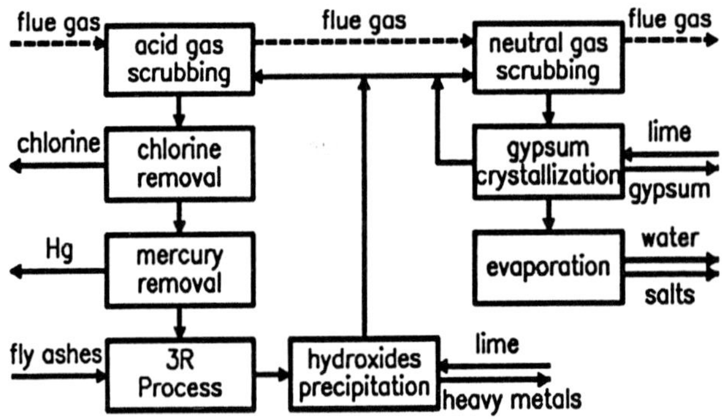

Fig. 1: Conceptual process flow sheet with outlets for major components

3 ANALYTICAL BASIS

3.1 Main Mass Streams

Modern MSWI generate about 10 to 15 kg of filter ash and between 5 and 8 kg of boiler ash per ton of waste. The amount of flue gas is in the order of 4000 to 5000 m^3/ton with the tendency towards 4000 m^3/ton waste. The chlorine input into the incinerator is around 8 kg/ton waste, approximately three quarters of which are released as hydrochloric acid and are removed from the flue gas in the APC system. The resulting chloride concentrations in the FGSS have to be kept high in order to reduce fresh water consumption and to improve chlorides recovery.

The sulfur content in the refuse of about 2.7 kg/ton [7] splits into one part remaining in the bottom ash and another being converted into SO_2. Normal flue gas concentrations of 200 - 400 mg/m^3 SO_2 cause sulfate ion inputs of 1.5 - 3 kg/ton waste into the neutral scrubber.

3.2 Acidity and Neutralization

The main input of acidity into the air pollution control system is represented by the HCl which is absorbed in the acid scrubbing unit. The pH value in the acid scrubber circuit has to be kept as low as possible for three reasons: for better mercury removal, for minimization of water consumption and for optimization of chlorides recovery. In practice the pH value is limited to 0.5 - 0.3 due to corrosion problems. This corresponds to a HCl concentration of 10 - 20 g/l and can be met with a typical water consumption of around 300 l/ton waste. In some incinerators this value is restricted down to or even below 100 l/ton. In these cases a partial neutralization of the FGSS is required. Hence, the chlorides concentration increases up to 60 g/l.

Laboratory tests and the results of semitechnical tests in the DORA pilot plant [2] have established that the 3R Process can be operated with satisfactory extraction efficiencies at a final pH value of 4. This corresponds to a consumption of about 30 % of the nominal HCl supply (approx. 2 kg/ton waste) and means that about 70% of the HCl input into the APC system (about 4 kg chlorine per ton of waste) are available for recovery and utilization.

Experimental work with 3R filtrates (pH=4) indicated that the required lime consumption for neutralization and heavy metals precipitation is around 0.04 - 0.05 kg/kg filter ash. On the basis of about 20 kg of fly ash per ton of refuse this results in a total lime consumption of approximately 1 kg/ton waste. The stochiometric amount of lime which would be needed to neutralize these 2 kg of HCl without passing the 3R Process is about 2.7 kg or 0.14 kg/kg filter ash. This significant difference clearly indicates the advantage of using the 3R Process as a partial neutralization step, on top of its primary objective of residue inertization.

A second important source for acidity input into the APC system is represented by the sulfur dioxide absorption in the neutral scrubber. Under normal operation conditions about 2.5 kg of sulfate ions per ton of refuse enter the neutral scrubbing stage. The neutral scrubber is fed with the basic mother liquor (pH=10) of the heavy metals hydroxides precipitation stage. However, this covers only a small part (less than 0.1 kg) of the total 2.5 kg lime per ton of refuse needed for complete neutralization.

3.3 Chlorides

Chlorides are the major constituent of the circulating solutions of the proposed process. The input sources are the HCl in the flue gas (approx. 6 kg/ton waste) and some water soluble chlorides of sodium, potassium and calcium present in the filter ashes at concentrations of about 0.05 kg/kg filter ash. Hence, the total chlorides input into the system is about 7 kg/ton waste. As stated above, some 30 % of the HCl (2 kg/ton waste) is consumed by the 3R Process, the remaining 4 kg/ton are available for recovery. It has to be proved, to what extent the chlorides present in the 3R filtrate (approx. 3 kg/ton waste) can be fed back into the acid scrubber and there become available for recovery too. Two methods of chlorides recovery have been investigated with respect to their applicability in the proposed process: the electrolysis and the extractive distillation. Basically both alternatives are suitable, hence, their implementation respectively gives rise to some minor changes in the flow scheme. In this paper only the electrolytic option is described. Regardless of the applied method a certain amount of the chlorides inventory, however, will be purged out of the system via the salts evaporator.

3.4 Sulfates

The average sulfate input into the APC system is around 2.5 kg/ton waste. An additional input from soluble sulfates in the filter ashes can be estimated on the basis of the DORA experiments to be in the order of 0.2 kg/ton waste. The main outlet for the sulfates is the gypsum. The nominal production of gypsum dihydrate is about 5 kg/ton refuse.

The stochiometric consumption of calcium ions for gypsum generation is around 1.2 kg/ton waste. Calcium ions are introduced into the system by the lime used for neutralization purposes. As stated above, the heavy metals precipitation consumes about 1 kg and the neutralization of the FGSS in the basic scrubbing unit needs another 2.5 kg of lime per ton of waste. This represents a total input of 1.4 kg of calcium ions per ton of refuse.

Another source for calcium ions are the filter ashes via the 3R Process. The DORA experiments point out that about 0.5 kg of calcium ions are dissolved per ton of waste. Taking this into account, a total of about 1.9 kg of calcium ions per ton of waste is present in the system. More than 60 % of the calcium goes with the gypsum, the remaining part is purged out of the system via the evaporator together with the other soluble salts.

3.5 Heavy Metals

Mercury is the most relevant metal in waste incineration with respect to ecotoxicity. The input with the refuse is about 3 g/ton waste [7, 8]. Approximately the total inventory appears in the flue gas and more than 80 % can be absorbed in the acid FGSS [9, 10]. The mercury is to be

Table 1: Recovery potential and residues in the combined process

	kg/ton waste
recovery potential	
chlorine / hydrochloric acid	4
gypsum dihydrate	5
mercury	0.003
zinc	0.5
cadmium	0.02
copper	0.005
residues	
3R Product	20
soluble salts	6

removed from the FGSS prior to the 3R Process. This is done with practically 100 % efficiency by ion exchange [2].

Other heavy metals enter the process along with the filter ash and are leached to a certain extent in the acid extraction step of the 3R Process [2]. The recovery potential of some metals can be estimated on the basis of published waste concentrations [9, 10] and is summarized in Table 1. The obvious contemporary method for their removal from the 3R filtrate is hydroxide precipitation which results in about 0.6 to 1 kg metal hydroxides per ton of refuse. If there should open up the potential of the reuse of special metals, other methods for recovery like ion exchange or electrolysis are feasible as well, however, probably more expensive. In that case sodium hydroxide has to be used for the pH control of the acidic scrubbing unit.

3.6 Water

The proposed process has no outlet for liquid residues. Hence, it consumes water only for replenishment of losses. A reasonable estimate of the consumption of water is therefore rather difficult. However, some loss items can be evaluated.

The input of water into the APC system with the flue gas is in the order of 0.13 to 0.15 kg/m^3 or about 700 kg/ton waste. The wet scrubbing system is operated at temperatures between 60 and 65 °C. The exhaust fumes are supposed to be water saturated at this temperature and thus remove around 0.2 kg/m^3 or 1000 kg/ton waste out of the system.

Minor amounts of water go with the 3R filter cake which has a moisture content of 30 %. This sums up to about 6 kg/ton waste. The heavy metals hydroxides cake contains about 90 % water, accounting for some 5 kg/ton refuse. The wet gypsum dihydrate will add another 2 kg/ton. If an aqueous HCl is recovered, its water content has to be replenished as well.

4 RESIDUE DISPOSAL

According to the proposed process, the waste streams are separated into groups, each having its characteristic handling technique. As far as this can be done in the present state of the concept the final disposal of the single residues will be discussed in the following.

Mercury and the other heavy metals are obtained as insoluble solid products which should be recovered due to their quality, but in all probability will be stored in underground sites.

The 3R filter cake is stabilized by the addition of a binder material and fed back into the incinerator, where it is inertized and combined with the bottom ash [2]. This mass stream can either be utilized or openly landfilled. The same final option goes for the gypsum. Its commercial value is doubtful because of the availability of cheap and high quality natural gypsum.

The major waste disposal problem still concerns the soluble salts. As mentioned above, the chlorides recovery is not applied to the total chlorides inventory of the system. A certain portion is emitted to the evaporator. Also the sulfates recovery is operated with some losses due to the solubility product of gypsum. The total salts residue stream can be assessed to about 5 - 6 kg/ton refuse. They comprise a mixture of mainly KCl, NaCl, $CaCl_2$ and minor quantities of bromides, fluorides or sulfates. These materials have no known commercial value. However, they contain no hazardous compounds and their quantity is only about 30 % of the total amount of normally emitted salts. Due to legislative restrictions underground storage seems to be the only option.

5 ELECTROLYTIC CHLORIDE RECOVERY

5.1 Fundamentals

The electrochemical recovery of chlorine from concentrated solutions of sodium chloride as well as hydrochloric acid is well-known and industrially established [5]. The anodic discharge of chlorine from dilute chloride solutions is not implemented, it is, however, chemically feasible [11]. Side reactions like chlorate formation can occur only at very low chloride concentrations due to simultaneous oxygen evolution.

The chemistry of the anodic chlorine electrode process was studied by Hine and co-workers [12]. Polarization data for acidified sodium chloride solutions and graphite electrodes indicate increasing current densities at increasing concentration, showing a reaction order between 0.6 and 1.0. Tafel slopes for chlorine (120 - 140 mV/decade) and for oxygen evolution (210 - 240 mV/decade) are specified, but no data are given concerning current efficiencies and overall cell voltage. Thus on top of the informations available from literature experimental verification at realistic conditions is needed.

5.2 Experimental verification

First experiments have been conducted with simulated FGSS containing 0 - 100 g/l NaCl. These solutions had been acidified with sulfuric acid to a pH range of 0 - 1.5 and were electrolyzed between two parallel graphite electrodes having a surface of 20 cm^2. All experiments were carried out at 25 °C. Current-voltage curves were analyzed in order to estimate the electrochemical reaction potentials and the different components of the ohmic voltage drop.

The voltage drop ΔV of the experimental cell is the sum of 4 components:

$$\Delta V = (V_a + \eta_a) + (V_c + \eta_c) + \left(\frac{d}{\lambda * A}\right) * I + R_0 * I \qquad (1)$$

The first two terms account for the electrochemical reaction potentials at the anode (a) and at the cathode (c) which are separated into the reversible potentials (V) and the overpotentials (η). The other terms describe the ohmic voltage drops and depend on the total current (I) through the cell. The third term takes into account the ohmic voltage drop within the electrolyte. It depends on the interelectrode distance (d), the conductivity of the electrolyte (λ) and the effective cross-section of the solution (A). The fourth term estimates the ohmic losses associated with the cell structure, R_0 accounting for the resultant resistance.

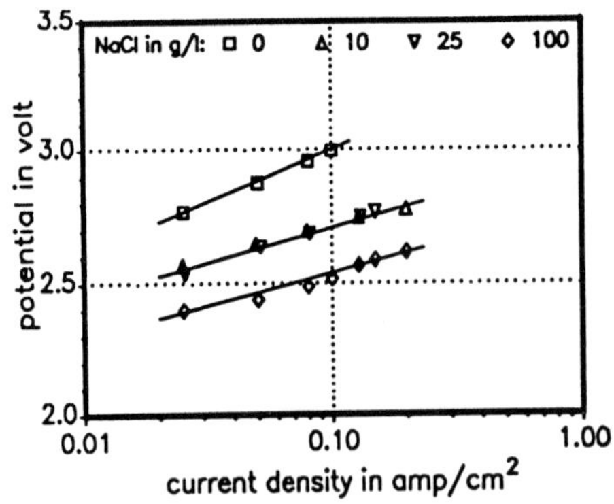

Fig. 2: Current-voltage curve - effect of NaCl concentration

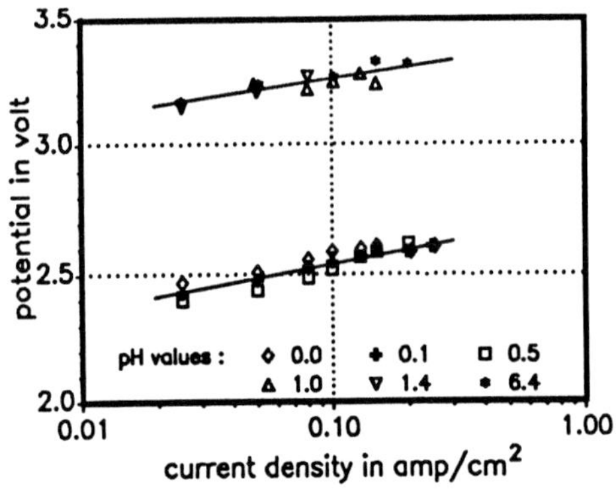

Fig. 3: Current-voltage curve - effect of pH value

The separate determination of the current depending components follows from measuring the resistance R versus the electrode distance according to the relationship:

$$R = R_0 + \left(\frac{1}{\lambda * A}\right) * d \qquad (2)$$

The reversible electrochemical potential is estimated by substracting the overall ohmic losses (R * I) from the measured overall voltage drop at the current I. The potential which is calculated in that way is the combined polarization voltage of both the anode and the cathode.

It has to be stated that this technique is unusual in basic electrochemical research. It is, however, very usefull for getting directly relevant technological information at minimum time and expenses.

The effect of NaCl concentration and pH value on the polarization voltage is shown in Fig. 2 and 3 respectively. The Tafel slopes vary from about 400 mV/decade for solutions containing no NaCl to about 200 mV/decade at 100 g/l NaCl. Fig. 3 demonstrates that the pH value actually controls the reaction mechanism, showing two different, however, parallel polarization lines at two different acidity ranges. In the more acid range ($pH<1$) the combined polarization potentials are decreased. This is an indication of the prominent influence of H^+ ions which affect the cathodic H_2 discharge and transfer most of the cathodic current.

The latter is confirmed by measuring the conductivity of synthetic FGSS which had been acidified with sulfuric acid. The operation temperature was 25 °C. Fig. 4 summarizes the results. At $pH > 1$ the conductivity is constant and is actually that of the NaCl, whereas it increases at $pH<1$ with increasing H^+ ions concentration. These general findings comply with data that were found in the literature [3].

5.3 Bench-scale demonstration

The expperimental verification of the electrochemical chlorides recovery was demonstrated in a bench-scale electrolyzer. It consisted of two parallel planar graphite electrodes and two perforated PVC plates between the anode and the cathode compartment which served as mechanical separators. A closed electrolyte circulation was maintained with a circulation rate of about 15 l/h. The cathodic hydrogen was vented, while the anodic gases were pumped out through a bubble column containing aqueous NaOH.

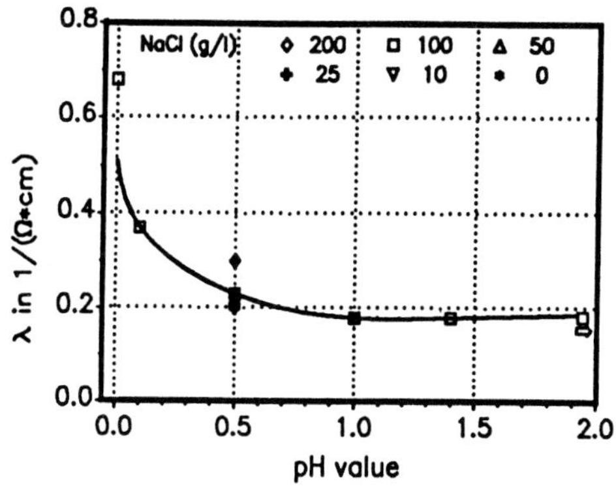

Fig. 4: Conductivity of simulated FGSS - effect of pH and NaCl concentration

The experiments were carried out batchwise at ambient temperature. The electrolyte hold-up was 0.7 l, a quantity of 20 g NaOH was used for chlorine absorption. After a predetermined operation time the electrolyte was heated up to about 90 °C to expel dissolved chlorine to the absorption column. The actual discharged amount of chlorine was analyzed using ion chromatography. Several test runs were carried out. The cell had unusual high solution voltage losses due to the simple cathode-anode separator, which nevertheless could not be removed because of safety reasons.

Typical quantitative results are summarized in Table 2. They clearly support the technological feasibility of the electrochemical recovery process. Chlorides can be electrolyzed with reasonable current yields even if their concentration is very low. The quantitative energy data have still limited practical significance because of the unusual ohmic resistance of the cell.

6 IMPLEMENTATION AND CONCLUSIONS

An example of a process flow-scheme which combines chlorides utilization with the 3R Process is given in Fig. 5. The electrolytic recovery of chlorine from FGSS is directly applied to a circulating side-stream of the acidic srubbing unit. The emitted FGSS is used for the 3R Process extraction stage after the mercury has been removed. The filtrate passes the metals recovery unit and is partly recycled to the acid scrubber, partly fed into the gypsum precipitator. The soluble salts accumulated in the circuit of the basic FGSS are evaporated. The condensates from the evaporator are used to wash the 3R filter cake.

Table 2: Results of bench-scale demonstration of the electrolytic chlorides recovery process

operation parameters		
NaCl (g/l)	10.0	100.0
initial pH	0.5	0.5
final pH	0.6	0.7
duration (h)	2.5	2.7
av. current (Amp)	1.2	1.6
av. potential (Volts)	10.0	10.0
results		
current efficiency (%)	42.5	62.5
energy (kWh/kg Cl_2)	31.0	21.0

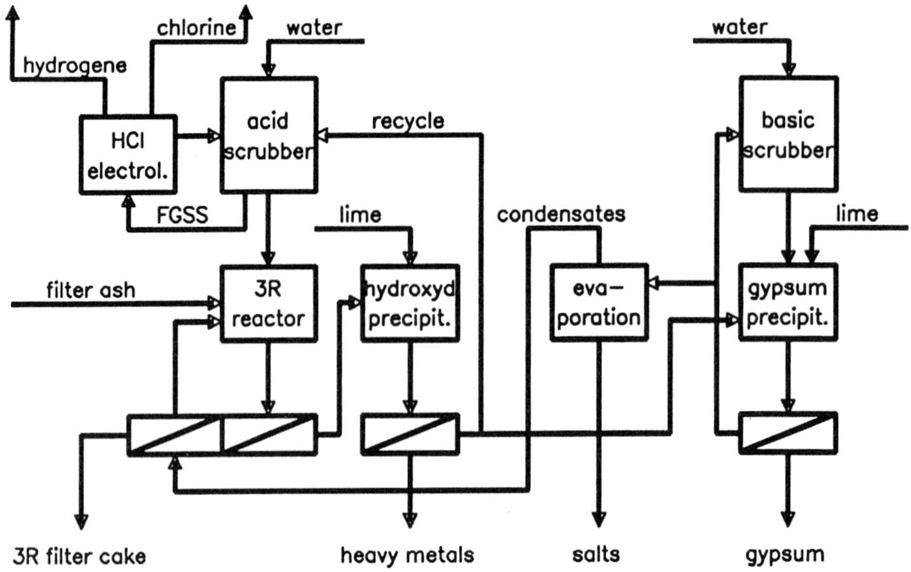

Fig. 5: Combined 3R - chlorides recovery process (gas flows and Hg removal are omitted)

Hence, the proposed combined process fulfills the following main objectives:
- the 3R Process inertizes filter ashes and saves considerable amounts of neutralizing agents,
- mercury, cadmium, zinc and some other metals are separated,
- about 70 % of the chlorides inventory can be recovered,
- the sulfates are utilized as gypsum,
- the closed cycles minimize water consumption and enable to operate without waste water.

The feasibility of several stages of this conceptual process have been tested in the laboratory, in bench-scale facilities or have already been implemented in other full scale facilities. The following steps will be the semi-technical demonstration of the eletrochemical chlorides recovery and finally of the entire combined process.

7 LITERATURE

1 Vogg, H., Chemie- Ingenieur-Technik, 56 (1984) 740 - 744
2 Vehlow, J., Braun, H., Horch, K., Merz, A., Schneider, J., Stieglitz, L. & Vogg, H., Waste Management & Research, 8 (1990) 461 - 472
3 Kürzinger, K. & Stephan, R. in: Müllverbrennung und Umwelt 3 (Thome-Kozmiensky, K.J., ed.), EF-Verlag, Berlin, 1989, pp. 343 - 348
4 Juritsch, V. & Rinn, G., ibid., pp. 349 - 358
5 Hölemann, H., Chemie-Ing.-Techn., 34 (1962) 371 - 376
6 Karger, R., AbfallwirtschaftsJournal 2 (1990) 365 - 375
7 Brunner, E.H. & Mönch, H., Waste Management & Research 4 (1986) 105 - 119
8 Schneider, J. in: Messen und Analysieren in Abfallbehandlungsanlagen (Thome-Kozmiensky, K.J., ed.), EF-Verlag, Berlin, 1987, pp. 283 - 290
9 Braun, H., Metzger, M. & Vogg, H., Müll und Abfall 18 (1986) 89 - 95
10 Reimann, D.O., Müll und Abfall Beihefte, 29 (1990) 12 - 16
11 Haber, F. & Grinberg, S., Z. anorg. Chem. 16 (1897/1898)
12 Hine, F. et al., Electrochem. Technol. 4 (1966) 555 - 559, and J. Electrochem. Soc. 121 (1974) 749 - 756 and 1289 - 1294

Waste Materials in Construction.
J.J.J.R. Goumans, H.A. van der Sloot and Th.G. Aalbers (Editors)

COMPOSITION AND LEACHING CHARACTERISTICS OF ROAD CONSTRUCTION MATERIALS

J.J. van Houdt*, E.J. Wolf* and R.F. Duzijn+

* Ministry of Transport and Public Works, Road and Hydraulic Engineering Division, P.O. Box 5044, 2600 GA Delft, The Netherlands.
+ TAUW Infra Consult B.V., P.O. Box 479, 7400 AL Deventer, The Netherlands.

ABSTRACT

The present paper discusses the environmental impact of various materials used in road constructions and the degree to which application methods are of importance in this context. Some twenty materials were selected for study after an inventory had been made of both commonly used road construction materials (primary materials) and waste materials considered to be suitable for such applications (secondary materials). Each material was sampled in duplicate and its composition and leaching characteristics determined using standard tests. It was found that generally primary materials pose less of a threat to the environment than secondary materials, regardless of whether the comparison is made on the basis of composition or leaching behaviour. A number of primary and secondary materials have been selected for further evaluation under practical conditions.

INTRODUCTION

At present, substantial quantities of quarried materials such as gravel, sand, clay and limestone are used in road and hydraulic engineering works. However, the Dutch government is anxious to preserve natural resources of the mentioned type [1, 2] and is actively considering whether waste products can replace such materials in (road) construction applications. Careful assessment of the likely impact bulk waste materials in their various applications can have on the environment is therefore essential.

In this respect, in 1987, research was initiated at the environmental section of the Road and Hydraulic Engineering Division of the Department of Public Works to determine the composition and leaching behaviour of typical road construction materials. The main aim of this research programme is to increase understanding about the effect road construction materials in their various forms have on the environment and particularly the soil. To facilitate implementation of the results, the research programme has been divided into discrete phases, namely:

1. collecting relevant data on road construction materials and compiling an inventory of potential applications;
2. determining the composition and maximum leachability of candidate road construction materials;
3. establishing the leaching behaviour of road construction materials in standard column tests;
4. conducting simulated practical tests;
5. performing and monitoring practical trials.

EXPERIMENTAL

Phase 1 of the research programme was principally concerned with collecting data about primary (traditional) and secondary (alternative) road construction materials from the literature and oral information. Particular attention was focused on the way in which such materials can be applied and the degree to which they are already used. Where possible, information was also compiled on the concentration of pollutants present in these materials and their susceptibility to leaching. After reviewing the data collected, some twenty virgin materials were selected for further research. At a later stage the effects of mixing and compounding waste products will be studied as well as the effects of compaction.

For each of the twenty virgin materials selected, two samples were taken from two different batches. In order to assess the variation in composition of particular materials, every effort was made to obtain samples from both seriously contaminated batches and less polluted material. Where this was not possible, samples were taken from two batches chosen at random.

In the experiments carried out in phases 2 and 3 of the research programme, particular attention was focused on the following contaminants:

- metals (As, Ba, Cd, Co, Cr, Cu, Hg, Mo, Ni, Pb, Sb, Se, V, Zn and Sn);
- sulphur (total);
- organic compounds (EPA-16 PAH).

These contaminants were chosen because of their relevance to environmental health policies and the fact that there are reasonable grounds for suspecting the presence of such pollutants because of the origin of the materials and/or evidence in the literature. It was also felt that analysing materials for a standard set of contaminants would allow the results to be compared in a more meaningful way. Where appropriate, additional tests were performed to detect the presence of other constituents.

After the initial series of tests to establish the composition of the various materials, due to high costs, leaching experiments were performed only on samples taken from the more heavily polluted batches of material. Leaching tests on bitumen tar had to be omitted due to its physical state (viscosity).

To assess the leachability of the inorganic constituents in each of the candidate materials, use was made of a standard test [3]. This test is intended to provide worst case data which allow estimates to be made of the long-term pollution threat of inorganic constituents. Where significant quantities of PAH's were known to be present, additional leaching tests were performed. For the purpose of these tests, the size of the particles was reduced to less than < 0.2 mm in diameter using cryogenic techniques. An L/S ratio of 5 was employed and after four hours shaking, the leaching medium was centrifuged and analysed.

In addition, standard column tests were performed on all the candidate materials to establish their short-term inorganic leaching characteristics [6]. A summary of the salient features of the different tests is given in Table 1.

Although the above mentioned leaching tests have not been developed specifically for PAH leaching, the results were considered sufficiently accurate for ranking purposes.

Table 1: Characteristic differences between the tests used to determine composition and leaching behaviour [3].

COMPOSITION particle size reduction: metals < 0,2 mm, PAH 0,5-0,1 pretreatment with aqua regia
MAXIMUM LEACHABILITY particle size reduction < 0.125 mm stir for 4 hours in distilled water (L/S = 100) pH maintained at 4 analysis after filtration over 45μm
COLUMN TEST particle size reduction < 3mm column inner centre line 50 mm; bed height 400 mm distilled water flow from bottom to top pH = 4 flow rate m/48 l/h, m= dry mass of the tested material analysis of 7 water fractions

Following the laboratory tests to establish the leaching characteristics of the materials under consideration, a number of materials were selected for further study under simulated practical conditions (phase 4). Candidate **base course** materials are, for instance, being subjected to leaching trials in a test bed measuring approximately 1 m^2. In an attempt to model practical conditions as accurately as possible, tests are being performed on a 0.5 m thick layer of the chosen material (either in a homogeneous layer or as part of a mixture) after realizing the optimum water content for road construction purposes. The leaching behaviour of the candidate base course materials is being studied by allowing water to percolate through the test bed from above. To investigate the leaching behaviour of **embankment materials** under practical conditions, tests are being carried out in 3 metre high columns with a diameter of 0.6 m. Layer thicknesses of between 1.5 and 3 m will be used to simulate the conditions in embankments. The leaching behaviour of the candidate materials will be studied by allowing water to percolate through the columns from the top. After completing the simulated practical trials, the leaching characteristics of the various materials will be compared with the data generated in previous phases of the research programme. In addition, detailed chemical analyses will be performed on samples of the residue materials to determine their precise composition.

RESULTS

The salient features of the results obtained in the present research will be discussed in this section. For a complete treatment of all the results generated to date, reference is made to [4, 5, 6, 7].

Table 2 assesses the importance of various secondary road construction materials. In order to assess the potential pollution threat secondary road construction materials pose, comparisons have been made with the limit values included in the Soil Protection Act and the Chemical Waste Act. In this context, "A" represents the reference or background concentration, "B" the limit value above which a soil survey is required and "C" is the concentration above which an immediate remedial action programme or clean-up investigation should be performed.

Table 2: Inventory of secondary materials of potential use in road construction.

I: Estimated annual production rate
- - : I< $0.5*10^6$ tonnes/y
- + : $0.5*10^6$ tonnes/y < I < $1.0*10.^6$ tonnes/y
- ++ : I > $1.0*10.^6$ tonnes/y

II: Estimated take-up for road construction purposes
- - : II < $0.05*10^6$ tonnes/y
- + : $0.05*10^6$ tonnes/y < II < $0.5*10.^6$ tonnes/y
- ++ : II > $0.5*10.^6$ tonnes/y

III: Estimated pollution threat
- - : III < b value Soil Protection Act
- + : III > b value Soil Protection Act
- ++ : III > c value Soil Protection Act
- +++ : limit value Chemical Waste Act

Secondary road construction material	Key selection criteria I	II	III
Incinerator slag	+	+	++
Incinerator fly ash	-	-/+	+++
Asphalt rubble	+	+	++/+++
Dredging spoil	++	-	-..+++
Dredged sand	++	?	?
Building rubble:			
* **Crushed concrete rubble**	+	++	+
* **Crushed brick rubble**	+	++	-
* **Screened sand**	-/+	-/+	+
Calcium sulphate waste	++	-	++
Electric furnace slag	-	-/+	++
Phosphoric slag	+	++	+++
Blast furnace slag:			
* **Granular slag**	++	+	++
* **Foamed slag**	+	+	++
* **Particulate slag**	+	++	++
* Pellet slag	-	-	++
Coal residues:			
* **Pulverised coal fly ash**	+	+	++/+++
* Fluidised bed ash	-	-	++/+++
* Brown coal fly ash	-	-	++/+++
* Coal gasification residue	-	-	?
* **Bottom ash**	-	+	-
Mining waste	-	-	+
Steel slag	-	++	++

In addition to the secondary road materials printed in bold in the table above, the following primary materials have also been included in the study: embankment sand, industrial sand, gravel, lava and bitumen tar. The composition and leaching characteristics of Zn and PAH's of the chosen road construction materials are given in Table 3. The range of values given indicates the likely variation in composition.

Reference to Table 3 shows that the concentration of zinc in brick rubble, pulverised coal fly ash and incineration fly ash can vary significantly. In the case of PAH's, significant differences were found in the concentrations present in asphalt rubble and brick rubble. It can also be seen from Table 3 that there appears to be a relationship between the composition of a given material and its maximum leachability rating for zinc. However, it should be noted that materials containing relatively few pollutants can have a procentually high leachability. Embankment sand is a particular case in point.

Table 3: The concentration (mg/kg dry matter) and the maximum leachability (μg/l) of zinc and PAH's. The values given in parenthesis are the results of repeat tests.

MATERIAL	ZINC		TOTAL PAH	
	concen-tration	leach l/s=100	concen-tration	leach l/s=5
Incinerator slag	2300- 2400	14000	0.1- 5.2	7 (2)
Incinerator fly ash	7200-11000	120000	<0.1	
Asphalt rubble	24- 62	63	0.9-29.0	170 (22)
Dredged sand	42- 70	320	0.3	
Concrete rubble	56- 78	310	3.0- 6.0	10 (7)
Brick rubble	70- 170	820	5.3-23.3	110 (56)
Screened sand	160- 230	1150	20.0-32.0	69 (14)
Electric furnace slag*	210- 215	1000	0.1- 1.1	
Phosphoric slag	3- 43	37	0.1	
Particulate slag	3- 6	48	0.3- 1.4	
Gran. bl. furn. slag	1	10	0.1	
Foamed slag*	3- 4	37	<0.1	
Pulv. coal fly ash	52- 155	86	<0.1	
Bottom ash	12- 18	68	0.1	
Steel slag	17- 29	53	0.2- 0.4	
Embankment sand*	7- 8	145	0.1- 0.2	
Industrial sand	7- 23	34	0.1- 0.8	
Gravel	18- 30	50	<0.1	
Lava	43- 44	40	<0.1	
Bitumen tar	39		150,000	
Bitumen	5		110	

* batches chosen at random

The five materials in the above list with the highest concentrations of total PAH's were selected for further leachability tests. In order to assess how representative the maximum leachability ratings for PAH's are when determined using water, repeat measurements were performed on samples that had already be tested. Reference to Table 3 shows that substantial quantities of PAH's are leached out from the various materials and that the leaching process had not been completed during the first series of tests. It should be noted that the highest leachability rating was not ascribed to the material with the highest PAH's concentration.

Table 4: The composition and leaching characteristics (column test) of road construction materials expressed in terms of the A, B and C limit values defined for inorganic constituents in soil and groundwater.

composition		A	B	C
	steel slag			electric furnace slag pulverised coal fly ash incinerator slag incinerator fly ash
C		asphalt rubble concrete rubble	lava/particulate slag dredged sand foamed slag slag sand	phosphoric slag screened sand
B		gravel brick rubble		
A		embankment sand industrial sand bottom ash		

leaching

To weigh the leachability data, comparisons were made with the limit values specified in the Dutch Soil Pollution Act. This allows materials to be classified on the basis of the three categories defined earlier (A, B and C). Failure to comply with a particular limit value results in a material being classified in a higher risk group, regardless of whether the limit value is exceeded by more than one constituents. This implies that there could be both quantitative and qualitative differences between the leaching characteristics of materials in the same group. For example, incineration fly ash and pulverised coal fly ash both contravene the "C" limit on composition defined in the Soil Pollution Act, but incineration fly ash contains 7 constituents that exceed the "C" value while pulverised coal fly ash has 2 constituents that exceed the "C" value.

It can be seen in Table 4 that materials that are highly susceptible to leaching also contain substantial quantities of pollutants. However, highly polluted materials are not necessarily susceptible to leaching as is evidenced by the results obtained for steel slag. This material has a high concentration of pollutants, yet is relatively insensitive to leaching. In general, it can be stated that primary materials pose less of a threat to the environment than secondary materials. The concentration data accord with the data found in literature [8].

After considering the leaching behaviour of the various materials, the following candidate embankment and base course materials were selected for testing under simulated practical conditions.

- pulverised coal fly ash - incinerator fly ash - sand	- embankment materials
- crushed concrete/brick rubble - particulate slag/foamed slag/slag sand - sand	- base course materials

This phase of the research programme is currently nearing completion.

CONCLUSIONS

The results of the leaching tests and chemical analyses show that secondary road construction materials are less acceptable from an environmental point of view than primary materials. Significant differences were found in the composition of the various materials. For instance, incinerator fly ash and bitumen tar were found to contain high concentrations of contaminants, while industrial sand was found to be relatively free of pollutants. It should, however, be noted that also significant variations were observed in the concentrations found in different samples of the same material.

Comparison of the leaching behaviour of the various materials showed that the portion of contaminants leached out of pulverised coal fly ash, screened sand, crushed blast furnace slag, industrial sand and incinerator fly ash was relatively high. In contrast, materials such as gravel, lava, steel slag and asphalt rubble are less susceptible to leaching.

Materials highly susceptible to leaching also contain substantial quantities of pollutants. However, highly polluted materials are not necessarily susceptible to leaching (steel slag).

Leaching in practical situations is thought to be influenced by parameters such as particle size, porosity, chemical retention (pH, Eh) [9], interaction with the surroundings [10], etc. The extend of the influence of these properties will have to be determined under (semi-)practical conditions. Research under these circumstances will have to teach the environmental risk of using contaminated materials in road and hydraulic engineering constructions.

REFERENCES:

1. Ministeries VROM/EZ/L&V/V&W, Notitie inzake preventie en hergebruik van afvalstoffen, 1988.
2. Ministerie V&W, Beleidsnota Gegrond Ontgronden, 1989.
3. NVN 2508: Bepaling van de uitloogkarakteristieken van kolenreststoffen, UDC 662.62/67:543.2, 1987.
4. RWS-DWW/TAUW Infra Consult b.v., Inventarisatie wegenbouwmaterialen: Notitie als onderdeel van fase 1 van het onderzoek naar emissies door uitloging van wegenbouwmaterialen, RWS-DWW rapportnr. MI-OW-88-57, 1988.
5. RWS-DWW/TAUW Infra Consult b.v., Rapportage fasen 2A en 2B: samenstelling en maximale uitloogbaarheid, RWS-DWW rapportnr. MI-OW-89-06, 1989.
6. RWS-DWW/TAUW Infra Consult b.v., Rapportage fase 3: uitloogonderzoek aan enkele wegenbouwmaterialen, RWS-DWW rapportnr.MI-OW-89-100, 1989.
7. RWS-DWW/TAUW Infra Consult b.v., Fase 4: semi-praktijkonderzoek naar het uilooggedrag van enige wegenbouwmaterialen. Deel 1: Poefopzet en karakterisering van de materialen, RWS-DWW rapportnr. MI-OW-90-37, 1990.
8. ECN, Elementsamenstelling van primaire en secundaire grondstoffen, Mammoet deelrapport 06, 1990.
9. NOVEM/RIVM, Beoordelen van bouwstoffen in het licht van het bouwstoffenbesluit, Workshopverslag 1-11-'89, Bilthoven. NOVEM, Utrecht.
10. ECN/McMaster University, Canada, Zelfvormende en Zelf-Herstellende Afdichtingen: Concept, Modellering, en Laboratorium resultaten, in publication.

Waste Materials in Construction.
J.J.J.R. Goumans, H.A. van der Sloot and Th.G. Aalbers (Editors)

MUNICIPAL SOLID WASTE COMBUSTION ASH AS AN AGGREGATE SUBSTITUTE IN ASPHALTIC CONCRETE

D.L. GRESS[1], X. ZHANG[1], S. TARR[1], I. PAZIENZA[1] and T.T. EIGHMY[1]

[1]Environmental Research Group, Department of Civil Engineering, 236 Kingsbury Hall, University of New Hampshire, Durham, New Hampshire 03824 (USA)

SUMMARY

A two year study is underway to evaluate the physical and chemical properties of the bottom ash process stream from the 500 TPD waste-to-energy facility in Concord, New Hampshire. The use of bottom ash as an aggregate substitute product in asphaltic base course is envisioned. Research is underway to characterize the time-dependent properties of the bottom ash for product acceptance, to develop both hot mix and cold emulsion formulations, and to evaluate the leachate release rate characteristics from various blends using a variety of batch and lysimeter leach tests. Results to date suggest that the bottom ash product stream is relatively constant, hot mix formulations meet State Department of Transportation specifications, and bitumen is effective in encapsulating bottom ash and reducing salt leachability.

1. INTRODUCTION

In the United States, consideration is being given to the use of bottom ash from municipal solid waste combustion as an aggregate substitute in construction materials (1). The anticipated hierarchy for use in the United States reflects regulatory concerns that certain waste products be encapsulated or stabilized before use. Consequently, the use of bottom ash is likely to be in bituminous base course, bituminous wearing course and concrete construction materials before it is used in granular sub-base, structural fill, or embankment applications. This hierarchy differs somewhat from typical uses of bottom ash in Europe as a granular, soil-like material (2-5).

Earlier work in the United States by Walter (6,7) showed that optimum hot mix formulations contained 50% bottom ash with asphalt cement contents of 5.5 to 6.5 weight percent. Marshall stabilities for strength and flow met typical specifications.

A number of demonstrations on bottom ash use in base course or wearing course were conducted in the 1970's and early 1980's (8-13). The results are summarized in Table 1. General observations from these studies (1) suggest that conventional asphalt mixing and paving equipment can be used, loss-on-ignition values for the bottom ash should be less than 10%, fly ashes should not be incorporated into the blends, vibrators on feed bins are necessary, and plant temperature control is important given the moisture content of the residues. Additionally, these studies suggest that optimum mixes for hot mix work can contain 50 to 75% residue though the absorption of bitumen may be high.

A few studies have looked as using fused (14) or cement-treated (15) residues as aggregate substitutes. Recent work by Chesner et al. (16) has shown that bottom ash from the Southwest Brooklyn, New York combustor can be used as aggregate substitute; they found a 30% bottom ash substitution to perform as well as controls in batch Marshall testing.

Recent evaluations of bottom ash use by Chesner (17) have examined economic, regulatory, and environmental concerns surrounding the use of bottom ash. Institutional issues may be the largest impediment to active use in the United States despite the fact that its use is technically and economically feasible.

The scope of this study was to expand on previous efforts by (i) evaluating the time-dependent physical and chemical properties of the bottom ash process stream for product acceptance, (ii) evaluating both hot mix and cold emulsion formulations for product development, and (iii) investigating the effects of bottom ash substitution and bitumen (or emulsion) content on elemental release rates from the materials. Our approach to evaluating bottom ash as an aggregate substitute is shown in Figure 1.

2. MATERIALS AND METHODS

The combustor that is being sampled for this project is the 500 TPD (454 tonnes per day) Concord, New Hampshire combustor. The facility is owned by Wheelabrator Concord L.P. and operated for the Concord Regional

TABLE 1

Combustion Residue Use Studies in Bituminous Applications in the United States.

Project	Date	Residue %	Asphalt %	Lime %	Length m	Thickness cm	Performance	(Ref.)
Houston, TX	1974	100[a]	9.0	2.0	60	15 BC[c]	Excellent	(8-10)
Philadelphia, PA	1975	50[a]	7.4	2.5	30	3.8 WC[d]	Acceptable	(11)
Delaware Co., PA	1975	50[a]	7.0	2.5	20	3.8 WC	Acceptable	(11)
Harris, PA	1975	50[a]	7.0	2.5	80	3.8 WC	Poor	(11)
Washington, DC	1977	70[a]	9.0	2.0	130	11.3 BC	Good	(12)
Lynn, MA	1979	50[a]	6.5	2.0	1500	3.8 WC	Excellent	(13)
Harrisburg, PA	1976	100[a]	6.7	0.0	60	3.8 WC	Excellent	(14)
Tampa, FL	1987	15[b]	7.0	-	400	2.5 WC	Excellent	(15)

[a]Largely bottom ash; possibly some economizer ashes.
[b]Bottom ash and fly ash.
[c]BC = base course.
[d]WC = wearing course.

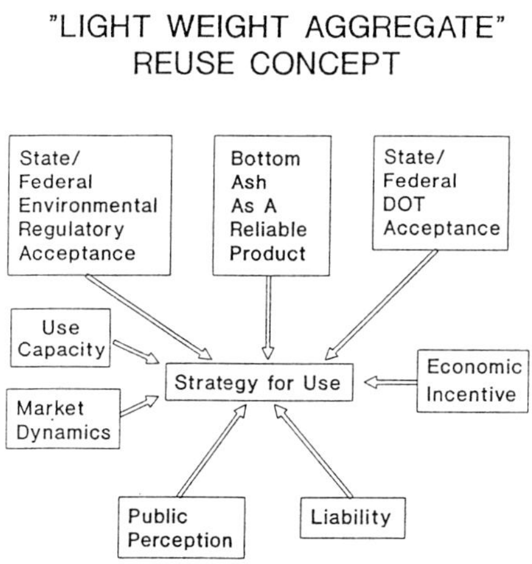

Fig. 1 Aggregate Substitute Concept.

Solid Waste/Resource Recovery Cooperative. The facility has two process trains consisting of Von Roll reciprocating stoker grates, Babcock and Wilcox boilers, and Wheelabrator Technology dry lime scrubber/fabric filters. The bottom ash from each train is quenched in its own quench tank.

Bottom ash is sampled from the drag chain conveyor according to the scheme shown in Figure 2. Arrangements are made to prevent economizer ash from entering the quench tank prior to and during sampling. The quench tank is also cleaned prior to sampling. Combustor performance is monitored to relate product quality to combustor operation.

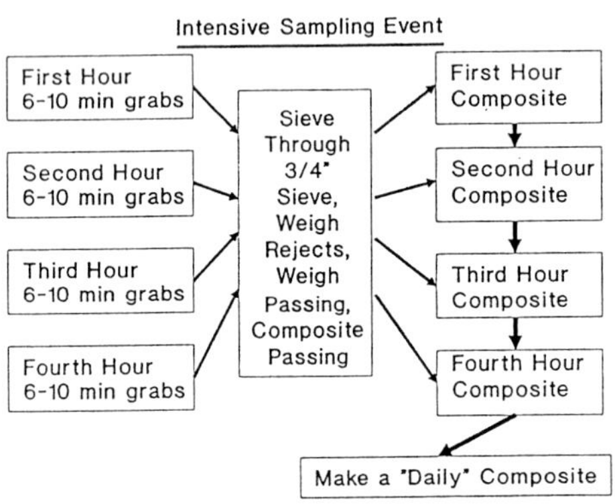

Fig. 2 Schematic showing sampling program. The 3/4 inch (1.9 cm) cut-off is used to meet New Hampshire base course specifications.

For evaluation of time dependent physical and chemical properties of the bottom ash, the tests shown in Table 2 are conducted at varying frequencies.

Leaching tests are being conducted on Marshall samples from both the hot-mix and cold-emulsion work. Additionally, a test patch containing 25%

TABLE 2

Testing for Bottom Ash Product Acceptance

Time-Dependent Physical Properties[b]	Time-Dependent Environmental Properties
- % 3/4 inch (1.9 cm) minus[a]	- Elemental composition
- Moisture (ASTM D2216)[a]	- Acid Neutralyzing Capacity (ANC)
- LOI (ASTM C114)[a]	- Static pH leach test (pH 7,4; L/S 100)
- Ferrous content	- Others
- Particle size distribution (ASTM C136)[a]	
- Absorption and specific gravity (ASTM C127,C128)[a]	
- Moisture Density (ASTM D1557)	
- CBR (ASTM D1863)[a]	
- Sodium sulfate soundness (ASTM C88)[a]	
- LA abrasion (ASTM C131)[a]	
- Unconfined compressive strength (ASTM D2166)	
- Marshall stability[a] (ASTM D1559)	

[a]Denotes generic State Department of Transportation preferred tests for lightweight aggregate use in bituminous road materials
[b]Data also useful for other potential applications

bottom ash and 9% asphalt cement was paved, cured for a week and then broken up and placed in a lysimeter for long term field leaching studies. For comparison, 3/4 inch (1.9 cm) minus bottom ash is also being leached in an adjacent lysimeter.

3. RESULTS AND DISCUSSION

Grain size distributions for 29 samples are shown in Figure 3. All the samples fall within the upper and lower limits for distribution for a New Hampshire Department of Transportation Type B base course. In theory, if 100% substitution of bottom ash for natural aggregate was specified, the bottom ash would meet gradation specifications.

Three parameters have been selected to present time-dependent changes in physical or chemical properties of the bottom ash. Figure 4 shows 3/4 inch (1.9 cm) minus material. With the exception of one event, the quantity passing is very constant. Figure 5 shows LOI. The LOI is relatively constant at 6 to 9%. Figure 6 shows total lead. The elemental lead compositional variability within an event is as variable as time dependent daily composite variability. Work is underway to relate combustion operation to bottom ash quality.

Typical physical and chemical properties of the bottom ash are shown in Table 3. The data were compiled from the product acceptance testing conducted to date. The quenched ash is a wet, but lightweight material with highly absorptive properties. It's durability, based on LA abrasion and Na_2SO_4 soundness appears acceptable. The finer fraction is more absorptive and friable. The material compacts well and has reasonable buffering capacity.

The data on time dependent physical and chemical properties are in general agreement with data from plants in Europe (18) and the United States (16) for product acceptance. The bottom ash is not excessively variable in its physical properties. Additional data are needed to assess chemical property variability. Further work is planned to evaluate product reliability, to further develop the data base, to identify surrogate parameters as gross indicators of product quality, and to intercorrelate combustor operational data with product quality data.

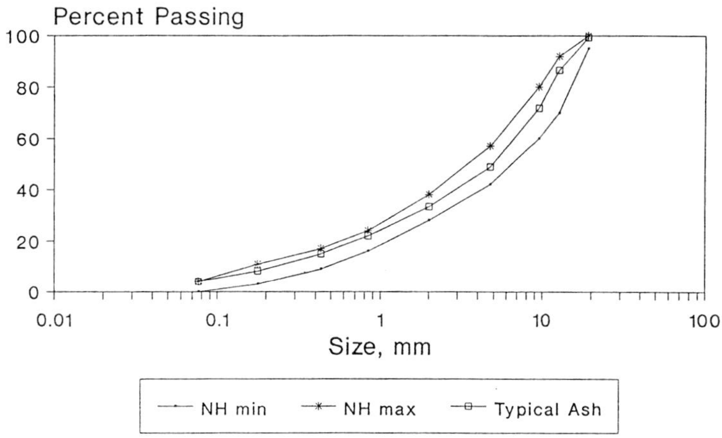

Fig. 3 Average grain size distributions of 29 hourly or daily composite samples of bottom ash are indicated by the typical ash plot.

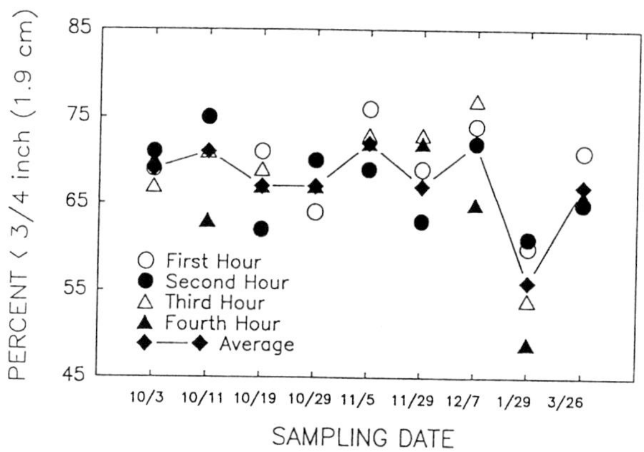

Fig. 4 Time dependent variation in % 3/4 inch (1.9 cm) minus material of the bottom ash.

TABLE 3

Bottom Ash Physical and Chemical Properties

Parameter	Range of Values
Water Content (%)	2.6 - 53
Uniformity Coefficient (D_{60}/D_{10})	11.6 - 38.0
Effective Size (D_{10}, mm)	0.155 - 0.762
Bulk Specific Gravity (<4.75 mm)	1.30 - 2.06
Bulk Specific Gravity (>4.75 mm)	2.03 - 2.43
Absorption (%, <4.75 mm)	7.66 - 21.23
Absorption (%, >4.75 mm)	1.74 - 7.80
LOI (%)	4.8 - 10.7
Ferrous Content (%)	15.6 - 39.9
Unit Weight (kg/m^3)	1,109 - 1,223
Optimum Proctor Moisture (%)	11 - 17
Proctor Dry Density (kg/m^3)	1,724 - 1,782
LA Abrasion (%)	46.4 - 48.2
Na_2SO_4 Soundness (Fine Fraction)	10.38 - 14.32
Na_2SO_4 Soundness (Coarse Fraction)	2.51 - 2.76
Acid Neutralizing Capacity (meq/g)	1.5 - 3.5

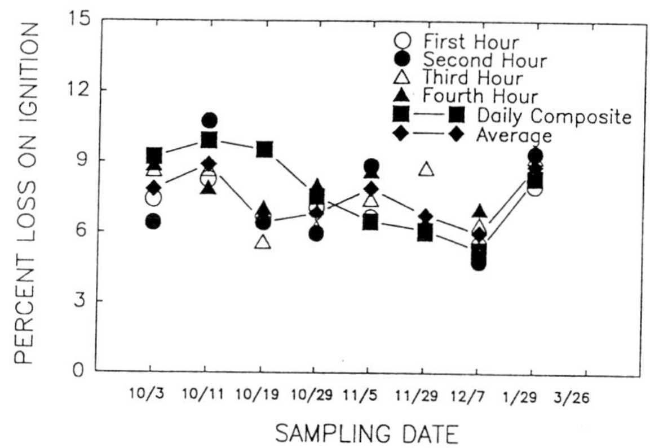

Fig. 5 Time dependent variation in LOI of the bottom ash

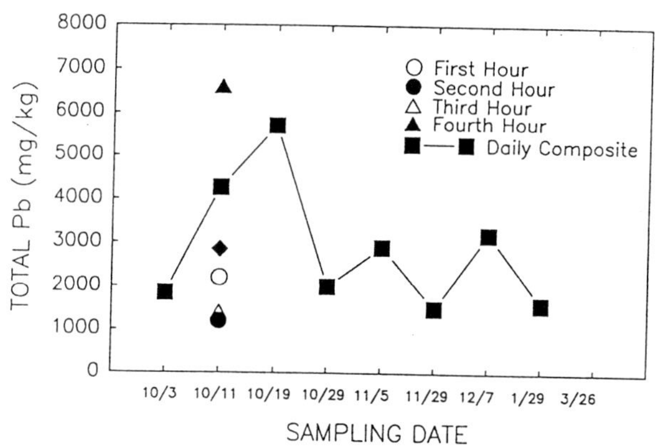

Fig. 6 Time dependent variation in total Pb of the bottom ash, Pb was determined by $HF/HCl/H_2O_2$ complete digestion

The data on optimizing the Marshall properties of the bottom ash as an aggregate substitute are shown in Figures 7-12. Various levels of bottom ash substitution with natural aggregate were tested (0, 25, 50, 75, and 100% bottom ash). A variety of percent asphalt cement levels were also investigated (4 to 12% by total weight).

Marshall stability is shown in Figure 7. Typical high strength values of 2000 lbs (908 kg) were seen. Percent air voids usually fell in the range of 3 to 8% (Figure 8). Flow values were somewhat high, even with the aggregate control (Figure 9). Typical flows were 12 to 20 1/100's of an inch (0.03 to 0.50 cm). Unit weight values of the blends were 125 to 140 lbs/ft^3 (2,003 to 2,244 kg/m^3), indicating that bottom ash is a lightweight aggregate (Figure 10). Percent voids in the mineral aggregate (Figure 11) were out of specification for the 100% aggregate blend. Percent absorbed asphalt (Figure 12) levels were high in all samples; indicative of the absorbtive and porous nature of the bottom ash.

The results from the Marshall stability work are similar to previous efforts (12,16). The results suggest that a 75% blend at asphalt cement contents of 6 to 9% are technically feasible. Higher percent substitutions may be achieved. Further work is ongoing to examine the economics surrounding utilization.

The evaluation of the lysimeter leachate characteristics from the excavated test patch and the control bottom ash is shown in Table 4. Despite the lesser quantity of the bottom ash in the bottom ash/asphalt blend, the encapsulating property of the bitumen is apparent. Na, K, Cl, and SO_4^{2-} levels are all dramatically lower in the asphalt blend leachate. Metals leachability from both materials is not problematic.

4. CONCLUSIONS

The results to date suggest that bottom ash is a suitable lightweight aggregate; the time dependent physical properties are reasonably uniform. These data suggest that bottom ash as a lightweight aggregate substitute or product is an acceptable continuously-produced material. Economic evaluations of percent substitution and relative use of asphalt cement are needed. Nevertheless, a 75% blend of 9% asphalt cement is a technically good hot mix blend. Lower asphalt contents will be considered. Work is

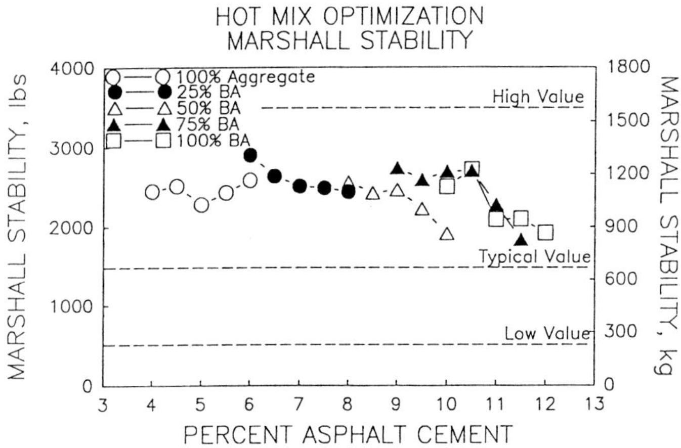

Fig. 7 Marshall Stability

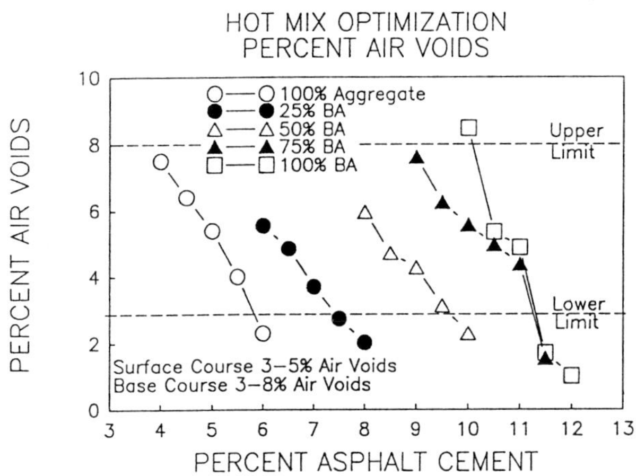

Fig. 8 Marshall Percent Air Voids

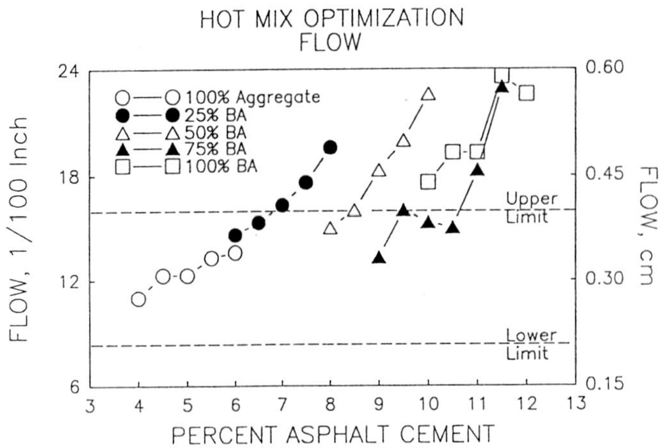

Fig. 9 Marshall Flow

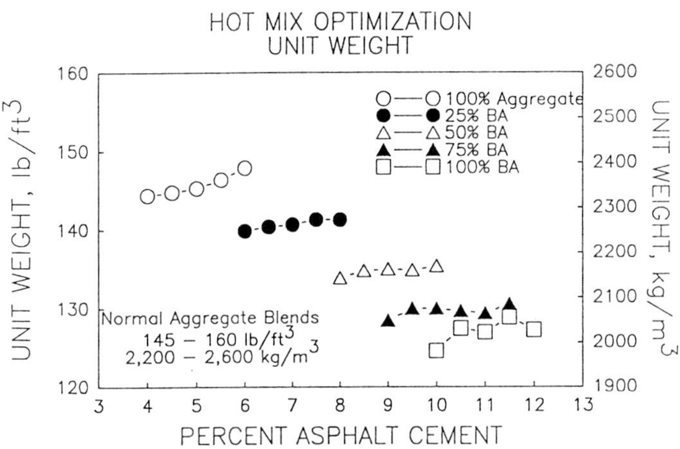

Fig. 10 Marshall Unit Weight

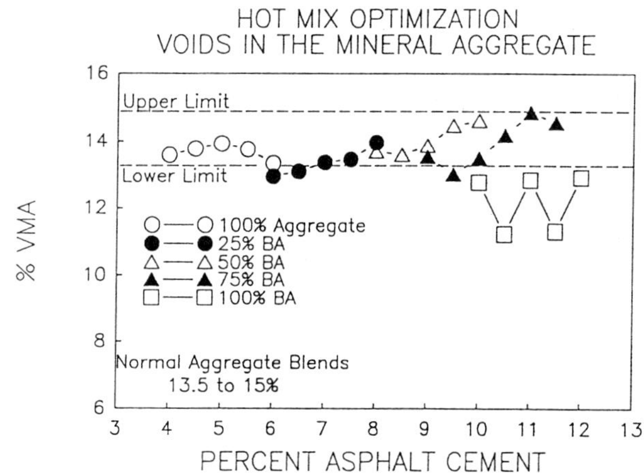

Fig. 11 Marshall Percent Voids in the Mineral Aggregate

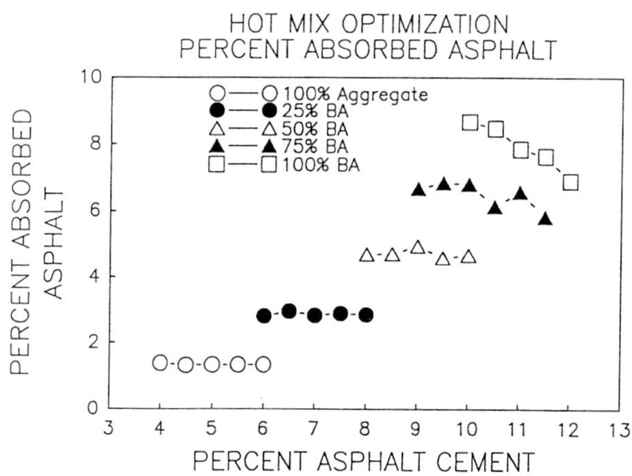

Fig. 12 Marshall Percent Absorbed Asphalt

TABLE 4
Leachate Characteristics From Bottom Ash and Bottom Ash/Asphalt Road Way Material

Parameter	EPA Method or Standard Method	Bottom Ash Total	Bottom Ash <0.45 μm	Bottom Ash/Asphalt Total	Bottom Ash/Asphalt <0.45 μm
L/S	-	0.069[b]	-	0.073 (0.22)[b]	-
pH	-	6.41±0.12	-	7.15±0.18	-
Alkalinity (mg/L)	-	29±3.0	-	23±0.3	-
Conductivity (mmhos/cm)	-	8,322	-	165	-
COD (mg/L)	410.4	310	-	22	-
NO_3^--N (mg/L)	300.0	<0.50	-	<0.050	-
NO_2^--N (mg/L)	300.0	<0.50	-	<0.050	-
NH_4^+-N (mg/L)	350.2	5.0	-	0.20	-
SO_4^{2-} (mg/L)	300.0	1,700	-	31	-
Br^- (mg/L)	300.0	30	-	<1.0	-
PO_4^{3-} (mg/L)	300.0	<0.01	-	<0.01	-
Cl (mg/L)	300.1	1,700	-	24	-
Al (mg/L)	3010,6010	0.20	0.19	<0.10	<0.10
Sb (mg/L)	3010,6010	0.063	0.068	<0.060	<0.060
As (mg/L)	7060	<0.010	<0.010	<0.010	<0.010
Ba (mg/L)	3010,6010	0.10	0.12	<0.10	<0.10
Be (mg/L)	3010,6010	<0.0050	<0.0050	<0.0050	<0.0050
Cd (mg/L)	3020,7131	<0.0050	<0.0050	<0.0050	<0.0050
Ca (mg/L)	3010,6010	590	640	22	26
Cr (mg/L)	3010,6010	<0.010	<0.010	<0.010	<0.010
Co (mg/L)	3010,6010	<0.030	<0.030	<0.030	<0.030
Cu (mg/L)	3010,6010	0.59	0.59	<0.020	<0.020
Fe (mg/L)	3010,6010	0.050	0.038	<0.030	<0.030
Pb (mg/L)	3020,7421	0.0053	0.0053	<0.0050	<0.0050
Mg (mg/L)	3010,6010	6.5	6.6	1.3	1.7
Mn (mg/L)	3010,6010	0.32	0.33	0.020	0.022
Hg (mg/L)	7470	0.0006	0.0004	<0.0003	<0.0003
Mo (mg/L)	3010,6010	0.34	0.36	<0.10	<0.10
Ni (mg/L)	3010,6010	<0.030	0.030	<0.030	<0.030
K (mg/L)	3010,6010	220	220	3.9	3.7
Se (mg/L)	7740	<0.010	<0.010	<0.010	0.010
Si (mg/L)	3010,6010	2.2	1.9	1.2	1.2
Ag (mg/L)	3010,6010	<0.020	<0.020	<0.020	<0.020
Na (mg/L)	3010,6010	740	680	8.1	8.3
Sr (mg/L)	3010,6010	4.4	3.9	0.15	0.14
Ti (mg/L)	3010,6010	<0.10	<0.10	<0.10	<0.10
V (mg/L)	3010,6010	<0.010	<0.010	<0.010	<0.010
Zn (mg/L)	3010,6010	0.068	0.34	<0.020	0.025
<u>VOCs</u>	624				
Tetrahydrofuran		Trace[a]		65 ppb[a]	
<u>ABN's</u>	625				
		None		None	

[a] Component of PVC cement used in the leachate collection piping system.
[b] L/S is weight of leachate collected divided by wet weight of ash. For BA/Asphalt, L/S is weight of leachate collected divided by weight of asphalt plus wet ash (or just wet ash).

underway to also evaluate cold emulsions. The benefits of the encapsulating properties of bitumen in controlling salt release are apparent. More work on the fundamentals of bottom ash leaching are ongoing. A field demonstration is planned.

ACKNOWLEDGEMENTS

We thank the Concord Regional Solid Waste/Resource Recovery Cooperative and Wheelabrator Environmental Systems for their support of this research.

REFERENCES

1 W.C. Ormsby, in T.T. Eighmy and W.H. Chesner (Eds), Proceedings of the First International Conference on Municipal Solid Waste Combustor Ash Utilization, UNH Printing, Durham, NH (1988), pp. 49-62.
2 J. Hartlen, in: T.T. Eighmy and W.H. Chesner (Eds), Proceedings of the First International Conference on Municipal Solid Waste Combustor Ash Utilization, UNH Printing, Durham, NH (1988), pp. 33-48.
3 O. Hjelmar, in: W.H. Chesner and T.T. Eighmy (Eds), Proceedings of the Third International Conference on Municipal Solid Waste Combustor Ash Utilization, UNH Printing, Durham, NH (1991), pp. 59-70.
4. J. Vehlow, in: W.H. Chesner and T.T. Eighmy (Eds), Proceedings of the Third International Conference on Municipal Solid Waste Combustor Ash Utilization, UNH Printing, Durham, NH (1991), pp. 85-96.
5 P. Leenders, VEABRIN, unpublished manuscript (1991).
6 C.E. Walter, U.S. Patent 3,907,582 (1973).
7 C.E. Walter, J. Env. Engineer. Div. ASCE 114 (1976) 139-140.
8 J. Haynes and W.B. Ledbetter, Incinerator Residues in Bituminous Base Construction, (1976) FHWA/RD-76/12.
9 D.J. Teague and W.B. Ledbetter, Three Year Results on the Performance of Incineration Residues in Bituminous Base, (1978) FHWA/RD-78/144.
10 J. Ederly and W.B. Ledbetter, Field Performance of Littercrete (Incinerator Residue) in a Bituminous Base, (1981) FHWA/RD-88/022.
11 R.J. Collins, R.H. Miller, S.K. Ciesielski, E.M. Wallo, M.J. Boyle, D. Pindzola, and J. Tropea, Technology for Use of Incinerator Residue as Highway Material, (1977) FHWA/RD-77/151.
12 R.D. Pavlovich, H.J. Lentz, and W.C. Ormsby, Transportation Research Record 734 (1979) 38-44.
13 G. Bastanza, Summary Update of Research Projects With Incinerator Bottom Ash Residue, Department of Environmental Management (1982).
14 R.R. Synder, Evaluation of Fused Incinerator Residue as a Paving Material, (1980) FHWA-TS-88-229.
15 D.G. Mehan and W.F. Hooper, in Proceedings of the Municipal Waste Combustion Conference, April 15-19, (1991) AWMA Publ., Pittsburgh, pp. 627-641.
16 W.H. Chesner, R.J. Collins, and T. Fund, Assessment of the Potential Suitability of Southwest Brooklyn Incinerator Residue in Asphaltic Concrete Mixes (1988) NYSERDA Energy Authority Rept. 90-15.
17 W.H. Chesner, in Proceedings of the Municipal Waste Combustion Conference, April 15-19 (1991) AWMA Publ., Pittsburgh, pp. 608.
18 Veabrin Kwaliteitskontrole van AVI-Slakken '87-'88, (1988) RAP-305/JJS/avd.

Waste Materials in Construction.
J.J.J.R. Goumans, H.A. van der Sloot and Th.G. Aalbers (Editors)

THE USE OF INDUSTRIAL BY-PRODUCTS WITH HYDRAULIC BINDERS: REFUSE INCINERATION ASHES AS AN EXAMPLE

Michael Schmidt, Oberklamweg 6, D-6906 Leimen, Germany
and
Paul Vogel, Heidelberger Zement AG, Research and Development Department, D-6906 Leimen, Germany

Summary

Recycling materials and industrial by-products can be used in construction if the structures built with these materials adequately withstand all loads occurring during use and if they have no unacceptable effects on the environment. The laboratory evaluation of such materials, the testing procedures applied, and the fixing of threshold values must adequately take into consideration the constructional and environmentally relevant properties of the material as well as the structural conditions of the structure itself. The constructional properties, e.g., strength and freeze-thaw-resistance, and deficiency in environmental requirements can be significantly improved by a consolidation with hydraulic binders. In some cases the leaching of certain pollutants is considerably reduced by physicochemical bonding and by sealing the grain structure from water.
Refuse incineration ashes are an example of the successful use of residual substances. Several experimental roads were built on the basis of comprehensive laboratory research. On the whole it could be determined that the use of these ashes as base material with hydraulic binders in road construction is possible in many cases. But they differ in their behavior from hydraulically bound road bases composed of natural minerals. Therefore, what is needed to prevent risks is an intensive pre-evaluation with stiffer frost testing conditions and special attention to the mixing technique, the particularly appropriate binders, the curing, as well as the attainment of the desired compactness of the grain structure. On the basis of comprehensive experiments, recommendations are provided here for the use of refuse incineration ashes.

1. General outline

Every year 220 tons of recyclable building materials and industrial by-products are produced in the FRG (West), e.g., receycled concrete, receycled asphalt, construction waste, and fly ashes. 3.5 million tons of combustion residues are left behind from refuse incineration. At present, these materials are only reused to a small extent and only for minor purposes in construction. Most of them are deposited in refuse dumps because of their insufficient constructional properties and in some cases on account of environmentally relevant pollutants. Often they are less firm than natural minerals, are more porous, and have a lower resistance to frost. Some of the materials in question also contain environmentally relevant pollutants that should not be mobilized. In each individual case, therefore, the following questions should be closely examined: what are the properties of a particular material, how can it be improved if necessary, and where should it be used.

The use of industrial by-products and recycling materials for road building, applied in an unbound state or bound with hydraulic binders, offers many benefits. The ever-diminishing space for waste dumps is less threatend, and the exploitation of natural resources reduced. Moreover, the use of these materials in many cases is profitable, because significant refuse costs can be avoided.

2. Appropriate Structural and Ecological Evaluation of Recycling Materials and Secondary Materials for Roads

2.1 Structural Evaluation

The evaluation of industrial by-products and recycling materials for use in road building can be carried out according to the scheme in fig. 1. In stage 1, the structural properties of the starting material, e.g., the strength, the freeze-thaw-resistance, and the structural constituents, are established and evaluated, after which an initial decision concerning its suitability is made. At best, it is suitable in an unbound state, e.g., for frost blanket layers. At worst, the material shows such poor technological properties that the only alternative is disposal.

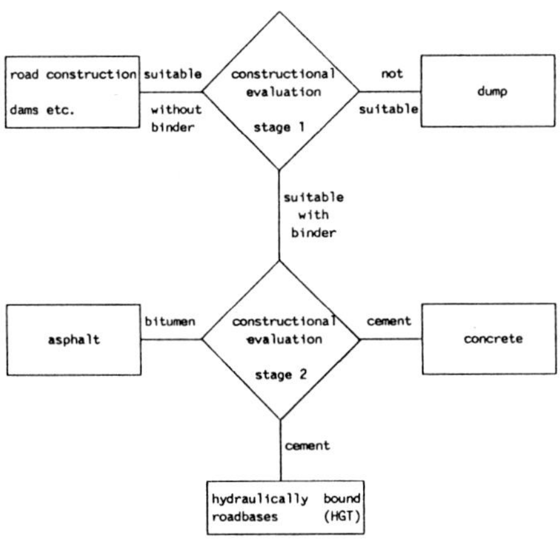

fig. 1: Evaluation of recycling- and secondary materials for use in structural engeneering with or without binders

In many cases the properties of recycling materials and secondary materials can be improved with hydraulic binders to allow for their subsequent use. The final decision in this regard, however, can only be made by an evaluation of the material in its bound state, i.e., in a hardened hydraulically bound mixture, designated here as stage 2 of the structural evaluation.

The engineering requirements comprise four basic properties:

* the strength must be sufficient
* adequate resistance to weathering influences
* volume stability
* sufficient workability

In order for hydraulically bound road bases [1] to obtain adequate solidity and durability, a minimum amount of binder is generally required, depending on the mineral mixture used. This is determined on the basis of a compressive stength test. If necessary, an additional frost test is carried out according to the TP-HGT [2]. This procedure has proved its efficiency with hydraulically bound road bases composed of natural mineral materials. Fig. 2, however, shows by the example of a hydraulically bound road base of refuse incineration ashes that this experience can not always be applied to other cases. After the 12 freeze-thaw-cycles that are usually required in frost tests for hydraulically bound specimens, the measured length alteration was less than the threshold value of 1 o/oo, even for the lowest binder content of only 4 wt.%. According to this criterion, the complete hydraulically bound road base should be frost-resistant. However, upon further freeze-thaw cycles extreme structural disorders occurred relatively quickly, which was reflected in crumbling of some of the specimens. Therefore satisfactory long-term frost resistance can not be expected in spite of apparently satisfactory test results, since the criteria proven for hydraulically bound bases of natural minerals are not sufficient to account for hydraulically bound bases of refuse incineration ashes. At present, the testing procedures is being intensified by an increase in the number of freeze-thaw-cycles and by enhanced moist curing of the specimens during the thaw phase.

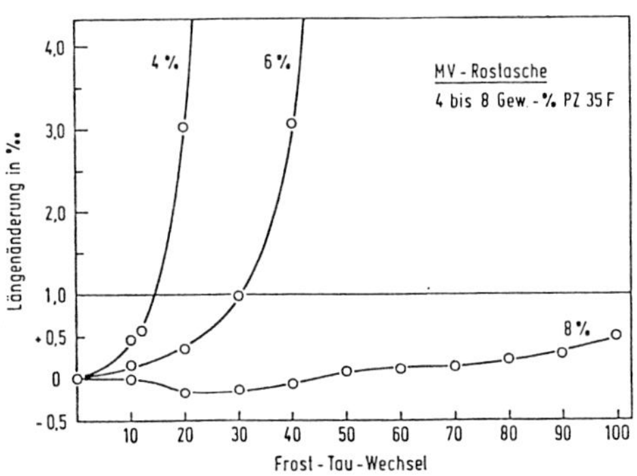

fig. 2: Length alteration with increase of freeze-thaw-cicles of hydraulically bound refuse incineration ash

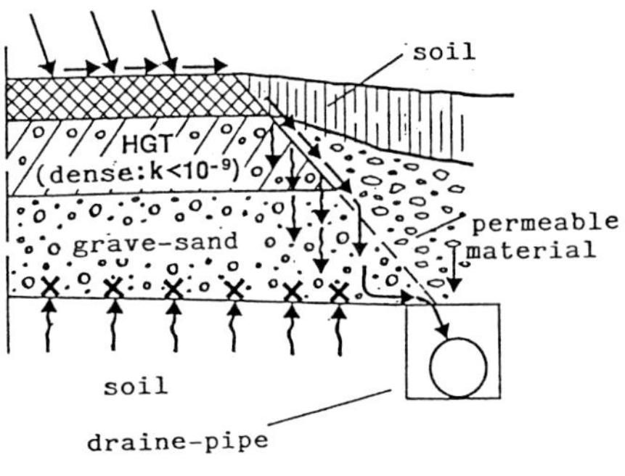

fig. 3: Waterproof pavement construction

2.2 Ecological Evaluation

Provided that a pollutant is a long-term fixed constituent of the material, it is not harmful to the environment. A matter of great importance, however, is the amount of soluble constituents present which in the course of time can make their way into the environment. Leaching is essentially determined by three influencing factors:

1. the structural conditions that determine whether moisture can soak through the structure and thereby reach the processed material.

2. the structure of the compacted by-product - it is practically impossible for materials contained in a waterproof structure to be dissolved.

3. the water solubility of the pollutant in question, which varies greatly and depends essentially on the ambient conditions, e.g., on the pH; in the course of hydration during consolidation with cement, a pronounced reduction in solubility can take place.

Fig. 3 schematically shows a typical road structure. A frost-protection subbase consisting of water-permeable natural gravel is first applied. This prevents solvent water and moisture from rising by capillary action into the structure from under the ground. The base above the subbase is made of secondary material and is hydraulically bound with binders in such a manner that it is practically waterproof. The structure is completed by an asphalt cover which seals off the hydraulically bound base from surface water. Most of the water is laterally diverted to the drainage system. In this case, far fewer pollutants can leach out than with a water-permeable base without an upper asphalt layer. This factor must be taken into consideration in a practice-oriented evaluation of compatibility with environmental requirements.

The positive influence of the impermeability of the grain structure can be determined by testing procedures. This is shown in fig. 4, which depicts the results of model tests at the Research Institute of the German Cement Industry [3]. Increasing amounts of cement were added to an equigranular sand. As a result, a greater number of hollow spaces were filled. The parameter for the permeability is the permeability value k, which was determined in a special testing installation described in [3]. A known quantity of chromium was added to the mixing water. The leaching is defined as the proportion of chromium that leaches out after water has flowed through a specimen for 24 hours relative to the entire quantity originally added. This rate dropped corresponding to a decreasing coefficient of permeability. A decrease of the permeability by approximately two powers of ten resulted in a drop in the leaching rate also by two powers of ten.

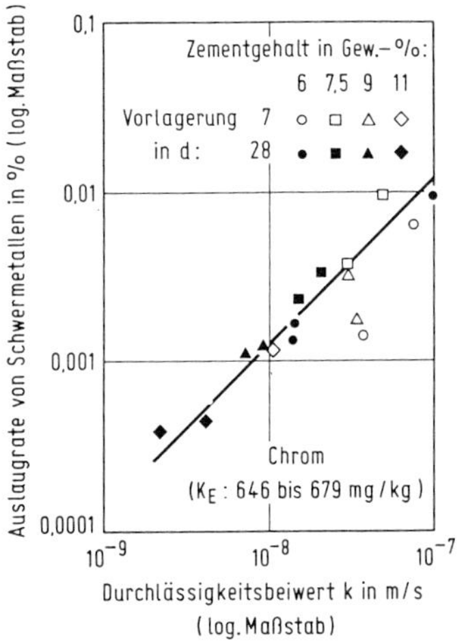

fig. 4: Decrease of leaching rate (chromium) depending on the permeability (k)- value of hydraulically bound sand

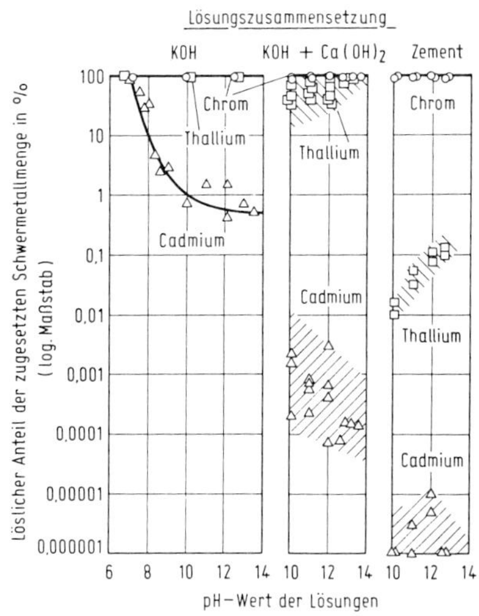

fig. 5: Decrease of the leaching rates of different heavy metals depending on the pH-value

The third essential influence is the chemical ambient conditions which particularly determine the solubility of some heavy metals. Fig. 5 shows the influence of the composition of the solution on the leaching rate of the elements chromium, thallium, and cadmium, which were added to the solution. Although the solubility of the chromium remained unchanged, the solubility of the cadmium receded in the cement paste up to the specification limit.

All three examples illustrate that a proper evaluation of the environmental compatibility is only possible with procedures that sufficiently take into consideration the type of use - bound or unbound - and the structure of the finished building material. A meaningful environmental test of bound substances can be carried out only with bound, hardened specimens. In this respect, the flow-through procedure described, for example, in [3,4] or an easier to handle so called "trough procedure" is more suitable. At present, experience is being gathered with the trough procedure, so that it might be described in a test specification in order to achieve uniform testing procedures for all test locations.

3. The Use of Hydraulically Bound Refuse Incineration Ashes in Practice

In a comprehensive research programm of the Research Institute of the German Cement Industry [1], the fundamental structural and ecological principles for using refuse incineration ashes in hydraulically bound building materials were established. The most essential parameters of the investigation were

- o determination of the binder content
  - compressive strength
  - frost testing
- o strength and deformation
- o durability according to different storage conditions
- o volume stability
- o environmental compatibility

Based on the experience gathered at the above-mentioned research institute refuse incineration ashes were tested for their actual use.

In a city street in Mannheim the originally planned 15 cm thick unbound gravel base was replaced by a hydraulically bound road base of the same thickness and made of refuse incineration ash. A structural suitability test was made in the research and development laboratories of the company Heidelberger Zement. In addition, orientating experiments on the leaching behaviour and the water permeability were carried out. A waterproofed hydraulic binder for bases of the strength class HT 35 according to DIN 18506 (Heidelberg Recycling Binder), which was specially developed to suit this purpose, was selected.

The grain distribution of the refuse incineration ash fulfilled the ZTVT requirements of a natural gravel mixture of 0/32 mm [1]. The water content of the ash delivered and used for the suitability test was 18 wt.%. Most of this water was absorbed by the material.

3.1 Suitability Tests

3.1.1 Structural Suitability

With the Proctor test, the optimal water content of the ash-cement mixture was determined to be 20 wt.%. In this case the Proctor density was 1.72 g/cm^3.

After 28 days, the compressive strength was determined on specimens with a diameter of 150 mm and a height of 125 mm according to the TP-HGT [2]. It increased nonlinearly with an increasing binder content of 5-9 wt.%. With higher compressive strength, the increase in strength is low because of the low grain strength of the refuse incineration ash. According to the test, the dry bulk densities of the mixture with 5 wt.% binders, however, were considerably lower than those of the other mixtures. As a consequence, the compressive strength of the mixture is too low. If the specimens had approximately the same bulk density, a binder content of 5 wt.% would be necessary in order to fulfill the requirements of the ZTVT [1] regarding compressive strength (7 N/mm^2). Previous investigations [4] showed that in most cases it is not the compressive strength which is important for the determination of a sufficient binder content of refuse incineration ashes, but the frost test. The intensified **frost-test procedures with 30 frost-thaw cycles, however, showed that a** sufficient frost resistance could be obtained with a binder content of at least 7 wt.%. It was therefore agreed that this amount would be used for the construction project.

3.1.2 Environmental Compatibility

For lack of another test instrument the environmental compatibility was investigated theoretically in pilot tests with consolidated compressive strength specimens, that is, in the same way as unbound incineration ash according to the standard DIN 18414 DEV S4 procedure.

Very small amounts of heavy metals were present in the starting material. With the addition of the hydraulic binder, it was possible to lower the total amount of solubles substances on the whole and to reduce considerably the soluble portions of sulphate, Na_2O, and K_2O. The permeability of the hydraulically bound specimens to water as an additional criterion for the leachability was very low with a k value of $1.2*10^{-9}$ m/s after 28 days. This corresponds to the requirements for sealing layers for refuse dumps. Thus the hydraulically bound refuse incineration ash should practically be impermeable to water.

3.2 Experience with the Test Section on an Urban Road

3.2.1 Placement of the Hydraulically Bound Base

The binder content of the mixture used at the construction site (7 wt.%) corresponded to approximately 120 kg/m^3. In accordance with the Proctor test the addition of water in the ready-mix plant was adjusted for constant semi-dry consistency, which was obtained. The placement was carried out in a single operation across the whole width of the road (approcimately 5.50 meters) with a standard asphalt road finisher. First, the hydraulically bound road base was compacted by the vibrating beam of the finisher. Next, it was further compacted two times by a rubber-wheel roller (fig. 6).

After running over it with a steel roller, no further compressive effect was observed. Afterwards, the surface of the completed base was found to be stable, compact, and free of cracks. In order to be able to observe the natural behavior of the road base produced in this manner, the hydraulically bound road base was not notched. According to the ZTVT, road bases are usually notched to control crack formation. A 10 cm thick bituminous base was applied two to three hours after the placement of the hydraulically bound road base. At this time, the solidity and the load-bearing capacity of the hydraulically bound road base had already reached such a level that trucks left no tire tracks (fig. 7). At the same time, the asphalt provided optimal protection against both drying out and frost effects.

fig. 6: Placement and compaction of the hydraulically bound base

3.2.2 Experiments and Results

During placement,samples were taken from every truck, and specimens were produced at the construction site (cylinder 15/12.5) for testing the compressive strength, the leachability, and the permeability. In addition to the specimens which were produced at the construction site, drill cores (ϕ 15 and 10 cm) were taken from the hardened road base after 28 days, to test compressive strength, leachability, and permeability. The compressive strength of the hydraulically bound specimens produced at the construction site came to an average of 14 N/mm^2 and was evidently higher than the requirements of the ZTVT [1] and the suitability test. This could be attributed to different storing conditions.

The leaching test, performed according to DEV S4 standards, demonstrated that the unbound refuse incineration ash in the field test contained somewhat lower amounts of pollutants than the material in the suitability test. This is why the pollutant-binding effect of the hydraulic binder can not be recognized as clearly in the environmental test of the specimens as in the suitability test. Regarding the leaching of pollutants, there are no significant differences between the drill cores and the specimens produced at the construction site. The k values of the specimens produced at the construction site, which were taken as a measurement for the permeability of the grain structure, were between $2*10^{-5}$ and $3*10^{-9}$ m/s. The drill cores of the finished base, however, showed with an average of $4.9*10^{-9}$ m/s a very constant and high degree of permeability. Presumably, a denser grain structure was obtained in the base than in the relatively small specimens, because it was compacted steadily and more intensively with the vibrating beam and the roller. Thus the base should be practically impermeable to water.

fig. 7: Surface of the HBB after ca. 3 hours

At present, the environmental compatibility and the compactness of the hydraulically bound road base are additionally tested with a field lysimeter. This method permits long-term assessment of leachability under natural conditions and most reflects what is found in practice. Moreover, in the range of the lysimeter - in contrast to the remaining area - the hydraulically bound road base was not covered by an asphalt coating, so that here more severe weathering effects are to be expected. The base has been in place for about one year without having shown evident changes.

4. Summary and Outlook

The previous experiments for using refuse incineration ash as mineral mixtures for hydraulically bound road bases can be summarized as follows:

1. Usually, the decisive criterion for properly determining the binder content is the adequate frost resistance of the hardened hydraulically bound road base. For sufficient frost resistance, a higher binder content may be necessary than the amount needed for attaining certain compressive strengths. Therefore testing of the frost resistance should always be included in the suitability test. The procedure must sufficiently allow one to ascertain the specific properties of the bases bound with secondary materials.

2. In order to obtain volume stability for hydraulically bound road bases composed of refuse incineration ashes, it is advantageous to store the residues for an extended period, before further processing. After the placement, it is also beneficial to cure the consolidated mixtures of the building materials with moisture for as long as possible. The use of the refuse incineration ash would be greatly facilitated, if coarser particles could be removed during the refinement.

3. Practically no soluble, environmentally-relevant pollutants are leached out of cement-bound refuse incineration ashes containing sufficient amounts of binder, if the grain structure becomes as impermeable to water as the layer for waste dumps (permeability value $k <= 10^{*-9}$ m/s).

4. The field test confirmed the results of the laboratory tests and of the suitability test: the applicability of refuse incineration ash as a hydraulically bound building material. Nearly two years after the placement of the hydraulically bound road base, no changes in the asphalt surface (cracks, spalls, vaultings, etc.) had been observed. The long-term behavior with respect to weather resistance, volume stability, and environmental compatibility will be observed further at the present test sections.

References

[1] Zusätzliche Technische Vorschriften und Richtlinien für Tragschichten im Straßenbau (ZTVT-StB 86), Ausgabe 1986. Der Bundesminister für Verkehr, Bonn 1986

[2] Technische Prüfvorschriften für hydraulisch gebundene Tragschichten (HGT) (TP HGT-StB 86), Ausgabe 1986. Hrg. Forschungsgesellschaft für Straßen- und Verkehrswesen, Köln 1986

[3] Sprung, S.; Rechenberg, W.: Einbindung von Schwermetallen in Abfallstoffen durch Verfestigung mit Zement. beton 38 (1988), H. 5, S. 193 - 198

[4] Schmidt, M.: Verwertung von Müllverbrennungsrückständen zur Herstellung zementgebundener Baustoffe. beton 38 (1988), H. 6, S. 238 - 254

Waste Materials in Construction.
J.J.J.R. Goumans, H.A. van der Sloot and Th.G. Aalbers (Editors)

INCINERATION SLAG IN ROAD CONSTRUCTIONS
example of an application

by Mank, J.A.M.[1] ; Brulot, J.[2] ; Janssen van de Laak, W.H.[1]

[1]Road and Hydraulic Engineering Division of the Min. of Transport and Public Works
[2]Regional office South Holland of the Ministry of Transport and Public Works

ABSTRACT
Application of incineration slag in embankments for roadbuilding meets uncertainties, both in material-technical and constructional and environmental aspects. The latter led to a large-scale-test in Highway 15 near Rotterdam. The results are given below.

1. INTRODUCTION

In Holland two important reasons are recognised to stimulate the use of secundary materials in roadbuilding and hydraulic engineering [1]. The first is that a number of these materials can replace primary minerals that become more and more scarce in this country. The second is that the dumping of industrial residues and waste can be prevented.
Incineration slag, produced when domestic waste is incinerated, belongs to that category. One of the possible solutions is to apply incineration slag in road-embankments in stead of the usual sand [2]. Experience existed moderately, but the possible environmental and hygienic consequences proved sufficient reason for the Rijkswaterstaat to realise a large-scale-test. More insight in the environmental risk and relevant engineering parameters may contribute to a larger use of this material.

2. INCINERATION SLAG AS BUILDING MATERIAL

Much industrial and domestic waste is dumped or incinerated in Holland. Globally is it separated at its sources, between heavily polluted waste (chemical waste e.g.) and lesser polluted waste (e.g. domestic). The former should be dumped in an acceptable manner or incinerated in specialized furnaces; the latter, when incinerated, is the basis for AVI-slag (the incineration slag referred to in this paper) [3].

AVI-slag is, in fact, the solid residual result of the incineration of domestic waste and similar light industrial waste.
The fly-ash resulting from this process is polluted so heavily that they have to be separated, and they may not be added to the AVI-slag in a later stage.

The market supplies annually 700,000 tons at this moment, increasing to over 2,000,000 tons at the beginning of the new century.

AVI-slag contains high concentrations of heavy metals, varying according to the nature of the waste to be incinerated. The spreading of this polluting matter through lixiviation is one of the greatest environmental risks.

In order to enable application of this slag in roadbuilding the Rijkswaterstaat determined the following (temporary) conditions:

- to prevent pollution of the soil, AVI-slag may not be applied in critial areas, such as soil-protection-areas and waterexploitation areas. Furthermore AVI-slag must be applied in such a way that they can be removed at any later stage; top and sides of the layer must be covered with a watertight layer; the AVI-slag must remain permanently at a level of 50 cms above the average highest groundwaterlevel;
- limitations are set to the composition of the slag (admissible percentage of iron-containing elements, non-incinerated parts and compostable parts);
- concerning the lixiviation certain conditions must be fulfilled, such as grain sizes and crushing factor.
- the actual use is subject to conditions regarding the compaction of the material in the construction.

The first two conditions are mainly of an environmental nature, the latter two concern the engineering. The environmental conditions meet the requirements laid down in the interim-policy of the administration, pending the Decree on Building Materials.

3. THE TESTPROJECT

3.1. General aspects

The application of AVI-slag in large-scale, constructive embankments is a fairly new field. Both in Holland and abroad the experience is modest. The Rijkswaterstaat made an inventory of the available knowledge and experience and decided to realise a testproject. This proved to be possible in 1988 in Highway 15 near Rotterdam.

3.2. Situation

Highway 15 connects the Rotterdam industrial and harbour area to the Ruhr in Germany (see figure 1). Industrial development in the western Rotterdam area caused serious circulationproblems at the end of the existing highway. As a solution, the Rijkswaterstaat decided to extend highway 15 to the west.

The first phase of the extension is a 4.5 km trajectory. Here a four-lane highway was planned. The complex infrastructure with many viaducts, and the relatively short distances between these viaducts, led to the decision to design the whole highway elevated.

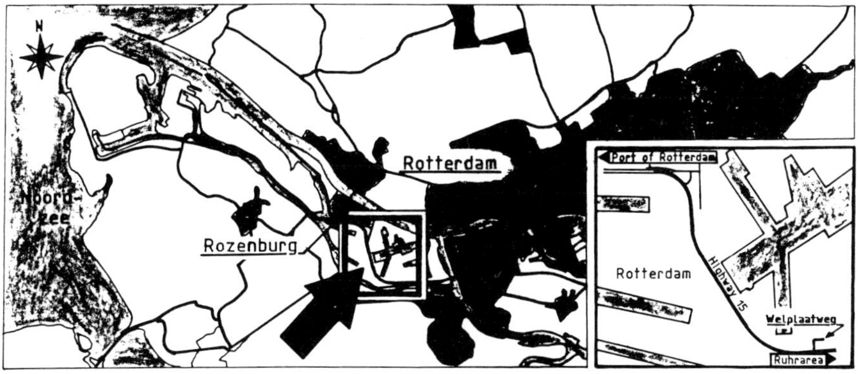

Fig. 1. Situation

At first, the embankment was to be made of sand. The contractor however suggested to replace sand by 400,000 tons of incineration slag, mainly available from the nearby furnaces. The Rijkswaterstaat agreed, considering the future policy and the industrial location. It was the first time for the Rijkswaterstaat to apply AVI-slag in such quantities in a roadconstruction.

3.3. Engineering aspects

The engineering design of the embankment with AVI-slag was highly influenced by environmental demands, i.e. the covering of top and sides with a watertight layer, so that rainwater does not perpetrate the slag and, in the long run, pollutes the underground and groundwater.

In agreement with the licensing authority it was decided to use for this purpose a mixture of natural materials, i.e. sand and bentonite (a swelling clay). To prevent frequent disturbance and perforations of the watertight layer by transportation activities during the work, or when installing road-furniture, a 1 meter layer of sand over the watertight layer was provided for. The design further reckons that the bottom of the body of AVI-slag, after setting of the subsoil, will remain at least 0.5 meter above the highest groundwaterlevel.

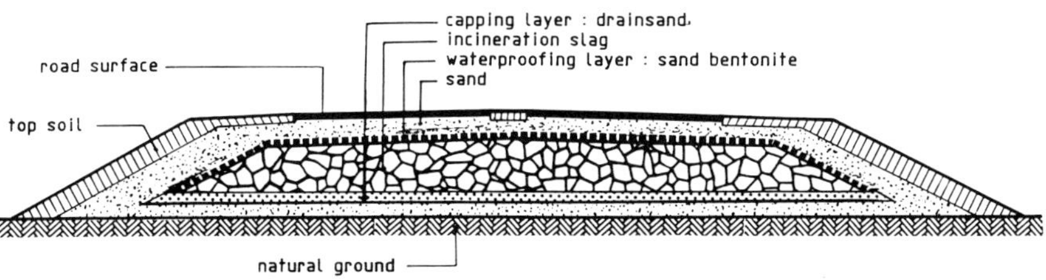

Fig. 2. The engineering design of the embankment

As to the permeability of the layer of sand/bentonite a limit of 10^{-10} meter per second is required by the administration.

3.4. <u>Planning the test</u>

Given the preferred construction a researchplan was made. Targets are:

- to obtain and increase general knowledge of AVI-slag and sand-bentonite;
- to obtain insight in engineering and environmental consequences as a result of the application of these materials in embankments.

Research would take place in the field and in laboratory-conditions.

3.4.1. <u>Civil engineering research</u>

Aspects requiring attention were:

- in the laboratory:
    * geotechnical properties of AVI-slag. The angle of friction and cohesion of the slag is determined through large-scale triaxial tests, to be used as input in stability-calculations. Compression-constants are determined by way of large-scale compression tests. This is to obtain knowledge of the behaviour in setting. Both tests use the material graded as used in the project
    * frictional properties of the sand-bentonite mixture by triaxial testing. Bentonite is a natural clayproduct, which came to be by sedimenting vulcanic ashes. The primary element is found in many countries, but not in Holland. It is marketed in powdery form. Its character is to swell greatly when water is added. In a correct mixture with sand it forms a watertight sealant.
    * permeability of sand-bentonite for water, also in the long run.

- in the field:
    * how AVI-slag sets in the long run;
    * the compaction of AVI-slag;
    * the quality of the watertight sealant; its permeability. With experimental tests, such as the mobile **ring-infiltration-meters** and the permanently built-in **lysimeter** the permeability of the sealant can be determined.
    * the quality of the sealing layer: it must be checked whether the added bentonite is sufficiently and homogeneously mixed in the actual construction; determination of the quantity through the methylene-blue test.

3.4.2. Environmental research

Special attention in this test-project will go to the environmental consequences of the application of AVI-slag in an embankment.

To follow this, synthetic sheets were placed under the layer of AVI-slag (testsites 1 and 2) as well as directly under the sand-bentonite layer (testsite 3). If water will percolate it will be caught. The length of the testsites is about 10 meter each.

The contents of pollution in the AVI-slag itself, the degree of lixiviation possible, and long-term developments in the contents of polluting elements in water samples, will be determined.

Important information is expected to result from this, concerning:

- quality and quantity of water emitted by the AVI-slag during the construction phase, before the sealant is applied, and which may result from:
    * compaction of AVI-slag, expelling present water; in order to determine this, moisturecontents are defined per layer carried into the construction, both before and after compaction;
    * intrusion of rain into the AVI-slag between the depositing and the sealing;
- effectivity of the sealant of sand-bentonite; how much water percolates through sealant and slag can be defined by a frequent sampling of testsite 3, where sheets are applied directly beneath the sealant;
- quality and quantity of water, coming from the layer of slag, during the actual use of the construction, i.e. after the sealant has been applied. This information can be obtained by registring during a long period of time of the quantity of water coming from testsites 1 and 2, where sheets are applied under the layer of slag.

Periodic sampling and analysis of this water, during a period of 3 years, leads to definition of the degree of pollution. Sampling frequency depends on the quantity of expelled water. Indicative, concentrations found will be checked against limitvalues as published in the Soil Protection Standard.

The research results have to be processed. That implies the necessity to have information on the quality and quantity of rain over the testing period. Near the testsite automatic rainmeters were installed for this purpose.

Slightly remote from the test sites, near to the road, sampling tubes were installed. Before the job was started, the begin situation was defined and reported on. From these tubes groundwatersamples will be extracted periodically, whereas samples of the soil in the near vicinity are taken as well.

It is to be analyzed whether possible diffusion of polluted percolation water outside the roadconstruction will affect the quality of soil and groundwater.
The concentrations to be found will again be checked against limitvalues as published in the Soil Protection Standard. This will eventually lead to answering the question whether the application of incineration slag according to the actual designs is acceptable.

In the laboratory the lixiviational behaviour and the actual composition of the slag, used at the testsite, will be analyzed. Three different lixiviation tests will be executed:
* maximal lixiviation at pH 7 and pH 4
* cascade test and
* column test, according to the Dutch standard NEN 2508

The samples will be analyzed to their contents of heavy metals cadmium, chromium, copper, molybdeen, lead, nickel, antimonium, zinc, mercury and arsenic. Furthermore on barium, fluorine, sulphate, chlorine and sulphide and the acidity (pH) and electrical conductivity.

4. DESIGNING TESTSITES

4.1. Where to situate testsites?

When choosing locations for the testsites, the expected settings were taken into consideration. A disturbance of the body of the road, a possible depression in the middle, makes it harder to interpret research results. Starting point of the calculations was that rain would flow sideways off the sealant. Therefore testsites are situated where, as a result of earlier **preliminary load** the primary setting was already effected. An added advantage is the lesser risk of damages to sampling systems.

4.2. Design

Testsite 1 and 2 under the core of AVI-slag

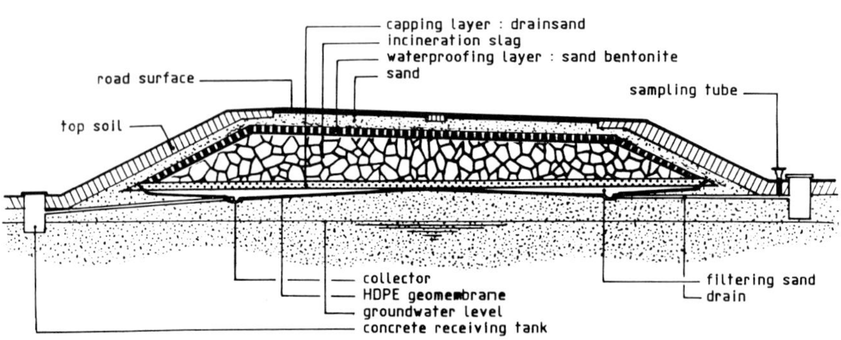

fig. 3. Testsites 1 and 2

After ample considerations, the choice was in favour of a system that buffers the percolate temporarily in the embankment construction itself (see figure 3).
Primarily the percolate is received from the body of slag in foil troughs filled with gravel. The foil, a geomembrane of 2mm HDPE, meets at its sides the sand-bentonite sealing layer.
The percolate is guided sideways to drains at the foil. These drains are made of HDPE-drain and are filled with filtering sand. To improve drainage, a slight difference in level is made. The draintubing (8 cm HDPE) transports the percolate to concrete containers, next to the slope.
The foil and the drains form together a temporary buffer. To guarantee this, valves were mounted at the end of the draintubes, inside of the concrete receiving tanks. Periodically the system is drained.

To check the level of the buffered percolate and to have an indication of the quantity to be drained, each testsite has a sampling tube connected to the drain. The expection was that during the first phase, when the sealant is not yet applied, the flow will be app. 4000 liters per hour. Once sealed, the flow will decrease to app. 0.1 liter per hour. The system is large enough to contain the water, received in each sampling period. The frequency of draining the system during the open period will be once every one or two days, decreasing to once every 2 or 3 months in the contained phase.

<u>Testsite 3, directly under the sealant</u>
This design is basically not different from the other two sites (see figure 4).

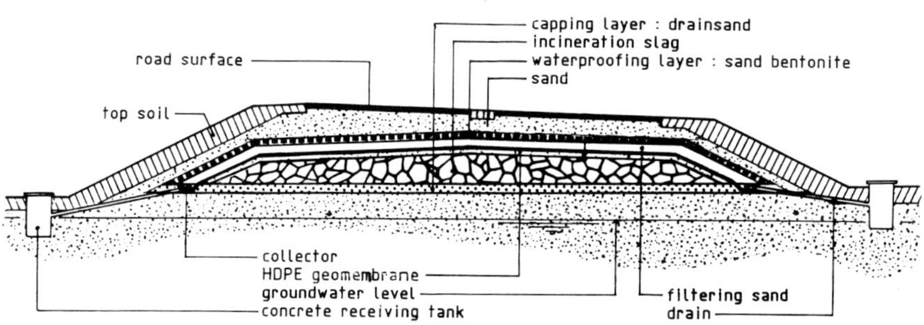

Fig. 4. Testsite 3

Here too, a foil is provided, this time directly under the layer of sand-bentonite; a (smaller) buffering capacity for the infiltrated water; and a draining system to concrete receiving tanks on either side of the slope, where the buffer can be drained periodically.

The shape however is different: where at sites 1 and 2 a trough construction is used, at site 3 a reverse trough construction was needed.

A general condition was that the situation at the testsites should differ as little as possible from the situation outside the testsites. Therefore the layer of draining sand, applied under the AVI-slag in general, is also applied at sites 1 and 2.
On the other hand, drainage of the percolate towards the drains had to be guaranteed, which called for a layer of filtering gravel under the draining sand. The HDPE-drains had to be provided with a poly-propylene jacket to prevent clogging in the long run; and the draincanals were filled with filtering sand for the same reason.

During the drainage, periodic partial samples are taken from the drainwater of the testsites 1 and 2. This is in favour of a qualitative analysis of the percolate. These partial samples allow for a general sample to be analyzed.

4.3. Sampling

The sampling provisions were made for longterm sampling. We think that sampling should take place over a period of 20 to 30 years. During the first three years, sampling will have to be intensive. With the advancement of time, the frequency will decrease noticeably. Sampling of the "open" phase started in september 1990, sampling of the "covered" phase began in october 1990. Sampling of testsite 3, directly from under the sealant, started early 1991.

5. RESULTS

Although the testing sites were completed only recently, and reports on the laboratory research are only partly finished, we can report the following on the actual constructing activities:

- application of AVI-slag did not lead to problems: the compaction that is measured complies with standards and does not decrease as the construction of the actual embankment progresses
- so far only a marginal setting of AVI-slag could be found;
- the temperature of the AVI-slag decreases with time;
- the mix-in-plant method led to a homogeneous product as to the sand-bentonite mix

As far as the environment is concerned, only modest sampling results are available. As far as the flow of percolate is concerned, the information can be given in figure 5.
It shows the cumulative quantities of percolate, as drained. It also shows the moment at which the layer of slag was covered with the sealant.

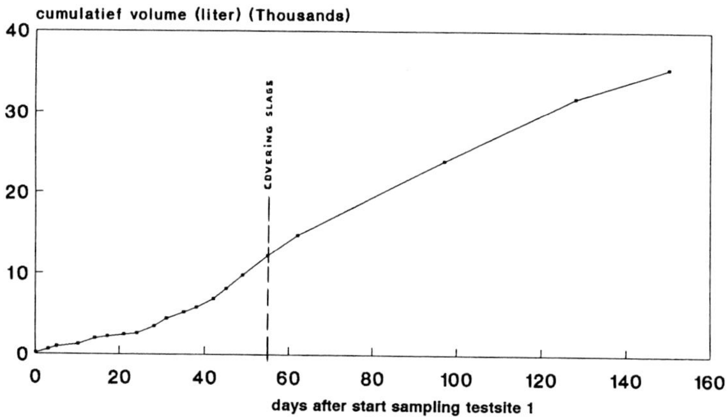

Fig. 5. The cumulative quantity of percolate after start sampling testsite 1.

The first analyses concerning the quality of percolated water have been made. Before publishing their outcome, and expressing views on the contents of polluting matter, we need the as yet not available information about the quality of AVI-slag, draining sand en rainwater.

It is expected that most results will become available halfway 1991. A definite judgement whether the use of AVI-slag in constructive road-embankments is acceptable can be expected in or about 1994.

6. LITERATURE

1 "Gegrond ontgronden", Landelijke beleidsnota voor de oppervlaktedelfstoffenvoorziening voor de lange termijn. Ministerie van Verkeer en Waterstaat, 's Gravenhage, 1987

2 "Resten zijn geen afval (meer): Afvalverbrandingsslakken", publikatie 15, Stichting Centrum voor Regelgeving en Onderzoek in de Grond- , Water- en Wegenbouw en de Verkeerstechniek (C.R.O.W.), oktober 1988

3 "Resten zijn geen afval (meer): Bijzondere ophoogmaterialen", publikatie 16, Stichting Centrum voor Regelgeving en Onderzoek in de Grond- , Water- en Wegenbouw en de Verkeerstechniek (C.R.O.W.), oktober 1988

Waste Materials in Construction.
J.J.J.R. Goumans, H.A. van der Sloot and Th.G. Aalbers (Editors)

UTILIZATION AND DISPOSAL OF SOLIDIFIED AND STABILIZED CONTAMINATED SOILS

Margareta Wahlström[1], Bob Talling[2], Jaakko Paatero[2], Esa Mäkelä[1], and Martti Keppo[3]

[1] Technical Research Centre of Finland, Chemical Laboratory, P.O.Box 204, SF 02150 Espoo
[2] Partek Corporation, Cement Division R&D, SF 21600 Parainen
[3] Partek Conrete Industry Corp., P.O.Box 61, SF 00501 Helsinki

SUMMARY

In this project optimal binding systems have been developed for soils contaminated with wood preserving chemicals (As, Cr, Cu) and lead respectively. Properties like compression strength, frost resistance and swelling have been studied. Environmental impacts have been evaluated by using a Dutch tank leaching test. This summer two roller compacted concrete pavements, about 500 m^3 each, will be laid using contaminated soil as aggregate. Immobilized soils contaminated with wood preserving chemicals and lead can be used in construction of roads or storage areas on landfills. In case of slightly contaminated soil, the immobilized soil can also be placed in urban areas, e.g. in constructions of parking places.

1. Introduction

In Finland the only available remedial action techniques of contaminated soils are microbiological treatment or transportation of soils to the hazardous waste treatment plant Ekokem Ltd for incineration or for disposal on a landfill designed for hazardous wastes. Slightly contaminated soils can also be disposed on municipal landfills. However, the treatment techniques mentioned above are not always suitable or the treatment costs are too high especially if the amount of contaminated soil is huge.

One treatment technique for contaminated soils is immobilization with cement or other binding agents so that the leaching of the toxic substances is reduced. The benefits of immobilization technique compared to other techniques are the simplicity of the method, the availability of equipment and the short treating time.

The aim of the co-operation project of the Technical Research Centre of Finland and the Finnish company Partek Corporation is to develop an immobilization technique for treatment of contaminated soils and to study the leaching behaviour of immobilizated soil and to give recommendation for utilization or disposal of immobilized soils. The study will be finished at the end of this year.

2. Materials studied

Soil samples contaminated with arsenic, chromium and copper were taken from a site of a wood preserving plant (which was in operation the years 1953-1990) near Turku, Finland. The samples consisted of humus, saw dust, sand and clay. The total amount of soil to be treated is about 1000 m^3.

Soil samples contaminated with lead were taken from the site of an old lead smeltery (1929-1984) in Vantaa, Finland. Slag, which contained about 9 % lead, was found almost all over the site. The slag layers had been covered with filling soil (sand, clay), which partly due to air emissions from the smeltery contained on average 1-2 % lead. Beneath the slag there was a thin layer of peat, which contained up to 7 % lead, below that sand and clay. The total amount of soil to be treated is about 14000 m^3.

From both sites highly contaminated soil samples were chosen for the solidification tests (Table 1).

Table 1. Contaminated soils stabilized.

Soil	Concentrations of contaminants
Wood preserving plant:	
soil 1	As: 2400 mg/kg, Cr: 1600 mg/kg, Cu: 1400 mg/kg
soil 2	As: 5200 , Cr: 3500 , Cu: 2500
soil 3	As: 7400 , Cr: 4000 , Cu: 4400
soil 4	As: 9300 , Cr: 7000 , Cu: 7300
Lead Smeltery:	
top soil	Pb: 43000 mg/kg
slag + soil	Pb: 240000
peat	Pb: 24000

3. Design of optimal binding system

The first step is to characterize the soils i.e. grading, humus, clay, moisture, water absorption, easily soluble matters. etc. The characterization will also give information of wheather additives (chemicals) are needed to minimize solubility of the contaminants or to improve the hardening properties of the binder system.

After the characterization a few alternative binder systems are chosen for the small scale tests. The composition of the binder systems are optimized from Portland cement, pulverized fly ash, blast furnace slag, silica etc. In less hazardous cases flue gas desulfurization ashes and bottom ashes can be used as well as steel slag and coloured metal slags. The mixes are from 2 to 50 l and the soil is crushed or sieved to < 16 mm. Depending on the type of soil and amount of contaminants the ratio binder to soil is from 1:2 to 1:8. Typically the ratio is about 1:4.

After choosing the optimal binder system the concrete studies are scaled up to about 150 l batches. In this case the maximum soil aggregate size will be about 50 mm. The fresh concrete is tested for workability by a shear-compaction device, Intensive Compaction Tester (IC-Tester) in case of a stiff concrete.

The soils contain in general much fines and in some cases the workability can be improved by using additional coarse aggregate. The amounts of cement and water is optimized to make alternative concrete placement possible. Concretes with extreme stiffness to self-levelling compositions are prepared.

All the concretes are tested for compressive strength up to at least one year. The drying and cracking behaviour as well as stability and swelling in water is checked. The minimum strength levels are about 5 MPa at 28d. In many cases the ultimate strengths will be as high as 50 MPa.

In applications where the concrete is subject to weathering freeze-thaw test are made. After 100 cycles from +20 to -20 centigrades the scaling is to be less than 1 % volume.

4. Full scale solidification tests

Homogenization of the contaminated soil is extremely important. Homogenization is made by using high-efficient mixing and/or sieving and crushing techniques.

The full scale test are made on site. The concrete is mixed in a movable twin shaft continuous mixing plant (ARAN ASR-250X). In a field test the minimum amount of treated soil is in the range of 300 m^3. The concrete is normally laid as roller compacted concrete. Quality control is made on site by using the IC-tester for controlling optimum compaction properties. In some cases pumpable concrete is used, especially for encapsulation of large blocks of contaminants like lead slag.

5. Environmental impacts

5.1. Leaching tests

The environmental impacts of stabilized and solidified soils has been evaluated using the Dutch tank leaching test and the Dutch batch leaching test for the determination of the maximum leachable quantity (Draft NVN 5432, 1990). In the tank leaching test a specimen with dimensions over 40 mmm and a minimum age of 28 days was immersed in water (here pH of the water was adjusted to 4 using sulphuric acid) and at certain time intervals the solution was renewed and analyzed. The amounts of contaminated soils in the tested specimens were 70-80 % for soils from the wood preserving plant and 50-80 % for the soils from the lead smeltery.

This far leaching tests have been conducted with laboratory specimen. During the autumn 1991 the same tests will be conducted with bigger samples of different storage age from the field test constructions.

5.2. Results and discussion

The results of the tank leaching tests are shown in Figures 1-4. The cumulative flux in terms of leached amount per exposed surface area of the specimen is plotted against contact time on a log-log-scale.

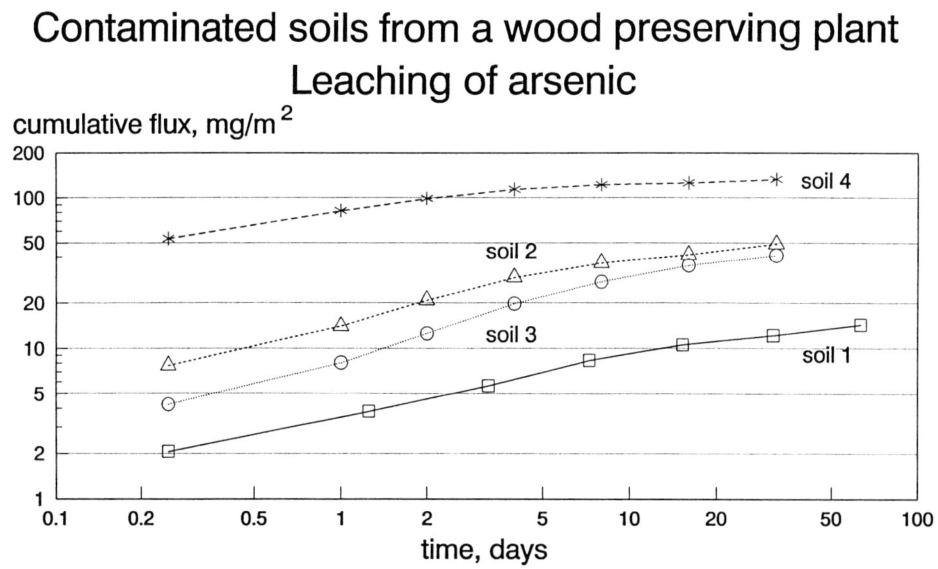

Fig 1. Leaching of arsenic from a solidified and stabilized contaminated soil.

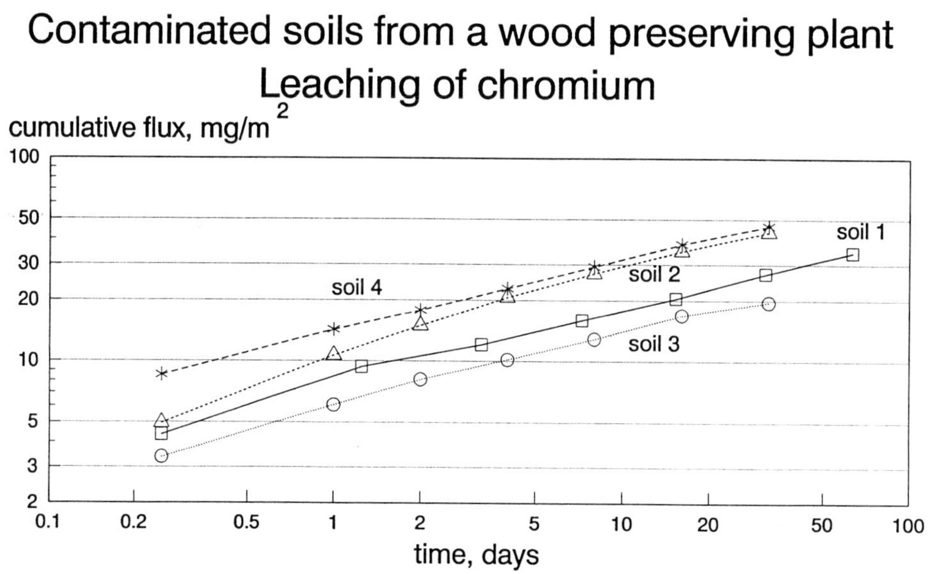

Fig 2. Leaching of chromium from a solidified and stabilized contaminated soil.

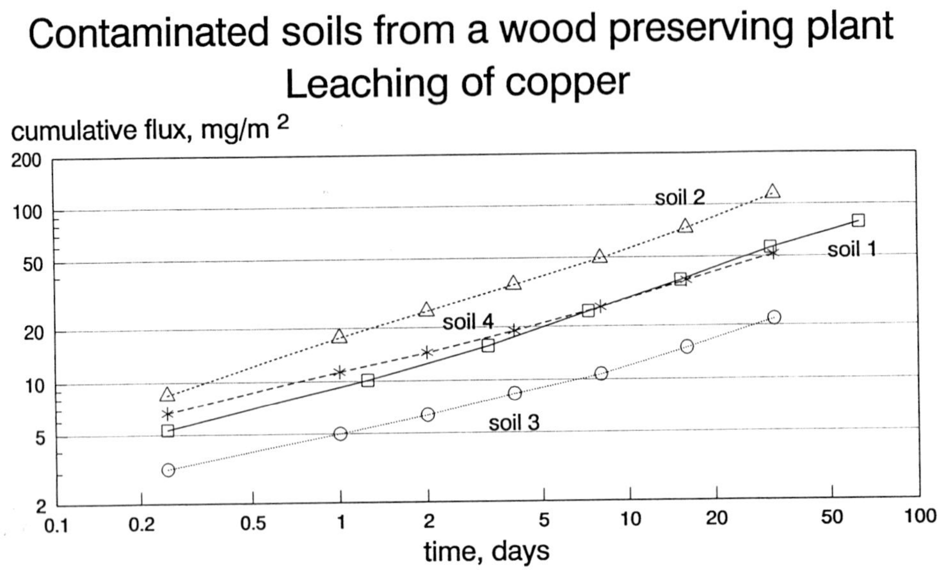

Fig 3. Leaching of copper from a solidified and stabilized contaminated soil.

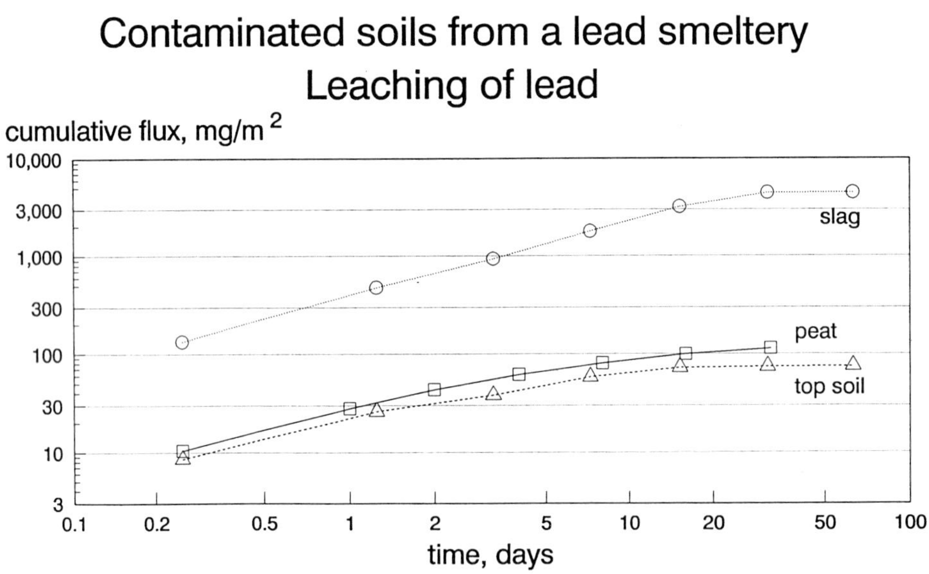

Fig 4. Leaching of lead from a solidified and stabilized contaminated soil.

5.2.1. Soils contaminated with wood preserving chemicals

The slopes of the lines in the plots are indicating that there were an initial wash-off of contaminants (slopes < 0,4) before the leaching was controlled by diffusion (slopes between 0,4-0,6). At the end of the leaching test there was a depletion of contaminants available for leaching (slopes < 0,4). The calculated effective diffusion coefficients for arsenic, chromium and copper were about 10^{-11} m^2/s, which means that the contaminants have intermediate mobility.

According to the results the leached amount of arsenic is related to the total concentration in the sample. The total chromium and copper concentrations didn't have a significant effect on the leached amount.

The solutions from the tank leaching tests were alkaline or neutral due to the calcium compounds in the binding agents. In the last leaching period in the tank leaching test the pH-values dropped from over 11 to 8. This may be due to the wash-off of alkaline components from the surface of the stabilized specimen.

5.2.2. Soils contaminated with lead

The leaching mechanisms of all samples studied are in the beginning diffusion controlled. At the end of the leaching period there was a depletion of lead available for leaching. The calculated effective diffusion coefficients of lead were less than 10^{-12} m^2/s, which means that the contaminant has very low mobility.

The leached amount of lead was related to the total concentration in the sample. The pH-values in the last leachate from the tank leaching test showed a drop from 11 to 8.

5.3. Evaluations of the test results

The results from the leaching test can be compared to airborne emissions in wet depositions, background concentrations in different water environments and to some reference materials normally accepted as construction materials.

For example in the tank leaching test which lasted 64 days the cumulative leached amounts calculated per the exposed surface area for two of the soils studied were the following:

- soil sample 1 from the wood preserving plant: 34 mg Cr/m^2, 14 mg As/m^2 and 73 mg Cu/m^2.
- top soil sample 1 from the lead smeltery: 75 mg Pb/m^2

The leaching from a soil construction would in the same time period surely be much smaller than this, due to lower temperature, and due to a shorter water contact, if the stabilized soil would be situated over the ground water level. The leaching for the first year can roughly be extrapolated from the plots, assuming that the leaching is diffusion controlled. A rough extrapolation gives following amounts for the first year: As: 50 mg/m^2, Cr: 100 mg/m^2, Cu: 200 mg/m^2 and Pb: 200 mg/m^2.

The airborne emissions in wet depositions were in the 80's the following: As: 0,3 $mg/m^2/a$, Cr: 0,1 $mg/m^2/a$, Cu: 1,3 $mg/m^2/a$ and Pb: 9,3 $mg/m^2/a$. This means, that the emissions of the stabilized soils contaminated with wood preserving chemicals and with lead are about 200-1000 and respectively 20 times greater than the airborne emissions.

Assuming that the net rainfall is about 300 mm/year, the theoretical average concentration of the leachate water from a soil construction can be calculated. The average concentrations in the leachate from the above soil types would be the following: As: 0,2 mg/l, Cr 0,3 mg/l, Cu 0,7 mg/l and Pb 0,7 mg/l. These concentrations can be compared to different background concentrations and toxicity values. For example the Canadian guidelines for the protection of aquatic life in fresh water are the following: As: 0,05 mg/l, Cr 0,002 mg/l, Cu 0,002 mg/l and Pb 0,001 mg/l. This means that the leachate from a stabilized soil has to be diluted about 500 times to have no adverse effect on the aquatic life. The quality of leachates from several Finnish municipal landfills have been analyzed and the median values were following: As: $<$0,006 mg/l, Cr: 0,02 mg/l, Cu: 0,02 mg/l and Pb: 0,003 mg/l. This means that the leachate from stabilized soil contaminated with wood preserving chemicals has to be diluted about 30 times and the stabilized soil contaminated with lead about 200 times not to differ from the mean leachate quality.

The results from the tank leaching tests can also be compared to the corresponding results from some reference materials. This study is still under work. According to preliminary results the leached amount from red and white bricks were around 1-3 mg Cr/m^2, for a piece of impregnated wood over 500 mg Cr/m^2 and over 200 mg As/m^2 and for a copper plate over 1000 mg Cu/m^2. These values are of course not acceptable emissions, but can be used as indicative values.

Laboratory tests give results for leaching when the immobilized contaminated soil is in contact with water all the time. However, in practical applications the immobilized soil material is practically taken in no contact with water. In such a case the diffusion phenomena can occur to an extremely low extent, only. Therefore, it is worth to observe that the laboratory tests and the above given comparative calculations only indiciate which kind of environmental impact can exist if the immobilized contaminated soils are of some unexpected reasons brought to a continuous water contact.

6. Conclusions

Stabilized soils will produce leachates with small concentrations of contaminants for a very long time. The leachate from a deposit can affect the ground water quality in the surroundings of the deposit. The concentrations are related to the optimal binding systems and the total concentrations of the contaminants in the soil.

The environmental impacts from a stabilized product has to be studied from case to case. This study should include leaching tests and also stability tests. In the Nordic countries it is also necessary to study the frost resistance.

The laboratory studies of soils contaminated by arsenic, chromium, copper and lead indicate that the stabilized soils can be used in the construction of roads or storage areas on landfills. The leached amounts from treated soils would have a small effect on the quality of the leachate from a landfill. In case of slightly contaminated soils, it is possible to construct parking places in urban areas, which are already polluted by traffic and industrial activities. However, results from the full scale tests are needed for the evaluation of environmental impacts in other surroundings than landfills.

References:

1. Ross, H., Monitoring of heavy metals in precipitation, National Swedish Environmental Protection Board, Report 3230, 46 pp. (in Swedish)

2. Results of air quality at background stations, Reports from the years 1980-1989, Finnish Meteorological Institute - Air Quality Department.

3. Assmuth, T., Poutanen, H., Strandberg, T., Melanen, M., Penttilä, S. & Kalevi, K., Environmental impacts from hazardous wastes in landfills, National Board of Waters and Environment, Publications of the Water and Environment Administration - Series A 67, 211 pp. (in Finnish)

4. Chemicals hazardous to the environment, Swedish National Chemicals Inspectorate, Report 10/89, 303 pp. (in Swedish)

5. Draft NVN 5432, Determination of the maximum leachable quantity and the emission of potentially hazardous components from construction materials, monolitic waste forms and stabilized waste materials, October 1989.

Waste Materials in Construction.
J.J.J.R. Goumans, H.A. van der Sloot and Th.G. Aalbers (Editors)

UTILIZATION OF INCINERATOR BOTTOM ASH - LEGAL, ENVIRONMENTAL AND ENGINEERING ASPECTS

J. Hartlén[1] and T. Lundgren[2]

[1] Swedish Geotechnical Institute, S-581 01 Linköping (Sweden)

[2] Terratema AB, FOA-huset, S-583 30 Linköping (Sweden)

SUMMARY
The legal, environmental and engineering aspects of utilizing incinerator bottom ash have been studied in a comprehensive project. The study shows that several laws may be applicable in Sweden when utilizing a residue. Present legislation is not geared to the new situation where such utilization is concerned.

Based on measured chemical and physical properties, a list of important parameters is presented and discussed in the paper. Impact on the environment is evaluated and related to other sources of environmental impact and the vulnerability of the environment. The conclusion is that graded bottom ash can be utilized, but with some limitations. Finally, the paper discusses how to regulate the use of incinerator bottom ash.

1. BACKGROUND

For several years now, waste management in Sweden has been aimed at increased reclamation and recycling. It is only natural that the ambition of plant owners is to recycle the residual products of incineration as well. In regard to the use of such products as high-grade fill and road-building materials, other reasons also speak in favour of such usage, chief among them being the increasingly shortage of natural materials like gravel, sand and crushed rock, as well as the need to spare these natural resources for reasons of nature conservation or as ground water reservoirs.

The consumption of natural materials in Sweden for roads and fill is at present about 90 million tonnes per year, mainly obtained from about 7,000 sources scattered all over the country. The potential for producing graded slag at the country's existing waste treatment plants corresponds to about 0.3 million tonnes per year. Concurrently, about 200,000 m^3 of landfill space would be saved annually.

Two fundamental conditions for utilizing slag are that the technical results of its use are satisfactory and that its effect on the environment can be accepted. It must be possible to describe and quantify its effect on the environment in absolute terms and in relation to the alternatives. From an environmental viewpoint, it is primarily the leaching of metals which must be taken into consideration. The leaching of salts may also have important local consequences. The problems of carrying out such an assessment arise from the fact that leaching is difficult to quantify and that general reference values are either unavailable, of doubtful accuracy or unrepresentative of the cases in question.

A major project has been carried out from 1987 to 1991 in order to determine the physical and chemical properties of incinerator bottom ash and coal bottom ash from grate firing and, on the basis of the results obtained, to define rules for facilitating official approval of re-use (Lundgren & Hartlén, 1991). The project included field and laboratory tests, as well as an evaluation of the environmental impact and current legal restrictions and/or prospects.

2. ENVIRONMENTAL LEGISLATION

The use of residual products such as slag and bottom ash from the combustion of different solid fuels for the construction of roads, for example, must be taken into consideration from the viewpoint of environmental protection. This means that the Environment Protection Act (ML) is applicable. In Sweden, however, it is not the utilization of the residual product in itself that requires a licence, but the use of "real property" in a specified manner when residual products constitute an element. This also applies to the controlled tipping of the residual product, described as "deposition of solid waste" in the Environment Protection Act and judged in accordance with this Act.

Uncertainty has existed in regard to whether or not a licence is obligatory under the Environment Protection Act for the utilization of slag and bottom ash. This has given rise to the need for a set of rules defining which products may be used, as well as where and how they may be used. With reference to such a set of rules, general licences can be granted in certain regions or for certain incineration plants.

3. ASH PROPERTIES

One requirement on the utilization of bottom ash will be that the ash is graded by using screens as well as drum magnets. After grading, about 70% can be regarded as a suitable substitute for coarse aggregate. About one-third of the remaining 30% from the grading operation consists of coarse scrap that can be sold and the other two-thirds will usually have to be disposed of in some way.

The constituents of the graded bottom ash have been identified (Jacobsson, 1989) and are shown in Table 1. The content is shown for different grain size intervals.

TABLE 1
Main constituents of graded bottom ash used in Malmö (Jacobsson, 1989).

	FRESH ASH FRACTION			AGED ASH FRACTION		
	5.6-8 mm	8-11.2 mm WEIGHT	11.2-16 mm	5.6-8 mm	8-11.2 mm WEIGHT	11.2-16 mm
Magnetic	-	-	-	-	-	-
Non-magnetic slag	45.0	55.6	50.6	43.4	37.9	28.5
Glass	44.9	40.0	35.9	49.0	49.7	56.6
Ceramic mtrl	1.3	2.4	3.4	2.3	3.9	5.0
Stone	8.7	2.0	10.1	5.3	8.5	9.9
Paper	0.1	-	-	-	-	-

As can be seen from the table, about 40% is vitreous material. It should also be noted that magnetic separation is extremely efficient as no magnetic materials at all were detected.

"Aged ash" is ash which has been kept in storage for a period of time before utilization. Experience in Europe is that such storage results in a more stable product (Hartlén, 1988). It is believed that oxidation of iron and other reactions take place during storage (Aubrey et al, 1986). Experience in Germany is that temporary storage prevents future swelling of the ash.

The water content of fresh ash (about one or two months old) before laying was about 23% and of aged ash about 16%. Water content varies widely, however. The lower water content of aged ash makes it easier to compact. Using a heavy vibrating roller (about 10 tonnes), a degree of compaction better than 90% modified Proctor was achieved (Hartlén & Rogbeck, 1989).

Graded bottom ash has been used on a trial basis in test roads. The results are so far promising and indicate that the properties when bottom ash is used are approximately the same as when natural aggregates are used. Leaching properties have been studied in the laboratory by means of sequential batch leaching and diffusion leaching tests.

4. ENVIRONMENTAL IMPACT

The leaching process in a fill of bottom ash above ground water level is highly complex. Since the porous material is unsaturated by water and almost completely dry during certain periods, the concentrations of the dissolved substances will not be constant in time and the rate of diffusion in the material will therefore vary (in time and space). The leaching process cannot therefore be simulated in a natural manner in the laboratory and particularly not in accelerated experiments.

Conventional shaking tests and special diffusion tests both indicate that the leachate content will decrease comparatively rapidly. There is not a lot of difference in this respect between the salts (sulphate and chloride) and metals in general. Some metals differ from the general pattern, however. Leaching tests cannot be verified until more leachate has been examined from the test roads and in particular from the road structure under the asphalt pavement (Kullberg, 1990).

A general assessment of the effect that graded bottom ash fill may conceivably have on the nearest (most sensitive) watercourse can be made if a number of assumptions are taken:

* The formation of leachate corresponds to the percolation through permeable fractured asphalt (100 mm/year).

* A sorption of transported elements corresponding to 50% takes place in the underlying soil.

* The leachate is diluted only with such surface water and ground water as have formed in its "own" runoff area.

* No sorption in the surface watercourse of the substances in question from the bottom ash is expected.

Dilution of the leachate in the internal runoff area, that is in the layers of soil and eventually in the local surface watercourse, will naturally depend on the size of the runoff area or its distance from the fill. In most cases, however, dilution corresponding to a factor of at least 100 within reasonable distance of a sizeable fill is likely to be achieved. If the background level is the same as that in unaffected watercourses, then increased contents and corresponding contamination factors calculated according to the National Environment Protection Board's "General Advice", Table 2, will then be obtained in the watercourse.

TABLE 2

The "normal" chemical composition of the leachate from graded slag is assessed on the basis of the project results as well as on the corresponding "unaffected" background content in surface waters and the calculated content resulting from the discharge of leachate into local watercourses.

Substance/ parameter	Content in leachate from graded slag	Content background surface water	Content local watercourse	Contamination factor
Chloride (mg/l)	120	4	5.1	1.3
Sulphate (mg/l)	300	15	17.8	1.2
Aluminium (μg/l)	50	30	30.2	1.01
Cadmium (μg/l)	0.5	0.02	0.025	1.2
Chromium (μg/l)	4	0.3	0.34	1.1
Copper (μg/l)	20	0.7	0.89	1.3
Nickel (μg/l)	25	0.5	0.74	1.5
Lead (μg/l)	10	0.2	0.30	1.5
Zinc (μg/l)	100	2.0	3.0	1.5

In spite of the conservative assumptions in the calculations, the contamination factor for nickel, lead and zinc would obviously reach the limit between "insignificant effect" (up to 50% increase) and "significant effect" (50-200% increase). However, it should be noted that this applies to an initial stage before dilution of the leachable part of the residual products starts to occur. A reduction of the leachate content by a factor of 5-10 is obtained after only a few years.

The environmental impact should not be evaluated solely on the basis of concentrations in leachate and thus in the affected ground water and/or surface water. The total amount in grammes, for example, is also of importance. Some data and comparisons are given in Table 3.

TABLE 3
Calculated quantities leached from a 1 km stretch of road constructed of different materials, expressed in grammes/year. The calculation is in respect of the initial stage of the leaching process (year 1).

Metal Leachate from	Al	Cd	Cr	Cu	Ni	Pb	Zn
Graded slag (Malmö)	16-250	0.2-1.2	3-9	13-100	20-80	0.6-27	10-800
Graded slag (Linköping)	7	<0.2	3	6	2	0.7	14
Natural gravel (Malmö)	800-3700	0.3-4	2-14	6-460	5-30	1-360	10-900
Natural gravel (Linköping)	8	<0.2	2	14	0.1	0.1	19
Run-off water from test road (Malmö)	1-320	1-3	3-12	38-165	5-120	1-29	13-220

In all probability, the leaching properties of most fill materials in urban environments are generally not more favourable than those of graded slag. In order to correspond to the background level in a natural environment (ground water), the leachate from such "graded slag" as has been studied here requires little dilution - in the order of a factor of 10. On the other hand, the amount of undiluted leachate necessary to achieve a significant effect on the environment corresponds to a factor 3-20 times higher than has been obtained in the leachate from "graded slag" and incinerator bottom ash.

Calculation of leaching from natural gravel in Malmö (Table 3) is based, like the other annual amounts, on the data obtained for the test road in Malmö. It should be pointed out that this data only represents a short period of leaching and may be influenced by such factors as the infiltration of surface water and dust from various activities in the area. In spite of the pronounced difference in leaching between natural gravel in Malmö and natural gravel in Linköping, leaching must still be regarded as slight in all cases.

5. PROCEDURES FOR UTILIZATION

The study and subsequent discussions with the Swedish EPA have shown that the utilization of screened bottom ash does not have a critical impact on the environment.

The materials should have the following properties:

- Pre-screened ash through the separation of magnetic materials
- Maximum grain size 50 mm
- No more than 10% of the particles smaller than 0.06 mm
- The content of metals in leachates from the laboratory must not exceed those in the leachates studied in the project
- Loss on incineration less than 4% - The ash must be kept in storage for at least 3 months.

The slag can be used chiefly for the following applications:

- Embankment fill for roads
- Reinforcement material for low-traffic roads and bicycle paths
- Fill under light buildings and floor structures

For the present, the guidelines below should normally be followed:

- The thickness of the fill should be limited to 3.0 m
- The bottom ash should be placed primarily above the ground water level and below pavement
- The bottom ash should be used primarily in urban areas and as fill.

In an initial stage, general advice will be prepared for a small number of products, including sorted bottom ash from incineration and cementstabilized coal fly ash, for example. When sufficient experience has been gained over a few years, a general system ought to be drawn up and made available. The responsibility for this rests with the National Environment Protection Agency.

The various suppliers ought to draw up procedures for the delivery of ash which include the following particulars:

- quantity delivered
- environmental quality assurance in respect of the delivered ash
- location
- plan and section
- control programme, if necessary

6. ACKNOWLEDGEMENT

The authors wish to thank Sysav AB, who has given permission to publish data from the on-going project. We aslo want to thank Stiftelsen Reforsk, who is financing a major part of the studies.

7. REFERENCES

Aubry, P., Poncelet, E. & Billard, H. (1986).Mâcheferes des usines d´aincineration d´ordures ménageries: Charactéristique et utilisations - A.N.R.E.D. Service Recherches, Etudes et Dévelopment.

Hartlén, J. (1988). Incinerator ash utilkization in some countries in Europe. Proc. Ash I-Conference. Philadelphia.

Hartlén, J. & Rogbeck, J. (1989). Sorted incinerator slag used as fill material. Proc. Int. Conf. on Municipal Waste Combustion, Vol. 1, p. 5B:1-13. ISBN 0-662 16891-7.

Jacobsson, T. (1989). Provväg med sopförbränningsrest i förstärkningslager. Sysav, Malmö. National Road and Traffic Institute. VTI Notat V 104. Linköping.

Kullberg, S. (1990). Miljöteknisk utvärdering av provvägar vid an vändning av sopslagg och kolbottenaska i Malmö och Linköping 1988-1990. Swedish Geotechnical Institute, Report No 2-482/88. Linköping.

Lundgren, T. & Hartlén, J. (1991). Slagganvändning - Teknik och miljö. Slutrapport. Stiftelsen Reforsk. Report (In print). Malmö.

Waste Materials in Construction.
J.J.J.R. Goumans, H.A. van der Sloot and Th.G. Aalbers (Editors)

PHYSICO-CHEMICAL AND MINERALOGICAL CHARACTERIZATION OF MINING WASTES USED IN CONSTRUCTION

E. VAZQUEZ[1], A. ROCA[1], A. LOPEZ-SOLER[2], J.L. FERNANDEZ-TURIEL[2], X. QUEROL[2] and M.T. FELIPO[3]

[1]Escola Tècnica Superior d'Enginyers de Camins, Canals i Ports, UPC. Campus Nord, Gran Capità, s/n, Mòdul B1. 08034 Barcelona. Spain.
[2]Institute of Earth Sciences "Jaume Almera", CSIC. C/ Martí i Franquès s/n. 08028 Barcelona. Spain.
[3]Fac. Farmacia, University of Barcelona. Av. Diagonal 643. 08028 Barcelona. Spain.

SUMMARY

Mining wastes from lead and lead-zinc mines of the Bellmunt de Ciurana and Vilaller areas (NE Spain) were characterized physico-chemically and mineralogically to assess the use of different industrial solid wastes in construction and in soil as part of a research program (in progress).

The high level of lead, zinc and occasionally iron sulfides (galena, PbS; sphalerite, ZnS; pyrite, Fe_2S), advised against their use as aggregates for concrete production. Nevertheless, the hardness of these materials, due mainly to the presence of the silicated mineral phases (quartz and feldspars), allow their use in road pavements. This study shows that the rock dump wastes of Bellmunt may be used in bituminous mixtures for surface, intermediate layer and basis of flexible road pavements for all types of traffic. The tailing pond wastes cannot be recycled in construction, but present work shows that bulking agents (Vilaller wastes), and fertilizers, amendments for acidic soils or substrates for horticultural crops (Bellmunt tailing pond wastes) have been found to be the best possibilities to recycle these mineral wastes on land.

1. INTRODUCTION

Large amounts of wastes are produced in areas where sulfides are mined. Most of them were dumped or disposed of without environmental quality criteria since they were considered to be inert materials. Currently, their disposal and use are restricted on environmental grounds. The possibility of recycling these wastes in construction and in soils (fertilizers, amendments, etc.) gave rise to this study.

With the aim of analyzing the application of sulfide mining wastes to construction we selected two areas in which lead and zinc sulfides (galena and sphalerite) have traditionally been worked by underground mining. These two areas are located in the south and in the north-east of Catalunya, NE Spain, in Bellmunt de Ciurana in the Priorat and in Cierco, near Vilaller, in the Alta Ribagorça (Fig.1).

Galena (PbS) rich veins hosted in Paleozoic granitic porphiries (Regia Mine) and / or slates (Eugenia Mine) were worked until the middle of this century in Bellmunt. In the mines of Vilaller, which ceased operation in the early 80s, three veins hosted in dioritic and quartz-dioritic porphiries and Devonian limestones and dolostones were worked. The mined veins are made up of mineralizations of argentifer galena, sphalerite(ZnS) and barite ($BaSO_4$).

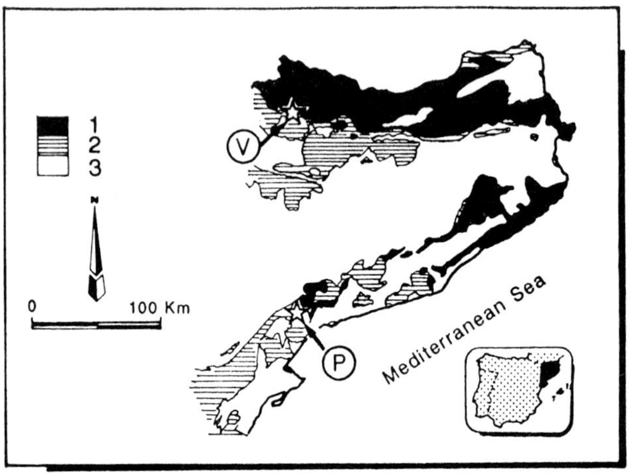

Fig.1. Geological location of the Bellmunt (P) and Vilaller (V) mining areas. 1, Paleozoic materials, mainly slates, shales, volcano-sedimentary rocks, marbles and granitoids; 2, Mesozoic sediments, mainly limestones, dolostones, marls, clays, sandstones and evaporites; and 3, Cenozoic sediments, mainly clays, sandstones, conglomerates, marls, limestones and evaporites.

2. SAMPLING AND CHARACTERIZATION

2.1 Sampling

Two types of mining wastes are found in the Bellmunt and in the Vilaller areas: 1) The waste rock dump of coarse and very coarse materials (generally between 5 and 10 cm in size) which comes from the hosting rocks of the worked veins; 2) The tailing pond wastes whith a fine sand grain size.

Two type 1 waste rock dumps (granitic porphyries and slates) and one type 2 tailing pond waste deposit were sampled in the Bellmunt area, in the vicinity of the Regia and Eugenia Mines. In the Vilaller area only tailing pond wastes were sampled (type 2).

2.2 Mineralogical and Chemical Characterization

The mineralogical composition of sampled wastes were determined by means of X-ray diffraction (XRD) analysis. XRD spectres were obtained by means of a powder diffractometer SIEMENS D-500 with a graphite monochromator and scintillation detector, using Cu radiation and operating at 40 Kv and 20 mA. Steps of 0.05 degrees of 2 theta were programmed with a counting time of 2 seconds per step.

TABLE 1
Mineralogical analysis of the mining wastes studied. Samples P-2 and P-3 are type 1 mining wastes and P-1 and V-1 to V-4 are type 2 mining wastes.

	SAMPLES						
MINERAL PHASES	**P-1**	**P-2**	**P-3**	**V-1**	**V-2**	**V-3**	**V-4**
CALCITE	-	-	-	31.4	50.3	27.8	41.5
DOLOMITE	18.6	36.4	24.1	3.3	5.8	4.0	2.5
KAOLINI./CHLORITE	18.9	18.5	11.2	7.5	6.2	7.9	6.0
ILLITE	11.1	11.3	33.4	11.6	4.5	13.8	12.2
QUARTZ	35.6	22.3	26.1	32.5	13.2	28.8	29.8
MICROCLINE	9.8	3.7	-	1.6	1.3	1.0	1.5
PLAGIOCLASE	-	-	-	2.8	1.0	1.1	1.7
GYPSUM	1.7	0.5	-	-	-	-	-
BARITE	1.0	1.8	<1	6.8	14.2	13.5	<1
ANGLESITE	1.8	1.1	<1	-	-	-	-
GALENA	-	1.4	1.7	-	-	-	-
PYRITE	1.4	1.3	-	1.4	1.8	-	1.0
SPHALERITE	-	-	-	-	-	-	1.0

The mineralogical quantitative analysis was carried out by the Reference Intensity Method (RIM) (1) for which the diffraction constants were determined by means of the intensity ratio of the problem pure mineral and the fluorite prior to the preparation of the binary mixtures at 50 %. The statistical data treatment of the results was carried out by means of the EVA program (2).

The chemical analysis was carried out by means of X-ray fluorescence spectrometry fitted with a RIGAKU S-MAX/E using a Rh tube at 50 Kv and 40 mA and a measuring time of 40 seconds. International standards were employed in calibration.

The quantitative XRD analyses of the studied samples (Table 1) show that there is a predominance of the silicate phases (quartz, SiO_2; illite, $(K,H_3O)Al_2Si_3O_{10}(OH)_2$; kaolinite, $Al_2Si_2O_5(OH)_4$; chlorite,

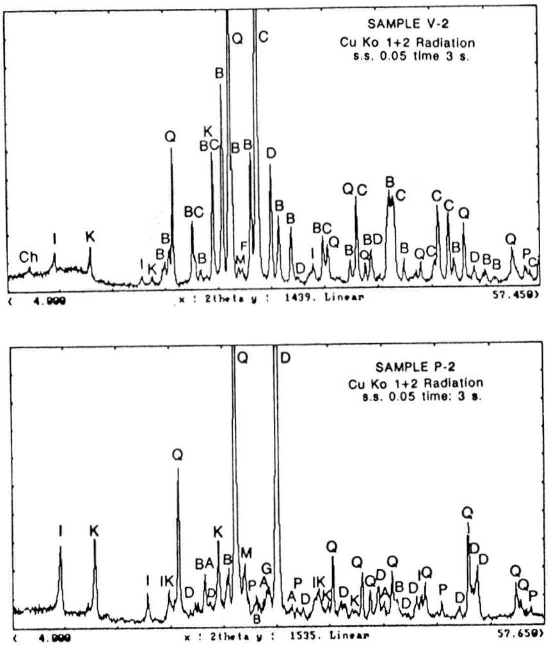

Fig.2. Representative X ray diffraction pattern of the mining wastes studied: P-2, Rock dump waste from the Bellmunt Mining Area; and V-2, Tailing pond waste from the Vilaller Mining Area. A, anglesite; B, barite; C, calcite; Ch, chlorite; D, dolomite; F, plagioclase; G, galena; I, mica - illite; K, kaolinite and/or chlorite; M, microcline; P, pyrite; and Q, quartz.

TABLE 2
Chemical analysis of the mining wastes studied.

	P-1	P-2	P-3	V-1	V-2	V-3	V-4
SiO_2	52.14	43.68	43.23	40.11	25.97	39.70	40.05
CaO	5.49	10.67	9.79	20.99	29.13	17.59	24.25
MgO	4.13	8.12	3.38	1.47	1.08	1.24	1.20
Na_2O	1.63	0.95	1.07	1.47	0.75	0.81	0.77
K_2O	5.41	3.29	4.16	3.98	1.97	2.70	2.13
Al_2O_3	15.53	14.75	19.67	10.12	7.83	9.37	9.51
Fe_2O_3	3.62	3.47	6.62	1.93	1.64	1.61	2.00
P_2O_5	0.13	0.11	0.20	0.17	0.13	0.14	0.17
TiO_2	1.16	1.20	1.57	1.54	1.96	1.97	1.15
MnO	0.18	0.20	0.21	0.12	0.11	0.10	0.14
LOI	8.21	11.05	9.25	15.12	21.58	17.81	17.59
Ba	5097	11824	3349	33013	70656	65759	2786
Co	22	21	36	20	11	14	19
Cr	25	30	98	34	47	30	33
Cu	200	406	287	223	253	182	186
Ni	23	22	79	21	18	8	20
Pb	13842	91183	15185	2229	2444	2256	1671
Sr	98	267	185	1319	2990	2550	329
Zn	6500	3794	693	2968	1705	2097	8292

$Mg_{4.7}Mn_{0.5}Fe^{3+}{}_{0.7}Al_{1.15}(Al_{1.3}Si_{2.7})O_{10}(OH)_8$; microcline, $KAlSi_3O_8$; and plagioclase, $NaAlSi_3O_8$ - $CaAl_2Si_2O_8$) and carbonate phases (dolomite, $CaMg(CO_3)_2$, and calcite, $CaCO_3$) both in the samples of Bellmunt and in those of Vilaller (Fig.2). Large amounts of barite are noteworthy in nearly all the samples, especially with reference to V-1, V-2 and V-3. The sulfides are mineral phases always dangerous when sulfide bearing materials are applied to the construction. In the present case, concentrations of galena, pyrite and/or sphalerite exceeding 1 % were found in all samples with the exception of V-3. These results were corroborated by chemical analyses (Table 2).

3. APPLICATION OF MINING WASTES IN CONSTRUCTION

The results of the chemical and mineralogical analyses of the mining wastes from the Vilaller area do not include the use of these materials in construction due to the presence of sulfides and to the narrow granulometry of the sand size which is excessively fine and difficult to stabilize.

The mining wastes from the Bellmunt area present favorable chemical and mineralogical patterns. Owing to the lack of exploitations of siliceous aggregates in the Province of Tarragona and the Administration requirements to use aggregates with an accelerated polishing coefficient (APC > 45) in road surfaces, this part of the study focuses on the use of these mining wastes for asphalt pavements in roads with heavy traffic.

3.1 Physical Characterization

The shape of granulometric fractions 5/12, 12/18 and 18/25 mm was studied and the flakiness and elongation ratios (NLT-354 specification) were determined (Table 3). The samples were prepared in a crushing plant set up by the collaborating company.

TABLE 3
Shape of the aggregates (NLT-354 specification).

Size	mm	Flakiness	Elongation
18/25	25 to 20	17	20
	20 to 12.5	10	33
12/18	20 to 12.5	31	17
	12.5 to 10	27	30
5/12	12.5 to 10	25	17
	10 to 6.3	25	39

TABLE 4. Composition of the bituminous mixtures for roads.

	Composition A	Composition B
Coarse aggregates	Sample P-3	Sample P-3
Fine aggregates	Limestone	Sample P-3
Sand	Beach sand	
Filler	Industrial calcium carbonate	
Bitumen	60/70	
Sp. gravity agg.	2.735	2.763

TABLE 5. Properties of the bituminous mixtures for roads. Marshall Test and Inmersion-Compression.

	Composition A				Composition B	
	D-12	S-12	D-20	S-20	D-12	S-12
Bitumen %	4.8	4.6	4.6	4.5	4.7	4.5
Sp.grav. g/cc	2.440	2.435	2.342	2.432	2.459	2.455
Stability Kp	1,250	1,380	1,300	1,420	1,800	1,900
Deform. mm	3.1	2.7	2.8	2.8	3.2	3.4
Voids aggreg. %	15.1	15.1	15.2	15.0	15.1	15.4
Voids mixt. %	3.7	4.2	4.3	4.3	3.9	3.7
Inm./Compress.	-	-	90.3	87.2	-	72.0

The mechanical resistance assessment of the mineral fabric of bituminous mixtures gives us an idea of the behaviour of the aggregate used. The Los Angeles Abrasion Test (NLT-149 specification) gives a result of 14.2 (mean of fifteen samples). The polishing resistance of the aggregate particles is directly proportional to the sliding resistance of the road surface. The average result of fifteen samples (NLT-174,175 specifications) yields mean values of 0.7 for the initial APC and 0.52 for the final APC. The bitumen adherence test (NLT-162,166 specifications) gives a covered area of 95 %. The specific gravity is 2.76-2.78 g/cc.

3.2 Study of Hot Bituminous Mixtures

Mixtures of D-12, S-12, D-20, S-20 types (specifications of the Dirección General de Carreteras de España) were prepared with A and B compositions (Table 4) which may be differentiated in accordance with the fine grain sizes of the mining wastes studied. Studies based on the Marshall and Immersion - Compression Tests (NLT-162 specification) were carried out to evaluate the mechanical properties and the effect of the water on the compacted mixture (Table 5).

3.3 Study of Cold Bituminous Mixtures

The coating test of the aforementioned aggregate fractions with bituminous medium curing emulsions with fluxants ECM-1 (NLT-145 specification) yields values exceeding 97 %.

3.4 Slurry Seal.

Crushed fractions of 3/8 and 3/6 mm from the mining wastes mixed with 0/3 mm slico-carbonate sand and Portland cement in slurries LB-2 and LB-3 were studied. High polishing resistance, good cohesion and consistency were obtained.

4. APPLICATION OF MINING WASTES IN SOIL

Mining wastes may be recycled through the soil as a promising method of disposal (fertilizers or amendments) even as bulking material for rehabilitation of degraded areas or as substrate for horticultural purposes. Nevertheless, this practice needs to be controlled in accordance with the soil characteristics (3-4), the physico-chemical properties of the wastes and their leachable toxic element content (5).

Three samples (V-2, V-4 and P-3) were selected on the basis of the aforementioned chemical and mineralogical patterns. In order to assess the recycling possibilities of these wastes the following analytical parameters were determined (Table 6): a) the most relevant physico-chemical parameters (pH, electrical conductivity and cation exchange capacity), b) the leachable toxic element content evaluated in accordance with the EPA extraction procedure (5); and c) a biological activity test.

TABLE 6. Physico-chemical properties and leachate element content according to EP procedure extraction (5).

	V-1	V-2	P-3	Ac. EPA (*)
pH	7.3	7.2	7.6	
EC dS m^{-1}	0.38	0.18	0.89	
CEC meq/100g	0.30	0.50	3.20	
Ag mg l^{-1}	<0.03	<0.03	<0.03	5.0
As mg l^{-1}	<0.50	<0.50	<0.50	5.0
Ba mg l^{-1}	0.75	1.40	0.06	100.0
Cd mg l^{-1}	0.17	0.22	0.56	1.0
Cr mg l^{-1}	<0.10	<0.10	<0.10	5.0
Hg mg l^{-1}	0.002	0.001	0.002	0.2
Pb mg l^{-1}	33.22	5.10	<0.10	5.0
Se mg l^{-1}	<0.50	<0.50	<0.50	1.0

(*) Acceptable element content in leachates for EPA (mg l^{-1}) (5).

From the physico-chemical point of view, these materials may be added to soil without encountering agronomical problems but they do not enhance soil characteristics. The main problem of this application is related to the environmental contamination caused by the leachability of some heavy metals by weathering. If the recommended EPA thresholds for leachates (Table 6) are taken into account, it may be said that the samples from Vilaller (V-1 and V-2) could present problems when used in soil recycling due to their high leachable lead content. It is likely that plants will not undergo any toxic effects but the lead content in soil and in groundwater may increase. This is not the case with the Bellmunt sample (P-3).

A respirometric test was performed to assess the effect of samples on soil biological activity. After twenty days' incubation of soil-waste mixtures at the rate of 20 % (W/W) no inhibitory effects were observed.

5. CONCLUSIONS

After laboratory testing to assess the potential recycling in construction and in soil of the Bellmunt and the Vilaller mining wastes, the following conclusions may be drawn:

1. The mining wastes from Bellmunt may be used in bituminous mixtures for surface, intermediate layer and basis of flexible road pavements for all types of traffic. The excellent abrasion behaviour should be noted. The results of resistance versus water are somewhat low although the other properties are satisfactory, when these materials are used as sand size aggregates. After the completion of the laboratory studies, some stretches of motorway and road using all the materials investigated in the present work were constructed. Approximately 100.000 tons of hot bituminous mixture and 600.000 m^2 of bituminous slurry seal were prepared with these materials. Their evolution under real conditions was studied and the results have been satisfactory to date.

2. Mining wastes from Vilaller may be used as bulking agents mixed with high cation exchange capacity materials to avoid lead contamination. Whereas, fertilizers (P and N contents are negligible although their exchangeable cation content is acceptable), amendments for acidic soils, or substrates for horticultural crops (after mixing with other materials to improve

their physical properties) have been found to be the best uses for the Bellmunt mining wastes.

6. ACKNOWLEDGEMENTS

We are grateful to Mr. M. Peralba for his useful contribution to the study of bituminous mixtures and slurry seals, and to Panasfalto S.A. for their helpful collaboration and material support. The present study was supported by the Spanish Ministry of Education and Science, Research Project nº PB87-0463-C02-01. Finally, we wish to express our gratitude to Mr. G. von Knorring for expert technical assistance.

REFERENCES

1 F.H. Chung, J. Applied Cristallography, 8 (1975) 17-19.
2 SOCABIM, EVA Program - Diffrac AT software, 1986.
3 R.C. Loehr, The role of soil science in the utilization, treatment and disposal of wastes. In: Utilization, treatment and disposal of wastes on land. SSSA, Madison, 1986.
4 R.N. Salcedo, F.J. Cross, R.L. Chrismon, Environmental impacts of hazardous waste treatment storage and disposal facilities. Technomic Publishing Co. Inc., Lancaster, 1989.
5 US EPA, Appendix II, EP Toxicity tests procedure. Fed. Register 5-19.80, vol.45, no. 98, Washington, 1980.

Waste Materials in Construction.
J.J.J.R. Goumans, H.A. van der Sloot and Th.G. Aalbers (Editors)

RECYCLING OF CONSTRUCTION WASTE

M.M. O'MAHONY[1] and G.W.E. MILLIGAN[2]

[1]Department of Civil Engineering, University of Dublin, Trinity College, Dublin 2, Ireland.

[2]Department of Engineering Science, University of Oxford, Parks Rd, Oxford, OX1 3PJ, United Kingdom.

SUMMARY

Although Britain is relatively rich in natural aggregate reserves, planning approvals to develop new quarries are running at about half the rate of aggregate extraction. This paper reports on a property study of crushed concrete and demolition debris with a view to using these materials in construction instead of natural aggregate. One of the objects of the research, on which this paper reports, was to accumulate data which would be a basis for the production of a British Standard for recycled materials in construction.

1. INTRODUCTION

Due to an increased awareness of the environment, planning approvals to develop new quarries in the United Kingdom are running at about half the rate of aggregate extraction. The use of secondary materials may not completely remove the problem of the resulting shortage of aggregate but it could alleviate it.

The increasing price of land in London and the South-East of Britain has caused the price of dumping to increase. Therefore demolition contractors have found that it is now more expensive to dump demolition waste then to recycle it. This encouraged the Institute of Demolition Engineers to sponsor research into the properties of aggregates produced from construction waste.

2. MATERIALS

The recycled aggregates included crushed concrete which was obtained from the breaking up and crushing of concrete slabs during a repair contract on the M25 orbital motorway. Demolition debris was the other recycled aggregate which was obtained from the demolition and crushing of various structures. It contained a random mix of materials including brick, concrete, glass, etc. This material was not cleaned and was included in the research as representing the worst condition in which a recycled aggregate might be used. The properties of these aggregates were compared with those of limestone when the materials were being considered for use in road construction. The aggregates were well graded and their particle gradings are included in O'Mahony (6). A further study was conducted which involved comparing recycled aggregate for use in concrete manufacture with the use of natural aggregate.

3. TESTS

For the unbound aggregate property study, the tests included CBR, shear strength, compaction and frost susceptibility tests. The CBR tests were conducted in accordance with BS 1377 (1) and were performed on samples at a range of densities and moisture contents, but particular attention was paid to the optimum moisture content and peak dry density condition.

The shear strength tests were performed in a 300mm x 300mm x 179mm shear box located at the Transport & Road Research Laboratory. The first series of shear strength tests was conducted at the same vertical stress of 50 kN/m^2 but the density of the material was varied whereas in the second series the density was maintained constant and the vertical stress was varied. The third series of shear strength tests involved testing the materials at the same conditions which Earland and Pike (2) describe in their test for the stability of gravel sub-bases.

The compaction tests were conducted to observe whether particles in the fraction 37.5mm-75mm affected the density of aggregates when placed in unbound layers in road construction. Although particles of this size are allowed, the maximum particle size normally present in natural aggregates is 37.5mm but in the demolition debris samples particles larger than this were evident. The frost susceptibility tests were conducted in accordance with Roe and Webster (3), as materials to be placed within 450mm of a road surface should not be frost susceptible. The materials were tested at three moisture contents including the optimum moisture content.

The tests conducted on the recycled aggregate concrete were compressive strength, modulus of elasticity, shrinkage and creep tests. Concrete at a range of water/cement ratios was examined.

4. **RESULTS AND DISCUSSION**

4.1 Tests on Unbound Aggregates

The CBR of crushed concrete at optimum moisture content and peak dry density was in the range of 400-500% and this compared well with the results for limestone. The CBR of demolition debris was considerably lower at 130% but this result was greater than that required by the Specification for Highway Works (4) for sub-base materials. Both recycled aggregates therefore could be used as sub-base aggregates with regard to bearing capacity.

Compaction tests conducted using a vibrating hammer on material contained in a 300mm diameter mould showed that the particles in the 37.5mm to 75mm range, which were present in the demolition debris samples, did not affect the maximum density which could be achieved. This suggests that the jaws on the crusher in a recycling plant need not be set as closely as those in crushers used in natural aggregate production.

It can be seen in Fig 1 that the recycled materials had friction angles similar to those of limestone although these angles were achieved at lower dry densities. The specific gravity of crushed concrete and demolition debris is lower than that of limestone and this accounts for most of the difference in density but it was found in compaction tests that the limestone particles pack more closely together than those of the recycled aggregates. The recycled aggregates would be useful as lightweight backfill to structures considering their high friction angles and low specific gravities.

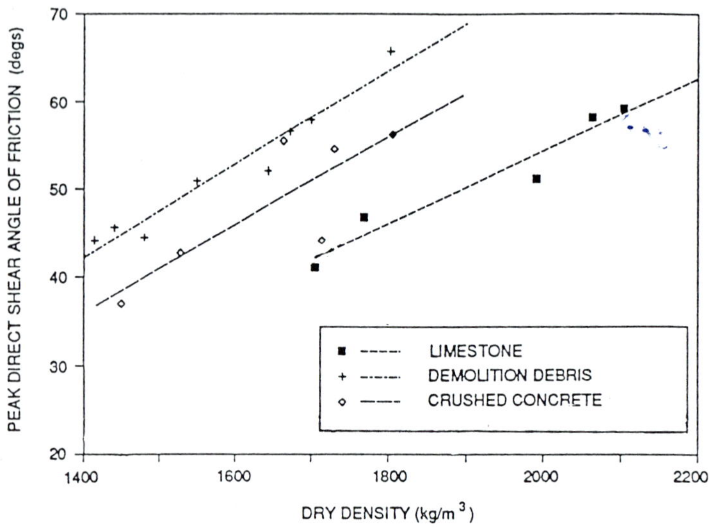

Figure 1. Influence of dry density on the angle of friction.

When the density of the limestone and recycled materials was maintained the same and the vertical stress was altered the friction angles of the materials were found not to be dependent on vertical stress, within the range $50kN/m^2$ to $200kN/m^2$.

The shear box test (2) for examining the stability of sub-base aggregates, which it is hoped will soon be included in a British Standard, involves testing the material at optimum moisture content and peak dry density at a vertical stress of $10\ kN/m^2$ and at a shearing rate of 1mm/min. The recycled aggregates, when tested in these conditions, were found to be in the medium strength category, as defined in (2), where a preliminary trafficking trial would need to be conducted on the material before it could be used as an aggregate in a sub-base layer. Limestone was found to be in the high strength category and therefore could be used without a preliminary trafficking trial in normal conditions.

Crushed concrete and demolition debris were found to be frost susceptible whereas limestone exhibited very little frost heave. This was the only test where the recycled aggregates did not perform well. Crushed concrete heaved 18mm at optimum moisture content and peak dry density and the heave exhibited by demolition debris was 12.3mm. The maximum frost heave allowed is 12mm. However, it is likely that the recycled aggregates would exhibit a self-cementing action with time similar to that which was found by Sweere (5). If this was the case then a lower frost heave might be measured if the materials were tested some time after placement. Further research is needed into the frost susceptibility of recycled aggregates.

4.2 Tests on recycled aggregate concrete

The mix proportions for the concrete mixes are listed in Table 1. The compressive strength of the recycled aggregate concrete was comparable with the strength of the conventional concrete after 28 days. The recycled aggregate concretes achieved the same or higher moduli of elasticity.

Shrinkage did not appear to be dependent on the type of aggregate used or the quantity of fine aggregate present but at high water/cement ratios, higher creep was exhibited in the recycled aggregate concrete than in the natural aggregate concrete.

Mix No.	CEMENT	FINE AGGREG	COARSE AGGREG	TOTAL WATER	FREE W/C RATIO	SLUMP (mm)
	Mass (kg) of constituents per m^3					
SET A						
C1	493	1004	670	212	0.36	10
R1	532	1004	670	255	0.35	50
C2	493	1004	670	233	0.4	115
R2	532	1004	670	229	0.3	10
C3	493	1004	670	246	0.45	190
R3	532	1004	670	279	0.4	190
SET B						
C4	580	744	839	236	0.35	35
R4	626	744	839	293	0.35	170
C5	580	744	839	265	0.4	190
R5	626	744	839	261	0.3	40
C6	580	744	839	293	0.45	235
R6	626	744	839	325	0.4	215

Note: C = control, **R** = recycled

Table 1 Concrete mix quantities and slump

5. CONCLUSIONS

It is clear from the above results that recycled aggregates could be considered for fairly demanding situations such as in sub-base layers, as granular fill to structures and as aggregate in concrete. However, further research is needed into the frost susceptibility of these materials and into the effect of varying degrees of contamination in recycled aggregates. Monitoring of contamination would be necessary if recycled aggregate was reused in industry to ensure that deleterious substances were not present.

The production of a British Standard on recycled aggregates and allowable levels of contamination would be useful for demolition contractors hoping to recycle construction waste on a large scale. To accomplish this recycling of construction waste must be taken more seriously by engineers and clients.

ACKNOWLEDGEMENTS

The research on which this paper reports was sponsored in part by the Institute of Demolition Engineers. Funding was also provided by the Geotechnical Consulting Group as part of a contract for the Transport and Road Research Laboratory.

REFERENCES

1 British Standard 1377 (1975) Methods of test for soils for civil engineering purposes. British Standard Institution. London.
2 Earland, M.G. and Pike, D.C. (1985) Stability of gravel sub-bases. RR64. Transport and Road Research Laboratory.
3 Roe, P.G. and Webster, D.C. (1984) Specification for the TRRL frost-heave test. SR 829. Transport and Road Research Laboratory.
4 Specification for Highway Works (1986) Department of Transport. HMSO.
5 Sweere, G.T.H. (1989) Structural contribution of self-cementing granular bases to asphalt pavements. Proc. 3rd Symposium on Unbound Aggregates in Roads (UNBAR 3). University of Nottingham.
6 O'Mahony, M.M., (1990) Recycling of materials in civil engineering. D. Phil Thesis. University of Oxford.

Waste Materials in Construction.
J.J.J.R. Goumans, H.A. van der Sloot and Th.G. Aalbers (Editors)

HIGH FREE LIME FLY ASH;CHARACTERIZATION AND USE

V.ROGIĆ[1], B.MATKOVIĆ[2], M.PALJEVIĆ[2], D.DIMIĆ[3], D.DASOVIĆ[1], C.W.ORMSBY[4] AND M.SELIMOVIĆ[1]

1) Faculty of Civil Engineering, University "Džemal Bijedić", 88000 MOSTAR, Yugoslavia, 2)"Ruđer Bošković" Institute, 41000 ZAGREB, Yugoslavia, 3)Institute for Research and Testing of Materials, 61000 LJUBLJANA, Yugoslavia 4) U.S. Department of Transportation Federal Highway Adiminstation 63000 Georgetown Pike McLean, Virginia 22101-2296 U.S.A.

SUMMARY

Fly ash from power plant Gacko (Bosnia and Herzegovina, Yugoslavia) contains up to 50 wt.% free lime and 13 wt.% anhydrite: the rest are Al, Si, Ca, Fe, Mg, Na, and K oxides in glassy phase. Fly ash is low radioactive therefore it was considered for the use in civil engineering.

Blend made with 70 wt.% fly ash and 30 wt.% silica fume corresponds to the quality of binder for soil grouting. Because of high free lime content in original fly ash, which causes unsoundness, fly ash is not suitable as ingredient of any hydraulic binder, like pozzolanic or masonry cement. Therefore original fly ash was treated with water for the hydration of free lime.

Obtained "hydrated fly ash" mixed with 30 wt.% of ordinary portland cement corresponds to the quality of masonry cement.

Blends, which correspond to the quality of binder for soil grouting, were tested at the construction site of a dam for hydro-power plant and at lignite mine for injection works.

Binder which corresponds to the quality of masonry cement was used as mortar for masonry work and for outdoor and indoor plastering of walls.

Satisfactory results were obtained in all these uses.

INTRODUCTION

Fly ash "Gacko" is waste material from power plant Gacko (Bosnia and Herzegovina, Yugoslavia) which uses indigenous lignite (it is burnt at approximately 1000 °C). This lignite contains marlaceous limestone; its amount depends upon the thickness and number of marlaceous limestone layers which intersect lignite layer. Therefore fly ash contains 30 to 50 wt.% free lime (total CaO from 70-80 wt.%) and approximately 13 wt.% $CaSO_4$ (anhydrite). Free CaO and $CaSO_4$ are crystalline components, the rest is glass composed of Al, Si, Ca, Fe, Mg, Na and K oxides.

ASTM standard (1) classfies fly ash into class F (from bituminous coal or anthracite) and class C (from subbituminous coal or lignite with CaO content < 10 wt.%). Both classes of fly ash have pozzolanic properties; class C also has some cementitious properties.

Another classification is made according to chemical composition (2). This classification devides fly ash into four types according to the content of SiO2, CaO and SO3. The corresponding composition of fly ashes is presented in Table I together with the composition of fly ash "Gacko" which (according to it's composition and properties) does not belong to any type or class.

Table I Classification of fly ashes according to chemical composition; wt.%.

Type	SiO_2	CaO (total)	CaO (free)	SO_3	Al_2O_3	Fe_2O_3	MgO
I	50	5-7	-	1-5	4-25	7-9	-
II	40-50	9-22	-	0.5-5	17-25	8-12	-
III	2-5	40-46	10	9	7-8	-	-
IV	3	33	23	26	4	-	4
Gacko	9	70-80	30-50	8	5	2	1

Types I and II correspond to ASTM classes F and C, respectively. Type III is basic, Type IV is sulfatic, while fly ash Gacko is high basic.

The production of fly ash "Gacko" is 400.000 t/year. Its transportation and storage causes technical, economical and ecological problems. Water at deposite site has pH=12.5 and the protection of the environment from possible contamination of groudnwater is wery expensive.

Fly ash "Gacko" is low radioactive. The concentration of radionuclides is below the limit allowed for civil engineering materials (3). Therefore fly ash "Gacko" was considered for commercial use as main ingredient in blends which could correspond to the quality of binder for grouting, masonry and pozzolanics cements.

COMPONENT OF THE BINDER

Fly ash (FA), Chemical composition: SiO2, 8.89; Al2O3, 4.81; Fe2O3, 2.59; CaO, 72.18; MgO, 1.28; alkalies as Na2O, 0.34; SO3, 7.68; CO2, 1.62 wt.%.

Hydrated fly ash (HFA), obtained by hydration of original fly ash in a factory for lime hydration, contains: Ca(OH)2, 50; CaSO4, 9; CaCO3, 5; CaO free, 5 wt.%; these components are crystalline, the rest is amorphous phase.

Silica fume (SF): SiO2, 95 wt.%; specific surface (BET), 19.6 m2/g. Lime: silica fume strength at 7 days (55°C) is flexural/compressive strength = 4.9/44.2 MPa.

Ordinary portland cement (OPC), calculated composition: C3S, 64.4; C2S, 9.9; C3A, 10.5; C4AF, 10.7 wt.%.

RESULTS AND DISCUSSION

1. Fly ash - silica fume mix

Blend was made by homogenizing 70 wt.% of original fly ash and 30 wt.% of silica fume. Blend was tested in paste and mortar.

Paste was prepared using water to solid ratio 0.6. Setting time was determined using Vicat needle; initial setting time was at 405 min and final at 855 min.

Soundness was determined using modified Le Chatelier and pat tests: samples were cured in moist air (R.H. > 90%, T=20 °C) for six days before boiling in water for two hours; (a) the difference between two measurements of indicator needle distance was 1.1 mm; (b) pate were sound. Some pats, after initial curing of six days in moist air, were kept 28 days in the air at 65% R.H.; they were distorted, but no cracks were visible. If initial curing of paste samples in moist air was shorter than six days, samples were unsound.

Heat evolution of paste at isothermic conditions reaches maximum at 930 min, Fig.1.; Original fly ash paste without silica fume has its maximum at 15 min.

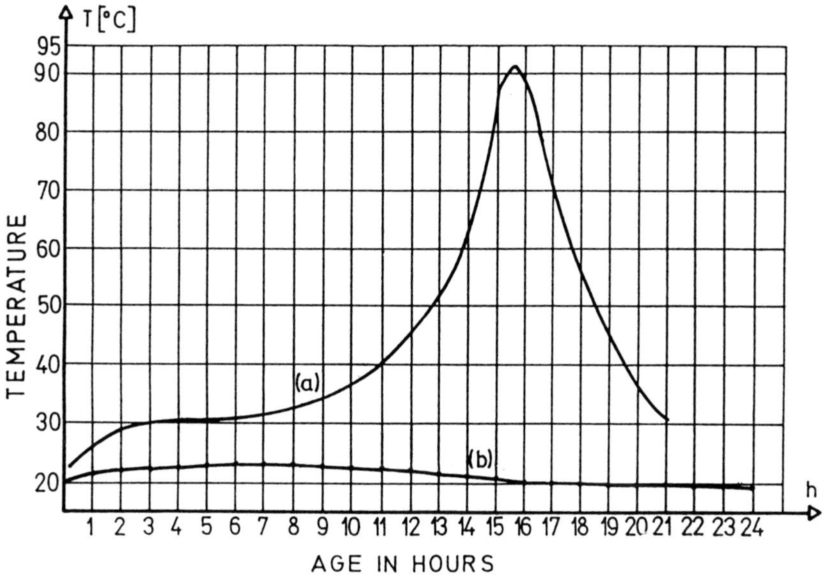

Fig.1. Temperature-time curves of paste samples : (a) fly ash:silica fume = 70:30; water:solid = 0.6; (b) hydrated fly ash:silica fume = 70:30; water to solid = 0.49.

Mortar specimens were prepared by mixing binder with graded sand (1;3) and water (water to binder ratio = 0.75) and cast in 40 by 40 by 160 mm molds. After 48 hours curing at 20 °C and > 90% R.H. the specimens were demolded and stored under water at 20 °C until tested. Results are presented in Fig.2.

Length change was determined in mortar specimens. After curing for 3 days at > 90% R.H. and 20 °C, specimens were demolded, reference length was read and specimens were expose to drying at 60% R.H. Results are presented ina Fig.3.

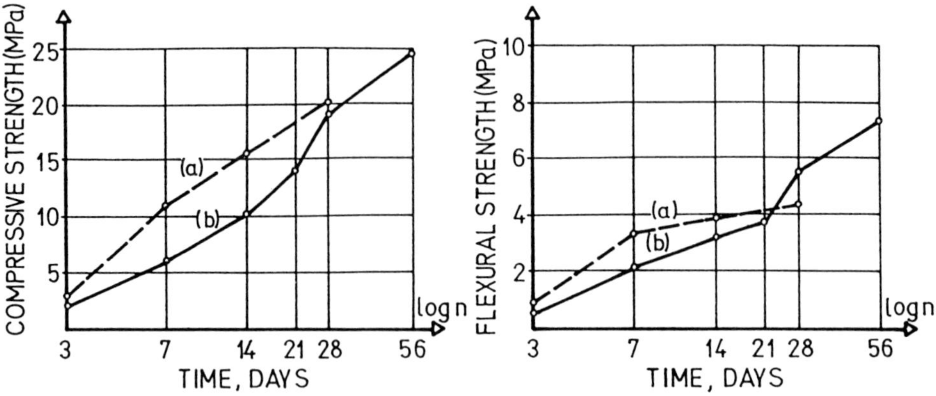

Fig. 2. Strength development in mortar specimans: (a) fly ash:silica fume = 70:30; (b) hydrated fly ash:silica fume = 70:30.

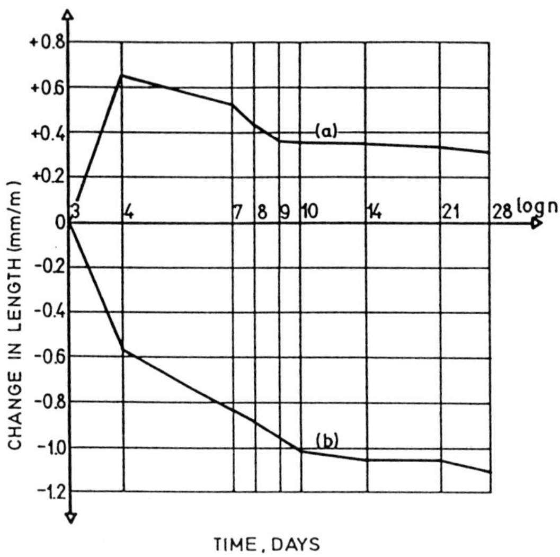

Fig. 3. Length change of mortars made with mixes of : (a) fly ash:silica fume = 70:30; (b) hydrated fly ash:silica fume = 70:30. Reference length reading was on demolded samples after curing three days in molds at 95% R.H.

The effect of superplasticizer addition (48%aqueous solution of naphtalene sulfonate condensate) in water content and strength developement in mortars is presented in Fig. 4.

The effect of silica fume content in fly ash mix on strength development is presented in Fig. 5. Amounts above 30 wt.% silica fume in the mix increase, while smaller (below 30 wt.%) reduce the strength. In spite of the fact, that binder with more than 30 wt.% silica fume gives higher strength, the blend with 30 wt.% was selected because silica fume is expensive (in comparation to fly ash) and strength development in mortars made with 30 wt.% silica fume content in the mix still corresponds to the quality of binder for grouting. The other results, like soundness, heat evolution (Fig.1.); and length change (Fig.3.) suggested also that this binder does not correspond to the quality of a pozzolanic cement and so could be used only for grouting where high water to binder ratios are used.

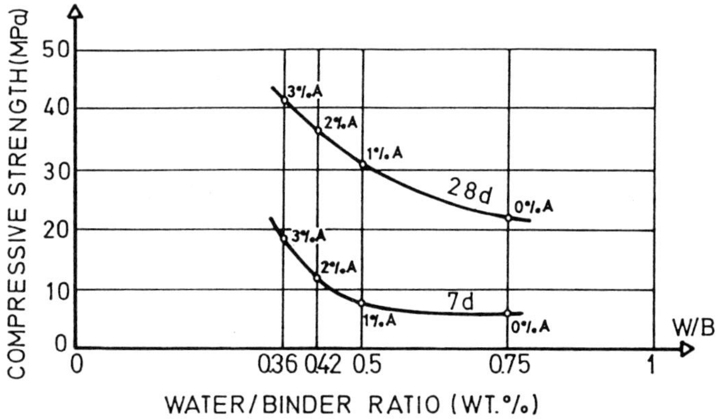

Fig.4. Compressive strength (at 7 and 28 days) in mortar samples (fly ash:silica fume = 70:30) as a function of water to binder ratio and superplasticizer addition. The amount (A) of superplasticizer is expressed in weight percent of dry superplasticizer by weight of total binder.

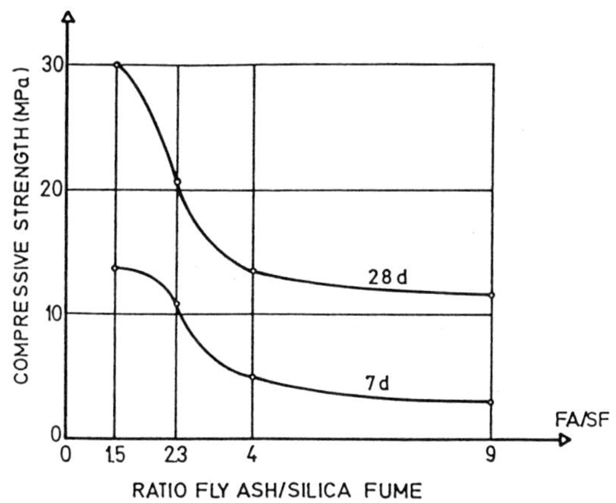

Fig.5. Compressive strength (at 7 and 28 days) in mortar samples as a function of silica fume contet. Fly ash:silica fume = 1.5 corresponds to the ratio 60:40: 2.3 to 70:30: 4 to 80:20; and 9 to 90:10.

1.1 Testing of blends for soil grouting

Blends made with fly ash (70 wt.%) and silica fume (30 wt.%) mix with solid to water ratios from 1:1 to 1:5 were compared with portland cement (95 wt.%) and bentonite (5 wt.%) standard blends regarding rheological properties.

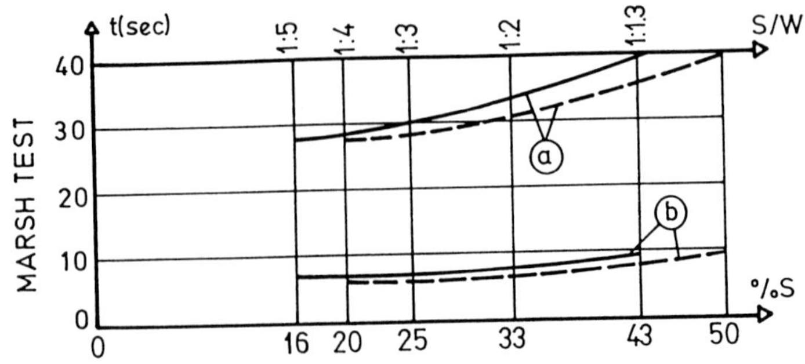

Fig. 6. Marsh test for viscosity of blends for grouting. Rate of flow through (a) ϕ = 4 mm ; (b) ϕ = 10 mm. Full line denotes fly ash blend; broken line that of portland cement blend. S = percentage of solid; S/W = solid to water ratio.

Viscosity was determined by using Marsh flow test, Fig.6., and plastic viscosity by Fann method, Fig.7. Stability of blends, expressed as a decantation volume versus decantation time, is presented in Fig.8. Fly ash mixes have better rheological properties than corresponding portland cement mixes according to results in Figs.6. to 8.

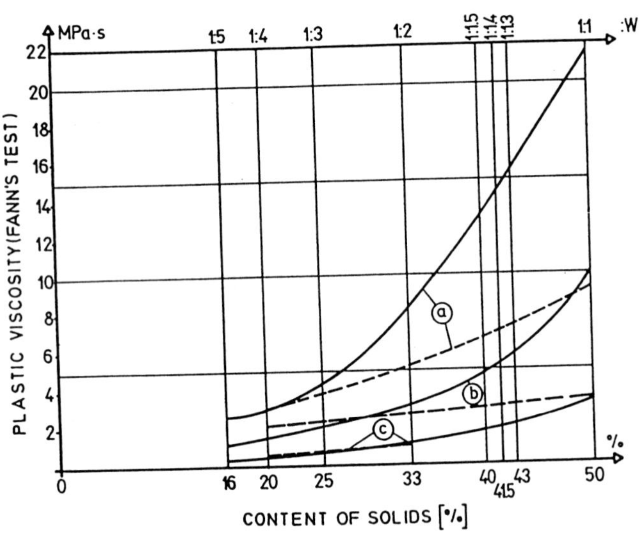

Fig. 7. Fann test for plastic viscosity of blends for grouting. Angular velocity of rotation (a) 600/300; (b) 200/100; (c) 6/3 cycles/min. Full line denotes fly ash blend, broken line that of portland cement blend. S:W = solid to water ratio.

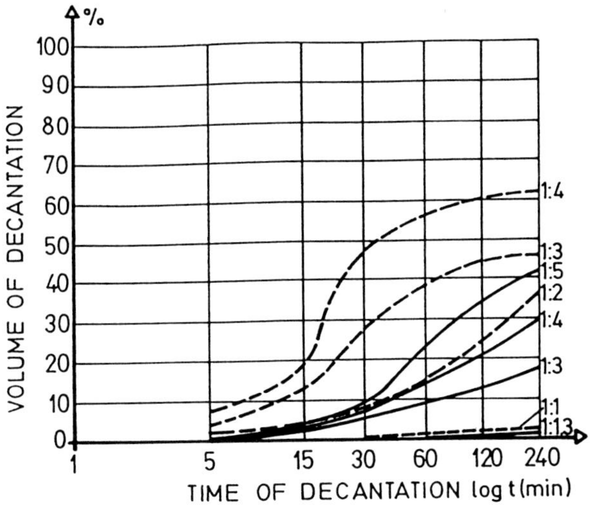

Fig. 8. Decantation volume of blends for grouting with different solid to water ratio (from 1:1 to 1:5) versus time of decautation. Full line denotes fly ash blend, broken line that of portland cement blend.

A comparison between setting time, strength and permeability between fly ash and corresponding portland cement mixes after filtration of execive water by vacuum suction is given in Table II.

Table II Setting time, strength and permeability of filtered fly ash and portland cement mixes for grouting.
S:W FIL(min) WC (wt.%)

Mix	Solid to weight ratio	Time of filtration (min)	Water content filtration (wt.%)	Setting time (min)		Compr. strength (MPa)		Permeability
				initial	final	7d	28d	(cm/sec)
95%	1:4	83	31.25	110	190	39.1	47.2	$5.5x10^{-8}$
OPC	1:3	83	31.79	165	230	36.3	44.1	$7.3x10^{-8}$
+	1:2	73	28.08	140	200	36.5	45.7	$6.4x10^{-8}$
5% B	1:1	63	21.09	85	120	38.9	45.3	$7.6x10^{-8}$
	1:5	83	31.49	100	210	3.6	12.4	$6.7x10^{-8}$
FA:SF	1:4	83	41.69	130	225	3.6	13.3	$6.2x10^{-8}$
70:30	1:3	73	32.69	190	245	3.1	12.3	$6.1x10^{-8}$
	1:1.3	43	35.18	250	290	4.2	15.9	$7.4x10^{-8}$

OPC = ordinary portland cement; B = bentonite; FA = fly ash; SF = silica fume; S:W = solid:water ratio; FIL = time of filtration; WC = water content in sample after filtration

Fly ash mix has lower strength than coresponding portland cement mix, but still corresponds to the quality required binder for watertight soil grouting.

Fly ash mixes for soil grouting with solid to water ratio 1:4, 1:3 and 1:1 were tested for injection works "in situ" at the construction site of hydropower plant dam and in lignite mine. The obtained results confirm that this mix satisfy and can be used as a binder for soil grouting.

1.2. Testing of blends in concrete for diaphragm construction

Mix proportions for concrete specimens (120 by 120 by 360 mm) are given in Table III.

Table III Mix design for diaphragm wall

Binder FA-SF /kg/m3/	Bentonite wt.% of binder:	Agregate* /kg/m3/	Water /l/m3/	Porosity /%/	Workability		Vol. weight /kg/m3/
					Slump /cm/	Flow /cm/	
85	0	2029	189	1.2	18.0	52.0	2250
	5	1969	208	1.0	18.0	51.0	2200
	10	1924	225	0.7	19.0	52.0	2180
125	0	2026	175	0.9	20.0	52.5	2270
	5	1831	246	0.6	19.0	52.0	2160
	10	1783	263	0.4	19.0	50.0	2170
175	0	1859	219	1.2	19.0	51.5	2180
	5	1738	263	0.6	18.0	51.0	2140
	10	1636	298	0.4	17.0	52.0	2080

FA-SF = fly ash:silica fume = 70:30;
** Limestone, Dmax = 31.5 mm.*

Specimences were cured in moist room (>90 % R.H., 20 °C) until tested (they were demolded after 48 hours). Strength development is presented in Fig.9. Results suggest that the binder without bentonite at the dosage of 85 kg per m3 concrete corresponds to the requirements for diaphragme wall.

During the construction of a dam for hydropower plant 6 m3 of portland cement concrete mix was supstituted by fly ash concrete mix. Mix design for concrete together with strength and permeability, determined on specimens cored from diaphragm wall after 28 days, are presented in Table IV.

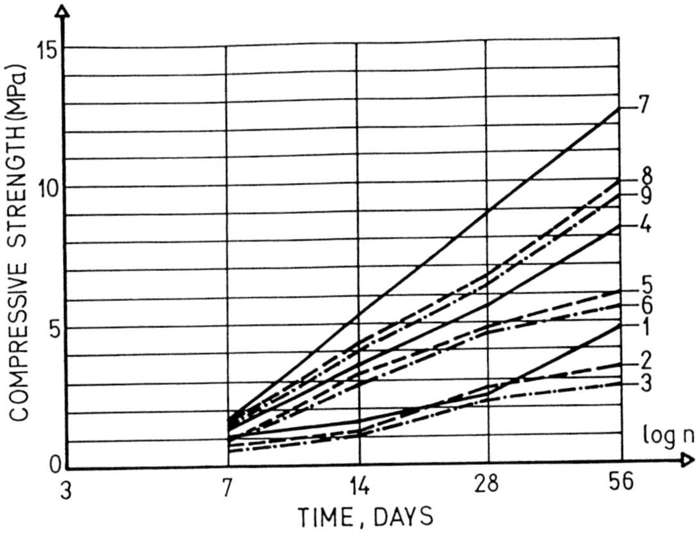

Fig. 9. Strength development in concrete specimens.
85 kg/m3 binder : (1),0; (2),5; (3),10 wt.% bentonite.
125 kg/m3 binder : (4),0; (5),5; (6),10 wt.% bentonite.
175 kg/m3 binder : (7),0; (8),5; (9),10 wt.% bentonite.

Table IV Mix design and properties of concrete for construction of diaphragm wall

Mix design	Portland cement	(kg/m3)	100	-
	Fly ash binder	(kg/m3)	-	85
	Bentonite	(kg/m3)	30	8.5
	Aggregate	(kg/m3)	1716	1924
	Water	(l/m3)	275	225
Fresh concrete	Slump test	(cm)	20.6	20
	Flow test	(cm)	53.2	53
Hardened concrete* at 28 days	Compressive strength	(MPa)	2.8	1.8
	Permeability	(cm/sec)	8.7x10-7	4.5x10-6
	Modulus of elasticity	(MPa)	2650	2500

* *Samples were obtaind by coring*

2. Hydrated fly ash-silica fume mix

Hydrated fly ash was considered for binder which could correspond to the quality of pozzolanic cement. Mixes with hydrobed fly ash : silica fume ratio = 70:30 were tested regarding heat evolution in pastes (Fig.1.), strength development in mortar specimens (Fig.2.) and during shrinkage (Fig.3.). Obtained results suggest that this mix could correspond to the quality of pozzolanic cement its strength development would be faster and during shrinkage smaller. This type of cement could be used in non - structural, low strength concrete, i.e. for highway sub-bases , but not as a binden for masonry or plastering (3).

3. Hydrated fly ash-portland cement mix

Reactions in pastes and strength development in mortars were investigated previously (3). There was suggested that this mixture corresponds to the quality of masonry cement. Therefore mortar was tested for masonry work and for outdoor and indoor plastering. Mortar used for masonry work and plastering had compressive strength 13 MPa, flexural strength 4.2 MPa. Adhesion in concrete wall without any pretrmeatment of the surface was 0.3 MPa. Plaster on the walls, after three years is sound. Because fly ash contains appreciable amount of anhydrate, it could be considered that it can cause unsoundness in the mixture with portland cement. But results (3) are quite opposite. The presence of anhydrite can be. considered as usuful component, because it reacts with alumina from glassy phase calcium hydroxide and water forming ettringite, which contributes to the strength.

CONCLUSION

Investigations presented in this report showed that some parts of the waste from power plant Gacko can be good for use as civil engineering material in spite of its unusual composition.

Original fly ash can be used only as the main ingreediant in binder for soil grouting. While pre-hydrated fly ash, which is stabilized through the hydratation of free CaO in $Ca(OH)_2$, can be used as binder for masonry and plastering work, if 30% of portland cement is added. There are also some possibilities of realizing the use of hydrated fly ash-silica fume mixes. To achieve that the effect of some admixtures and additions has to be investigated.

References

1. ASTM C 618-89. Standard specification for fly ash and raw - or calcinate natural pozzolans for use as a mineral Admixture in portland cement concrete. Annual Book of ASTM Standards, Vol. 4.02. Am.Soc. for Testing and Materials,(1989) pp 296-298 Easton, MB, USA.

2. S.Droljc, D.Dimic, M.Ferjan, The use of fly ash in production of bricks, mortars and lightweight aggregates for concrete. Presented at 1st International conference in the use of Fly Ash, Silica Fume, Slag and Other Mineral Byproducts in Concrete, Montebello, Quebec, Canada 1989.

3. B.Matković, Ž.Gržeta, M.Paljević, V.Rogić, D.Dasović and D.Dimic, Hydrated fly ash with SiO2 fume and/or portland cement addition. Reaction in pastes and strength development in mortars, Cem. Concr. Res. 20, 475-483 (1990).

Waste Materials in Construction.
J.J.J.R. Goumans, H.A. van der Sloot and Th.G. Aalbers (Editors)

CHEMICAL PROCESSES AT A REDOX/pH INTERFACE ARISING FROM THE USE OF STEEL SLAG IN THE AQUATIC ENVIRONMENT

ROB N.J. COMANS, HANS A. VAN DER SLOOT, DIRK HOEDE AND PETRA A. BONOUVRIE
Netherlands Energy Research Foundation (ECN), P.O. Box 1, 1755 ZG Petten, The Netherlands

SUMMARY

The environmental impact of the use of steel slag in a fresh water system has been studied using a combination of laboratory tests, a pilot scale experiment and field measurements. The results obtained at the different scales are consistent and indicate the development of a relatively sharp discontinuity in the composition of interstitial water in the steel slag emplacement, below which highly alkaline and reducing conditions were measured. The mobility of major and trace elements on opposite sides of the redox/pH interface differs markedly. Consequences for the emission of these elements from steel slag are discussed.

INTRODUCTION

The use of solid waste materials in construction, both in the terrestrial and aquatic environment, leads to a reduction in the volume of waste requiring disposal and to a conservation of natural building materials. In addition to the testing of construction properties, increasing efforts are necessary to establish the environmental consequences of the application of waste products. Although standardized laboratory testing methods are being developed and improved for this purpose (1), accurate prediction of the impact of waste utilization in the terrestrial or aquatic environment is impossible without a better knowledge of the underlying processes active in the field.

The present combined laboratory and field study focuses on the application of steel slag in the aquatic environment as an alternative material for shore protection along rivers. Laboratory testing of the steel slag materials included availability and tank leaching experiments. In addition, leaching of major and trace elements was studied in a large (pilot) scale experiment simulating the actual situation in the field. Field data were obtained by monitoring the composition of surface water and interstitial water in a steel slag emplacement used for shore protection along a small river in the Dutch Biesbosch area. Results from laboratory and pilot scale experiments and field data will be compared and the underlying chemical processes are discussed below.

MATERIALS AND METHODS

Steel slag

The steel slag used for the present study was formed during the production of steel according to the Lintz Donawitz process and is therefore often referred to as LD-slag. The size of the individual pieces was 40-160 mm. Five different samples were used for total concentration measurements and laboratory tests (BG0, BM0, BG2, BM2, BRED). More porous (codes "BG") and more dense (codes "BM") slag samples could be distinguished visually. The BRED sample was used for laboratory tests under a nitrogen atmosphere.

Laboratory methods

Laboratory testing of the slag samples included total concentration measurements and availability and tank leaching experiments. Total concentrations were obtained from chemical analysis after destruction of the solids in a $LiBO_2$- (Na, K, Ca, Mg, Ba) or Na_2CO_3-melt (SO_4, Cl, F). Vanadium in the solids was analyzed directly by Instrumental Neutron Activation Analysis. The availability test (2) was developed to assess the potential release of constituents from a waste material on the very long term and consists of extraction of a fine grained sample (95% < 125 μm, after crushing of the sample) at a liquid/solid ratio of 100 and a constant pH of 4. This test considers the fraction of the element bound in poorly soluble mineral phases to be environmentally inert. The tank leaching test (3) was developed for the assessment of leaching mechanisms from intact waste products. The sample is immersed in water that is refreshed and analyzed at regular time intervals. The cumulative flux of constituents is calculated for each time interval in mass per unit surface area. A plot of this value against time reveals different release mechanisms such as dissolution, surface wash off and diffusion. The effective diffusion coefficient calculated from these experiments can be used to estimate long term releases from the waste materials.

Pilot scale experiment

The pilot scale leaching experiment was developed as an intermediate between laboratory tests and field measurements and consisted of a 1 m^3 polyethylene container (approximately 1 x 1 x 1 m) with five sampling taps mounted vertically at different depths. The container was filled completely with tap water. Approximately 1350 kg of steel slag was added, filling the container almost to the first tap ($\approx$ 0.2 m from the top). Water was pumped horizontally in a laminar flow between five inflow and outflow taps located across the top of the container. The flow rate was adjusted at 240 l/day. The container was coated on the outside with black foil in order to prevent algal growth. Water samples were taken regularly from the inflowing and outflowing water and from the interstitial water at different depths in the steel slag emplacement.

Field measurements

The field site was located along a small river in the Dutch Biesbosch area. A steel slag layer with a thickness of 20-25 cm was placed along the shore embankment (see Fig. 1) over a distance of approximately two kilometers. About 1000 kg of slag was used per meter. One month after the layer was placed, four Teflon (PTFE) sampling tubes were positioned following the slope of the embankment to a depth of 1.5 m in the steel slag (Fig. 1). A Teflon (PTFE) bailer was used to sample the interstitial water in the tubes at regular intervals.

Surface water was sampled at two locations in the river approximately 1-2 m from the slag emplacement. A third set of samples was taken upstream as a reference not influenced by the steel slag.

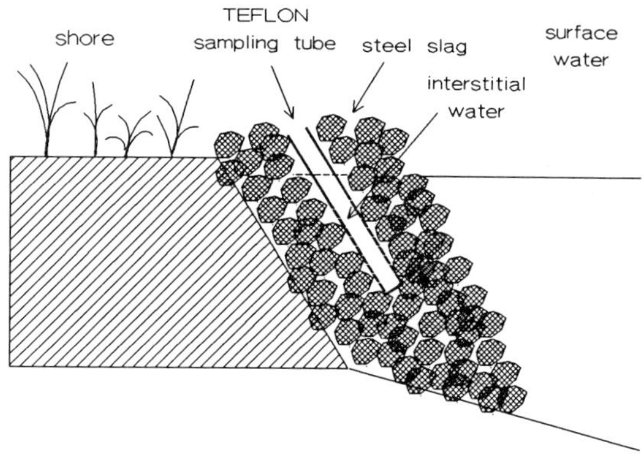

Figure 1. Schematic representation of the steel slag emplacement in the field and location of the tubes for interstitial water sampling.

Water analysis

All water samples were filtered through 0.45 μm membrane filters and stored in acid cleaned polyethylene sample vessels until analysis. Measurements of pH and E_h were taken directly after sampling. The following analytical methods were used: Flame Atomic Absorption Spectrometry (Na, K); Inductively Coupled Plasma Emission Spectrometry (Ca, Mg, Ba, V) and Ion Chromatography (SO_4, Cl, F). The low V levels in the surface water samples were measured using Instrumental Neutron Activation Analysis after preconcentration on activated carbon (4).

RESULTS

Laboratory experiments

The results of the total concentration measurements and laboratory tests are summarized in Table 1. The very high leaching of Ca from sample BG2 is immediately clear and is probably related to insufficient mixing of Ca added during the Lintz Donawitz process. Hydration increases the volume of the Ca-components and leads to deterioration of the slag by internal tension. A significant deterioration was indeed observed after 30 days of tank leaching with sample BG2.

The leaching behaviour of the redox sensitive elements Ba and V is also noteworthy. Because barium solubility is limited by $BaSO_4$, the mobility of this element increases at redox potentials low enough to reduce SO_4^{2-}. In contrast, vanadium is often observed to be immobilized under reducing conditions, possibly by the very strong adsorption of V(IV) on oxide minerals (5). When brought into contact with water, slags have been observed to develop reducing conditions, a property which has been ascribed to sulphide species leached from the solid (6). It seems that the deteriorated sample BG2 has developed more pronounced reducing conditions in the tank leaching experiment than the other samples, as evidenced by the lowest V and highest Ba emission. Surprisingly, the Ba and V emissions in the tank leaching experiment under a nitrogen atmosphere (sample BRED) seem to indicate that reducing conditions have not developed any further than during the other experiments performed under air. High sulphate levels from the oxidized surface of the slag may have prevented a further increase of the Ba concentration.

Table 1. Summary of parameters from the laboratory tests

Element	Code	Total conc. (mg/kg)	Availability (mg/kg)	pDe (m^2/s)	Emission 64 days (mg/m^2)
Ca	BM0	285600	51400	14.1	20600
	BG0	345700	82600	14.7	11300
	BG2	246200	80900	12.7	142700
	BM2	241600	53670	13.7	34500
	BRED	279800	67200	14.2	25200
V	BM0	2045	3.1	9.7	193
	BG0	2878	14.8	10.9	180
	BG2	2570	67	13.5	24
	BM2	3980	4	10.6	100
	BRED	2870	22	12.2	63
Ba	BM0	45	1.2	12.9	3
	BG0	156	15	14.5	3
	BG2	96	3.1	11.6	9
	BM2	148	7.8	13.7	5
	BRED	133	6.9	13.4	6
F	BM0	196	2.8	12.2	10
	BG0	356	7.2	12.3	17
	BG2	285	6.7	12.4	17
	BM2	95	3.3	12.6	7
	BRED	280	5	11.8	31
SO_4	BM0	7200	224	13.5	171
	BG0	5780	174	12.8	232
	BG2	4700	163	12.7	100
	BM2	2790	173	13.5	123
	BRED	5900	183	12.6	440
K	BM0	4700	87	13.9	42
	BG0	3400	68	13.2	80
	BG2	2600	52	13.4	70
	BM2	3000	137	13.7	45
	BRED	3000	86	12.8	170
Na	BM0	2900	47	12.0	200
	BG0	6700	64	12.2	150
	BG2	7000	67	12.0	270
	BM2	6000	40	12.3	180
	BRED	6600	55	12.2	200

pDe = -log(De) (effective diffusion coefficient); sample codes are explained in the Materials and Methods section.

Pilot scale experiment

Figure 2 shows the development of interstitial water composition over time, during the pilot scale experiment, at different depths in the simulated steel slag emplacement. Within one day, large changes are observed in pH and redox potential (E_h), leading to alkaline and reducing conditions. After a few days, a sharp discontinuity was formed at a depth of 15-35 cm below the steel slag surface. Below this depth the pH stabilized at strongly alkaline values of 12.5 to 13.5 and E_h values stabilized at -150 to -200 mV. Above the interface, oxic conditions were maintained at pH values of 8 to 9.5. The redox/pH interface remained stable during the whole 210 days of the experiment. Apparently the overflowing water kept the top 15-35 cm of the system oxidized, whereas the leaching of alkalinity and sulphide species from the slag maintained the highly alkaline and reducing conditions below the interface.

The chemistry of the interstitial water is strongly influenced by the redox/pH changes at the interface. Calcium concentrations reach values of up to 1200 mg/l below the interface, but remain identical to those of the inflowing water (50-100 mg/l) in the oxidized zone. Equilibrium calculations with the geochemical speciation code MINTEQA2 (7) indicate that, in equilibrium with atmospheric CO_2, calcite ($CaCO_3$) controls Ca solubility above the interface. Large quantities of this mineral were indeed observed as a coating on the steel slag and the container wall in the region of the oxidized zone. Mass balance calculations indicate that probably all dissolved carbonate introduced with the aerated inflowing water reacts with dissolved calcium in the oxidized interstitial water layer. No equilibrium with atmospheric CO_2 exists, therefore, in the zone below the interface. MINTEQA2 calculations predict portlandite ($Ca(OH)_2$) to control Ca solubility at the high pH values in this zone. After termination of the experiment, crystals found on steel slag samples from the lower parts of the container were indeed identified as portlandite by X-ray Diffraction Analysis. Large quantities of Ca are leached from the slag because of the high total levels obtained during the LD-process and the high availability of Ca in the slag (Table 1).

Magnesium, sodium, potassium and chloride are not or only insignificantly leached from the steel slag. The latter three elements behave conservatively and are not influenced by the redox/pH interface, whereas Mg is almost completely removed from solution below the interface. Equilibrium calculations indicate that brucite ($Mg(OH)_2$) lowers dissolved Mg to μg/l levels at the extremely alkaline conditions.

Below the interface, Ba is strongly released from the steel slag and reaches concentrations of over 2 mg/l, whereas above the interface dissolved Ba is not significantly higher than in the inflowing water. Sulphate, leached from the slag in only insignificant quantities, shows concentration profiles opposite to those of Ba. Equilibrium calculations indicate that both above and below the interface, Ba solubility is controlled by barite ($BaSO_4$).

Initially, and below the interface, F reaches concentrations 1.5-2 times those of the inflowing water. After approximately two months, the concentrations at all depths, but especially below the interface, are lower than in the inflowing water. It is presently not clear which process controls the solubility of F. Because the solutions are undersaturated with respect to CaF_2, apatite ($Ca_5(PO_4)_3(OH,F)$) solubility, or sorption processes remain among the possibilities.

Another trace element of importance is V, especially when considering its high total concentration in steel slag (Table 1). Two distinct stages of V release from the slag can be distinguished from Fig. 2; a rapid leaching of V from all depths in the container to very high concentrations of up to 400 μg/l and a release from the oxidized zone, noticeable over the entire experimental period. The initial release occurs while the interstitial water at all depths is still oxic and may be caused by wash off from the surface of the steel slag or by rapid leaching from fine particles at the steel slag surface. The general tendency of V to be mobilized under oxic conditions and immobilized under reducing conditions is in agreement with observations that reduced V(IV) is very strongly adsorbed on oxide minerals (5), which are ubiquitous in steel slag.

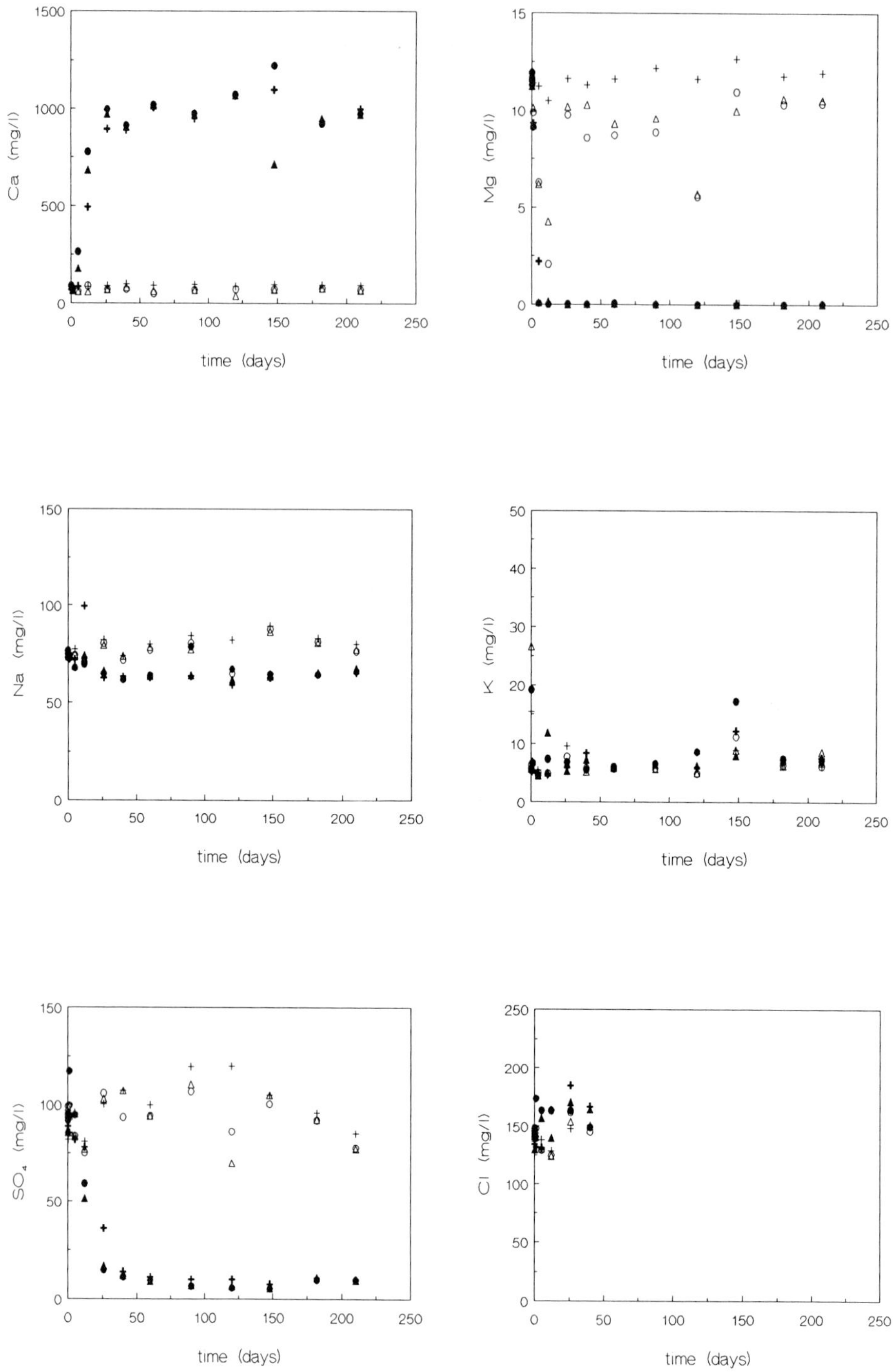
Ca (mg/l)
time (days)
Mg (mg/l)
time (days)
Na (mg/l)
time (days)
K (mg/l)
time (days)
SO4 (mg/l)
time (days)
Cl (mg/l)
time (days)

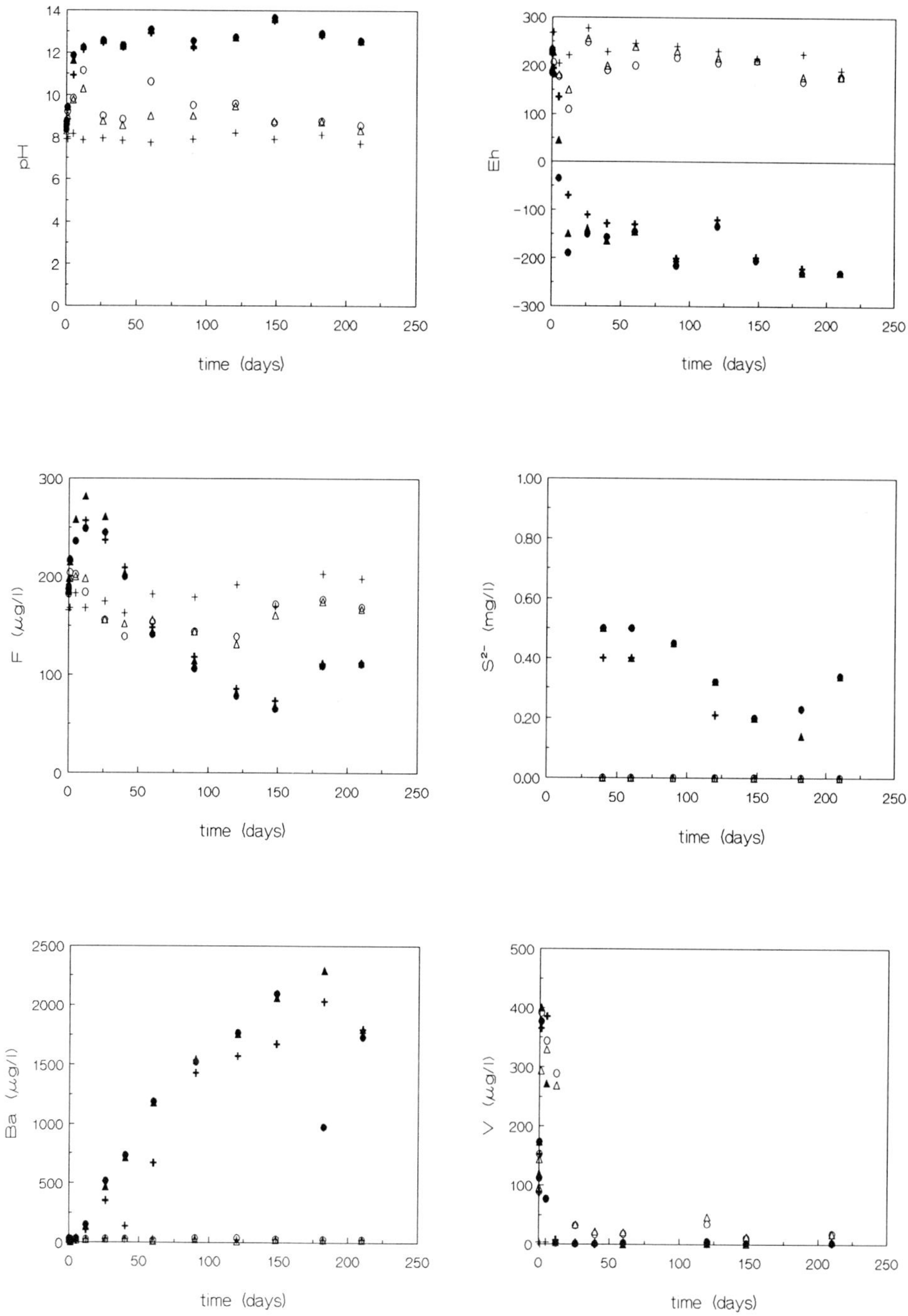

Figure 2. Composition of interstitial water in the simulated steel slag emplacement during the pilot scale experiment. (+) inflowing water; (Δ) outflowing water; (○) 15 cm depth; (+) 35 cm depth; (▲) 55 cm depth; (●) 75 cm depth.

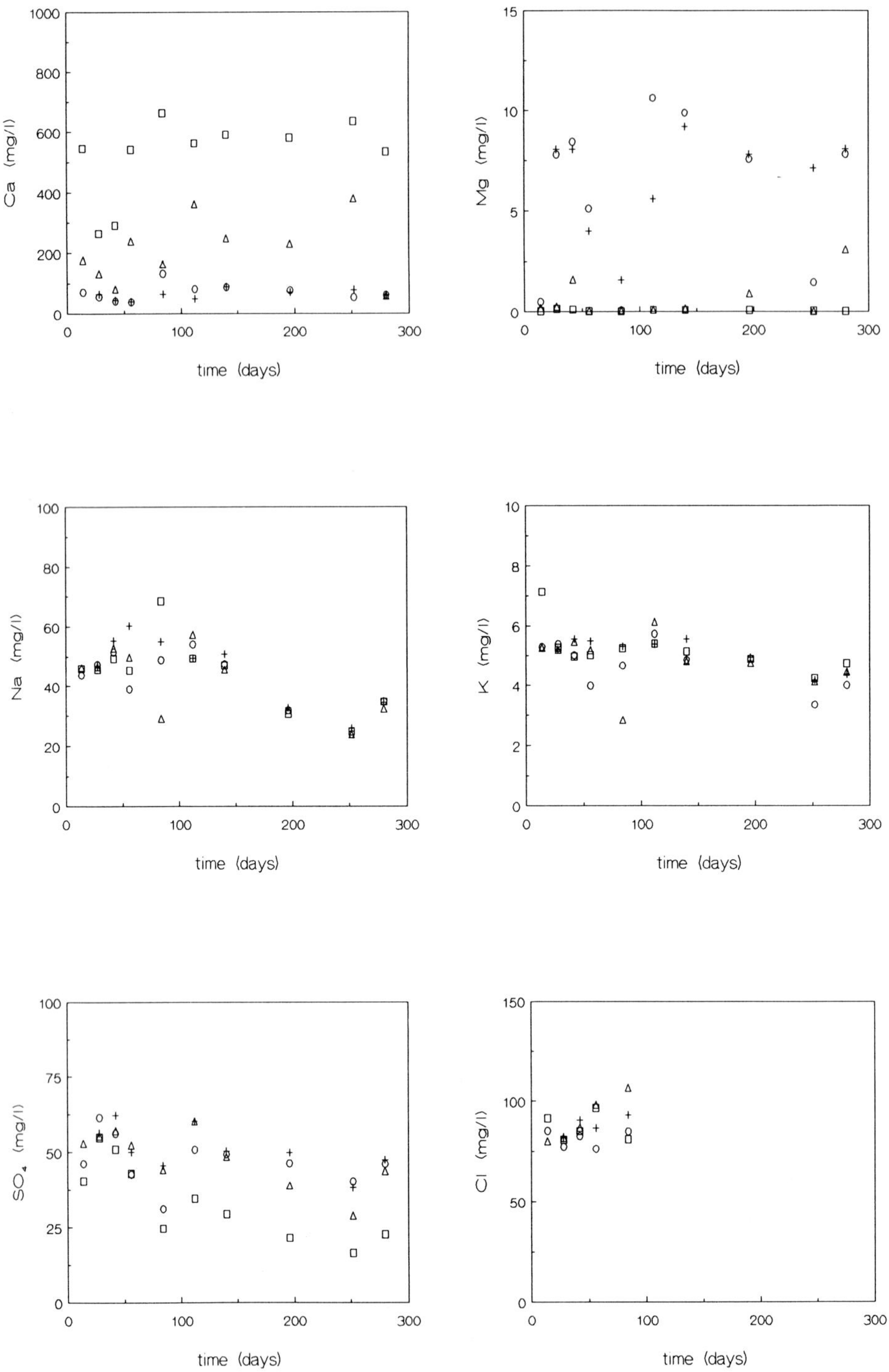
Ca (mg/l)
Mg (mg/l)
Na (mg/l)
K (mg/l)
SO_4 (mg/l)
Cl (mg/l)
time (days)

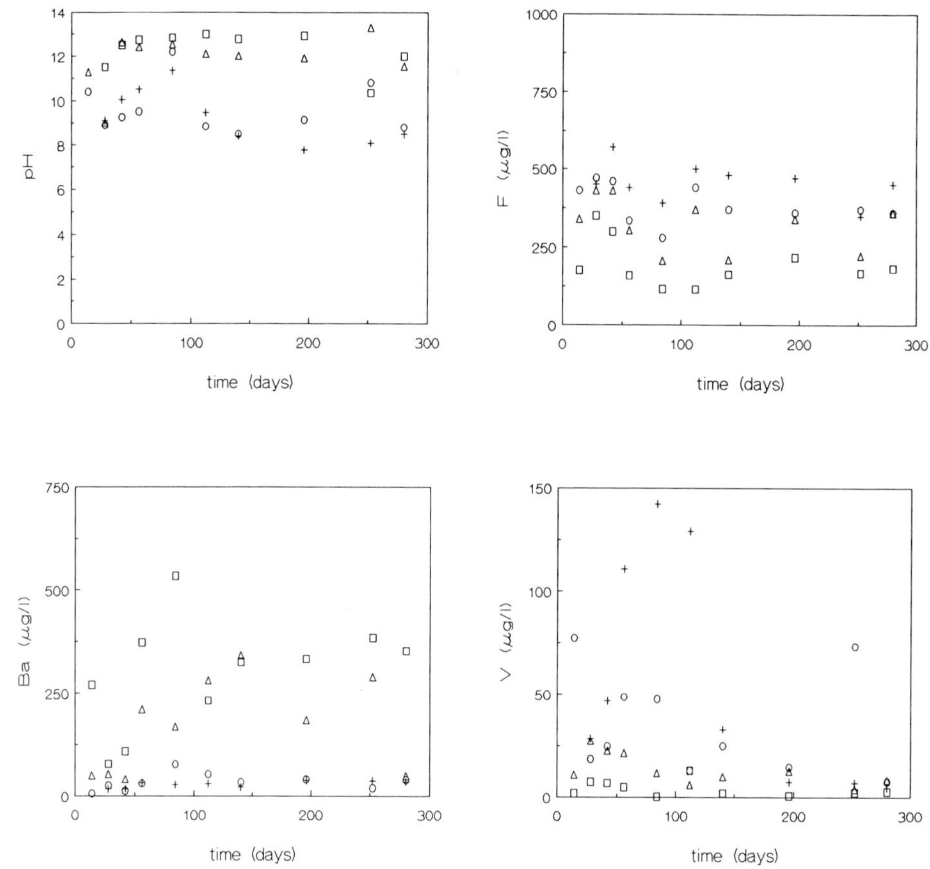

Figure 3. Composition of interstitial water in the steel slag emplacement in the field. (+) tube I; (Δ) tube II; (○) tube III; (□) tube IV.

Field measurements of interstitial water composition

The chemical composition of interstitial water sampled from the steel slag emplacement in the field is comparable to the pilot scale measurements (Fig. 3). Although no E_h measurements are available from these samples, the increase of pH values up to 13, together with the behaviour of redox sensitive elements such as Ba and V, indicates the development of highly alkaline and low redox conditions as observed in the pilot scale experiment. Sampling tubes I and III show, however, lower pH values over the entire monitoring period than tubes II and IV. Possibly, the former tubes were in more direct contact with the surface water. Tubes II and IV seem, therefore, more representative for the interstitial water composition.

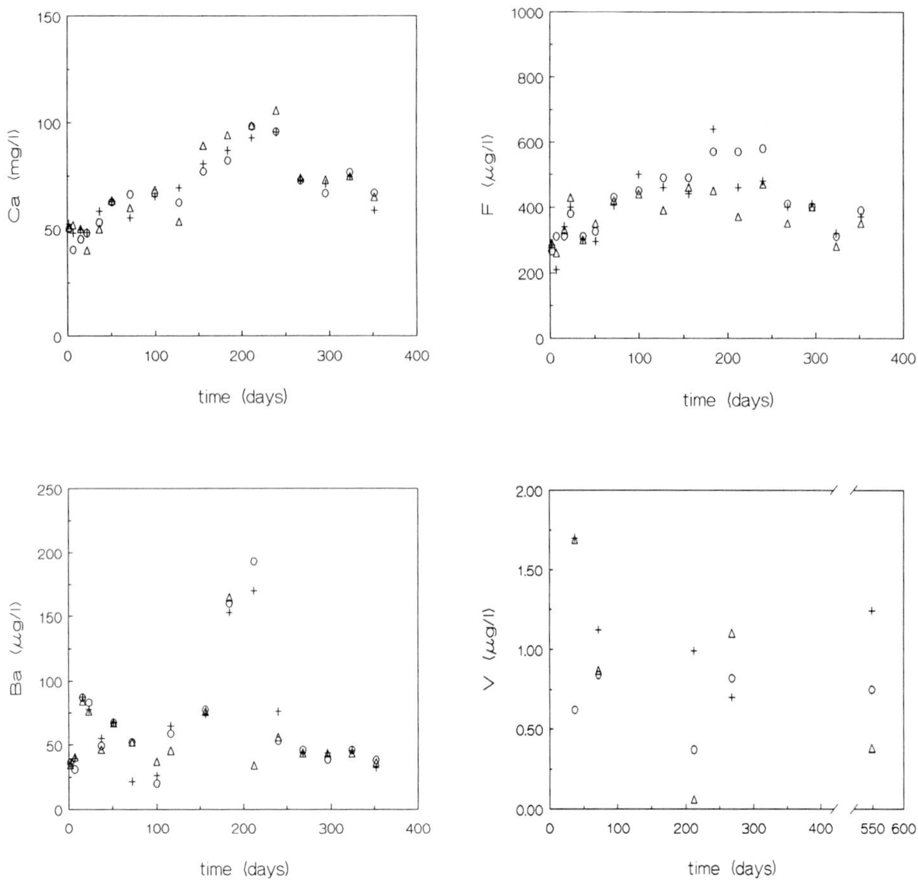

Figure 4. Concentrations of Ca, F, Ba and V in surface water at site I (+) and site II (Δ) near the steel slag emplacement in the field and at the reference site III (○), located more upstream.

Interstitial Ca concentrations reach an order of magnitude higher values (up to 600-700 mg/l) than those in the surface water (Fig. 4), and compare well with values measured in the pilot scale experiment. After approximately one year, when the tubes were removed from the emplacement, calcite coatings were observed over almost the entire length of the tubes. Magnesium is almost completely removed from solution in tubes II and IV, probably by precipitation of brucite, as indicated by the pilot experiment. Barium and SO_4 show the barite solubility control discussed above. Maximum barium concentrations are, however, about four times lower than measured at the pilot scale. Interstitial F concentrations are generally identical to those in surface water (Fig. 4), but are clearly lowered in tube IV. This behaviour may be related to the high Ca levels in this tube. The behaviour of V compares well with the observations in the pilot scale experiment. The highest concentrations are measured in the initial stage and are comparable within a factor of 2.5. Concentrations in tubes I and III are generally higher than in the other tubes which are more alkaline and probably reducing. These observations are again consistent with the observed immobility of V under reducing conditions.

Surface water measurements in the field

The composition of surface water was monitored at two locations (I and II) near the steel slag emplacement and reference samples were taken from location III. With the possible exception of V, none of the measured elements showed concentrations above those measured at the reference site. Observed concentration changes in time were consistent at all three sites and reflect seasonal fluctuations in the river. Figure 4 shows the surface water concentrations of Ca, Ba, V and F. Very low V concentrations were measured and large fluctuations were observed. However, the fact that concentrations at location I were consistently higher by a factor of 1.5-2 may indicate a measurable release from the steel slag emplacement.

DISCUSSION

The highly alkaline and reducing character of steel slag appears to determine its behaviour in the aquatic environment. Of the elements which are leached from the slag in significant amounts, Ca is controlled by pH whereas Ba and V are controlled by redox potential. The pilot scale experiment clearly revealed a sharp discontinuity in pH and redox potential within the interstitial water in the steel slag emplacement. This discontinuity is formed by the leaching of alkalinity and reducing sulphide species from below, and the mixing with near neutral and oxidized fresh water from above. This system seems to have reached a steady state in the pilot scale experiment.

The redox/pH interface plays an important role in controlling the leaching of Ca, Ba and V from the steel slag. These processes are illustrated in Figure 5. Calcium is leached from the steel slag in the reduced/high pH zone and diffuses upwards in response to the steep concentration gradient across the interface. The flux of Ca from below and dissolved CO_2 from above the interface lead to precipitation of calcite and a corresponding lowering of dissolved Ca in the oxic zone. The precipitation zone prevents, therefore, the leaching of large amounts of Ca to the overlying water. A similar process prevents a massive leaching of Ba from the system. Barium is mobilized in the reduced zone with low SO_4 levels and diffuses across the interface into the oxic zone. The flux of dissolved SO_4 from the overflowing water into the oxic zone causes Ba to precipitate as barite, leaving only low residual Ba concentrations in the oxic interstitial water. The behaviour of V at the redox/pH interface is opposite to that of Ca and Ba in that it is mobilized in the oxic zone. Dissolved V diffuses into the reduced zone and is immobilized, possibly by adsorption of V(IV) on oxide minerals. A portion of V(V) may coprecipitate with $Fe(OH)_3$ in the oxic zone. Some freshly precipitated $Fe(OH)_3$ was observed in the oxic zone, but this process does not seem important enough to significantly lower dissolved V concentrations. Part of the V may, therefore, have diffused into the oxic overlying water.

Although our sampling of the field emplacement was not detailed enough to demonstrate the existence of a sharp redox/pH interface in the interstitial water, there must be a similar transition zone between reducing conditions observed in the sampling tubes and the oxic surface

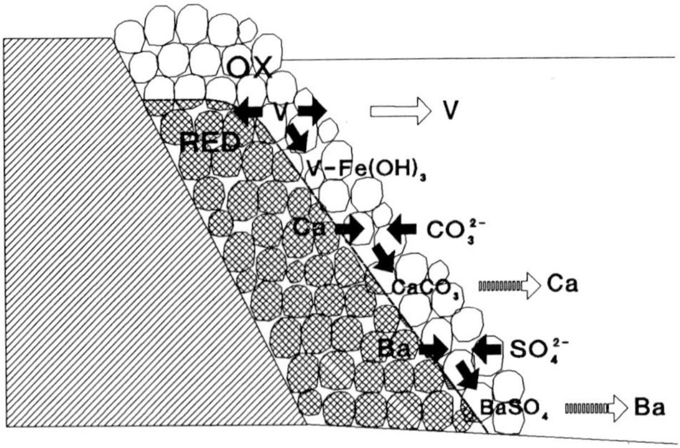

Figure 5. Schematic representation of processes involving Ca, Ba and V at the redox/pH interface in the steel slag emplacement.

water in the river. We believe that the location of this interface depends on the hydrodynamic conditions in the river and the thickness of the steel slag layer. The processes observed in the pilot scale experiment are therefore believed to reduce the emissions of Ca, Ba and V from the emplacement in the field. Surface water measurements near the emplacement support the hypothesis that, with the possible exception of V, emissions of the measured elements from the steel slag are negligible.

REFERENCES

1. H.A. van der Sloot. Leaching behaviour of waste and stabilized waste materials; characterization for environmental assessment purposes. Waste Management and Research, 8 (1990) 215-228.
2. Determination of leaching characteristics of coal combustion wastes. Dutch pre-standard NVN 2508, 1st edition February 1988.
3. Determination of the maximum leachable quantity and the emission of potentially hazardous components from construction materials, monolithic waste materials and stabilized waste products of mainly inorganic character. Concept Dutch pre-standard NVN 5432, edition March 1990.
4. H.A. van der Sloot. Neutron activation analysis of trace elements in water samples after preconcentration on activated carbon. ECN-1 (1976).
5. B. Wehrli. Vanadium in der Hydrosphäre; Oberflächenkomplexe und Oxidationskinetik. Ph.D Dissertation, Eidgenössische Technische Hochschule (ETH) Zürich, 1987.
6. M.J. Angus and F.P. Glasser. Mat. Res. Soc. Symp. Proc. 50 (1985) 547-556.
7. D.S. Brown and J.D. Allison. MINTEQA1, An equilibrium metal speciation model: User's manual. US EPA Report No. 600/3-87/012 (1987).

Waste Materials in Construction.
J.J.J.R. Goumans, H.A. van der Sloot and Th.G. Aalbers (Editors)

The Leaching Behaviour of Some Primary and Secondary Raw Materials Used in Pilot-scale Road Bases

E. Mulder

TNO Environmental and Energy Research, Dep. of Environmental Technology, P.O. Box 342, 7300 AH Apeldoorn, The Netherlands

SUMMARY

At TNO a two years lasting pilot-scale research was carried out into the leaching behaviour of some eight primary and secondary raw materials, used in road bases. Objective of the study was comparison of leaching characteristics between primary and secundary raw materials and verification of the suitability of standard leaching tests. The research was performed in test bins, measuring 1 x 2 meters, containing a sand layer, a foundation layer and an asphalt layer with grass verges. The materials that were investigated as a foundation layer were slags of municipal waste incineration, bottom ash, concrete destruction debris, lavalite, phosphorus slags, a mixture of blast-furnace and steel slags, a fly ash-cement stabilization, a sand-cement stabilization, and sand.

1. INTRODUCTION

From 1986 to 1989 the joint project 'Environmental implications of useful applications of primary and secondary raw materials' (Mammoth '85) was carried out by four institutes, i.e. the National Institute for Public Health and Environment (RIVM), the Netherlands Energy Research Foundation (ECN), INTRON, and the Netherlands Organization for Applied Scientific Research (TNO). The main objective of this project was to make an estimation of the environmental implications of replacing primary materials (natural materials) by secondary materials (waste materials) in road constructions and building materials.

Within the framework of this Mammoth project TNO Environmental and Energy Research (formerly TNO's Division of Technology for Society) carried out a pilot-scale research into the leaching of some eight primary and secondary raw materials, used in road bases. The objective of this pilot-scale leaching research was twofold, namely:

- to make a comparison between primary and secondary raw materials, when used in road bases;
- to verify the suitability of the Standard Leaching Test for predicting leaching in practice.

The research was performed on a pilot scale because of on the one hand the similarity with road constructions in practice and on the other hand the relatively low costs, making it possible to investigate more materials.

Chapter 2 describes the set-up of the research and the investigation methods. Chapter 3 presents the results, which are subsequently discussed in chapter 4. Finally, chapter 5 closes the paper with some conclusions.

2. SET-UP and INVESTIGATION METHODS

The following primary and secondary raw materials were investigated as a foundation layer in the road bases:

- slags of municipal waste incineration (MWI-slags);
- bottom ash (originating from a power generation plant);
- concrete destruction debris (unwashed);
- lavalite;
- mixture of phosphorus slags and furnace slag sand;
- mixture of blast-furnace slags, steel slags and furnace slag sand;
- fly ash-cement stabilization (the fly ash originating from a power generation plant);
- sand-cement stabilization.

Besides, sand was included as a reference (blank).
In chosing the primary and secondary raw materials, attention was paid that the selected materials are used or can be used in practice as road-base materials. Furthermore only those materials were chosen that were to be investigated in the Mammoth project on a laboratory scale as well with the help of leaching tests.

The pilot-scale research, that lasted for two years, was carried out in coated metal test bins, measuring 1 x 2 meters. Figure 1 shows a cross-section of such a test bin.

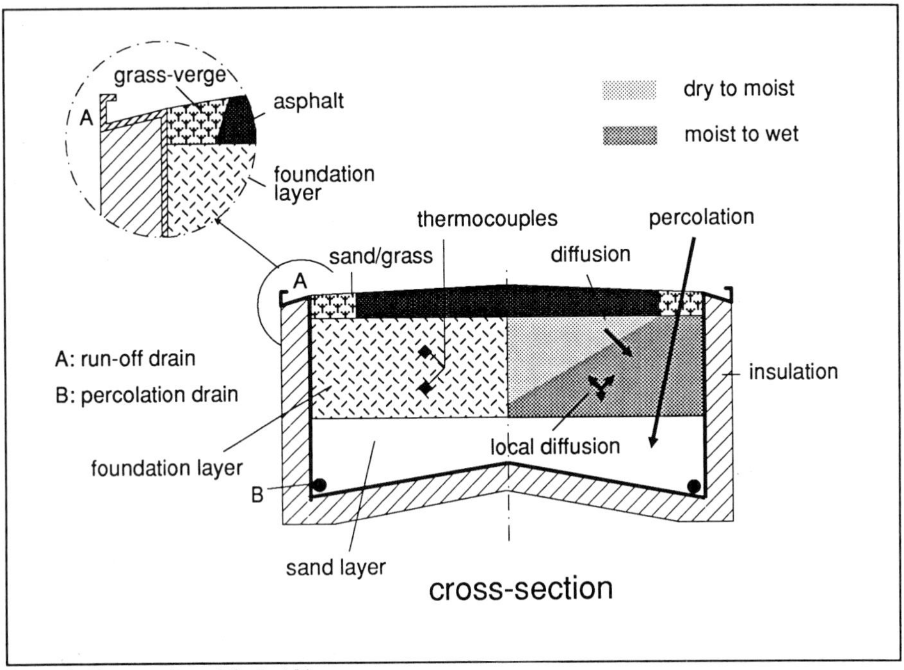

Figure 1: Cross-section of the test bins

The road-base construction in the test bins consisted of:

- A sand layer (drainage sand) of approximately 20 cm. This sand was constructed in a moist condition. A percolate drain was fixed at the bottom of this sand layer.

- A foundation layer of approximately 20 cm. The road-base materials were constructed in a moist condition too. In constructing a mixture of (secondary) materials, these materials were previously mixed in a concrete mixer. Eight thermo-couples were fixed in the foundation layer for on-line measurement of temperatures. (The blank had a sand layer of approximately 35 cm only).
- A 5 cm thick asphalt upper layer with a grass verge on both sides. Next to the verge, a drain was fixed for the run-off.

The test bins were placed outside under normal weather conditions.

During the two-year monitoring programme, on-line run-off and percolate quantities were monitored. Apart from this, the run-off and percolate flows were sampled intermittently, according to a fixed scheme; the samples were chemically analyzed as to main elements, anions, and trace elements. In order to be able to pronounce upon the (average) humidity level of the foundation layers, the test bins were weighed under different weather conditions. As, in particular in the beginning, many percolate samples contained sediment and/or fine particles, all samples were filtrated, while a number of filter-residues were subjected to further investigation.
Figure 2 presents an overview of the nine test bins.

Figure 2: Overview of the nine test bins

3. RESULTS

Starting from the measured quantities of run-off and percolate, a liquid balance of ingoing and outgoing flows was calculated for each test bin. These balances did not show large differences between the various test bins. It turned out that from the quantity of rain that entered the bins, only approximately 5% had drained away as run-off, while approximately 10% had evaporated.

The percolates from the test bins containing MWI-slags, bottom ash, lavalite, phosphorus slags en sand (blank) were neutral (pH 7-8). The (small quantities of) filtrated brown sediments from these percolates probably consisted of manganese and/or iron complexes, that were washed away from the sand layer. The percolates from the other test bins (containing concrete destruction debris, a mixture of blast-furnace and steel slags, fly ash-cement stabilization and sand-cement stabilization) were strongly alkaline (pH 12-12.5). The large quantities of sediment that were found in these percolates, appeared to consist, for the greater part, of humic acids washed away from the sand layer by the strongly alkaline percolate.

Weighing of the test bins showed that their humidity varied only slightly under different weather conditions. However, during a long, dry period a certain drying-up occurred. The average humidity per test bin varied from 8 to 23%. The humidity level of the different road-base materials was calculated on the basis of the average humidity, the calculated porosity of the material, and a number of assumptions. It turned out that the calculated humidity levels varied between 40 and 90%.

In the case of laboratory leaching tests, measured concentrations of main elements, anions and trace elements are generally related to the liquid-solid (L/S) ratio. In order to do this for the pilot-scale results as well, the real time scale was converted into an L/S scale with the help of the aforementioned liquid balances and humidity levels. Multiplication of the measured concentrations by these L/S ratios resulted in emissions in mg/kg. For the comparison of the pilot-scale results of the different materials among themselves and with the results of standard leaching tests cumulative emissions were calculated at L/S = 5 l/kg. Table 1 presents the cumulative emissions for the main element calcium, for the anion sulphate and for the trace elements arsenic, chromium, copper, manganese, molybdenum, vanadium, and zinc. The table also presents average pH values for all test bins.

Table 1: Cumulative emissions (mg/kg) at L/S = 5

	Ca	SO_4	As	Cr	Cu	Mn	Mo	V	Zn	pH
MWI-slags	1530	3650	<0.030	<0.028	0.25		0.284		0.36	7.6
Bottom ash	766	1439	0.053	<0.009	0.04		0.280			7.6
Concrete destruction debris	709	170	0.100	0.039			<0.089	0.12	0.16	12.0
Lavalite	394	697	0.061	<0.010		0.59	0.056		<0.08	7.8
Phosphorus slags	2238	4820	0.104	0.013			0.147			7.8
Blast-furnace/steel slags	947	1887	0.260	0.021			0.120			12.0
Fly ash-cement stabilization	244	600	1.120	1.750			4.350	1.14	0.07	12.3
Sand-cement stabilization	1794	234	0.178	0.018		<0.11	0.065	0.18	<0.12	12.4
Sand (blank)	555	663	0.022	<0.005		<0.38	0.033			7.5

4. DISCUSSION

4.1 Comparison between primary and secundary raw materials

The various materials investigated, both primary and secondary, were compared as to leaching, on the basis of cumulative emissions at L/S = 5, as presented in table 1.
The primary road-base materials (sand, lavalite, and sand-cement stabilization) generally show a small release of main and trace elements. An exception to this is the release of calcium and arsenic from the sand-cement stabilization, which is rather high. The higher arsenic release is probably the result of a higher arsenic content in the stabilization and/or a higher availability for leaching of arsenic from the cement particles. The higher release of calcium (and chromium) is linked up with the high pH of the percolates.
Compared to the reference materials (primary raw materials), concrete destruction debris is the only secondary road-base material investigated with a release of components in the same order of magnitude as that of the reference materials. The other secondary materials show a considerable (more than a factor of 5) to a high (more than a factor of 20) release of these components:

- *MWI-slags:* a considerable release of sulphate, copper, molybdenum, and antimony;
- *Bottom ash:* a considerable release of molybdenum;
- *Phosphorus slags:* a high release of fluoride, a considerable release of sulphate;
- *Blast-furnace and steel slags:* a considerable release of arsenic;
- *Fly ash-cement stabilization:* a high release of chromium and molybdenum, a considerable release of arsenic and vanadium.

It can be concluded that fly ash-cement stabilization, overall, has the highest release of trace elements. Phosphorus slags have a high release of fluoride only. The aforementioned elements could, in general, be used as guide elements in assessing these kind of materials as road-base materials.

4.2 Verification of laboratory leaching tests with the help of the pilot-scale research

For this reason, the results of the pilot-scale research were compared with those of the laboratory leaching tests (column test [3] and tank leaching tests [4]) and of large-scale column experiments [5]. All investigations were performed within the framework of the Mammoth project. The large-scale column experiments were carried out in order to demonstrate the effect of size reduction (as a necessity for small-scale laboratory testing) and direction of flow (upflow or downflow) in column leaching tests.
The leaching results were compared again on the basis of the cumulative emissions at L/S = 5. The results of the tank leaching tests (expressed in mg/m^2) were converted to mg/kg through multiplication by the leaching area (the surface between the foundation layer and the sand layer) and division by the moistened amount of road-base material. The conversion of the time-scale to the L/S-scale was carried out in the same way as in the pilot-scale research.

Table 2 presents the results of the different investigations for two secondary raw materials (MWI-slags and fly ash-cement stabilization).

Table 2: Cumulative emissions of laboratory and large-scale investigations at L/S = 5

	Ca	SO_4	As	Cr	Cu	Mo	Zn	pH
MWI-slags								
laboratory column test	3560	2790	0.02	0.09	4.29	0.63	0.35	11.7
large-scale column UP	2290	4650	0.03	0.05	3.50	0.52	0.25	9.6
large-scale column DOWN	1800	3940	0.04	0.26	5.35	0.42	0.25	9.9
pilot-scale road base	1530	3650	0.03	0.03	0.25	0.28	0.36	7.6
Fly ash-cement stabilization								
laboratory tank leaching test	91	105	0.03	0.61	0.01	0.72	0.02	11.2
laboratory column test	53	275	0.09	1.76	0.02	1.40	0.05	12.3
large-scale column DOWN	227	427	0.21	1.99	0.25	5.56	0.25	12.6
pilot-scale road base	244	600	1.12	1.75	0.34	4.35	0.07	12.3

In comparing these numbers, one should take into account that many measured concentrations are just above the detection limits, which may cause the cumulative emissions to deviate up to a factor of 2. When considering the results in table 2 in this way there is one systematic difference, i.e. between the laboratory tank leaching test and the other tests and large-scale researches. This aspect will be discussed in section 4.3.
Other, specific differences between leaching on laboratory and on large scales will be discussed now with the help of some aspects that affect the leaching behaviour of primary and secondary raw materials. These aspects are direction of flow, size reduction, pH, and the presence of a sand layer underneath the leached materials.

Direction of flow

In the standard leaching test, the choice has been made for upflow in the column leaching test to reach a better reproducibility. In practice, however, there is downflow, as well as in the pilot-scale road bases, and in one of the large columns. Especially the large column experiments showed that in the case of downflow pH's were higher and that oxidation reactions occurred. In general, leaching behaviour was not affected by this phenomenon, but it was, in the case of specific elements (e.g. chromium).

Size reduction

For the laboratory column test, particles have to be reduced to 3 mm. This means first an increase in specific area, which in turn will affect the release of components. Eventhough, this has not been proved in the large column experiments. Also an other research [6] stated that the coarse fraction (which was reduced) did not significantly contribute in the specific area.
On the other hand, the results in table 2 show that size reduction may give rise to increasing pH values. For MWI-slags, for instance, this may be caused by impermeable skins going into pieces when the slags are broken.

pH
As mentioned before, size reduction (for laboratory leaching tests) may give rise to an increasing pH. This is linked up with the leaching of (earth)alkali-metal (hydr)oxides. Differences in pH value may also be the result of carbonation, a conversion of an oxide into a carbonate because of a reaction with carbon dioxide from the environment [7]. This effect may appear more easily in longer lasting research and in practice than in laboratory tests.
The solubility (and also the leachability) of many trace elements depends on the pH of the liquid phase. In this way, via the pH, the release of trace elements (arsenic, copper, molybdenum) is affected by the release of, for instance, calcium oxide.

The presence of a sand layer
The presence of the sand layer underneath the road base affects the leaching of the various components in at least three different ways:
- An initial effect caused by the replacement of water that is initially present in the sand layer. This effect cannot be observed equally well for all components and test bins.
- Adsorption of components leached from the road-base materials to sand particles in the sand layer. This effect was observed for, for instance, copper.
- Domination of the leaching from the sand layer, thus masking the leaching from the road-base materials. This effect was observed for a number of components, like sulphate and manganese.

The third effect can be illustrated starting from figure 3. This figure gives the cumulative emission of sulphate (as a funcion of the L/S-ratio) for the bin with sand only (bin 9) and for the bin with bottom ash (bin 2). As the figure shows, both leaching curves are similar, from which may be concluded that the leaching from the bottom ash is an apparent leaching only. In reality, it is a leaching from the sand layer underneath the foundation layer.

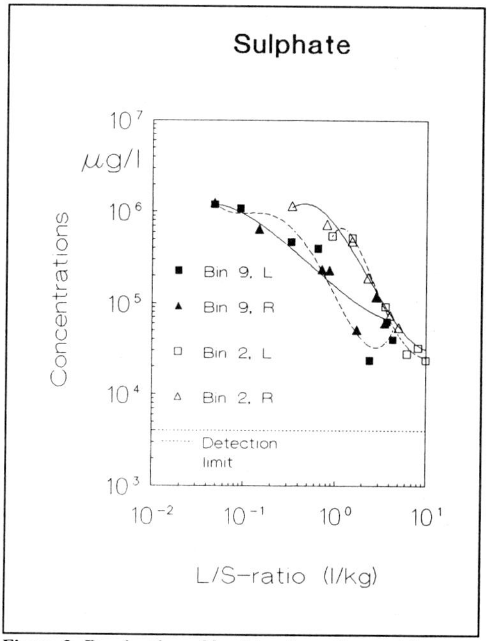

Figure 3: Domination of leaching from the sand layer

4.3 Leaching mechanisms in road bases

The previous section mentioned a systematic discrepancy between the extrapolated results of the tank leaching test and the results of the pilot-scale road-base research. The two different leaching machanisms that play a part in leaching in practice (i.e. diffusion and percolation) account for this discrepancy.

Diffusion
This leaching mechanism occurs for instance in all kind of products, especially porous products. When the material is moistened, the pores are filled up with water, which causes many metal oxides to dissolve. The dissolved components, however, can reach the environment only after diffusion through the pore structure of the material [4]. In this leaching mechanism diffusion is the rate determining step. The diffusion rate depends, among others, on the structure (matrix) of the product and on the distance to be covered to the water phase outside the product. In the tank leaching test only this type of leaching is measured and quantified.

Percolation
This mechanism occurs when there is an enforced flow of the liquid phase through a granular material. Also in this case, of course, there is diffusion, but because of the small distances to be covered, diffusion is not the rate-determining factor in most cases. The overall emission in this leaching mechanism depends primarily on the rate of percolation and is therefore higher than in the case of diffusion only. In the column leaching test, percolation is the main leaching mechanism.

In the pilot-scale road bases both diffusion and percolation will play a part in the overall leaching. This is schematically shown on the right-hand side of figure 1. Starting from an impermeable asphalt layer, no percolation will occur in the middle of the road base. Leaching is here a result of diffusion (to the percolation zone). Rain water, flowing from the asphalt layer, and percolating through the grass verge is enforced to percolate through the foundation layer as well (because of the fact that the foundation layer continues as far as the wall). As mentioned in chapter 3, the major part of rain water entering the test bins percolated through the foundation layer. All this explains that percolation was the dominating leaching mechanism in the pilot-scale road bases. This also follows from the results, presented in table 2.
In practice, however, the situation is rather different. Road bases in practice are much wider and there is no percolation enforcement. In practice diffusion will occur under the asphalt layer, as it does in the pilot-scale road bases. (If the asphalt layer is really impermeable, there may even be no leaching at all in the middle of the roadbase.) The borders of road bases in practice are only partly covered with asphalt in most cases. In these borders percolation will occur, but not to the same extent as in the pilot-scale situation, as a result of the possibility of flowing round about (and not through) the roadbase. Though leaching is a mixture of diffusion and percolation in a road base in practice, overall leaching goes more in the direction of diffusion. In view of the substantially lower release rates in diffusion-controlled systems, it is advised to account for this aspect in road construction designs (and cover the whole border with asphalt).

5. CONCLUSIONS

* The reference (= primary) materials investigated (sand, lavalite, and sand-cement stabilization) show a small release of trace elements. Compared with these primary materials, the investigated secondary materials show:
  - a release in the same order of magnitude for *concrete destruction debris*;
  - a reasonable release (more than a factor of 5) of some elements for *MWI-slags, bottom ash and blast-furnace and steel slags*;
  - a high release (more than a factor of 20) of one or more elements for *phosphorus slags and fly ash-cement stabilization* respectively.

* A comparison between the results obtained in this pilot-scale research and those obtained during the laboratory leaching research, leads to the following statements:
  - For some specific materials and/or elements, differences are observed between leaching on laboratory scale and on pilot scale, due to differences in flow direction, size reduction, pH, or the presence of the sand layer, but
  - Generally, the laboratory leaching tests, developed by SOSUV, are uesful instruments in simulating, within a short period of time, the leaching in practice over a long period of time.

* The leaching research in pilot-scale road bases shows that two mechanisms of release play a part in a road-base application: percolation (simulated in the column leaching test) and diffusion (simulated in the tank leaching test).

6. REFERENCES

[1] E. Mulder, The leaching behaviour of some primary and secundary raw materials used in pilot-scale road bases (in dutch),
Mammoth report 11, TNO-report ref.no. 90/006721, Apeldoorn, May 1990.

[2] C.W. Versluijs a.o., Mammoth '85, integral evaluation of the subprojects (in dutch),
Mammoth report 13, RIVM-report ref.no. 738504008, Bilthoven, July 1990.

[3] Th.G. Aalbers, J. Keijzer and R. Gerritsen, Leaching behaviour of primary and secundary raw materials (in dutch),
Mammoth report 7, RIVM/TNO-report ref.no.738504010, Bilthoven, July 1990.

[4] G.J. de Groot a.o., Characterization of the leaching behaviour of intact products (in dutch),
Mammoth report 9, ECN-report ref.no. C--90-007, Petten, March 1990.

[5] I.H. Anthonissen a.o., Research into the practical suitability of the standard leaching test by means of scaling up (in dutch),
Mammoth report 10, RIVM-report ref.no. 738504007, Bilthoven, April 1990.

[6] R. Gerritsen and E. Mulder, Influence of particle size reduction on the leaching behaviour of fly ash/cement prismas and phosphorus slags (in dutch), TNO-report ref.no. 88-064, Apeldoorn, February 1988.

[7] R. Gerritsen a.o., Fly ash in coarse-ceramic products (in dutch), TNO-report ref.no. 90-014425, Apeldoorn, October 1990.

Waste Materials in Construction.
J.J.J.R. Goumans, H.A. van der Sloot and Th.G. Aalbers (Editors)

Standardization of terminology, characterization methods, acceptance procedures and leaching tests for waste materials

M.J.A. van den Berg, P.M. Eckhart and W.P. Bijl
Nederlands Normalisatie Instituut, P.O. Box 5059, 2600 GB Delft

SUMMARY

For the appropriate handling and treatment of waste, standardization of terminology, classification, acceptance procedures and test methods for determination of composition and leaching behaviour of waste materials is very important. Chemical and physical characterization of the composition of waste materials by sampling and analysis is essential for effective decision making upon the kind of treatment and final disposal of the various wasteflows. Furthermore performance of leaching tests is necessary in order to predict and control potential contamination from landfills and re-used waste materials. Within various countries definitions of waste materials, classification methods and test procedures with respect to the treatment of waste, have been developed. These definitions and methods, however, are not always comparable and therefore standardization is necessary to establish unambigious criteria, both on a national and international level, to ensure environmentally safe treatment of wastes. The Netherlands Normalization Institute (NNI) set up a national standardization programme in the field of waste management, part of it in close collaboration with the Netherlands Agency for Energy and Environment (NOVEM) and took the initiative to propose a similar programme on the European level.

1. Introduction

1.1 The problem of waste

Considering the affection of the environment, the production and handling of waste is an increasing problem in the industrialized world. One of the most difficult aspects is the final destination of waste. In the past landfilling was considered as the best and cheapest way of waste disposal. In the last few years, however, it has been learned that landfills caused in many cases very serious and unacceptable soil and groundwater contamination.

The amounts of the waste produced and its contamination aspects are distressing to such an extent that ways of disposing other than landfilling have to be found.

1.2 Dealing with waste in the Netherlands

In the Netherlands landfilling is still the mostly applied

'solution' for the waste problem (60%), but incineration and recycling of waste are becoming more and more important. The policy of the government is that by the year 2000 the amounts of waste to dispose of should be reduced by prevention (10%) and recycling (55%). Landfilling of waste should be reduced to only 10% of the total amount, which is still about 3 million tons a year.

1.3 Classification of wasteflows

When separating waste at its source it is possible to categorize wasteflows in e.g. domestic, demolition or chemical waste. After a thorough investigation of the constituting compounds - by chemical analysis of samples taken from the waste - a proper way of treatment can be chosen. A good classification system gives the people who are handling the wasteflows the opportunity to transport and allocate the waste in the appropriate way without analyzing it each time, which is far more efficient with respect to time and money.

1.4 Leaching characteristics

In the case of re-using or landfilling of waste, knowledge of the leaching characteristics of the material is necessary. In order to predict and control potential contamination, the availability of standardized leaching tests is essential. Several institutes in countries like the USA, Canada, Germany and the Netherlands are developing leaching tests for various waste materials. However, these tests, nor the terminology or classification methods regarding waste, are comparable.

To ensure safe treatment of waste - from an environmental point of view - throughout the world, unambigious definitions, criteria for classification and acceptance and standardized methods for sampling, analysis and leaching tests are needed.

2. Standardization

2.1 The aim of standardization

Standardization institutes, like NNI in the Netherlands, develop unambigious standards for terminology and methods for measuring physical and chemical characteristics of diverse objects. In the process of standardization representatives of all interested parties combine their knowledge and experience in order to reach a consensus. In case of environmental issues interested parties are industrial organizations (producers of waste as a by-product of an industrial process and the organizations which transport or treat the waste), commercial laboratories as well as public research institutes and the government. The consensus they are trying to reach will focus on several questions: "What are the most commonly used definitions for the various waste materials?"; "What are the most suitable and applicable methods for classification and characterization?"; "What are the most desirable procedures for acceptance of waste?" and "What are the most applicable test methods for determination of the leaching characteristics of the various compounds that the waste consists of?"

The consensus on the diverse items results into standards that are published by national or international organizations like ANSI (USA), DIN (Germany), CEN (Europe), ISO (Worldwide) and, of course, NNI (The Netherlands). These standards are important tools for governmental agencies that have the task to control the appropriate (legal) handling of waste, since these standards are unambigious and they are generally accepted by all parties involved.

2.2 Reaching consensus

The various parties that are involved in the development of standards on environmental issues, in this case on the problem of waste, represent different interests. Governmental controllers generally prefer to know the precise concentrations of contaminating compounds in waste, which are obtained by the application of precise and probably expensive methods. Commercial laboratories, however, prefer efficient, easy and cheap methods for routine use. Industrial organizations do not want to take great economical risks, therefore e.g. methods of analysis and leaching tests should not be too expensive from their point of view, but should - above all - be enforced to all their competitors and leave no room for disputing the obtained results by controlling agencies.

2.3 Benefits of standards

When a government wants to protect the environment against further pollution she may set legally bounding maximum acceptable values for contaminating compounds. However, without standard methods for measuring these contaminating compounds, enforcement of legal regulations will be difficult and the results of measurements may be questioned. When standard methods are available there will be less disputes about the meaning and

precision of certain measured values. Furthermore all parties involved will be treated equally by applying standardized procedures for methods and tests.
But there are more benefits. The communication on waste problems is less difficult when everybody uses the same terms for the same subjects.
With standard procedures for classification and characterization of waste, its acceptance or refusal at treatment facilities can be based on commonly accepted rules. Also, proper destinations for the various waste materials can more easily be chosen and govermental controllers have less problems with assessing the environmental risks for transportation, landfilling, re-using etc. of waste.

3. Standardization programme in the Netherlands

In the field of environmental issues the working programme of NNI is executed by technical committees in the following sectors: water, soil and air quality, radio-activity and waste materials. The environmental standardization projects are financed by the Dutch government mainly.
For the time being there are two NNI technical committees on waste materials. One has a broad scope with respect to standardization in the field of waste materials; the other has the specific task of standardizing leaching tests (see Figure 1).

Figure 1: The organization for standardizing with respect to waste materials

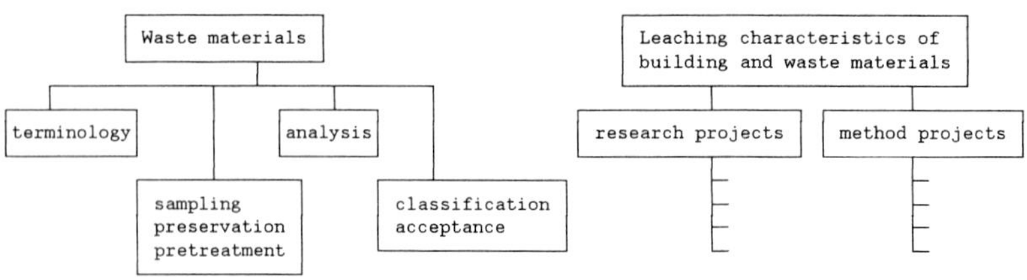

The reason for establishing two committees instead of one was a practical one. Within the terms of the Act on Soil Protection, the Dutch government has planned to publish in 1992 a decree on building materials, in which the government wants to refer to standardized leaching tests. In order to provide for these tests at short term a large project was set up, financed by the

government. The part of this project devoted to the development of standard methods for leaching tests (on building and re-used waste materials) is supervised by the NNI-committee.
The other committee is responsible for all other standardization issues with respect to waste materials. Due to pressure of time and to prevent retardation of the leaching test project the two technical committees were not combined. In the future the organizational structure of standardizing issues in the field of waste materials will probably be reconsidered.

3.1 NNI Technical Committee on waste materials

In 1991 the technical committee on waste materials was restructured to be able to develop, in a systematical and efficient way, standards that are related to the Netherlands' Acts on Waste. Four subcommittees have been established that are responsible for setting up or continuation standardization projects on specific subjects. The projects of these subcommittees are described in the next paragraphs.

3.1.1 Terminology

The subcommittee on Terminology started its work about 10 years ago. At this moment, the national standard on waste terminology (dating from 1985) is being revised. The existing terms and definitions will be updated and new sections on e.g. hospital waste, building and demolition waste and dredging sludge are added. This terminology is used in all sorts of organizations and improves the communication on waste management.

3.1.2 Classification and acceptance

The subcommittee on Classification and acceptance will establish a classification system for waste flows. These should be classified into unambigious, logically coherent groups, subdivided on basis of the way of disposal or on basis of analogous treatment in testing procedures (like sampling or characterization of the compounds of the waste).
It is preferred to standardize a classification system that can be used by processing industry, transporting industry, waste treatment facilities and examinating agencies, controlled by the government
A second task of this subcommittee is to develop unambigious acceptance procedures for wastes by treatment and disposal facilities, including a sampling strategy and analytical procedure. For well-known, properly classified wasteflows it is not necessary to apply the complete procedure of sampling and

analysis every time the waste is presented to a waste disposer. Mostly the composition of such wasteflows does not vary much from time to time and therefore one thorough investigation is sufficient, followed by random tests.

Furthermore the subcommittee should make a guideline on the use of standard methods for testing and analysis, giving guidance on which sampling, sample pretreatment and analytical methods should be applied for the various types of waste.

3.1.3 Sampling, preservation and sample pretreatment

According to the standardization programme of the subcommittee on Sampling, preservation and sample pretreatment, standard methods will be developed for sampling of solid and fluid waste, for preservation of samples that are not readily analyzed and for pretreatment of samples. E.g. samples should be taken and homogenized in such a way that the subsample that is analyzed is a representative part of the bulk material. Of course the standard methods for sampling and sample pretreatment developed by the NNI technical committees on water and soil quality will be very useful as a starting point.

3.1.4 Analysis

The subcommittee on Analysis will develop standards for the various aspects of analysis of organic and inorganic compounds in waste flows. Important aspects are the partial or total destruction of a sample and the instrumental analysis. Much attention will be paid to the development of rapid screening methods. Again the standards already developed by the NNI technical committees on water and soil quality for analysis will be helpful.

One of the most important stages in the development of a standardized method is the validation. Therefore interlaboratory tests should be performed.

3.2 NNI Technical Committee on leaching characterization of building and waste materials

As mentioned before, the working programme of the committee on leaching characterization of building and waste materials is strongly connected with the Dutch decree on building materials. In a former research project on the leaching characteristics of coal combustion slag, much knowledge on leaching tests has been gathered. The projectteam responsible for this project was of the opinion that for standardization of leaching tests for various materials, a methodology comparable to the one followed for testing the leaching behaviour of coal combustion slag, could be used.

To obtain the necessary knowledge on leaching behaviour, an extensive programme of projects has been set up, to support the work of the NNI technical committee. This programme is financed by the government mainly and it is managed by the Netherlands Agency for Energy and Environment (NOVEM). Two types of projects can be distinguished: research projects and projects with the aim to develop test methods (see also figure 1). In the next paragraphs the two types of projects are described further.

3.2.1 Research projects

In the research projects fundamental knowledge will be obtained on the leaching behaviour of several constituting components of waste. This scientific information will form the basis for future development of standards on leaching test methods.

3.2.2 Method projects

The goal of the method projects is to develop standardised leaching tests within a short time, based on already available knowledge of leaching behaviour. To perform these tests, standardised procedures for sampling, sample pretreatment, destruction and analysis of the soil and water polluted by leaching of building and waste materials are needed as well. Already available standard methods (e.g. from the NNI technical committees on water and soil quality) will be evaluated with respect to their applicability for determination of the leaching behaviour of waste, and, if necessary, new standard methods will be developed. Much attention will be paid to the validation of existing methods for destruction and analysis of inorganic compounds in solids and solid residues, and to the development and validation of methods for destruction, extraction and analysis of organic pollutants. Polluting inorganic compounds that can leach from building and waste materials like flyash, granulate, dredging sludge etc. are metals and metal salts. Organic contaminants are e.g. polyaromatic hydrocarbons, polychlorobifenyls and other halogenated organic compounds.

Besides methods of analysis for eventually polluted soil and water, methods for analyzing the physical and chemical constitution of building and waste materials itself, to assess their potential leaching behaviour, will be obtained.

A standard method for the determination of the maximum

leachability of inorganic compounds is already in an advanced stage of development. An existing draft standard on the characterization of leaching behaviour of non volatile organic compounds will be evaluated on its applicability.
In another method project a standard method will be developed to investigate the correlation between the geometrical surface and leaching behaviour and the influence of particle size reduction on leaching properties (because the building and waste materials that are (re-)used often are crushed in order to be applicable).
Furthermore it is known, that certain slags have during a given time a reducing effect on the leaching behaviour of materials they are mixed with. This reducing capacity, however, depends on the amount of oxygen that is present where the slags are applied. In one of the method projects a standard test method will be developed for the determination of the reducing capacity of these slags under various conditions.
In another project the goal is to develop validated standards for 'condensed' leaching tests: these tests should rapidly give information about the potential environmental contamination aspects of new (building) materials. For the validation of these tests the availability of appropriate reference materials is required. Application of these tests should simplify the environmental quality control of new materials by the government and give the industry the opportunity of acquiring certificates.

4. International standardization on waste materials

Within the International Organization for Standardization (ISO) there is yet no technical committee on waste materials, but in Europe the Netherlands have put forward a proposal for a Technical Committee in CEN (Comité Européen de Normalisation). The reason for doing this is that the problem of waste is not restricted to a national level and that it will be much more efficient to develop standards directly on an international level than to wait until all European countries have already developed their own. As these standards will be the result of a consensus process they will be generally accepted in Europe and therefore the implementation of EC-directives, referring to such standards, will be easier and far more effective. Several EC-directives on waste materials have been published already or are in preparation, for example a proposal for a Council Directive on the landfill of waste (October 1990), a proposal on the

Waste Materials in Construction.
J.J.J.R. Goumans, H.A. van der Sloot and Th.G. Aalbers (Editors)

LEACHING TESTS FOR CONCRETE CONTAINING FLY ASH - EVALUATION AND MECHANISM

R. H. Rankers and I. Hohberg
Institut für Bauforschung, RWTH Aachen, Schinkelstr. 3, D-5100 Aachen (Germany)

SUMMARY

Various leaching tests were carried out with mortar mixtures containing fly ash. Mortar specimens using one bituminous coal fly ash and one fly ash from municipal waste incineration were prepared. The results of the leaching tests were compared and classified. In addition, diffusion rates for certain elements were determined by means of tank tests with mortar specimens.

1. INTRODUCTION

Fly ash has long been used as an concrete addition or cement component. In Germany, interest has hitherto centred solely on the technological properties of fly ash, but environmental impacts are now of increasing imortance.

Fly ash contains some soluble salts and trace elements with probable environmental effects.

In order to evaluate the environmental impact resulting from the use of fly ash in concrete, it is necessary to aquire an understanding of the leaching properties of the contaminants. These data can be obtained only by means of leaching tests carried out in the laboratory.

The leaching tests determine the total amounts of leachable contaminants (leaching rates). These leaching rates subsume a number of superimposed effects, of differing degrees of importance. Depending on the concentration of a trace element in the pore water and in the contact (elution) solution, diffusion may take place from the solid to the eluate or vice versa. Diffusion is accompanied by solution phenomena, consisting of either (a) initial contact dissolution, in which soluble components are adsorbed on the surface at the first contact with the eluate or (b) continous dissolution, when the entire material (matrix) is highly soluble. The leaching rate is extremely dependent on the concentration and solubility of the contaminants and on the permeability of the matrix (1).

When fly ash is incorporated in cement, the surface area exposed to the environment and hence the leaching of toxic ocomponents are reduced. Toxic components are immobilized by (a) chemical reaction, in which the form of the contaminant is altered (e.g. through precipitation, changes in the pH value) and (b) complexing or sequestering the contaminant in a solid matrix (2).

Fixation does not normally result in the total immobilization of contaminants, but slows down the contaminant leaching rate. In the course of time, the matrix may become less alkaline, allowing insoluble salts to become soluble and resulting in potential environmental stresses due to toxic materials.

The evaluation of leaching data on fly ash products is extremely difficult, owing to the large number of different leaching processes involved. The present paper aims to compare and classify results for a number of different leaching procedures.

2. PROCEDURE

2.1 Materials

Two fly ashes were selected for use in the tests: a bituminous fly ash approved as a concrete addition (SFA) and a fly ash from municipal waste incineration (MVA). Mortar mixtures were prepared according to German standard DIN 1164 Part 5 (3) or to certification regulations (4), using a German PZ 45 F. Mixture A is a reference mixture

TABLE 1

Characteristics of Mortar Specimen

	mixture A	mixture B	mixture C
dimensions of specimen [cm]	4 x 4 x 16	4 x 4 x 16	4 x 4 x 16
density [kg/m^3]	2218	2224	2232
compr. str 28d [N/mm^2]	48,3	38,4	29,2

TABLE 2

Elemental Composition of Materials

	Na mg/kg	K mg/kg	Ca mg/kg	Cl^- mg/kg	Cr mg/kg	Cu mg/kg	Cd mg/kg	Zn mg/kg	Pb mg/kg
cement	1287	7353	389352	650	99	83	2	505	70
SFA	6908	26876	6417	44	203	203	2	530	283
MVA	269345	39289	73939	104900	238	108	415	29153	1071
Mix. A	1548	5808	210975	425	52	43	<1	262	36
Mix..B	2097	6625	90216	245	42	44	<1	214	47
Mix. C	2556	6390	118074	4999	43	91	30	1757	608

containing no fly ash. Mixture B contains SFA and Mixture C contains MVA, with 20 % replacement of cement by fly ash (w/(c+f) = 0.50). Mortar specimens prepared from the mixtures were removed after 24 hours' curing time and stored for 28 days at 20 °C and 100 % humidity prior to testing. Table 1 lists the dimensions and characteristics of the mortar specimens. Table 2 indicates the chemical compositions of the materials employed.

2.2 Leaching-tests

2.2.1 DEV-S4-Procedure

The leaching procedure generally used for elution in Germany, the DEV-S4 batch test (standardized German procedure for water, wastewater and sludge testing, sludge and sediments (Group S)) (5) was employed. The test procedure is extremely simple: 100 g of specimen material with a grain size <10 mm are placed in a 2 l wide-necked polyethylene bottle and 1l of demineralized

TABLE 3

Results of the DEV-S4-Procedure

sample	pH	cond. µS/cm	Cr mg/l	Cu mg/l	Cd mg/l	Zn mg/l	Pb mg/l	Na mg/l	K mg/l	Ca mg/l	Cl^- mg/l
cement	12.8	13190	<0.05	<0.05	<0.02	<0.02	<0.05	78	751	879	21.3
SFA	10.4	1402	0.2	<0.05	<0.02	0.02	<0.05	38	56	242	1.7
MVA	10.8	23660	0.1	<0.05	<0.02	0.10	0.25	2539	4536	1162	8167
Mix A	12.7	7960	<0.05	<0.05	<0.02	<0.02	<0.05	22	130	714	6.3
Mix B	12.1	4890	<0.05	<0.05	<0.02	<0.02	<0.05	25	153	421	3.0
Mix C	12.3	4240	<0.05	<0.05	<0.02	<0.02	<0.05	58	146	363	135

water is added. The bottle is agitated at room temperature for 24 h, avoiding further crushing of the specimen material through friction. The specimen is then filtered through a 0.45 μm membrane filter and the composition of the resulting eluate is analyzed.

The parameters determined in the test are:

1. the concentrations by mass of environmentally relevant elements;
2. the pH value of the leachate;
3. the specific electrical conductivity of the leachate.

Table 3 shows the results of the DEV-S4 tests.

The advantages of this method are its simplicity, its rapidity and the extremely good reproducibility of the results.

The DEV-S4 procedure prescribes a liquid/solid (L/S) ratio of 10, a value which is rarely reached under landfill conditions, and then only after an extremely long period (possibly several centuries). The test results should therefore be judged critically.

An additional disadvantage of the DEV-S4 procedure is that it was originally designed for testing sludges rather than compact building materials. In consequence, only limited allowance is made for stabilization factors such as the fixation of toxic components in cement. Nor is it sensible to crush material to a grain size of <10 mm after covering it with cement or mortar. The crushing of incorporated or fixed material inevitably produces new reaction surfaces at which leaching can occur. A leaching test other than DEV-S4 should therefore be used to validate the success of any modifications carried out after fixation or incorporation.

2.2.2 Determination of Maximum Leachability

Another batch test is the maximum leachability determination procedure described by Sloot and Bijen. The test determines element availability and total trace element content (S_{0i}). A 5 g specimen of the materials (<125μm) is added in small portions to 500 ml of demineralized water adjusted to a pH value of 4 by means of HNO_3 and stirred for five hours. The pH value of 4 is maintained by adding 1n HNO_3.

This test was also performed on all specimen materials. The test is intended to determine the proportion of the total trace element concentration which can be mobilized. The maximum leachable concentration is calculated using the following equation:

$$S_{ai} = \frac{(100 \times G2 + a) \times C_{0i}}{G2} \quad [mg\ kg^{-1}] \qquad [1]$$

S_{ai} = the maximum available contration of element i [mg/kg]
C_{0i} = the concentration of element i in leachate [mg/l]
G2 = the weight of the sample [kg]
a = the volume of added HNO3 [l]

The quotient f = $S_{ai}/S_{0i}*100$ (where S_{0i} is the total concentration of the element i in the sample) yields the percentage of the relevant element which can actually be mobilized, i.e. leached, by a given leachate. Table 4 shows the results of the analyses.

TABLE 4
Maximum Availability of Elements

sample	Cr mg/kg	f %	Cd mg/kg	f %	Cu mg/kg	f %	Zn mg/kg	f %	Pb mg/kg	f %	Na mg/kg	f %	K mg/kg	f %	Cl mg/kg	f %
PZ	41	40	<2	<1	43	52	202	40	17	24	1292	100	8573	117	645	100
SFA	13	6	<2	<1	23	11	77	15	<10	<3	401	6	1404	5	43	97
MVA	6	3	355	86	303	27	19146	66	519	46	21099	78	38390	97	94854	90
Mix A	22	44	<2	<1	28	64	184	72	<10	<1	728	47	4313	74	420	99
Mix B	12	30	<2	<1	12	12	16	8	<10	<3	531	30	3184	48	239	99
Mix C	8	20	24	82	42	62	1640	93	133	22	2024	79	4900	77	4948	99

The test is simple to perform and results show good reproducibility. Its importance lies in its ability to estimate the fraction of toxic elements which can be leached from a fly ash product under extreme conditions, i.e. the worst case environmental stress due to the waste product.

2.2.3 Column test

In a column test the material is brought into a glass/polyethylene column and than perculated with the leachant.

A normal column test was performed for Mixtures B and C. Material crushed to a grain size < 10 mm was percolated with demineralized water adjusted to a pH of 4 by means of HNO_3. The L/S ratio was 5. The leachate was collected in five fractions.

Results for the column test were not reproducible, probably due to the form of elution. A better procedure would be an inverse column test. In general, the concentrations of toxic elements are higher than in the eluates obtained using the DEV-S4 method.

A comparison of column and batch tests indicates that batch tests are simpler to perform and achieve greater reproducibility. It should, however, be noted that the column extraction test involves a continuous flow of liquid through a fixed bed of solid waste. This form of liquid-solid contact may be more representative of waste leaching under field conditions than the batch method.

3 DETERMINATION OF COEFFICIENTS OF DIFFUSION

The leaching process is approximately described by Fick's law if (i) dissolution is extremely rapid as compared to diffusion, (ii) leachant imbibition attains a fixed equilibrium value very soon after commencement of the elution test and (iii) the embedded salt is fully soluble in the leachant. Given these preconditions, the following equations will apply (7):

$$J = C_0 \left(\frac{D_e}{\pi t}\right)^{1/2} \quad \text{and} \quad D_e = \pi t \left(\frac{Ji}{f C_0}\right)^2 \qquad [2]$$

$$J_i = \sum_{i=1}^{n} (C_i - C_0)\, V/At_i \qquad [3]$$

J = the flux of diffusion [mmol m^{-2} s-1]
D_e = the effective coefficient of diffusion [m^2 s-1]
C_i = the concentration in leachate at the ith interval [mmol l^{-1}]
C_0 = the original concentration in leachate [mmol l^{-1}]
V = the volume of leachate [l]
A = the surface area [m^2]
t_i = the total contact time after the ith interval [s]
n = the number of cycles
f = S_{ai}/S_{0i}

Since diffusion through adsorption phenomena and exchange reactions with the pore walls is reduced in the constrained geometry of a porous medium, measurement in the tank leaching tests is restricted to determining an effective diffusion coefficient.

Tank leaching tests were carried out to determine diffusion coefficients. The mortar cubes were placed in polyethylene containers so as to allow direct contact with the leachate on all sides. The leachant was demineralized water adjusted to a pH of 4 using HNO_3. The volume of leachant was 5 times the surface area of mortar specimen (the surface to volume ratio is 5).

The leachant was renewed at several intervals. The schedule for leachant renewal was based on the assumption that bulk diffusion is the dominant leaching mechanism. If this is the case, the leaching rate decreases with time, and it is necessary to increase the time between renewals gradually in order to ensure that the leached species can be measured analytically in the leachate. The renewal intervals are determined according to the equation:

$$t_n = n^2 t1 \qquad [4]$$

n = the leaching period
t_1 = the end of first leaching period
t_n = the end of the nth leaching period

The diffusion fluxes were determined using Equation [3]. In a log-log plot of cumulative ion flux J_i over contact time, diffusion-controlled leaching results in a slope -1/2 (8). Table 5 lists the slopes for the log-log plot of element flux over t and the coefficients of diffusion. These approximate to -1/2 plots, indicating that leaching is in fact diffusion-controlled.

TABLE 5
Slopes of the log-log plot of J_i against contact time and diffusion coefficients

	mixture B		mixture C	
	slope	D_e [m2/s]	slope	D_e [m2/s]
Na	-0.75	3.59 E-12	-0.66	3.54 E-16
K	-0.72	1.40 E-15	-0.66	7.31 E-16
Cl	-0.69	2.95 E-17	-0.57	2.91 E-19
Cr	-0.61	6.47 E-19	-0.68	7.25 E-18
Cu	-0.65	3.64 E-22	-0.71	5.14 E-22
Zn	-0.51	6.98 E-20	-0.49	2.58 E-19
Pb	-	-	-0.73	3.44 E-21

Tank tests indicate the dynamics of a leaching process over time, permitting description of the leaching mechanisms. A further advantage is that the specimen material need not be crushed, suiting these tests for the evaluation of compact building materials or modification measures. Leaching rates can be predicted from the calculated coefficients of diffusion. The reproducibility of tank tests is excellent. In combination with batch tests, they provide valuable information on the environmental impacts of building materials containing fly ash.

4 CONCLUSIONS

The leaching tests carried out in the study are simple to perform and (with the exception of the column test) yield reproducible results. Appraisal of the environmental impacts of building materials which contain fly ash on the basis of a single leaching test is not feasible. Batch tests fail to reflect real conditions adequately, while column tests, though more representative of real conditions, are not sufficiently reproducible. Neither test permits appraisal of modification measures.

Tank tests are extremely suitable for determining leaching mechanisms. In the case of cement products, the dominant leaching mechanism is diffusion. Leaching rates can be predicted from the calculated coefficients of diffusion. Tank tests can also be used to evaluate the success of contaminant fixation.

The 45 F portland cement used in the tests possesses an extremely high contaminant-fixing capacity. All contaminant concentrations in the leachates from the various tests (except the maximum leachability test) are below the permissible limit prescribed by the German Drinking Water Order.

The grain size of the material and the pH of the leachant likewise the liquid/solid respectively volume/surface ratio have a great influence on the contaminant concentrations in the leachant.

REFERENCES

1 Cotè. P.L.; constable. T.W.; Moreira. A.; nuclear and waste management 7(2). 129 (1987).

2 Bishop. P.L.; Gress. D.L.; New York. ASCE. 1982. National conference on environmental engineering, 423

3 German standard. DIN 1164 part 5. 1986

4 "Richtlinie für die Erteilung von Prüfzeichnen für Steinkohlenflugasche als Betonzusatzsoff nach DIN 1045" (1986)

5 German standard. DIN 38 414, part 4. 1983

6 Sloot. H.A. van der; Piepers. O.; Kok. A.; Omschriving van de Standaarduitloogtest voor Verbrandingsresiduen. BEOP-25. 1983.
Bijen. J.M.J.M.; Sloot. H.A. van der; leaching test for fly ash containing produkts. Quelle unbekannt.

7 Amarantos. S.G.; Papadokostaki. K.G.; Petropoulos,J.H.; comparison of leaching tests and study of leaching mechanims. Athen: National research centre for the physical sciences domocritos. 1985.

8 Sloot van der. H.A.; Groot de. G.J.; Wijkstra. J.; leaching characteristics of construction materials and stabilization products containing waste materials. Petten: Netherlands engergy research foundation (ECN). 1987. restricted distribution ECN-87-093.

Waste Materials in Construction.
J.J.J.R. Goumans, H.A. van der Sloot and Th.G. Aalbers (Editors)

EFFECT OF PARTICLE SIZE DISTRIBUTION ON LEACHING PROPERTIES OF BUILDING MATERIALS

D. GOETZ and W. GLÄSEKER
Institute of Soil Science, Allendeplatz 2, 2000 Hamburg 13 (Germany)

SUMMARY

Particle size dependent leaching behaviour of 2 building recycling materials were investigated. The results are showing, that a decrease of particle size is not automatically connected with an increase of the total surface area. There is a good correlation between surface area and amount of components leached.

1. INTRODUCTION

Secondary building materials for road construction can have variable particle size distributions. The particular fractions often are containing a distinct material composition. Following the view of governemental authorities, a test of grained material is preferred to simulate a worst - case scenario for leaching of contaminants. On the other hand, an investigation of unaltered samples is preferred by the users of these building materials.

The tests should be adapted to the distinct building materials and practical conditions to get an objective examination for materials. The alteration of particle size distribution during sample preparation is often discussed controversely. To obtain a data basis for leaching of distinct particle size fractions, different building materials were used for a column and a shaking test, which are representing two various leaching mechanisms.

2. MATERIALS

The following materials were used for the leaching tests:

- Cement-stabilized fly-ash , composed with

coal fly-ash	78,14 %	by weight,
cement PZ 35F	6,25 %	-"-
water	15,61 %	-"-

The material was stored for 4 month after setting.

- Secondary building material (debris), containing bricks, clinker, tiles and conrete with remains of mortar.

The materials were crushed into pieces < 10 mm, dried at 105°C for 24 hours. The dried material was sieved to gain the follwing particle size fractions:

10 - 6,3 mm	0,8 - 0,4 mm
6,3 - 2,0 mm	0,4 - 0,1 mm
2,0 - 0,8 mm	< 0,1 mm

3. METHODS

Four tests were used for leaching the several particle size fractions of the two examined materials:

- The regular g erman shaking test (DEV S4) using a liquid to solid ratio (L/S) of 10 and 3 replicates.
- Column - test: Upstream percolation through a PVC - tube, length 20 cm and 5 cm diameter, using a vessel mounted into the circuit for compensation. The leachate was circulating at a speed of 30 ml/min for 24 hours. Similar to the shake - test, the materials were leached with a L/S ratio of 10 and 3 replicates.
- Leaching with Ammoniumacetate for examination of exchangeable cations: 10 g material, used in a column - test with a L/S of 3, was extractet for 4 times over a total time of 13 hours with a pH 7 Ammoniumacetate solution. The L/S ratio of this test was 20 and 3-5 replicates were conducted.

Total concentrations were determined after hydrolysis with nitric acid, HF and boric acid. Total surface area was determined by Fa. Ströhlein, using the BET Method.

4. RESULTS AND DISCUSSION

The experiments showed interesting relationships between particle size distribution and leaching. The dependence of some leaching parameters from the particle size and surface area are presented in Fig 1 - 7 and 9 - 10.

The replicates of the experiments are showing a good conformity, Therefor it could be said that random effects are playing a minor role.

The leachate pH of the distinct particle size fractions is varying within one pH grade and resulting in a characteristical pH curve for each material (Fig. 1a,b). In the column - test, the leachate pH of debris (HBS) is increasing from the largest particle size fraction (> 6,3 mm) to the 2-0,8 mm particles, declining in the next fraction and rising with the smaller particles again. In contrast to that, leachate pH of the shaking test is decreasing with the smaller particle size fractions.

The corresponding curves of the fly ash cement (FAZ) are analogous to HBS, but the maximum pH was observed in the 0,8-0,4 mm fraction and there was no pH increase in the leachates out of the smaller fractions. The curves of shaking test leachates are comparable to the HBS material.

Leachate conducticity of both tests are presented in Fig 2a,b. Column - tests are also showing maxima curves with the highest conductivity in the 2-0,8mm (HBS) and the 0,8-0,4 mm (FAZ) fraction respectively.

The Ca content (Fig. 3a,b) is comparable with the conductivity curves because Calcium is a dominant ion in the leachates of both materials. But the maxima of the column test are less distinct and the leachates of the shaking test are showing no (HBS) or slight (FAZ) increase of Ca concentration.

The concentrations of well soluble Na- and K salts (Fig. 4a,b and 5a,b) are lower by a factor of 10. Therefor they are playing a minor role in leachate element composition. The same can be mentioned for the heavy metals. To give an example for these elements, the Chromium concentrations are presented in Fig 6a,b.

The Na, K and Cr concentrations of HBS leachates obtained in column and shake tests are increasing with decreasing particle size, as expected. Opposite to HBS, the corresponding figures of FAZ are showing a clear maximum curve with the highest values at the 0,8-0,4 mm fraction.

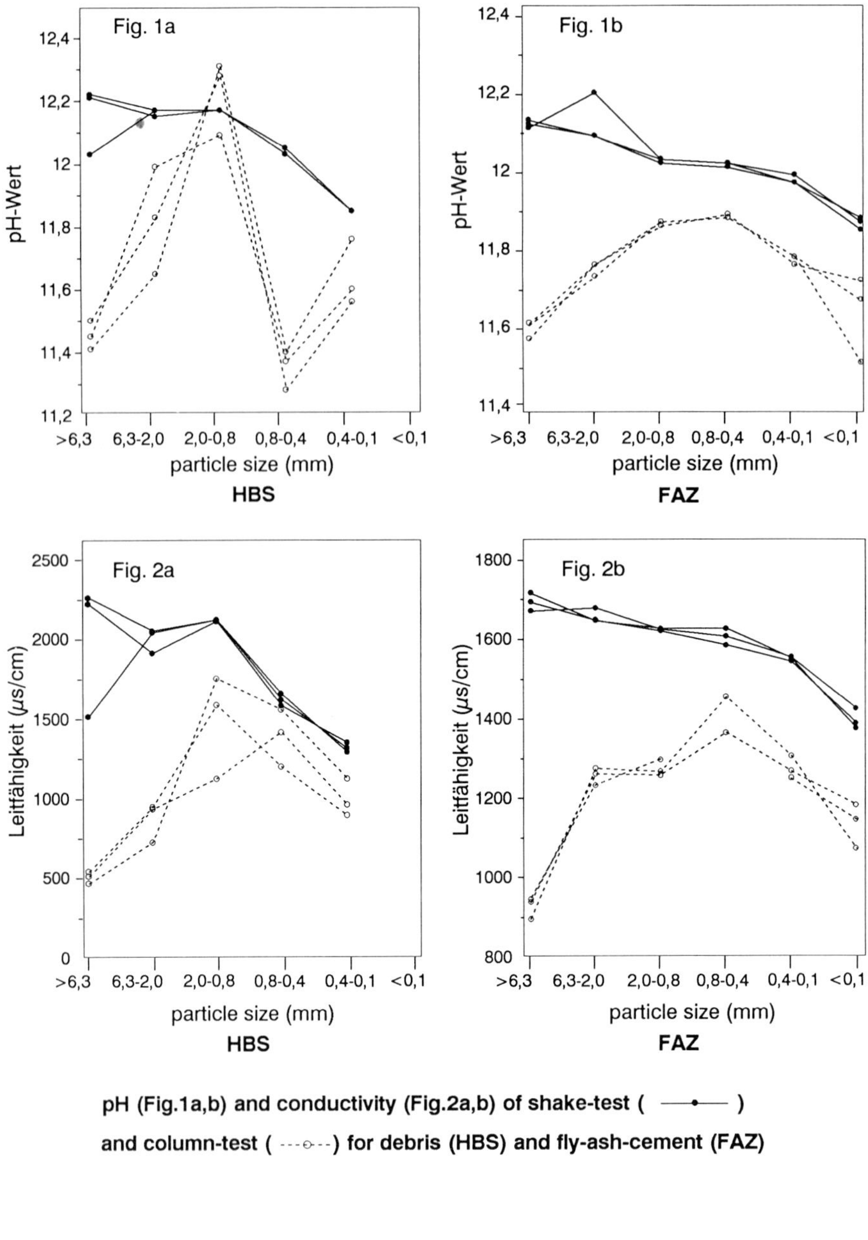

pH (Fig.1a,b) and conductivity (Fig.2a,b) of shake-test (—•—)
and column-test (---o---) for debris (HBS) and fly-ash-cement (FAZ)

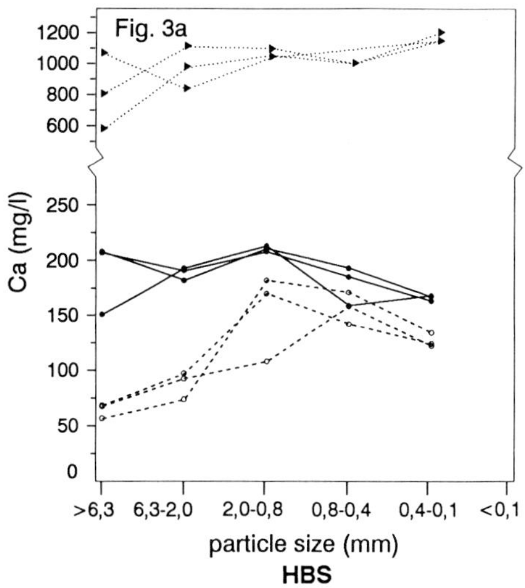

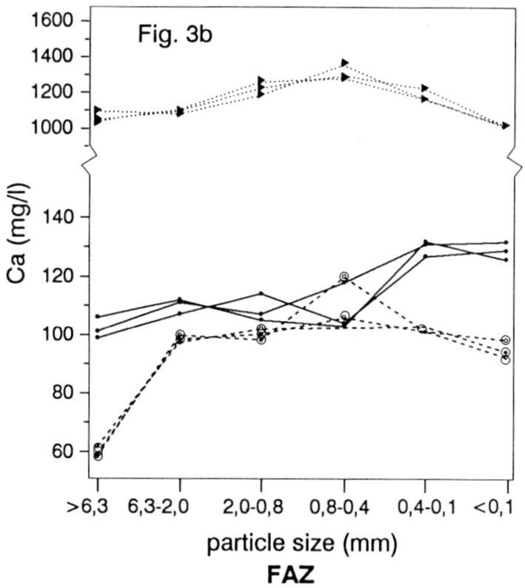

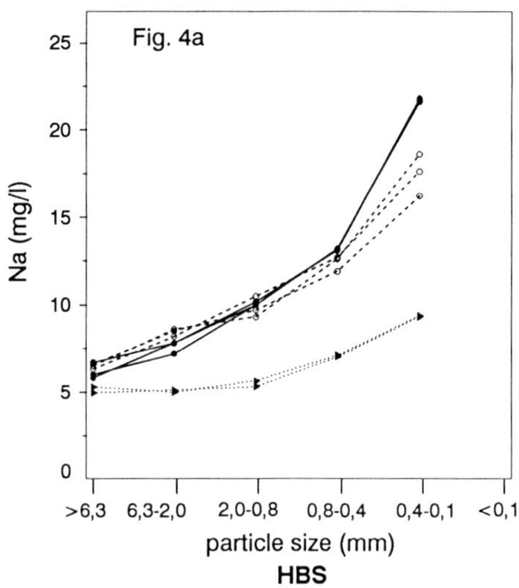

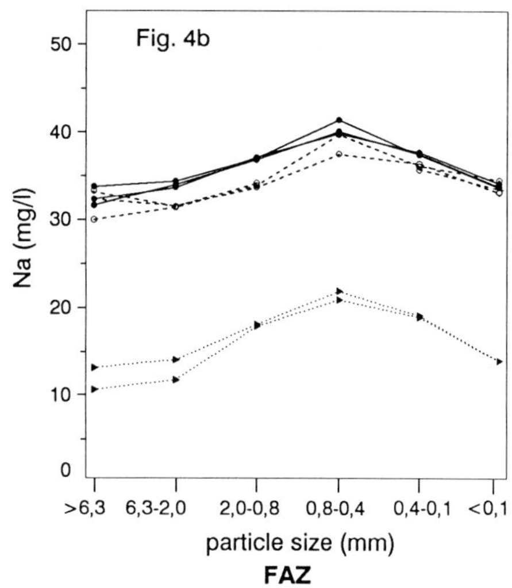

Ca-concentrations (Fig.3a,b) and Na-concentrations (Fig.4a,b) of shaking-test (—•—) column-test (---o---) and NH_4 Ac-solution after extraction (·····▸·····) for debris (HBS) and fly-ash-cement (FAZ)

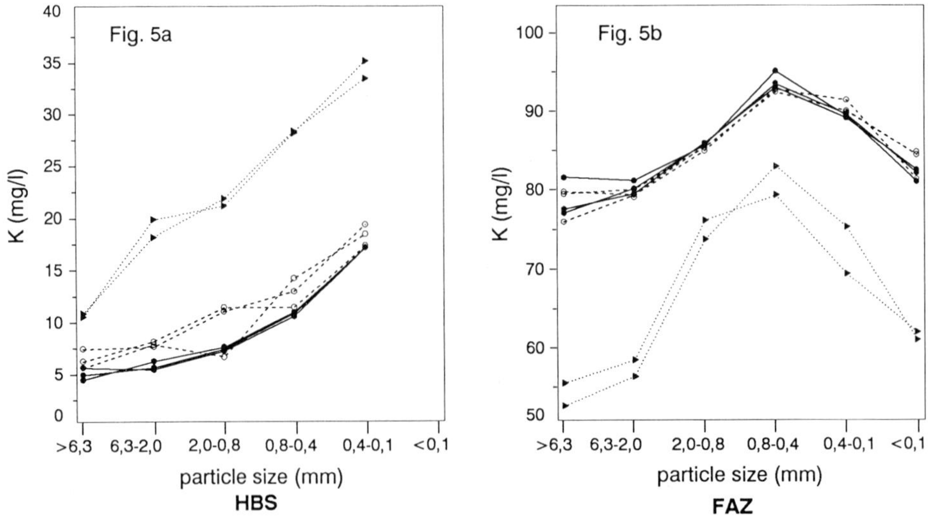

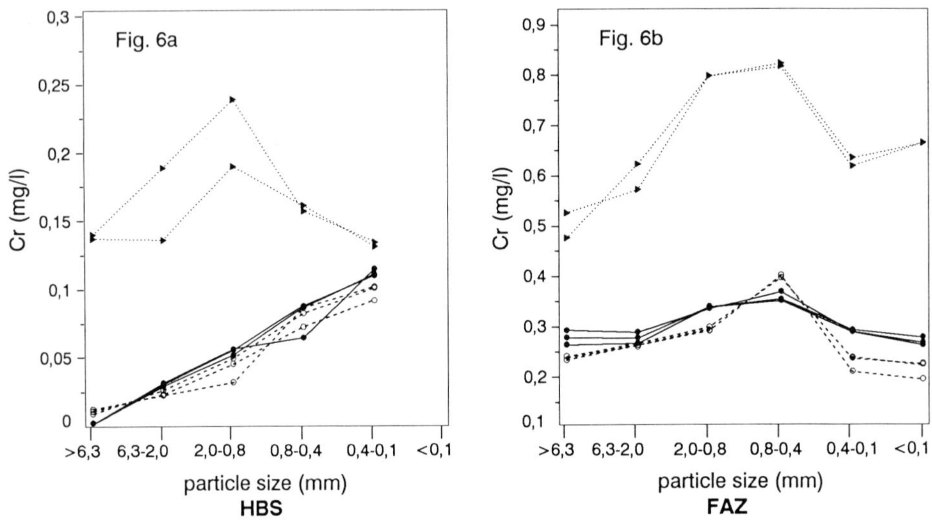

K-concentrations (Fig.5a,b) and Cr-concentrations (Fig.6a,b) of shaking-test (——•——) column-test (---○--) and NH_4 Ac-solution after extraction (·······▸·····) for debris (HBS) and fly-ash-cement (FAZ)

Cation exchange capacity (CEC) curves of HBS and FAZ particle size fractions (Fig. 7) show the same shape like the above mentioned figures, CEC of HBS material increases from 40 to 130 mmol/kg with decreasing particle size and the CEC of FAZ show a maximum of 200 mmol/kg at the 0,8-0,4 mm fraction.

The ion concentrations in the Ammoniumacetate solutions (Fig. 3-6) after extraction can be parallelized with the element concentrations mentioned before except for Chromium. Although the Ammoniumacetate treatment of the material occured after leaching in a column - test, it is likely, that not only exchangeable ions but also remaining salts were soluted. This effect is clear visible by the large amount of Ammoniumacetate - mobilized Ca ions, which exceed the total CEC by several times.

The total surfaces of the particular particle size fractions are presented in Fig 8. As we expected, the surface area of the HBS materials is increasing with decreasing particle size. But FAZ is showing a maximum in the 0,8--0,4 mm fraction.

For comparison, surfaces of basalt particle size fractions were determined. In this case, surface area is increasing until a particle size of 2-0,8 mm. Then, this curve shows a minimum at the 0,8-0,4 mm fraction.

The results of our experiments were a bit surprising. It is visible, that some parameters as pH, conductivity and solubility of Ca are determined by the chosen leaching procedures. Each test showed complete different curves. In oppposite, the composition of the material plays a minor role.

Particle size analyses of material, which are used in a shaking test, indicated that mechanical abrasion led to a disintegration of sample particles. This effect was notable seen from the coarser particles. However, leaching behaviour of Cr, Na and K was not influenced by the grinding effect of the shaking test.
These Elements show nearly the same solubility in column- and shaking tests.

But there ist a difference between the particle size dependent curves when both examined materials are regarded. Due to heterogeneous composition of the material, consisting of several compounds as mortar,

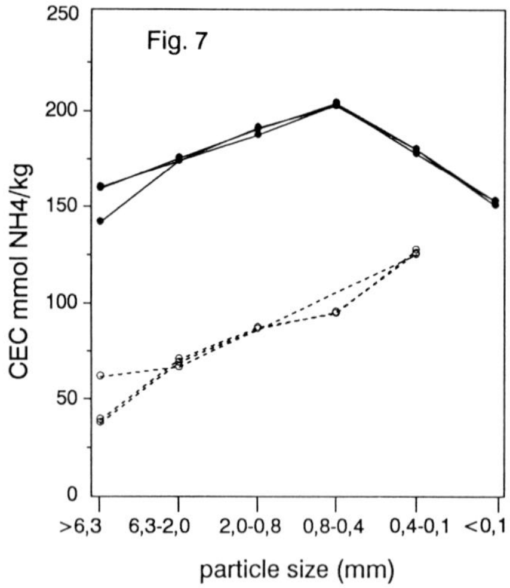

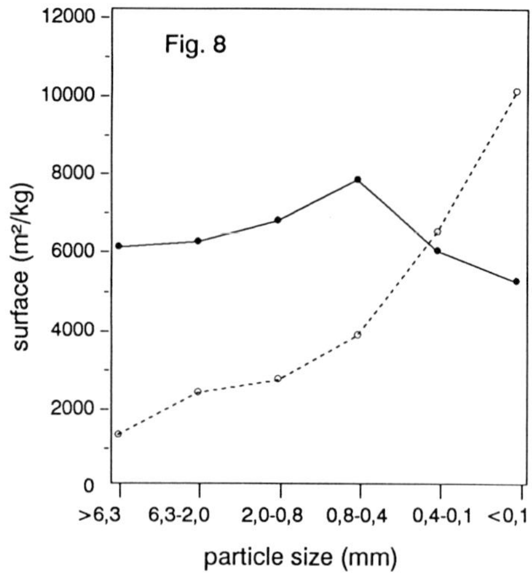

debris (HBS) fly-ash-cement (FAZ)

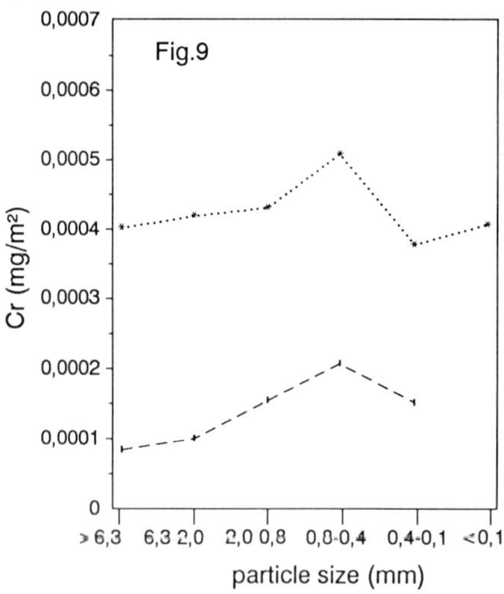

column-test

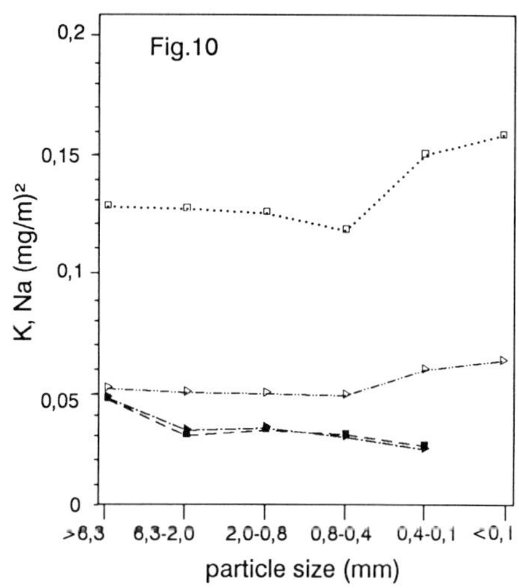

column-test

fly-ash-cement (FAZ): K Na Cr

debris (HBS): K Na Cr

bricks or cement, a rather unsteady concentration curve of HBS leachates was expected. But a continuous increase of Cr, Na and K concentrations with decreasing particle size was detected.

Opposite to the HBS material, leaching of particle size fractions of the very homogeneous FAZ was resulting in Cr, Na and K - concentration curves with a maximum in the 0,4-0,8 mm fraction. The same shape is visible from figures which show the dependence of CEC from particle size. By regarding our results avaiable up to now, this coincidence can be explained with the fact that the elements were in an exchangeable state before leaching out of the investigated materials.

Change of CEC with particle size can be explained with the total surface area, whose curve show the same shape. Taking the value for the theoretic spherical surface as basis, an expotential increase of total surface can be expected. Leachate analyses of the HBS material were fulfilling these expectations.

Decrease of total surfaces of particle size fractions 0,4-0,4 mm and < 0,1mm of FAZ can only be explained by chemical transformations on surfaces of the fine particles. Fly - ash is consisting to the greater extend of small glazed globules which can be destroyd in the fine fractions. With that, new reactive surfaces were formed which lead to a agglutination of the filigree structure elements. In this way, the avaiable inner surfaces could have been reduced. Chemical transformations at particles surfaces are probable, because the particle size dependent pH values of this homogenous material are showing relatively strong variations.

The strong variations of particle size dependent HBS leachate pH can be most easily explained with inhomogenous distribution of material in the distinct particle size fractions. There seems to be less influence of the high pH values on the leaching behaviour of the examined materials. A coincidence ist only visible between pH and electric conductivity of the FAZ.

The experiments are showing that there are no simple relations between particle size and leaching parameters. Even total surface is not automatically increasing with decreasing particle size. This phenomenon is even visible at a apparent homogenous and pore - free material like basalt.

Also, there can be a particle size dependent mobilization of the elements to be investigated. E.g. a liniear liberation of Sulfate with decreasing FAZ particle size or a maximum curve, describing the K and Na - content of FAZ leachate was detected. Other investigations, dealing with the particle size dependent leaching cf steel-works slag, lead to a miniumum curve for Vanadium and an increase of Chromium with decrease of particle size.

But these experiments show that the total surface of the materials is often determining the leaching of elements. The surface to amount of element leached - ratio is nearly equal for all particle size fractions (Fig. 9,10).

This means under the aspects of practical uses, that grinding of sample material must not lead automatically to a increased leaching. A convention for the particle sizes to be investigated is useful to assure a comparability of different materials.

Waste Materials in Construction.
J.J.J.R. Goumans, H.A. van der Sloot and Th.G. Aalbers (Editors)

FRENCH APPROACH TOWARDS THE EVALUATION OF MONOLITHIC AND SOLIDIFIED WASTE:
Development of a new leaching procedure

Jacques Méhu, Yves Perrodin, Bernard Sarrazin, Jean Véron

Laboratoire de Chimie Physique Appliquée et Environnement
Institut National des Sciences Appliquées
Bat 404 - 20, av Albert Einstein
69621 Villeurbanne Cedex - FRANCE

I - SUMMARY - INTRODUCTION

In France the problem of landfill as a means of waste disposal and the delicate operation of setting criteria for waste qualification to landfill led the different partners concerned (authorities, industries, either producers or eliminators, research laboratories,...) to work together on processes that would firstly render pollutants as insoluble as possible, and secondly fix them in a mineral or organic matrix to ensure structural integrity and protection against leaching phenomena.

The SRETIE of the French Ministry of the Environment initiated a joint research programme between IRH (Institute of Hydrological Research) and INSA of LYON (National Institute of Applied Science) with the aim to elaborate and validate an experimental protocol for a leach test adapted to monolithic or solidified waste.

The year before, a first coordinated work(1), based on a literature survey of leaching tests, including procedures concerning solidified nuclear waste, provided the guideline of a protocol.

Then, further work was conducted which concerned the selection and preliminary characterization of a wide range of wastes of different nature and behavior, and the definition and application of a preliminary selection test for monolithic or solidified waste.

Apart from the design and application of this preliminary test, the specific work of INSA was to study the behavior of monolithic or solidified wastes with respect to different leaching tests.

- leaching test for crushed material (4 mm), i. e. french standard test NF X31-210.
- washing test on a monolithic sample in the case of materials which can be cut to 4x4x8cm or 4x4x4cm specimens, or moulded into cylinders (height 5cm,Ø 5.5cm)
- washing test carried out on a granular fraction in the case of materials which cannot be cut to 4x4x8cm or 4x4x4cm specimens.
- test by percolation under pressure in the case of materials having appropriate size and permeability.

To date, in France there is one standard leaching test, which is the NF X 31-210

However, this test is not adapted to monolithic and solidified waste because it involves crushing to particles of 4mm.

In this study, the results of the washing test and the under pressure percolation test have been compared to the standard test. The results have been interpreted in order to assess the relevance and validity of the proposed tests, as well as the details concerning their experimental application.

A final selection of evaluation methods was carried out which led to the elaboration of a common experimental protocol (INSA of LYON-IRH) including all the verification of structural integrity as well as the leach test itself. In a first approach, threshold values were proposed for the different decision levels involved in the application of the protocol.

A wide use of this protocol would allow the validation and justification of these threshold values. The protocol could then be submitted for approval to AFNOR. It could also be the french contibution to a european procedure.

II - EXPERIMENTAL PROGRAM ON THE FEASIBILITY OF LEACH TESTS FOR MONOLITHIC OR SOLIDIFIED WASTE.

II-1 - The tests

The final protocol for evaluation of monolithic or solidified waste will consist in two sets of tests : 1/ mechanical strength, wet-dry, water absorption capacity, and 2/ the leach tests. However since this procedure would be long and costly, a preliminary selection test was proposed in order to avoid applying the complete procedure to a waste which would have a very poor structural integrity.

A - PRELIMINARY TEST

* Aim

To select waste which can be classified as "monolithic" for which the complete protocol should be applied.

* Means

Granulometric study after soaking in demineralized water for 16 hours.

* The waste

A minimal number of monoblocks are taken from a representative sample of the waste to make a sample of $m0 = 100 \pm 20g$

Any waste, of which the representative sample would contain more than A% by mass of fragments <10mm would be excluded from the application of the following step in the protocol and would undergo the standard AFNOR X 31-210.

* The test

1. Make a specimen of approximately 100g (as above)

2. Eliminate the fines from the surface of the waste by turning it over delicately on all sides on a flat clean surface.

3 Weigh the specimen accurately.

4. Evaluate in parallel the dry mass (which will be noted m0s), on another sample of the same waste, by drying in an oven at 103°C±2°C to constant mass or 24 hours maximum.

5. Immerse the specimen in demineralized water for 16 hours (liquid/solid ratio equal to 10). The waste is placed on a plastic grid (Ø4mm), so that all sides are in contact with water. The flask containing the waste is open to the air at 20°C±2°C.

6. After 16 hours the waste is removed and placed on a sieve of circular mesh(Ø10mm) itself placed on another sieve of circular mesh (Ø4mm)

The product is then washed delicately, first with its own leachate, and then with the same volume of dimineralized water.

7. Recover the extracts from each of the two different sieves on filter paper and dry in an oven (103°C ± 2°C) to constant mass or for 24 hours maximum.

8. Weigh each fraction :

Ø > 10mm fraction	--> m10s
4 < Ø < 10mm	--> m4s

* Measured parameters

Soluble and fines Ø < 4mm fraction	Fsf=100 x [m0s-(m10s+m4s)]/m0s
Large particles Ø>10mm fraction	Fg=100 x m10s/m0s
Medium particles (4<Ø<10mm) fraction	Fm=100 x m4s/m0s
Large/medium particles ratio	R=m10s/m4s=Fg/Fm

The fractions will be expressed in %

From these parameters, it will be possible to define threshold values to classify a waste as "monolithic", to which the complete protocol should be applied.

Fsf *max* : maximum tolerated for the "soluble+fines" fraction
Fg *min* : minimum required for the "large particles" fraction
Fm *max* : maximum tolerated for the "medium" fraction
R *min*: minimum required for the "large/medium" ratio.

In this study the protocol will be carried out regardless of the values obtained for the different parameters.

B - ORIENTATION OF WASTE FOLLOWING THE PRELIMINARY TEST.

Four situations are possible :

* The product is in the form of a moulded cylinder (Ø:5.5 cm; H:5cm)

* The product is in the form of blocks which can be cut to 4x4x8 or 4x4x4 specimens. The specimens are then cut with an accuracy of 20mm maximum.

* The product is in pieces and specimens can not be cut. Therefore it is partially crushed and the particles between 10 and 20mm are retained for later testing.

For those three possibilities the washing test can be carried out.

* The product is in the form of blocks which can be cut to specimens of 7.5cm in diameter by 3cm thick, for the under pressure percolation test.

C - LEACHING TESTS

* AFNOR X 31-210 : this test is carried out to serve as a reference for all wastes having passed the preliminary test.

* Washing test on a 4x4x8cm or 4x4x4cm specimen or a cylinder, or the fraction between 10 and 20 mm, using demineralized water with a liquid/solid ratio of 10 for each of the 3 successive washing cycles. Comparison with the results of the X 31-210 test.

The specimen is fixed in a plastic basket, which is itself attached to the flask. Mechanical backwards and forwards stirring of the flask (60 movements/min) ensures the washing of the specimen by the water.

* Under pressure percolation test with demineralized water (5 extracts), with a liquid/solid ratio of 10 for the first extract and 5 for the four following extracts. Comparison with the results of X 31 -210.

The three leach tests

Standard test AFNOR X31-210

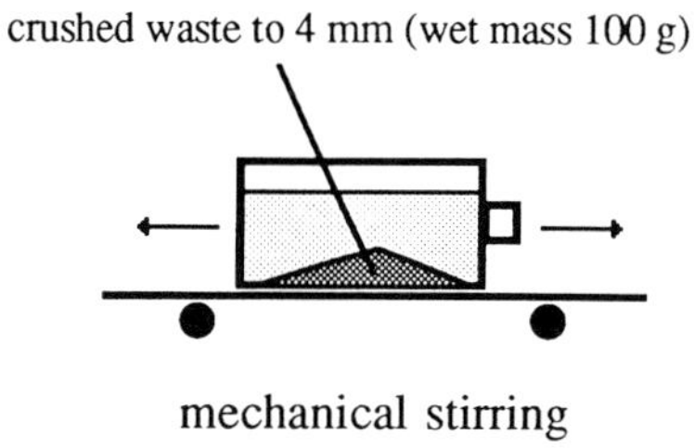

- mass of demineralized water / mass of wet waste = 10
- stirring frequency : 60 per min
- contact time : 16 h
- filtration and leachate analysis (L1)
- 2 other extractions (L2 et L3)
- cumulative extracted soluble fraction for each parameter (mg per kg of wet mass).

Washing test (flow-around test)

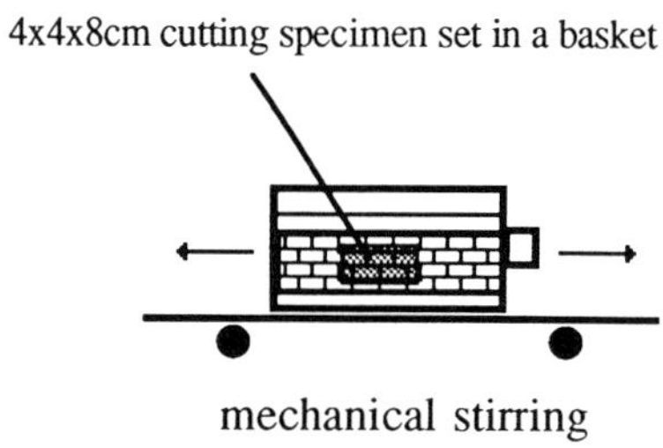

- mass of demineralized water / mass of wet waste = 10
- stirring frequency : 60 per min
- contact time : 16 h
- filtration and leachate analysis (L1)
- 2 other extractions (L2 et L3)
- cumulative extracted soluble fraction for each parameter (mg per kg of wet mass).

Under pressure percolation test

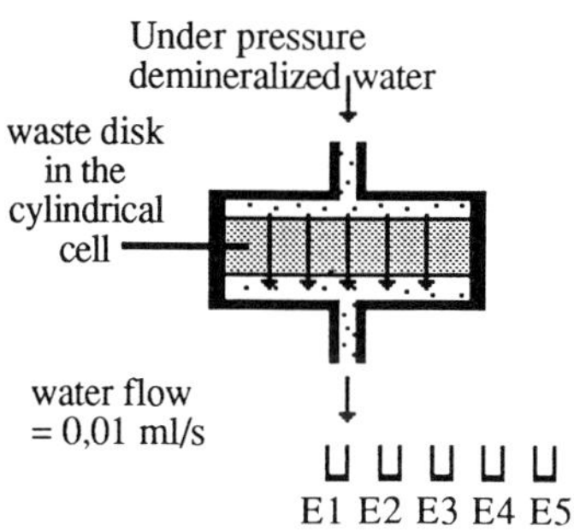

- mass of water E1 / mass of wet waste = 10
- mass of water E2 / mass of wet waste = 5
- mass of water E3 / mass of wet waste = 5
- mass of water E4 / mass of wet waste = 5
- mass of water E5 / mass of wet waste = 5
- analysis of the 5 extracts,
- cumulative extracted soluble fraction for each parameter (mg per kg of wet mass).
- potentially extractible soluble fraction. (calculated by computation of experimental data)

* Leachate parameters measured :
  - global: COD, TOC, conductivity, red-ox potential, pH.
  - specific parameters for the tested waste

* Number of tests :

For each waste the following tests are carried out :

- X 31-210	3 leachates
- washing	3 wash waters
- Percolation under pressure	5 extracts

II 2 Wastes Tested

11 types of different wastes were tested by the different experimental procedures

Each type of wastes was separated into subspecimens (12 maximum)
and distributed according to the best adapted type of test.

N° 1 N° 2	Preliminary test	carried out on 11 types of wastes
N° 3 N° 4	X31-210 test	
N° 5 N° 6	Washing test on 4 x 4 x 8 cm specimen or fragments 10 < ø < 20 mm	
N° 7 N° 8	under pressure percolation test	carried out on 5 types of wastes
N° 9 N° 10	Washing test on 4 x 4 x 4 cm specimen	carried out on 2 types of waste solidified in the laboratory
N° 11 N° 12	Washing test on a cylinder	

The tables below present the form of the waste to which the washing test was applied : 4x4x8cm specimens or fragments between 10 and 20mm. In the case of the laboratory solidified wastes, two forms were added to the study : 4x4x4cm specimen and cylinders of Ø 5.5cm and H 5cm.

Solidified waste	Industrial solidified waste	Laboratory solidified waste containing NaCl	Laboratory solidified waste containing Pb
Crushed to 4 mm for X31-210	X	X	X
4 x 4 x 8 cm specimen	X	X	X
4 x 4 x 4 cm specimen		X	X
Cylinder Ø 5,5-H 5cm		X	X
Specimen for the under pressure percolation test	X	X	X

Monolithic waste	Foundry sands			Desulfuration slag	Steel works slag	Metal hydroxide sludge	Distillation residue	Paint residue
	1a	1b	2					
Crushed to 4 mm for X31-210	x	x	x	x	x	x	x	x
4x4x8 cm specimen	x	x					x*	x*
Fragments 10 < Ø < 20 mm			x	x	x	x		
Specimen for the under pressure percolation test	x	x				x		

* the specimens were not cut out of the monolithic waste but were moulded.

The following national protocol was elaborated taking into account the experimental results obtained.

Acknoledgement:

The authors wish to thank IRH for its collaboration in this work and the SRETIE of the French Ministry of the Environment for funding.

References:

(1) Evaluation des dangers et des risques des déchets solides massifs. Choix d'un protocole national. (1ère phase bibliographique).
SRETIE Ministère de l'Environnement - n°88 292

III - FRENCH APPROACH TO THE EVALUATION OF MONOLITHIC AND SOLIDIFIED WASTES: Proposed protocol by IRH and INSA de Lyon (June 1991)

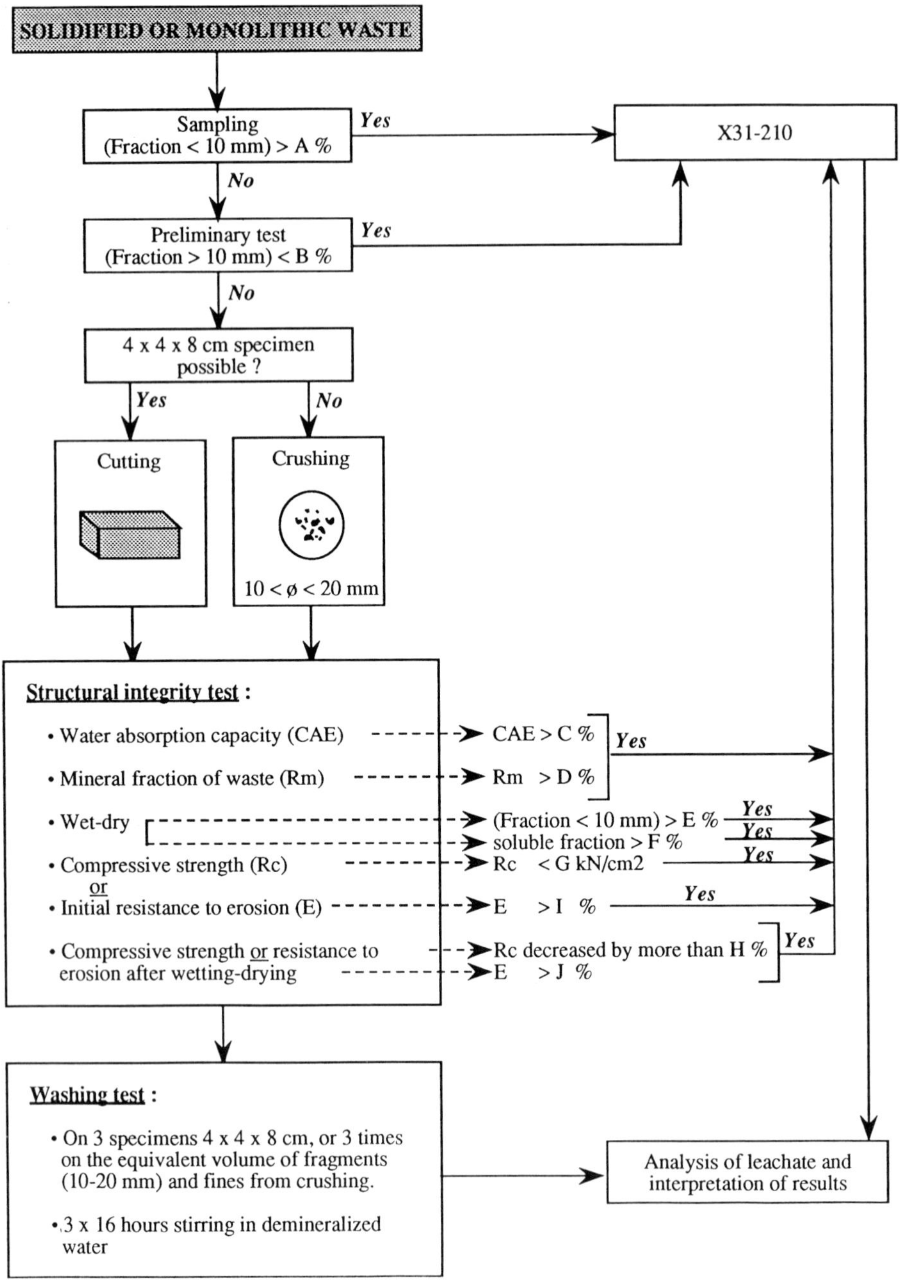

Waste Materials in Construction.
J.J.J.R. Goumans, H.A. van der Sloot and Th.G. Aalbers (Editors)

A TEST METHOD FOR THE DETERMINATION OF THE LEACHABILITY OF TRACE ELEMENTS FROM WASTES BOUND WITH CEMENT

W. Rechenberg, S. Sprung and H.-M. Sylla

Forschungsinstitut der Zementindustrie, P.O. Box 301063,
D-4000 Düsseldorf 30 (Federal Republic of Germany)

SUMMARY

Some wastes may be used, provided that constituents of environmental concern are not leached in an unacceptable magnitude. Cylindrical cement stabilized materials were tested in a flow through apparatus. Water was pressed through the specimens with a pressure of 1 bar. The leachate was analysed by AAS.

The examination showed a maximum leaching rate of 0,2 % of the content. The leaching rate decreased with increasing cement content and time. If the specimens were crushed the leachability increased, depending on the particle size. The results have shown that tests by which the structure of the material is destroyed during preparation, cannot be taken as suitable for an assessment of the leachability under application conditions.

1. INTRODUCTION

The German Waste Removal Act (1) requires the extensive re-use of wastes. Residues from power plants or from waste incineration may be used for example in road construction, provided the materials satisfy the required load capacities, and the incorporated constituents of environmental concern are not leached in an unacceptable magnitude under field conditions.

Therefore, it seemed to be necessary to develop a test method for the leachability of constituents of environmental concern from wastes stabilized with cement under conditions of normal application in practice.

2. TEST SPECIMENS

The leachability was tested on cylindrical test specimens. These specimens were made from several fly ashes and incineration ashes as well as from sand (model mortars) with a maximum grain size of 2 mm (table 1). As binding agent an OPC was used. The test specimens were prepared with cement contents ranging from 5 to 16 % by wt. and compacted in steel moulds (2-4). The water content was chosen to give an optimal compactability of the mixed material sample. In the case of the model mortars prior to mixing, soluble salts of lead, cadmium, chromium, mercury, thallium and zinc in different concentrations had been dissolved in the mixing water. The test specimens were cured for 7 and 28 days, respectively, at 20 °C and approximately 100 % rel. humidity in a fog chamber. Cylindrical test specimens made of cement-bound materials - called cement-bound Proctor cylinders - are normally used to assess and optimize the physical properties of subbases in road constructions (1), (5), (6).

Quartz Sand **Cement** **Mixing Water** **Elements**	**0/2 mm** **5 to 16 % by Wt OPC** **Workability** **Pb, Cd, Cr, Hg, Tl, Zn**
Compaction	**Proctor Method** **DIN 18 127**
Curing	**7 or 28 Days respectively** **20 °C, 100% rH (Fog Chamber)**

Table 1: Preparing, compacting and curing of cement mortar specimen

3. TESTING THE LEACHABILITY

3.1 <u>German Standard Procedure</u>

The only method standardized for testing the leachability for instance of heavy metals from solid materials is the procedure DEV-S4, described in the German standard DIN 38 414, part 4 (7). This method is comparable to the recently published draft NVN 5432 (8-10) according to the determination of the maximum leachable quantity. Originally the DIN method was developed only for the evaluation of sludges and sediments.

According to this standard, the sludge should be tested unchanged in size. In practice 100 g of a crushed material (< 2 mm) is constantly shaken for 24 hours in a volume of 1 l of deionized water. The solution is filtered off and the trace elements are determined in the leachate. If necessary, the leaching has to be repeated several times.

At present the method is specified for the examination of slags and wastes of different types (11), (12). If the test is performed on uncrushed samples, that means in a condition as produced or dumped, the assessment of the leachable amount of the total trace content seems to be justified. But it must be taken into account that the leachability depends on the particle size of the sample. Crushing or milling prior to testing will therefore simulate a higher and more theoretical leachability. In particular this will be the case if samples are chemically treated with acids or bases (8-10), (12), (13).

In literature the grain size is considered only in a few cases. In the following tests according to DIN 38 414, part 4 (DEV-S4) a particle size distribution from 0 to 2 mm was used. To indicate the influence of size, additionally samples of crushed solid materials ranging from 5 to 10 mm were tested as shown in table 2. The results of these leaching methods were compared with those of a Flow-Through-Test (14).

Designation	Preparation	Test
Method 1 DEV S 4	Sample broken	0/2 mm 100 g/l, 24 h
Method 2 DEV S 4 modified	Sample broken	5/10 mm 100 g/l, 24 h
Method 3 Flow through method	Sample unchanged	Proctor cylinder H/D = 120/96

Table 2: Methods for testing leachability

3.2 Flow-Through-Test

With this method the leachability was tested on cement-bound Proctor cylinders with a height of 120 mm and a diameter of 96 mm. The photo (fig. 1) shows the sequence of the preparation steps. The humid Proctor cylinder is placed and centered into a cylindrical acrylic glass tube. The marginal gap between the inner tube wall and the cylinder surface is sealed with a two-component-resin in order to avoid water flow in this zone during measurement (14).

Fig. 1: Preparation of cylinders for testing leachability

The thus prepared specimen is placed on the bottom part of the testing device, fig. 2, which is hermetically closed with a cap. The device is filled with deionized water through the feed

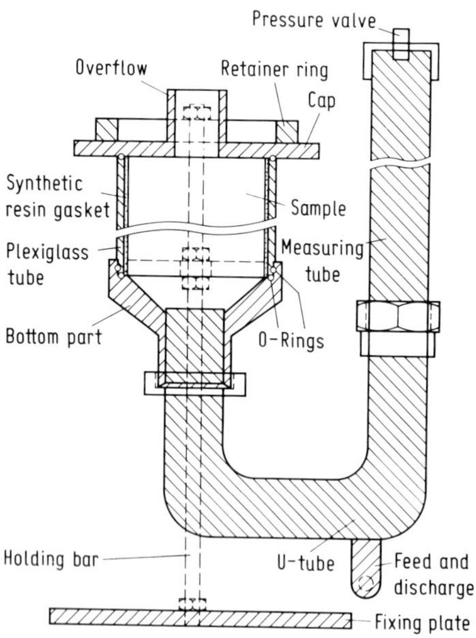

Fig. 2: Device for testing leachability of cement-bound materials

and discharge stop-cock. Hereby the air entrained in the cylinder is pushed out. To the measuring tube, which is made from acrylic glass, a meter is fixed. When the water reaches a level above the zero point of the meter, the stop-cock is closed. Air from a pressure bottle is introduced through the pressure valve. The pressure is raised to 1 bar and maintained. The measurement starts when the first water droplets appear on top of the cylinder. The flow velocity of the water through the sample cylinder is determined by measuring the difference in hight of the water level in the measuring tube over a certain time. As a result the permeability coefficient of the sample can be calculated. The water from the overflow is collected and analyzed.

The figs. 3 and 4 only show the results of the model mortars under investigation. Each specimen contained 16 mg of thallium and 679 mg of chromium per kg of mortar, added as water-soluble salts. The different curing time of 7 and 28 days is indicated by open circles and closed circles, respectively. The ordinates of the diagrams show the leaching rate of the elements in % of the total content. The abscissas give the cement content of the mortars in % by wt. In the left part of both diagrams the leaching rate of the crushed mortar samples with a maximum grain size of 2 mm tested according to DEV S4 is shown. The diagrams in the middle part show the results obtained with the modified leaching procedure S4 using coarser fragments from 5 to 10 mm of the cylinders. The results of the Flow-Through-Test are presented in the right part.

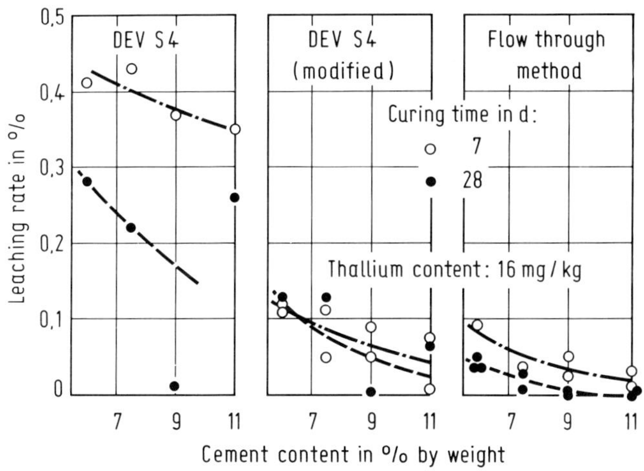

Fig. 3: Leaching rate of thallium depending on method, cement content and time of hydration

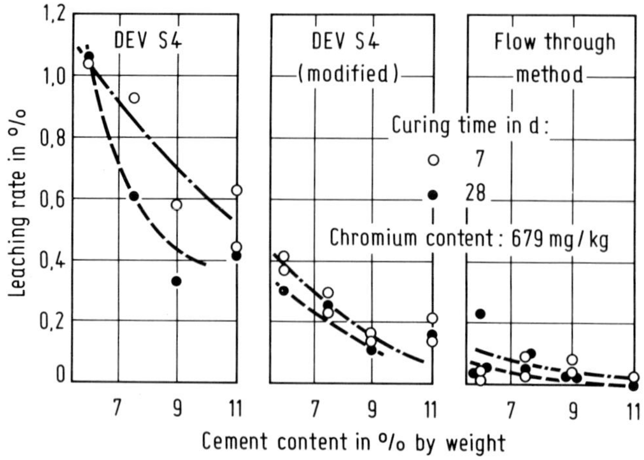

Fig. 4: Leaching rate of chromium depending on method, cement content and time of hydration

Independent of the test procedure used the leaching rates of thallium and chromium were low and practically did not exceed a value of 1 % of the total content. Besides this, the leaching rate generally diminished with rising cement content and curing time. The interpretation of this behaviour is that the structure of the specimens tested becomes increasingly more dense with the curing time by the formation of a growing amount of hydration products from the cement. Additionally, the hydrated cement is increasingly able to bind thallium and chromium.

Another result which has to be emphasized is that the leaching rate largely depends on the procedure test. In comparison the procedure according to DEV S4 leads to the highest leaching rate on the basis of equal cement content, curing time and density. This is due to the mechanical destruction of the material structure prior to the examination of leachability. By far the lowest rates were found with the Flow-Through-Test, using the original solidified and uncrushed specimen. Altogether, the test results, also with other elements, lead to the conclusion that the leaching behaviour of solidified materials after crushing does not reflect the real properties under the conditions of application.

The decisive correlation between the leachability and the structure density of a cement-bound material is indicated in figs. 5 and 6. The ordinates show the element content in the leachate after a leaching time of 24 h in mg/l and the abscissa the different permeability coefficients of the specimens in m/s. The varying cement contents and curing times were marked seperately.

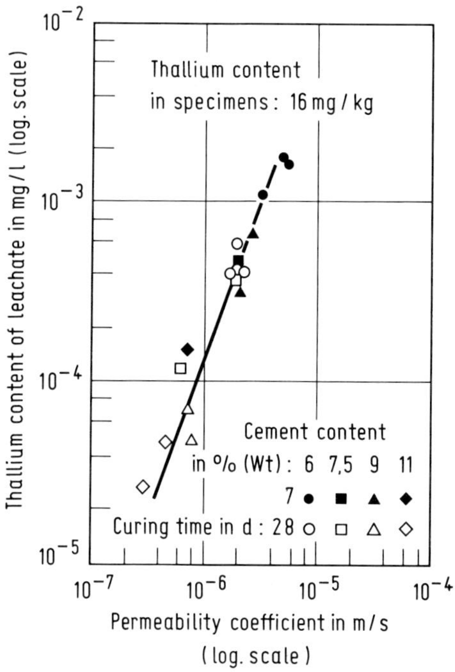

Fig. 5: Concentration of thallium in the leachate as a function of the permeability coefficient of cement-bound materials

The figures demonstrate the distinct dependence of the leached element concentration on the permeability coefficient, which reflects the predominant influence of the structure density. Comparably dense specimens with a permeability coefficient ranging from $1 \cdot 10^{-6}$ to $1 \cdot 10^{-7}$ m/s already led to low concentrations of thallium or chromium in the leachate of less than 10^{-4} or 10^{-2} mg/l, respectively. In the case of chromium the concentration remains under the German limit for drinking water, if the coefficient of permeability is less than $1 \cdot 10^{-6}$ m/s.

From these results can be concluded that the investigated water-soluble trace elements in wastes can be bound almost completely if the wastes are solidified with the necessary amount of cement. The finding is even valid in the case of abnormally high element concentration. This is due to the fact that the dense structure of a specimen with a comparably small surface decreases

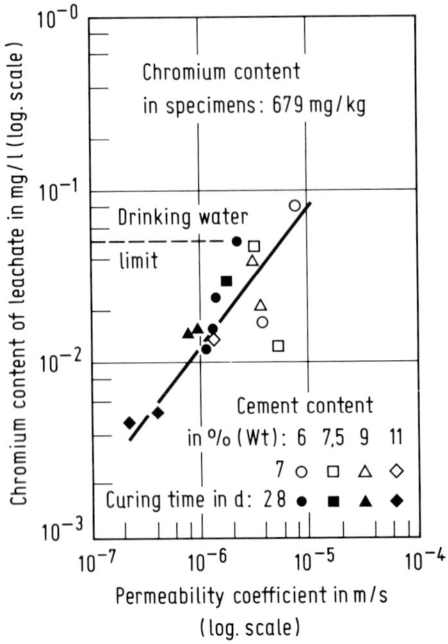

Fig. 6: Concentration of chromium in the leachate as a function of the permeability coefficient of cement-bound materials

the leachability to a minimum. Besides this the physical effect is improved by a chemical binding capacity of the hydration products of the cement. Transferred to slags and ashes from power and waste incineration plants these materials therefore may be reused for subbase road constructions without a negative effect on the environment, as other investigations have proved (15-19).

REFERENCES

1 Hösel, G., und H. v. Lersner: Recht der Abfallbeseitigung des Bundes und der Länder - Kommentare. Bd. 1, E. Schmidt Verlag, Berlin 1972-1984.

2 DIN 18 127: Baugrund, Versuche und Versuchsgeräte, Proctorversuch. Beuth-Verlag, Berlin und Köln 1987.

3 Proctor, R. R.: Fundamental Principles of Soil Compaction. Engin. News Rec. Vol. 111 (1933) August/September.

4 British Standard 1924: Methods of Test for Stabilized Soils. BSI, London 1957.

5 Forschungsgesellschaft für das Straßen- und Verkehrswesen, Arbeitsgruppe Untergrund-Unterbau: Merkblatt für Eignungsprüfungen bei Bodenverfestigung mit Zement. Köln 1975.

6 Forschungsgesellschaft für das Straßen- und Verkehrswesen, Arbeitsgruppe Betonstraßen: Technische Prüfvorschriften für hydraulisch gebundene Tragschichten (HGT), TP HGT-StB 86, Köln 1986.

7 DIN 38 414, Teil 4: Deutsche Einheitsverfahren zur Wasser-, Abwasser- und Schlammuntersuchung, Schlamm und Sedimente (Gruppe S), Bestimmung der Eluierbarkeit mit Wasser (S 4). Beuth Verlag, Berlin und Köln 1984.

8 NVN 5432 (Draft 1990): Waste Materials, Construction Materials and Stabilized Waste Products. Determination of the Maximum Leachable Quantity and the Emission of Potentially Hazardous Components from Construction Materials, Monolithic Waste Materials and Stabilized Waste Products of Mainly Inorganic Character.

9 Sloot, van der H. A., G. J. de Groot and J. Wijkstra: Leaching Characteristics of Construction Materials and Stabilized Products Containing Waste Materials. Amer. Soc. Test. Mater., Stand. Techn. Publ. 1033, P. 125/149. Philadelphia, PA 1989.

10 Groot, de G. J., J. Wijkstra, D. Hoede and H. A. van der Sloot: Leaching Characteristics of Selected Elements from Coal Fly Ash as a Function of the Acidity of the Contact Solution and the Liquid/Solid Ratio. Amer. Soc. Test. Mater., Stand. Techn. Publ. 1033, P. 170/183, Philadelphia, PA 1989.

11 Stahl-Eisen-Prüfblatt 1784. Verein Deutscher Eisenhüttenleute, Düsseldorf 1980.

12 Straub, H., G. Hösel und W. Schenkel (Hrsg.): Handbuch über die Sammlung, Beseitigung und Verwertung von Abfällen aus Haushaltungen, Gemeinden und Wirtschaft: Richtlinie für das Vorgehen bei physikalischen und chemischen Untersuchungen im Zusammenhang mit der Beseitigung von Abfällen, EW/77. 46. Lieferung 1977, Erich Schmidt Verlag, Berlin.

13 Gruber, H.: Elutionsverhalten von Blei- und Cadmiumverbindungen in Feststoffrückständen aus Rauchgasreinigungen. GIT Fachz. Lab. 28 (1984) H. 7, S. 603/605.

14 Rechenberg, W., und S. Sprung: Ein neues Verfahren zur Prüfung der Auslaugbarkeit umweltrelevanter Spurenelemente bei der Verwertung zementgebundener Abfallstoffe. In: B. Welz (Hrsg.): 4. Colloquium Atomspektrometrische Spurenanalytik, S. 295/302. Bodenseewerk Perkin-Elmer, Überlingen 1987.

15 Sprung, S., und W. Rechenberg: Einbindung von Schwermetallen in Sekundärstoffen durch Verfestigen mit Zement. beton 38 (1988) H. 5, S. 193/198.

16 Schmidt, M. Verwertung von Müllverbrennungsrückständen zur Herstellung zementgebundener Baustoffe. beton 38 (1988) H. 6, S. 238/245.

17 Sprung, S., und W. Rechenberg: Bindung umweltrelevanter Sekundärstoffe durch Verfestigen mit Zement. Zement und Beton 34 (1989) H. 2, S. 54/61.

18 Rechenberg, W., und S. Sprung: Probenvorbereitung zur Beurteilung der Auslaugung umweltrelevanter Spurenelemente aus zementverfestigten Stoffen. Abwassertechn. 41 (1990) H. 3, S. 24/27, und H. 4, S. 33/35.

19 Rechenberg, W., und S. Sprung: Umweltsicheres Deponieren von Abfallstoffen durch Verfestigen mit Zement. In: B. Böhnke (Hrsg.): Gewässerschutz-Wasser-Abwasser, Bd. 118, S. 178/203. Gesellschaft zur Förderung der Siedlungswasserwirtschaft an der RWTH Aachen, Aachen 1990.

Waste Materials in Construction.
J.J.J.R. Goumans, H.A. van der Sloot and Th.G. Aalbers (Editors)

THE NETHERLANDS LEACHING DATABASE: A USEFUL TOOL FOR PRODUCT QUALITY CONTROL, ENVIRONMENTAL CERTIFICATION AND EVALUATION OF LEACHING TEST RESULTS

G.J de Groot

Soil and Waste Research, Netherlands Energy Research Foundation (ECN), P.O. Box 1, 1755 ZG Petten, The Netherlands

SUMMARY

This paper describes a database developed to collate, organize and analyze information about the leaching of contaminants from waste and waste-containing materials. The organization of leaching test results and related information into database form is intended to allow systematic trends in leaching behaviour to be identified. The systematic information will, in turn, facilitate regulatory activities, certification activities and further research and can ultimately reduce costs associated with environmental testing by directing emphasis to the most important parameters.

1. INTRODUCTION

In recent years, the limited availability and high cost of waste disposal sites has led to increased emphasis on the reuse of bulk waste materials. A number of 'leaching tests' have been developed to predict the environmental impact of waste material reuse. In the Netherlands, standard leaching test procedures have been developed which reflect a systematic approach based on a fundamental understanding of the mechanisms controlling the release of contaminants from waste and waste-containing materials. These tests are described in the Dutch standards NVN 2508 [1] and NVN 5432 [2].

A comprehensive project was carried out in 1985-1990 ('Mammoet 85') by several institutes in the Netherlands (RIVM, TNO, INTRON and ECN) to investigate the leaching characteristics of a wide variety of bulk waste and building materials. The results from this project show systematic leaching behaviour within groups of materials [3,4].

Within the framework of the Mammoet project, ECN has developed a database for compilation and organization of the leaching test results. The database is called *UITLOOG*, the Dutch word for 'leach'. The database is incorporated in the 'Netherlands Leaching Database'. This paper discusses the purpose, structure and usage of *UITLOOG* in detail.

2. OBJECTIVES

The experimental procedures used to obtain systematic leaching information [1,2] are more extensive than the single extraction tests common in earlier work [5,6,7] and therefore are more expensive and time consuming. To take full advantage of the results, it is important that they be organized in a *relational database*, i.e. a structure in which data can be stored in separate tables that can be related to each other and/or connected to standard data processing software. The database approach greatly enhances the accessibility of the data, standardizes data processing and presentation, and aids in the identification of mechanisms controlling leaching. After leaching mechanisms for a specific type of material are identified, future testing can focus on the most relevant parameters; thereby reducing testing costs.

The specific objectives for the *UITLOOG* database are:

(a) to make the results of systematic leaching test accessible to research institutes and industry,
(b) to allow standardized processing of systematic leaching data,
(c) to enable statistical evaluation of leaching test results,
(d) to provide a data source for external software.

The database is located at ECN and presently contains results from the Mammoet project. In accordance with objective (a), it will be accessible to registered institutes and industrial users through telephone connections. The usage and further development of the database will be directed by a steering committee to be formed by the end of this year. Most of the major research institutes in the Netherlands will be represented in this committee.

A *standard application* has been developed to meet objective (b). It provides a user-friendly interface for processing data and generating reports. Available reports include raw data tables, calculated leaching parameters and information about leaching mechanisms. Graphical output can also be obtained through the standard application.

In accordance with objective (c), the database has been structured to allow statistical evaluation of leaching test results. The contribution of different sources of error, such as sample taking, sample preparation, leaching test and chemical analysis, as well as the accuracy and reproducibility of the calculated values can be determined.

Objective (d) requires that results in the database be readily available for external software, e.g. chemical reaction and transport models. This objective also requires coordination between database development and the development of external software or *database extensions*.

2.1 Systematic description of leaching behaviour

Although results from virtually all types of batch, column and tank leaching experiments can be accepted by *UITLOOG*, the database is based on the standardized Dutch leaching tests NVN 2508 and NVN 5432. These tests are fully described in references [1] and [2]. A discussion of their use and interpretation is provided in reference [8].

To briefly review, these tests measure the release of contaminants from waste and waste-containing products as a function of time. A distinction is made according to the size of the material. Test NVN 2508 is designed for granular or powdered materials from which the release of contaminants is percolation-dominated. Test NVN 5432 is designed for monolithic waste materials (with a diameter greater than 40 mm) and waste-containing products from which the release of contaminants is controlled by diffusion through internal pores.

The leaching tests allow the following characteristics of the materials to be identified:

- the maximum amount of each contaminant that can be released from a unit mass of the material,
- the release of each contaminant as a function of time, expressed as mass per unit mass for granular materials and as mass per unit surface area for monolithic materials.

For monoliths and waste-containing products the following additional parameters can be determined:

- the *physical retardation* or *tortuosity*, a parameter which quantifies the decrease in diffusion rates due to the tortuous path taken by components diffusing through the pore structure of the monolith or product,
- the *chemical retardation* of a contaminant, a parameter which quantifies the chemical interaction between the contaminant and the matrix of the monolith or product, by which the diffusive release of the contaminant is further decreased.

Information about the mechanism controlling the release of each contaminant can also be obtained from the leaching protocol. This topic is discussed in more detail in reference [8].

3. OUTLINE OF DATABASE

3.1 General description

The leaching database was originally written in FORTRAN on a Cyber main-frame computer [9]. That structure worked reasonably well but suffered from a number of deficiencies:

- Storage of multiple test results for the same sample was not possible without creating a completely new data set for the same sample including all data that were identical. Statistical analysis to quantify different sources of errors, such as sample taking, sample preparation, test performance and chemical analysis, was severely hampered by this deficiency.
- Additions to and changes in the database program required extensive reprogramming.
- Direct access to the database by external programs on a PC was not possible. Data on the Cyber had to be exported in a standard text file and transferred to the PC for further processing.
- The user interface did not meet current requirements in terms of convenience. A special stripped database version for the PC had to be written to create a 'user-friendly' interface, necessitating two implementations of every change.

In 1989, a conversion to a ORACLE RDBMS system was initiated. ORACLE is a truly relational, multi-user database system, that can run on many computer systems, including the PC. Several tools are available to create a convenient user-interface and to access data from PC applications, such as LOTUS and DBASE and programs written in high-level computer languages, such as C and PASCAL. One of the major advantages of the new system is its ability to handle complex branched database structures. This change allows development of a new structure for the storage of multiple test results of different data types. The system is installed on a COMPAQ-486 server (UNIX) located at ECN and accessible through telephone connections.

The tree-like structure of the database is clear in Fig. 1. The figure shows the data structure for a single sampling of a single material. Each box can be regarded as a sub-database or table in which some type of data from the sample is stored. For simplicity, only the four main types of data are shown: sample preparation; physical/chemical characterization; test conditions/parameters; and test results.

More than one set of information, called *records*, can exists in a table for a sample. Multiple records are represented in Fig. 1 by stacked boxes. Three stacked boxes are shown for each table; in reality this number is virtually unlimited. Furthermore, each record (one of the stacked boxes) in a table can be linked to multiple records in another table. In the box labelled 'SAMPLE DETAILS', there is only one record for each sampling. This box is the entry point for all other tables with information from that sampling.

Fig. 1. Schematic representation of the database structure

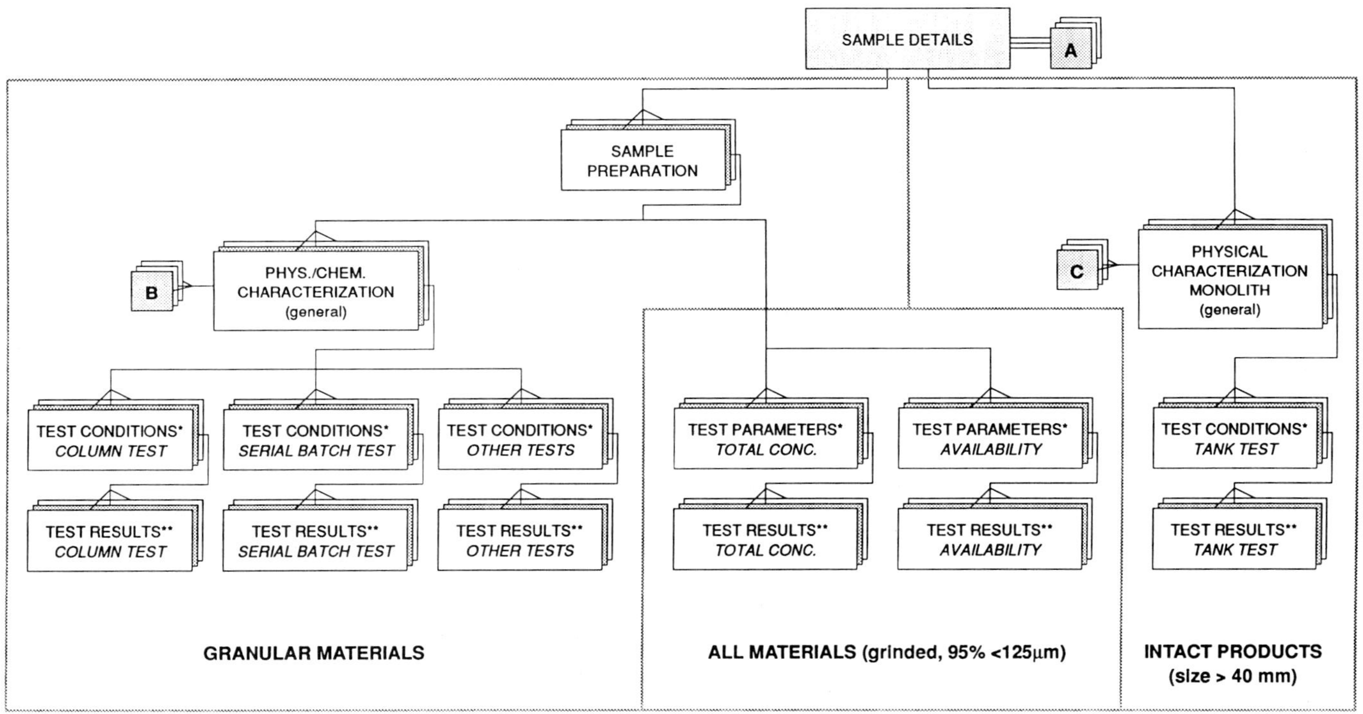

A: Connection to other databases
B: Size distribution and chemical speciation
C: Large number of sample-specific physical parameters

* *Linked with the LITERATURE sub-database (not shown)*
** *Linked with the sub-databases CHEMICAL ANALYSIS and DESTRUCTION METHODS (not shown)*

3.2 Types of data in database

The information stored in the database for each sampling of each material can be divided into four main types:

(1[a]) general information about the origin, production process, manufacturer, etc.,

(1[b]) information about the sampling procedure and sample preparation,

(2) results of the physical and chemical characterization of the sample,

(3) information about the type of test performed and the general leaching conditions,

(4) total element concentrations and leaching results including analytical parameters, e.g. methods, detection limits, standard deviations.

Each type of data is discussed in more detail below.

3.2.1 General information

The general information collected for each sample includes classification according to the size, type, usage, and overall leaching properties of the materials, and information about the origin of the sample including a description of the process by which the waste or waste-containing product is created, the name of the factory or plant, and the sampling procedure. In addition to the general protection of database information (see below), 'sensitive' items like the factory name and the process conditions are given extra protection to prevent unauthorized usage. General information is stored in box labelled 'SAMPLE DETAILS' in Fig. 1. Additional information about the procedures of sub-sampling and sample preparation prior to testing is stored in the 'SAMPLE PREPARATION' box.

3.2.2 Physical and chemical parameters

For granular materials, physical and chemical parameters measured on the sub-sample, such as percentage dry weight, percentage ash, percentage volatile, homogeneity, bulk density, chemical element speciation and particle size distribution, are stored in the boxes labelled 'PHYS./CHEM. CHARACTERIZATION' and 'B' in Fig. 1. For monoliths and solidified waste products, physical parameters such as apparent density, permeability, porosity and compressive strength are stored in the boxes labelled 'PHYSICAL CHARACTERIZATION MONOLITH' and 'C'. In addition, a virtually unlimited number of sample-specific parameters can be stored in the boxes labelled 'C'. New parameters can be defined on-line by the database administrator.

3.2.3 Leaching test conditions

The boxes labelled 'TEST CONDITIONS' in Fig. 1 contain general information about the test conditions, for example the pH, conductivity, and electrochemical potential of the leachates. This part of the database is also linked to the 'LITERATURE' sub-database.

Test conditions during the availability and total concentration tests are included for all materials. For granular materials, conditions of the column, serial batch, and single extraction tests are included, for example, the liquid to solid ratio and the column dimensions. For monoliths and stabilized waste products, general conditions for the tank leaching test are included, for example, the extraction volume, the water renewal times, the size of product, and the size reduction method, if required.

3.2.4 Leaching test results

The boxes labelled 'TEST RESULTS' in Fig. 1 contain the concentrations measured in the leaching test. A virtually unlimited number of duplicate analyses can be stored. Extensive information about the chemical analysis and sample destruction methods is also included.

3.3 Protection

Since the database is intended to be accessible to many research institutes and industrial groups, it is essential that unauthorized use of information is impossible. Therefore, a comprehensive protection scheme based on username and password combinations has been incorporated into the database. Each record of a sub-database can be protected separately. Before any information is retrieved, user privileges are checked for all records that the user tries to access. Extra protection is available for information that can relate leaching behaviour to product and factory names. Moreover, these names and other sensitive information can only be accessed by the owner of the sample from which the data originates, or by users that are authorized by the owner.

3.4 Communication with users

Registered users can make connection with the database system through a PC with modem using normal terminal emulation software (VT100 protocol). The standard *UITLOOG* application software on the host computer is available to add, retrieve and process data. The application program is menu oriented, and extended context-sensitive help is available at all times. Access of users to the different functions of the database is controlled by the steering committee. As mentioned above, users also require authorization from owners before 'private' information can be accessed.

4. STANDARD DATABASE APPLICATION

4.1 Standard functions

The main functions available in the standard application are:

(a) View/Add/Change/Erase data
(b) Generate standard reports, numerically and graphically
(c) Definition of a selection filter to retrieve data
(d) Import from and Export to a standard text file on disk
(e) General utilities

Access to the different functions depends on the privileges of the user (see section 3.4).

The user can formulate a large number of selection criteria (*selection filter*) to extract data related to a specific subject of interest. Calculated parameters can also be used in defining the selection filter. The filter is operative for most actions, such as extraction of raw data, calculation of derived parameters (e.g. percentage available for leaching, diffusion coefficient, tortuosity, chemical retention), plotting combined results, extended statistical analysis and exporting data to a disk.

4.2 Statistical evaluation of data

As discussed above, the tree-like structure of the database allows storage of multiple test results of different data types for the same sample. Extensive statistical evaluations can be performed to identify and quantify sources of error. For example, the variability in chemical analysis can be determined by calculating the mean and standard deviation of all concentrations for a chemical component ('TEST RESULTS') belonging to one record of 'TEST CONDITIONS/PARAMETERS'. The differences in these calculated means and standard deviations for different records of 'TEST CONDITIONS/PARAMETERS' show the variability between multiple tests for one sub-sample. In the same way, variability resulting from sub-sampling and sample preparation can be determined.

4.3 Determination of leaching systematics

The standard interpretation of leaching tests results described in [1] and [2] is relatively straightforward. However, the results obtained by the leaching test for intact products allow a more extensive evaluation regarding the leaching mechanisms involved and calculation of derived parameters. A calculation method for the determination of the leaching mechanisms has been reported earlier [8]. With this method, which is also included in the database application software, diffusion-controlled release can more easily be identified in the presence of other initial processes such as surface wash-off or slow wetting of the material. The *UITLOOG* program allows a standardized evaluation of the release mechanisms and presents a detailed report in tabular and graphical form.

4.4 Interface with other applications

To take full advantage of the *UITLOOG* database, it is necessary to allow external software and database extensions to access the data in a convenient way. Two methods are available. Firstly, data can be exported to an ASCII text file in a standardized format, called a *neutral file*. Secondly, interface libraries exist to access the database from PC software, such as LOTUS, DBASE and C programs.

The first option is coupled to the database selection filter described in section 4.1. In this way, only data that are requested by the user are included in the neutral file. The neutral file have to be transferred to the PC first (by file transfer through telephone lines or by diskette) before the PC software can process the data.

Using the second method, the application program with the interface library runs on a PC, which accesses the *UITLOOG* database directly through a modem connection. This option will only be available for a limited number of users. Some of the database extensions discussed below use this method.

5. DATABASE EXTENSIONS

Other databases and calculation modules will be linked to the main database. Two databases and one calculation module are now being developed.

5.1 Mobility database

This database will contain contaminant mobility measurements from diffusion tube experiments. These measurements are particularly useful for calculating contaminant distribution coefficients at in-situ liquid/solid ratios for different types of soil. The distribution coefficients can be used to describe contaminant sorption, an important process in determining the transport of contaminants in the environment surrounding the waste or waste-containing product.

5.2 Certification database

The certification database will collate results of periodic certification tests on products from waste stabilization and reuse processes. This database will play an important role in determining the long term variability in leaching characteristics of products from a continuing process. The frequency of testing required to determine if a material meets certification requirements can be reduced when the variability proves to be low. Since physical and chemical parameters of the product and its raw materials, and the relevant process conditions, will be stored in the database, correlations with leaching results can be studied. If a suitable correlation exist, further simplification of routine testing procedures will be possible, leading to a further reduction in costs.

5.3 Source term model

A project has recently been initiated to develop a release model for contaminants in waste-containing building materials, using *UITLOOG* as a data source. The purpose of this model is to provide a contaminant source-term in which laboratory leaching test results are translated to actual field conditions. Chemical and physical parameters that can directly influence the release of contaminants from the material are included.

One important aspect that is generally not recognised but which is being incorporated in the new model is the change in contaminant release caused by interactions at the interface between the material and its environment. These interactions are discussed in more detail in reference [10]. Results from diffusion tube experiments, including those contained in the mobility database, play an important role in quantifying many interface effects, such as precipitation reactions, changes in pH and element uptake from the surrounding environment.

The source term model will be modular in form. Standardized modules addressing specific parameters or processes are being developed separately and then incorporated into the overall structure of the model. In this way, a functional model can be obtained in a short time. Although the model will be quite simple at first, it can easily be enhanced by adding new modules or adapting old ones when more is known about specific processes.

6. CONCLUSIONS

The *UITLOOG* database allows information about the leaching of contaminants from waste and waste-containing products to be collated, organized and analyzed. Four types of information are included: general information about the material and the sampling procedure; results of physical and chemical characterizations; information about test methods; and test results.

The database has extended password-protection capabilities to prevent improper use of 'sensitive' information. Registered users can access a convenient interface for entering and reviewing data and for the calculation and presentation, in tabular and graphical form, of standard leaching parameters. Statistical analysis of the data is also supported. Extensions are now being prepared to incorporate contaminant mobility data, to follow a time series of certification tests, and to predict the net release of contaminants in different environments.

The use of a database to store leaching data from different wastes and waste-containing products in a database offers opportunities for a more efficient use of costly leaching studies. The knowledge gained by gathering this information into an accessible whole may prove particularly useful in selecting new directions for the reuse of waste materials.

7. REFERENCES

[1] NVN 2508 : Het bepalen van de uitloogkarakteristieken van kolenreststoffen. UDC 662.62/67:543.2, 1987. Also published earlier in English as: H.A. van der Sloot, O. Piepers and A. Kok. A standard leaching test for combustion residues. Technical report Bureau Energy Research Projects BEOP-31, 1984.

[2] Concept voornorm NVN 5432: Bepaling van de maximaal uitloogbare hoeveelheid en afgifte van potentieel schadelijke componenten vanuit constructiematerialen, monolitische reststoffen en gestabiliseerde reststofprodukten. September 1989.

[3] G.J. de Groot, H.A. van der Sloot, P. Bonouvrie, J. Wijkstra. Mammoet report no.09: Karakterisering van het uitlooggedrag van intacte produkten. ECN report no. ECN-C--90-007, 1990.

[4] Th.G. Aalbers and R. Gerritsen. Mammoet report no.07: Uitlooggedrag van anorganische parameters uit primaire en secundaire grondstoffen. RIVM and TNO, 1990.

[5] Hazardous waste proposal guidelines and regulations. U.S. EPA, 1978, (section 250.13, paragraph C to E).

[6] *Toxicity characteristic leaching procedure*. Federal Register, Vol.51, no.114, 1986.

[7] German standard method for water, wastewater and sludge investigation. Group S, DIN 38414, part 4: '*Auswaschbarkeit mit*

wasser' (S4). Deutsche Institut für Normung, Berlin, 1984

[8] G.J. de Groot and H.A. van der Sloot. *Determination of leaching characteristics of waste materials leading to environmental product certification*. Presented at '2nd International Symposium on Stabilization/Solidification of Hazardous, Radioactive and Mixed Wastes', May 29 - June 1, 1990 Williamsburg, Virginia, USA

[9] G.J. de Groot. Mammoet report no.12: Opzet en inrichting database uitlooggegevens. ECN-report no. ECN-C--90-008, 1990.

[10] D. Hockley and H.A. van der Sloot. *Modelling of interactions at waste-soil interfaces*. This proceedings, 1991.

Waste Materials in Construction.
J.J.J.R. Goumans, H.A. van der Sloot and Th.G. Aalbers (Editors)

ENVIRONMENTAL CERTIFICATION OF CALCIUM SILICATE

P.D. RADEMAKER[1] and G.J. DE GROOT[2]

1 Research Centre of Calcium Silicate Industry, P.O. Box 145, 3770 AC Barneveld (The Netherlands)

2 Netherlands Energy Research Foundation ECN, P.O. Box 1, 1755 ZG Petten (The Netherlands)

SUMMARY

In the near future the Dutch Government will introduce new legislation within the framework of the soil protection that will put limits to the composition and leaching behaviour of building materials. Enforcement of this legislation will require environmental certification of building materials. The calcium silicate industry has decided to participate in a pilot project to obtain such a certification. The reasons for certification of calcium silicate and the system of certification are discussed.

1. CALCIUM SILICATE PRODUCTION

Calcium Silicate bricks and blocks are produced in various European and Asian countries. The units are used as masonry. Raw materials are silicious material, usually sand, lime and water.

In the Netherlands calcium silicate is used in 60 % of the buildings. Particulary the larger elements (60 x 90 cm) allow for rapid and economically attractive building of separating and cavity walls. Application is in inner and in outer leafs of the building. Total annual production in the Netherlands is 3.5 million tons. There is a large variety of shapes and dimensions (see Fig. 1).

2. OTHER RAW MATERIALS

The majority of the calcium silicate bricks are produced with natural occurring sand, with lime and with water without any other addition. There is a growing tendency to study the possible use of other raw materials. This partially to find a substitute for the sand or lime but also to improve the product quality. Some examples of the use of other raw materials and of developments for such uses are given below.

Since many years the use of fly-ash from coal-burned power

Bricks

Blocks

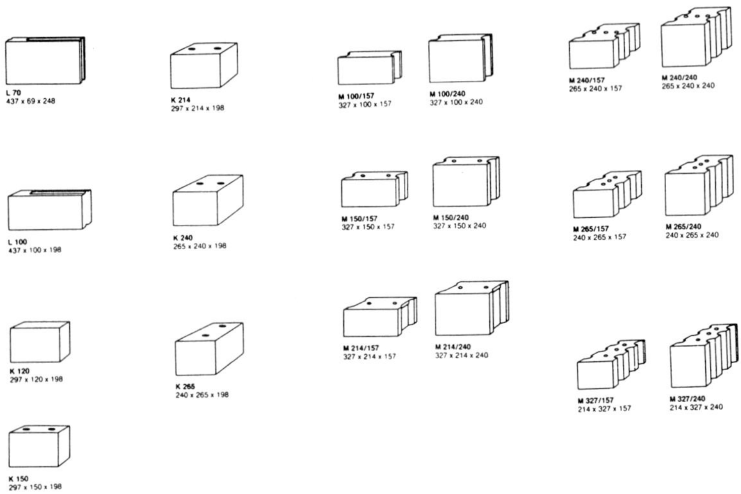

Elements

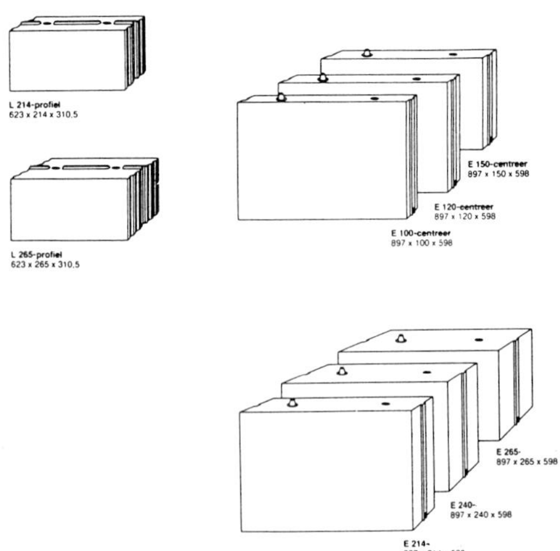

Fig. 1.

stations in calcium silicate has been investigated. One of the problems was the colour changes due to the residual carbon in the fly-ash. A new development in the Netherlands is a fluid-bed reactor that also uses limestone (calcium carbonate) and coals as fuel. The final product is lime (calcium oxide) mixed with fly-ash. This fly-ash now is carbon free. This mixture is suitable for the calcium silicate production. The fluid-bed reactor is situated on the premises of one of the Dutch calcium silicate plants.

Another possible addition to calcium silicate is pigment. A large majority of calcium silicate is white, but a growing interest is seen in yellow, red and grey coloured bricks. For this purpose the material is pigmented with the same colorants used in the concrete block industry. These pigments are based on iron (III)-oxide, but some pigments for other colours such as blue and green contain other base-metals.

Various development projects concerning the use of secondary raw materials in the calcium silicate production are in progress. An example of such a project is the use of burned sludge from the paperindustry. The Netherlands has a very high degree of re-use of paper. In the process to recycle paper, the old paper, is pulped in order to separate the fibre from the ink and the fillers. This sludge can be dewatered and burned. From the burning process an ash remains containing mainly china clay, alumina, calcium carbonate and calcium oxide. Such a mixture seems to be very suitable as an admixture for calcium silicate production.

Growing interest is generated in the Netherlands for the use of sea sand in the production of building materials. Although much work remains to be done, the option is worth a serious consideration.

In the production of calcium silicate a small part of the final product has to be considered as waste. From sawing the larger units and from breakage during handling a material that is unsuitable is formed. Rising disposal costs and better knowledge about possibilities to use this waste makes it more attractive to introduce recycling techniques.

In all these and in other comparable projects the effects of these raw materials on the production process, on the mechanical properties of the end-product and on the environmental consequences are studied. Of course this input of other material must also have a positive effect on the economics of the production and must be acceptable to the user. No negative impact on health or environment whatsoever can be accepted.

3. CERTIFICATION

All eleven calcium-silicate producing factories have acquired a product-certificate. In the Netherlands a building material can have a uniform certificate under the approved trade-name "KOMO".

Eight different independent certifying bodies are recognized to grant this KOMO-certificate. Each of those bodies works under the supervision of a central body, the Council of Certification. If a producer wants to apply for a certificate, a quality guidance document should be prepared. In this document the relevant procedures and quality requirements are described. For calcium silicate this document refers to the Dutch Standards NEN 3836 (Calcium silicate bricks and blocks) and NEN 3837 (Calcium silicate elements) for the product quality. Sixteen times per year samples from the product are taken at the factory by the certifying organisation (IKOB) and tested for mechanical, visual and physical properties.

The other aspects of the quality guidance document are the procedures to be followed by the producer. Daily and weekly tests and inspections of raw-material, intermediate products, and end-products are required. Four times per year the certifying organisation will control the documents from the producer concerning his own testing procedure.

If the new European CEN standard for calcium silicate will come into effect in 1993 or 1994 a very similar system of quality control is to be expected. Both factory production control and product control by producer and possibly by a third party are mentioned in the EC Directive for Building Products.

4. BUILDING MATERIALS ENVIRONMENTAL REGULATION

In 1988 the Dutch government started the preparation of a legislation for building materials concerning the protection of soils and ground-water. This legislation will put limits on the content of heavy metals, anions and organic contaminants in all building materials that can have contact with ground or surface water or rain. A number of leaching tests have been developed and according to the new legislation there will be limits to the total concentration of contaminants in the material and to the leaching rate from the material.

For preformed building products that do not exceed those limits there will be a classification, based on leaching behaviour, into two categories. Category I will be allowed without any major restrictions, for Category II several preventive measures are required.

It is intended that this legislation will take effect in 1992. The Dutch government indicates in its Explanatory Note to the legislation that it sees an important role for certification in maintaining the new law.

5. ENVIRONMENTAL CERTIFICATE

The trade organisation of the Dutch calcium silicate industry, VNK, has decided to start a pilot project to acquire an environmental certificate. The main consideration was the desire to have all calcium silicate in one category, notably category I. The actual certificate for calcium silicate considers quality aspects related to mechanical, physical and visual properties only. A new aspect will be the environmental properties related to soil and ground water protection.

Other aspects such as suitability for construction workers can become subject to legislation as well. In order to avoid confusion in the market, the calcium silicate industry prefers to combine all quality aspects into one certificate.

6. PREPARATORY ACTIVITIES

In a joint effort with NOVEM as governmental agency that promotes the use of secondary material, with IKOB as certifying body, with experts from ECN and MBN who have knowledge on environmental testing and on data basis for this, an addition to the existing Quality Guidance document is elaborated.

In this addition the system of sampling and testing of raw materials and of the end-product will be described. The system should satisfy two important requirements. First of all it should assure that the products comply with the environmental requirements, but also important it should allow that the number of samples to be taken and the number of analysis to be performed are limited. Too much sampling and testing would be prohibitive from an economical point of view.

To accommodate these two conflicting goals various decisions have been made. It is foreseen to have both an initial type-testing of the calcium silicate and a factory production control. From the results of the initial type-testing it will become evident what are the critical components and what is the risk to exceed the limits. Based on this information a frequency for sampling and testing as part of factory production control is selected. This frequency is variable and will be governed by the results from the regular analysis of raw materials and end-product. The variance in the test results can be assumed to be relatively high. For some components the legal limits are close to the analytical detection limit. The inspection plan will be based on the ISO standard 3951 "Sampling procedures for inspection by variables". With a choice of producer's and consumer's risk and of the Acceptable Quality Level (AQL) the number of analysis can be restricted and the appropriate quality level can be assured.

According to the governmental regulation more than twenty metals and anions should be tested for. From results of the initial type-testing and from the producers administration of raw materials this number of components can be reduced to some of the more critical. This will limit the analytical costs substantially. A further reduction in analytical costs can be attained by developing a short leaching test. The standardized leaching test takes 64 days which is undesirable for production control but is also rather expensive.

Another feature in the Quality Guidance document is the use of historical data on the composition and leachability of the calcium silicate. In order to evaluate these historical test-data a date base is required. This date-base can also serve to establish relations (if they exist) between for example:

- the source of raw materials and their composition
- composition of calcium silicate and its leaching behaviour
- process conditions and leaching behaviour.

It can also help to determine the natural occurring variation in composition of raw materials. A description of the structure and features of the leaching date-base called UITLOOG, is given by De Groot (this proceedings). Some details about sampling and sample preparation in relation to reference material and certification are discussed by Lamers et al (this proceedings).

In this "self-learning" system a set of raw materials and process conditions that assures products complying with all environmental requirements, can easily be identified. If a producer wants to experiment with other raw materials the system will be such that irresponsible behaviour causes frequent testing at high costs. On the other hand the quality assurance system does allow for new development if a justified use of secondary material is intended. Such a use of secondary material could help to reduce the input of natural raw material and also could avoid a potential disposal problem.

The complete description of the system of sampling and testing of raw materials and of end-product as part of factory control and of product control, together with conformity criteria will be a new paragraph to the Quality Guidance document for calcium silicate. If the producer can comply with all the criteria, he is allowed to bring the product to the market with a certificate that covers all quality aspects, including the environmental aspect.

It is not yet completely clear how this addition of an "Environmental paragraph" to the quality guidance document will fit in the new European standardization for Building Products. In the Interpretative Documents to the EC Construction Product Directive the technical specifications for building products are listed. One of the specifications is the release of pollutants to outdoor soil and water. For this specification a measurement of leaching of pollutants is required. The Interpretative Documents are unclear about requirements on concentration of pollutants in the building product, they do not exclude the possibility of such requirements. It remains to be decided what levels of requirements are allowed on a national level once a free flow of products across the European boarders should take effect.

7. CONCLUSION

The Dutch producers of calcium silicate have decided to investigate the possibility of an environmental certificate as part of the existing product certificate. This in view of new legislation for building materials to protect soil and ground water which sets limits to heavy metal content and leachability. Preparations are on schedule and an environmental section in the certificate can be expected immediately after the new legislation takes effect in 1992. Particular attention is given to the statistical treatment of all test results. Each analysis is relatively costly and the variance in the results is high. Only by appropriate use of statistics and by establishing a "self-learning" system, supported by a date base, costs can be kept within acceptable limits.

REFERENCES

1. G.J. de Groot, this proceedings
2. F.J.M. Lamers e.a., this proceedings
3. Directive for building materials and their effect on soil and ground water (Dutch legislation in preparation)
4. Dutch standards NEN 3836 and NEN 3837 (Calcium silicate bricks, blocks and elements
5. The Construction Product Directive of the European Communities, December 1988
6. EC Council Directive 89/106/EEC, Construction Products, Essential Requirement Hygiene, Health and the Environment
7. RCK Report CK-11, The use of papermill-ash in calcium silicate production (May 1991)
8. International Standard ISO 3951 "Sampling procedures and charts for inspection by variables for percent nonconforming"
9. International Standard ISO 3534 "Statistics-Vocabulary and Symbols"

Waste Materials in Construction.
J.J.J.R. Goumans, H.A. van der Sloot and Th.G. Aalbers (Editors)

TOWARDS A NEW APPROACH IN MODELING LEACHING BEHAVIOR

M. HINSENVELD

Department of Civil and Environmental Engineering, University of Cincinnati, 741 Baldwin Hall (ML 71), Cincinnati, Ohio 45221-0071 (USA)

SUMMARY

Modeling of leaching behavior from immobilized wastes is still in its infancy. The present approach is to curve-fit leaching data to a simple model that shows the required trends in leaching behavior. Many models, however, even though they summarize short-term leaching data reasonably, prove to be poor representations of the actual situation when concentration profiles in the specimen are observed. Some major models are discussed. Research suggests that a shrinking core type model for leaching of cement stabilized wastes may be more adequate than the bulk leaching models used to date.

1. INTRODUCTION

Short-term tests conducted under aggressive conditions have been widely used to predict the long-term leaching behavior of contaminants from cemented material. Since in many cases the short-term leaching behavior seems to be following some sort of diffusion curve, it is thought that the long-term behavior can be estimated by extrapolating the results of these short-term tests, without an in-depth understanding of the leaching mechanisms. In general, the indicated approach is valid for materials that do not react with the surrounding solution (acid rain, groundwater) and contain only limited amount of heavy metals. A completely different picture, however, is obtained when the matrix material contains components that can dissolve in the surrounding solution, like cemented materials and concrete. A short-term method in this case can seriously under- or over-estimate the actual leaching behavior.

Obtaining a mechanistic model of leaching behavior from immobilized waste is considered extremely difficult. Immobilized waste is heterogenous and mass transfer is complicated by numerous processes that can occur simultaneously [1]. Despite this complexity, presently used models assume the waste form to behave like a simple porous non-reactive system. In section 2, a few major models are briefly described to provide background information on the most important assumptions underlying these models.

At present, calibration of models is largely based on leaching data. It should be realized, however, that many fundamentally different models can easily be fitted to the same short-term leaching data. Fitting models to leaching data, therefore, is not a guarantee that they are adequate represention s of the actual situation as far as prediction of concentration profiles in the waste form is concerned. Internal investigation of acid leached specimens revealed that a shrinking core type model would be more appropriate than a bulk diffusion model. A bulk diffusion model, however, fits the short-term leaching data generated in the experiments equally well. This indicates that realistic leaching models should be based on leaching patterns as well as concentration profiles in the waste form and not on leaching patterns only.

In section 3, a simple shrinking core model is described. A shrinking core type model which also includes adsorption and internal diffusion is currently being developed at the University of Cincinnati [6].

2. PRESENT APPROACH TOWARDS MODELING LEACHING BEHAVIOR

2.1 Bulk diffusion

The simplest and most widely used model to describe the leaching process is based on bulk diffusion, as described by Fick's second law of diffusion:

$$\frac{\partial C}{\partial t} = D_e \frac{\partial^2 C}{\partial x^2} \tag{1}$$

where C is the concentration in the pore solution [kmol/m^3], t is the time of leaching [s], D_e is the effective diffusion coefficient [m^2/s] and x is the distance from the surface into the monolith [m]. Assuming that, due to the presence of the matrix, the path of diffusion is elongated by a tortuosity factor τ [-] and the area through which diffusion occurs is proportional to the porosity ϵ [m^3/m^3], the effective diffusion coefficient can be calculated from:

$$D_e = \frac{D_m \ \varepsilon}{\tau} \tag{2}$$

where, D_m is the molecular diffusion coefficient [m^2/s].

Apart from the obvious assumptions (flat surface, infinite medium, one-dimensional transport, equimolar diffusion, surface concentration constant), some very important assumptions underlie this model. Firstly, as mentioned, the model assumes that the leaching process does not alter the waste form (D_e is constant). A second assumption relates to the fact that, in practice, the concentration of the contaminants in the pore solution is not measured. Instead, the bulk concentration, C_t, defined as number of kmoles of contaminant in the monolith divided by the bulk volume of the monolith, is used. The use of C_t introduces another assumption, namely the concentration of the contaminants in the matrix equals the concentration in the pore solution at all times. More precisely stated, the amount of contaminants in the waste matrix and the pore solution is distributed as $(1 - \epsilon)/\ \epsilon$.

An analytical solution for the differential equation can easily be obtained. With boundary conditions: $x = 0$, $C = C_i$ for all t; $x = \infty$, $C = C_o$ for all t and $x > 0$, $C = C_o$ for $t = 0$, the dimensionless concentration profile as a function of x and t and the total amount leached [kmol/m^2] at $t = t$, respectively, are given by Crank [2]:

$$\Theta(x,t) = \frac{C - C_i}{C_o - C_i} = erf\,(x/\sqrt{4D_e t}\,) \tag{3}$$

$$M(t) = -2\sqrt{\frac{D_e t}{\pi}}\,(C_o - C_i) \tag{4}$$

The many assumptions underlying the bulk diffusion model have not impeded its wide used for fitting leaching data. There are three reasons for this. Firstly, there is an analytical solution available for the model and the model fits the limited range of data usually generated in laboratories (typically less than one year). Secondly, some of the assumptions do not influence the form of the equation on the short term, but merely change the numerical value of the "effective diffusion coefficient". The latter argument in fact is the most important, since adapting the parameters leads to a very good fit in terms of total residuals (but not always in terms of trends) between experimental values and the values computed from the fitted one parameter (D_e) equation.

In reality, the leachability is a complex function of many parameters, such as porosity, tortuosity, distribution of contaminants between pore solution and matrix, geometric form of the monolith, type of contaminant and degree of saturation of the matrix voids. A strong dependency of the effective diffusion coefficient on the number of elutions in batch tests is also reported [3,4], indicating that even process conditions are lumped in the effective diffusion coefficient.

2.2 Bulk diffusion with linear adsorption

The bulk diffusion model can be "enhanced" by introducing a linear adsorption isotherm to incorporate physical immobilization of contaminants onto the matrix [5]:

$$Kd = \frac{C_{mo}}{C_{im}} \tag{5}$$

where C_{mo} is the concentration of the contaminant in the pore solution and C_{im} is the concentration of adsorbed contaminants. This ultimately leads to an equation of the form:

$$\frac{\partial C}{\partial t} = \frac{D_e}{(1 + Kd)} \frac{\partial^2 C}{\partial x^2} \tag{6}$$

The equation is comparable to equation (1), the effect of the equilibrium being to slow down the leaching process. Based on leaching results only, no distinction can be made between this model and the bulk diffusion model.

2.3 Bulk diffusion with internal mass transfer resistance

This model may be used if the contaminants are not very soluble (which generally is aimed at in immobilization) or abundantly present in the waste form. Part of the contaminants will then be present in a precipitated form. This has been accounted for by introducing a dissolution constant [5]. This dissolution constant can also be interpreted as the rate of internal mass transfer or the rate at which non-precipitated contaminants are transferred from the matrix into the pore solution. Dissolution of the contaminants is assumed to be proportional to the difference in the concentration in the pore solution at equilibrium and at the true concentration. It is further assumed that the dissolution is a continuous process, meaning that the amount present in precipitated form is infinite (which is quite unrealistic). As before, a partition coefficient $Kd = C_{mo}/C_{im}$ can be defined. A mass balance across a differential section then leads to (writing C for C_{mo}):

$$\frac{\partial C}{\partial t} = D_e \frac{\partial^2 C}{\partial x^2} - \frac{\partial C_{im}}{\partial t} + k(C_s - C) \tag{7}$$

with boundary conditions: x > 0, t = 0, C = C_s; x = 0, t > 0, C = C_i; x = ∞ t > 0, C = C_s. On putting:

$$\Theta(x,t) = \frac{C_s - C}{C_s - C_i} \tag{8}$$

and substituting $D_e^* = D_e/(1 + Kd)$, the concentration profile and the total amount leached are given by Crank [2]:

$$\Theta(x,t) = \frac{1}{2} e^{-x\sqrt{k/D_e^*}}\, erfc\left(\frac{x}{\sqrt{4D_e^* t}} - \sqrt{kt}\right) + \frac{1}{2} e^{x\sqrt{k/D_e^*}}\, erfc\left(\frac{x}{\sqrt{4D_e^* t}} + \sqrt{kt}\right) \tag{9}$$

$$M(t) = -(C_s - C_i)\sqrt{D_e^*/k}\left\{\left(kt + \frac{1}{2}\right) erf\sqrt{kt} + \sqrt{kt/\pi}\; e^{-kt}\right\} \tag{10}$$

It is interesting to observe the limiting case for large kt, when erf $\sqrt{kt}$ approaches unity and (9) and (10) become:

$$\Theta(x,t) = \exp(-x\sqrt{k/D_e^*}) \tag{11}$$

$$M(t) = -(C_s - C_i)\left(t + \frac{1}{2k}\right)\sqrt{D_e^* k} \tag{12}$$

which illustrates the fact that the concentration profile approaches a constant value and leaching is infinite! When on the other hand kt is small, we find by expanding erf$\sqrt{(kt)}$ and exp(-kt) and neglecting powers of kt higher than the first [2]:

$$M(t) = -2(C_s - C_i)\left(1 + \frac{1}{2}kt\right)\sqrt{D_e^* t/\pi} \tag{13}$$

which reduces to the well known solutions for diffusion without reaction when k = 0. For small kt values (small k and/or small times), the total amount leached, M(t) is a function of $t^{0.5}$ and $t^{1.5}$. This model might be fundamentally different from the bulk diffusion model and the total amount of contaminants leached might not follow a square root of time relationship; it is difficult to discriminate this model from the bulk diffusion model using short-term leaching tests only.

3. TOWARDS A NEW MODEL FOR CEMENT STABILIZED WASTE

3.1 Introduction

In relatively inert specimens containing few buffering species, acid may penetrate the specimen deeply, solubilizing contaminants on its way. This is not so in specimen containing a large fraction of cement, where the acidity can be consumed by the available hydroxides.

In this section, a new model for acid leaching of cement stabilized specimen, the so-called shrinking core model will be presented. Shrinking core models are widely used in mining technology and chemical reaction engineering. They have, however, found little application in modelling leaching from wastes. A new model which combines the concept of a shrinking core with diffusion and adsorption is currently being developed at the University of Cincinnati [6].

3.2 Formation of different layers in acid leached cement

In an acid environment, acid soluble components such as Ca^{++}, H_4SiO_4, Al^{+++}, Fe^{+++} and OH^- will be leached out, resulting in a friable outer layer of acid soluble depleted material and leaving a fairly unaltered core [7]. Leaching experiments on cement specimen with acetic acid indicated the existance of three different layers [8]: (1) An unaffected core from which contaminants may partly be released but in which the matrix is unaffected by the leaching solution; (2) A relatively small core boundary layer in which reprecipitation of matrix material or gelformation has occured; and (3) A leached layer in which the leachable matrix components as well as the contaminants are in equilibrium with the leaching solution.

The core boundary layer separating the leached friable layer from the unleached core seems to be extremely small in acetic acid leached cement [8]. The sharpness of this core boundary layer may be attributed to the fact that portland cement has a large neutralization capacity and immediately consumes the acid penetrating the outer layer. However, there is also an indication that a number of reactions might be responsible for a closure of the pores at the interface, hence preventing a rapid penetration of acid into the core. Belouschek and Novotny [9], report that only a small decrease in pH value (to about pH 10) is needed to start the formation of insoluble silicate gels from the naturally in cement occurring monomeric silicates. They also report that complexation of acetate ions further enhances the gel formation. Van Dijk *et al.* [7] indicate the importance of $CaCO_3$ (CO_2 from ambient air) to form an impermeable layer. They found a strong reduction in leaching rate if cement was leached with CO_2 saturated acid as apposed to acid that was free of CO_2. It may be hypothesized that formation of this impermeable gel strongly limits metals from diffusing from the unleached core. It then may be expected that leaching of calcium and leaching of metals from the specimen are strongly correlated.

3.3 Shrinking core model

The shrinking core model is based on the assumption that the core is reduced in size by dissolution at a sharp core boundary, see figure 1 for the general principle and the notations used. In case an impermeable layer is formed at the core boundary, not only the release of calcium but also the release of solidified heavy metals from the core is described by the shrinking core model. Starting point for the model is the equilibrium reaction of calcium hydroxide in an acid environment, realizing that calcium hydroxide is the major acid consuming species: $Ca(OH)_2 + 2H_3O^+ = Ca^{++} + 4H_2O$.

The dissolution reaction of calcium hydroxide in acid is generally written as being first order in the hydrogen concentration:

$$R''_{bc} = -k''_s\left(c_{ac} - \frac{c_{bc}}{K}\right) \tag{1}$$

Where R''_{bc} = the surface reaction rate [kmole/m^2•s], k''_s = reaction rate constant [m/s], c_{ac} and c_{bc} = the hydrogen and calcium concentration concentration respectively at the core boundary [kmole/m^3] and K = linearized equilibrium constant [-]. The mass transport equations for hydrogen at the surface solute interface, the core boundary and in the leached layer respectively can be obtained from Fick's laws:

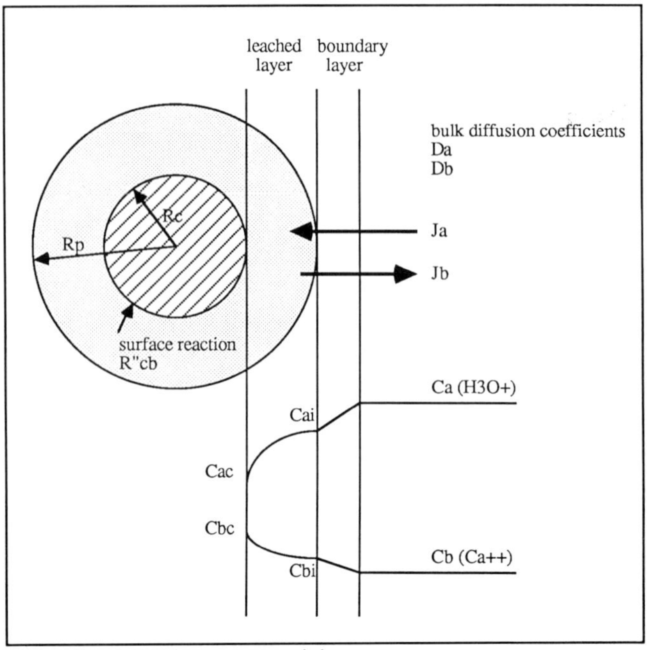

Figure 1: Shrinking core model

$$D_{ia}\left.\frac{\partial c_a}{\partial r}\right|_{r=r_p} = k_L(\bar{c}_a - c_{ai}) \tag{2}$$

$$D_{ai}\left.\frac{\partial c_a}{\partial r}\right|_{r=r_c} = \nu_a R''_{bc} \tag{3}$$

$$\frac{\partial^2 c_a}{\partial r^2} + \frac{2}{r}\frac{\partial c_a}{\partial r} = 0 \quad for \quad r_c \le r \le r_p \tag{4}$$

Where D_{ia} and D_{ai} = bulk diffusion coefficient and diffusion in the leached layer respectively of hydrogen [m^2/s], r_p = outside radius of the specimen [m], k_L = mass transfer coefficient [m/s], c_b = bulk concentration of hydrogen [kmole/m^3] and v_a = stoechiometric coefficient [-]. With the appropriate substitutions, the same holds for the calcium flux.

Realizing that the total flux of hydrogen and calcium respectively remain constant through any shell at $r \ge r_c$ for a fixed time, we obtain for the hydrogen flux at the interface, and in the leached layer respectively:

$$J_a 4\pi r_p^2 = k_{La} 4\pi r_p^2 (\bar{c}_a - c_{ai}) \tag{5}$$

$$J_a 4\pi r_p^2 = \frac{D_{ia} 4\pi r_p r_c}{r_p - r_c} (c_{ai} - c_{ac}) \tag{6}$$

Where J_a = flux of hydrogen through the solid-solute interface [kmole/m^2] and C_{ai} = hydrogen concentration at the interface [kmol/m^3]. For the calcium flux at the core boundary (reaction) through the leached layer and at the solid solute interface we obtain:

$$J_b 4\pi r_p^2 = -k''_s 4\pi r_c^2 \left(c_{ac} - \frac{c_{bc}}{K} \right) \tag{7}$$

$$J_b 4\pi r_p^2 = -\frac{D_{ib} 4\pi r_p r_c}{r_p - r_c} (c_{bc} - c_{bi}) \tag{8}$$

$$J_a 4\pi r_p^2 = k_{Lb} 4\pi r_p^2 (c_{bi} - \bar{c}_b) \tag{9}$$

The hydrogen and calcium fluxes are coupled by:

$$J_a = J_b \nu_a \tag{10}$$

By introducing the dimensionless variables: $r^* = r_c/r_p$, $Sh = k_L \bullet r_p/D$ and

$$A^* = \frac{2\nu_a}{Sh_a} + \frac{D_a}{D_{ia}} \frac{(1 - r^*)}{r^*} + \frac{D_a \nu_a}{k_s r_p (r^*)^2} + \frac{D_a}{KD_{ib}} \frac{(1 - r^*)}{r^*} + \frac{2}{Sh_b} \frac{D_a}{KD_b} \tag{11}$$

and eliminating the unknown local concentrations c_{ai}, c_{ac}, c_{bi} and c_{bc} from equation (5) to (10), an expression for calcium flux as a function of the bulk concentations of hydrogen and calcium and r^* is obtained:

$$J_b = \frac{D_a}{r_p} \left(\bar{c}_a - \frac{\bar{c}_b}{K} \right) \frac{1}{A^*} \tag{12}$$

The leaching process reduces the core diameter according to the relation:

$$J_b 4\pi r_p^2 = -4\pi r_c^2 \frac{dr_c}{dt} CA \tag{13}$$

Where CA is the solid concentration of $Ca(OH)_2$ in the core. Substitution of equation (12) in (13) leads to a differential equation for the core reduction:

$$\frac{dr^*}{dt} = -\frac{D_a}{r_p^2\,CA}\left(\bar{c}_a - \frac{\bar{c}_b}{K}\right)\frac{1}{A^*}\frac{1}{(r^*)^2} \tag{14}$$

For constant values of the concentrations of hydrogen and calcium in the bulk of the leaching solution, equation (14) can easily be integrated to give:

$$\begin{aligned}&\frac{D_a}{k_s r_p}(1-(1-\zeta)^{1/3}) + \left(\frac{2\nu_a}{Sh_a} + \frac{2D_a}{Sh_b K D_b}\right)\frac{1}{3}\zeta \\ &+ \left(\frac{D_a}{D_{ia}} + \frac{D_a}{KD_{ib}}\right)\left(\frac{1}{6} - \frac{1}{2}(1-\zeta)^{1/3} + \frac{1}{3}(1-\zeta)\right) \\ &= \frac{D_a}{r_p^2\,CA}\left(\bar{c}_a - \frac{\bar{c}_b}{K}\right)\cdot t \\ &\textit{with } \zeta = 1-(r^*)^3 \quad (\textit{conversionfactor})\end{aligned} \tag{15}$$

Unfortunately, the length of this paper does not allow an extensive analysis of the shrinking core model. Investigation of the conversion for three limiting regimes, however, provides some idea about the general behavior of the model for comparison with the bulk diffusion model described earlier.

1. The conversion is limited by diffusion through leached shell:

$$\zeta = \left(\frac{1}{\tau}\right)\cdot t \qquad \textit{with: } \tau = \frac{\nu_a CA r_p}{3k_L\bar{c}_a} \tag{16}$$

2. The conversion is limited by internal mass transfer:

$$1 - 3(1-\zeta)^{2/3} + 2(1-\zeta) = \left(\frac{1}{\tau}\right)\cdot t \qquad \textit{with: } \tau = \frac{r_p^2\,CA}{6D_{ia}\bar{c}_a} \tag{17}$$

3. The conversion is limited by chemical kinetics:

$$1 - (1-\zeta)^{1/3} = \left(\frac{1}{\tau}\right)\cdot t \qquad \textit{with: } \tau = \frac{\nu_a\,CA\,r_p}{k''_s\bar{c}_a} \tag{18}$$

The conversion for the three limiting regimes as a function of dimensionless time is depicted in figure 2, the x-axis being scaled by a square root. The conversion does not follow a square root of time relationship as in the the bulk diffusion models, but from the figure it is clear that up to a conversion of about 0.2 (20% leached) the curves could easily be fitted to a simple diffusion equation (straight line through first two points). The diffusion coefficient obtained from this fit, however, would have no sensible meaning and use of it for prediction of leaching behavior would lead to very erronneous results on the long term.

3.4 Discussion

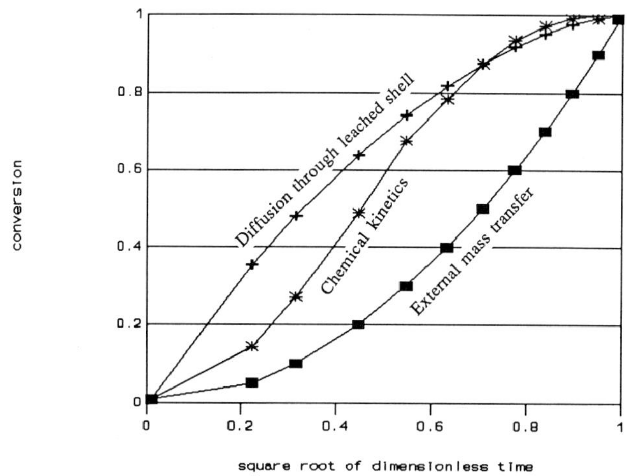

Figure 2: Conversion for three limiting cases.

Investigation of leached cement samples has shown that simple diffusion models are not adequate for leaching of calcium. Calcium release could be represented by a $t^{1/3}$ relationship rather than a square root of time relationship as in the case of diffusion dominated processes [10,11]. Research further indicated that release of calcium is in some cases unaffected by the size and consequently the geometric surface of the sample [11]. This is consistent with the shrinking core model. Zamorani, *et al.* [11] found that the time dependence of mass losses of calcium was lower for large-scale tests. They attributed the difference to the formation of $CaCO_3$ which increases in the large-scale tests. In case of formation of carbonates immobilization of calcium will occur. Calcium carbonate exhibits a low solubility resulting from the low value of the solubility product, $k_s = 9.3.10^{-9}$, which is about three orders of magnitude less than the solubility of $Ca(OH)_2$, $k_s = 6.10^{-6}$, and much less as compared to a soluble $Ca(HCO_3)_2$ compound [12,13].

The shrinking core model described in this paper, mainly relates to the release of calcium. In case an impermeable layer is formed at the core boundary, the model might in principle also be used for leaching of heavy metals. It should be taken into account, however, that adsorption of metals, leached from the core, onto the active sites in the leached layer may strongly reduce the rate of leaching into the leaching solution. In cases where the core boundary layer is not impermeable for heavy metals, the shrinking core model needs to be coupled to a diffusion model. Both aspects are currently being investigated.

4. CONCLUSIONS

- Leaching data from cement stabilized wastes are not sufficient to determine leaching mechanisms.
- It is possible to fit leaching data to a bulk diffusion model (square root relationship) when using short-term tests, even if the model is not an adequate representation of the system.
- In case the shrinking core concept proves to be an adequate concept for cement stabilized waste, the presently used tests in which surface to volume ratios are used, need to be abandoned.

Ackowledgement
This paper contains preliminary results of a research project on theoretical aspects of leaching behavior, financed by Center Hill Research Laboratories, Cincinnati, Ohio (USA).

REFERENCES

1 Godbee, H.W., D.S. Joy (1974), Assessment of the loss of radioactive isotopes from waste solids to the environment - Part 1: Background and theory, Oak Ridge National Laboratory, Reportnr. ORNL-TM-4333 (February 1974).
2 Crank, J. (1975), The mathematic of diffusion, Oxford University Press.
3 Côté, P.L., T.W. Constable (1982), Evaluation of experimental conditions in batch leaching conditions, *Resources and Conservation* **9** pp 59-73.
4 Kimmell, T.A., D. Friedman (1984), Model assumptions and rationale behind the development of EP-III, Oak Ridge National Laboratory (ORNL), uncoded.
5 Côté, P.L., T.W. Constable, A. Moreira (1987b), An evaluation of cement-based waste forms using the results of approximately two years of dynamic leaching, *Nuclear and chemical waste management* **7** pp 129-139.
6 Hinsenveld, M. (1991), Developing a mechanistic leaching model, Thesis proposal, March 1991 (draft).
7 Dijk, J.C., P.J. de Moel, W.F.J.M Nooyen, P.C. Nuiten (1986), Diffusiemodel voor zure aantasting van cement gebonden materialen, H2O 19(20) (1986) 482-487 (in Dutch).
8 Cheng, K.Y. (1991), Controlling mechanisms of metals release from cement-based waste form in acetic acid, PhD thesis, University of Cincinnati, 1991.
9 Belouschek, P., R. Novotny (1989), Zur chemie von pulverförmigen Wasserglass und seinen Folgeproducten: Kieselsäure und -gele in Wasser als Augangsmaterial für die Herstellung einer hochwertigen mineralishen Abdichtungsschicht aus bindigen Böden, Müll und Abfall 12 (1989) 636-643 (in German).
10 Zamorani, E., G. Serrini (1990), Evaluation of cement matrix preparation and leaching procedures usend in radioactive waste management, Radioactive Waste Management and the Nuclear Fuel Cycle, 14(3) (1990) 239-251.
11 Zamorani, E., G. Serrini, H. Blanchard (1986), Calcium release from cement samples of different size, Cement and Concrete Research, 16 (1986) 394-398.
12 Brodersen, K., K. Nilsen (eds) (1989), Characterization of radioactive waste forms. Report from the Commission of the European Communities, EUR 12077 EN, Vol. 1 (1989).
13 Brodersen, K., B.M. Pedersen, A. Winther (1981-1982), Comparative study of test methods for bituminized and other low- and medium-level solidified waste materials. Progress report for CEC contract WAS 235 DK (G), 1981-1982.

Waste Materials in Construction.
J.J.J.R. Goumans, H.A. van der Sloot and Th.G. Aalbers (Editors)

MODELLING OF INTERACTIONS AT WASTE - SOIL INTERFACES

D.E. Hockley & H.A. van der Sloot

Soil and Waste Research, Netherlands Energy Research Foundation (ECN), P.O. Box 1, 1755 ZG Petten, The Netherlands

SUMMARY

A "waste - soil interface" is created whenever waste disposal, storage, or reuse activities result in a waste or waste-containing material being placed on the soil. This paper discusses a systematic modelling and experimental study of waste - soil interfaces. A number of interfaces are discussed where "interactions" between the waste and soil phases lead to phenomena which are not predicted by independent consideration of either phase. In many cases, these interactions control the net release of contaminants to the surroundings. However, present regulatory assessments treat waste and soil phases independently. As a result, both the benefits and problems arising from interactions are overlooked.

1. Introduction

In this paper, the term "waste - soil interface" refers to the plane of contact between a waste, a treated waste, a stabilized waste, a solidified waste, or some other waste-containing product and an unconsolidated porous material (a "soil" in civil engineering terminology). The wide scope of this definition means that waste - soil interfaces arise in many systems. For example, waste disposal in landfill sites brings waste into direct contact with native soil materials or with clay liners. Waste solidification and stabilization create useful products which are brought in contacted with soil through construction activities. Bulk uses of wastes in embankments and dykes cause direct contact of waste with native soil or with a clay liner.

We are particularly interested in waste - soil interfaces where the behaviour of one medium is affected by the adjacent medium in ways that are not predictable from the way each medium acts independently. The mechanisms of these "interactions" will be made clear through examples presented below. A simple analogy can be drawn between these interactions and a "two-factor interaction" familiar to most scientists. It is well known that heat has an accelerating effect on some chemical reactions. Catalysts can also cause many reactions to accelerate. In some cases however, the combination of heat and catalyst causes a much greater acceleration than would be expected on the basis of their independent effects. Similarly, the combination of waste and soil at an interface can lead to unexpected results.

This paper introduces research conducted over the last three years into the general topic of waste - soil interfaces. Other manuscripts, now in preparation, will discuss in detail work on some of the individual interfaces [1-4]. The primary objective of this paper is to present the research as a coherent whole in the hope that, like the interfaces under discussion, it exceeds the sum of its parts.

2. Methods

2.1 Interface types

At most waste - soil interfaces, many processes occur simultaneously: contaminants and major elements leach from the waste into the soil, soil components move in the opposite direction, reactions occur which can alter the chemical and physical properties of both media. In order to systematize our study of these interfaces, we classify them according to the difference existing between the two media at the time they are brought into contact. For example, when two media with very different pH are brought into contact, a "pH-interface" is created.

When many properties change across the interface, it is necessary to focus on the behaviour of individual components, say, a particular contaminant or class of contaminants. For these components, only a limited number of the changes across the interface are important. For example, an interface between paint waste and a clay liner might be "seen" by trace metals as primarily a pH-interface, since that is the property that most affects their mobility. Organic contaminants might see the same interface as a "K_d-interface", since it is the change in sorption properties that most affects their mobility.

With this classification system in mind, it is possible to define a number of simple interface types. Characterization of processes at simple interfaces has been a primary goal of our work to date. Recent publications and manuscripts now in preparation discuss this work in detail [1-6]. Four examples are presented below. When the behaviour of a component at an interface requires that changes in more than one property be considered, we call the interface "complex". The number of possible complex interface types is very large. In our work to date, we have investigated only a few examples [5-7], two of which are discussed below. A further limitation of our work to date needs to be mentioned. The nature of the transport across the interface can have a significant influence on other processes. So far, we have focused on diffusion-dominated interfaces, under the reasoning that most waste disposal and reuse technologies have the reduction of advection as a fundamental objective, leaving diffusion as the dominant transport process. In future, we plan to extend our findings to systems with significant advection.

2.2 Mathematical models

An advantage of classifying interfaces into categories defined by changes in one property is that each interface type can then be expressed as a mathematical problem. When the problem can be solved, which is not always the case, the result is a mathematical model of the interface type. The model can then be tested against experimental results, and, if it passes the test, used for a number of purposes.

The first step in the modelling process is to formulate in mathematical terms the transport and reaction processes taking place at the interface. A general approach to the coupling of differential transport equations with algebraic reaction equations is presented by Rubin [8]. A number of generalized numerical methods for solving the resulting set of coupled equations have been proposed. Yeh and Tripathi provide an excellent review [9]. Unfortunately, the sharp changes in conditions at waste - soil interfaces limit the applicability of these methods. For some diffusion-dominated interfaces, analytical solutions presented by Crank [10,11] are applicable. Since diffusion equations are analogous to heat flow equations, analytical solutions presented by Carslaw and Jaeger [12] are also applicable in some cases. Solution methods for particular interface types are discussed in more detail in the individual manuscripts [1-6]. In most cases, some combination of approximate and analytical methods was necessary.

Once the solution to the mathematical problem has been obtained and tested against experimental data, the model is used to determine the behaviour of the interface under a range of conditions. A further use of the model is to test when the assumptions implicit in current practice are justified. Examination of the interface model often leads to simple criteria by which other workers can decide whether their waste - soil system can or cannot be treated as independent phases. Note that this type of result is impossible without mathematical models. To prove the limits of an assumption by experimental means would require an infinite number of tests.

2.3 Experiments

The experimental component of this study had two purposes. The first was to indicate which processes may be important at various waste - soil interfaces. The second was to provide data for testing of the mathematical models. Diffusion-tube methods were used for both purposes. These methods are similar to "self-diffusion" techniques used by earlier researchers [e.g. 13]. Their application to interface studies is more recent [14,1].

Most of the first type of experiments were carried out in conjunction with studies of specific waste disposal problems [14,15]. Media selected as soils in these experiments included soil minerals, such as peat, sand, kaolinite and bentonite, and realistic soils and sediments. Waste materials included contaminated soils, fly ashes, slags, and stabilized wastes. One limitation of the diffusion tube method is that only particulate media can be used. Bulky wastes like slag and stabilized wastes products require grinding. Results for those material were complemented with results from standard leaching tests.

In the second type of experiment, conditions were selected to come as near as possible to the modelled interface type. Artificial media, such as acid-washed sands or homogenized soil horizons, were used to attain precisely controlled conditions. To insure that the experimental data provided an independent test of the model, additional experiments were carried out to obtain parameter estimates.

The use of radiotracers for both major and minor elements allowed for inexpensive and accurate analyses. Contaminant tracers such as ^{99}Mo, ^{65}Zn, ^{75}Se, ^{74}As, and ^{210}Pb were used in the first type of experiment. Because the behaviour of major elements is better understood, tracers such as ^{45}Ca and ^{59}Fe were used in many of the model testing experiments. Methods were also developed to measure physical restrictions at interfaces using tritiated water and to conduct combined diffusion tube and leaching studies. Details of a number of experiments are included in the individual manuscripts [1-7,14,15].

3. Results

3.1 K_d-interface

For compounds which react with immobile solid phases, and, if those reactions can be described by linear sorption isotherms, the ratio of the contaminant's concentration in the mobile aqueous phase to that in the immobile sorbed phase is a constant "distribution coefficient" or K_d. As mentioned above, contaminants which are strongly affected by sorption see the interface as a "K_d-interface" [1]. K_d-interfaces are of practical interest because of the great number of matrix-contaminant combinations which can be described by a distribution coefficient. Rai and Zachara discuss the applicability of the K_d-model and list distribution coefficients for a number of elements in different matrices [16]. Hydrophobic organic contaminants are also well described by a K_d-model, at least at the low concentrations typical of environmental contamination [17].

Mathematical modelling of K_d-interfaces is relatively simple because the sorption reaction can be accounted for by dividing the transport rates by a constant "retardation factor". The retardation factor for each matrix is directly dependent on the K_d-value. This mathematical simplicity has an intuitive explanation. The transport processes are retarded because only a portion of the contaminant is mobile. Since K_d, the ratio of mobile contaminant to immobile contaminant, is constant, the retardation factor must also be constant. The retardation approach leads, in the case of diffusion-dominated K_d-interfaces, directly to analytical solutions [1].

Figures 1a and 1b show, respectively, predictions from the K_d-interface model and results from a diffusion tube experiment. The diffusing contaminant in the experiment was

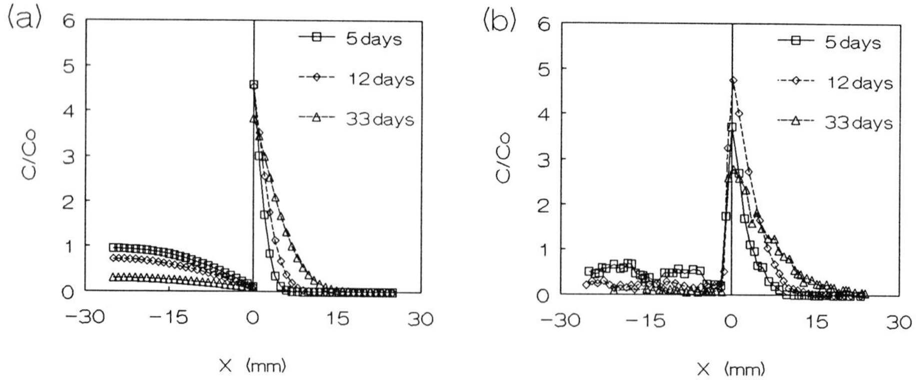

Figure 1. K_d-interface. (a) Model predictions. (b) Experimental results.

^{14}C-labelled naphthalene. The matrix on the left contained very little natural organic matter and hence did not strongly sorb the naphthalene. In that respect, it can be thought of as representative of a great number of wastes and stabilized wastes in which organic contaminants exhibit low K_d-values. The matrix on the right contained sufficient natural organic matter to cause relatively strong sorption of the naphthalene, leading to K_d-values similar to those observed in many surface soils. The naphthalene was originally mixed evenly throughout the left matrix. As predicted by the model, it diffused across the interface and immediately formed a concentration peak. In this case, the naphthalene is drawn out of the "waste" and concentrated in the first few millimetres of the "soil". As can be rigorously shown by comparing the mathematical models, this effect is not predictable without considering the presence of both media. [1]

3.2 pH-interface

A pH-interface is formed when a waste with a certain pH is placed adjacent to a soil with a different pH. Since soils and soil materials fall into a relatively narrow pH range, while wastes and stabilized wastes cover a much wider range, pH-interfaces are very common. In relatively few cases, the acid or base from the waste is itself a pollution problem. More commonly, it is the metals or anions mobilized by pH-changes that are the primary source of concern.

A general approach to the mathematical modelling of H^+ and OH^- diffusion is discussed by Lewnard *et al.* [18]. Their comments are also applicable to the problem of diffusion-dominated pH-interfaces, except that the boundary conditions are more complex. When only aqueous species are present, the multicomponent transport problem can be reduced to relatively simple analytical solutions. However, solid buffers are nearly always present in realistic waste disposal systems. The analytical solutions must then be replaced by numerical forms. Lewnard *et al.* present solutions for systems wherein the solid buffers can be described by distinct reaction sites [18]. Systems where the solid buffers are described by laboratory titration curves can also be accommodated in the numerical solution [5,2].

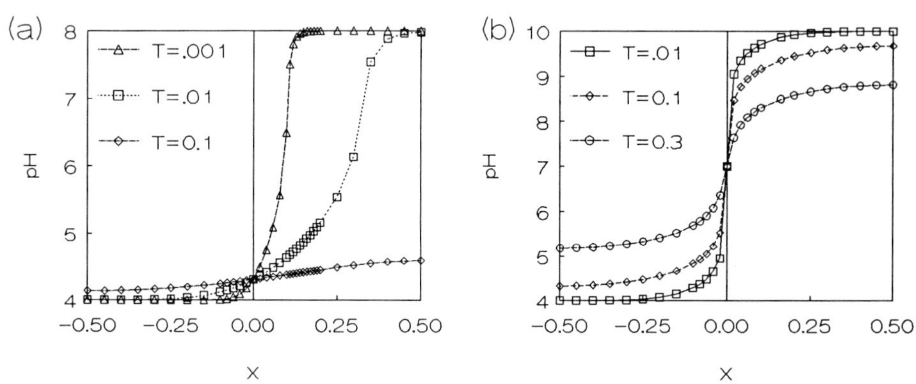

Figure 2. pH-interface models. (a) pH 4 vs. pH 8. (b) pH 4 vs. pH 10.

Figures 2a and 2b show model predictions of behaviour at two simple pH-interfaces without solid buffering. In Figure 2a, the acidity on one side of the interface exceeds the basicity on the other side. The result is that the basic side is quickly acidified. In Figure 2b, the acidity on one side balances the basicity of the other, resulting in a very stable pH-interface. A more realistic system which includes solid buffering is discussed in section 3.5 below.

3.3 Dissolution interface

A dissolution interface is formed when a waste matrix containing a solid phase is placed next to a soil wherein the solubility of the solid phase is higher, leading to dissolution. Like the pH-interface, dissolution interfaces are ubiquitous. They may be of practical interest for two reasons. First, when the dissolving phase is a major component of the waste matrix, the physical integrity of the waste may be jeopardized. Second, when contaminants are a component of the dissolving phase, their release to the surroundings may be controlled by the dissolution.

Dissolution of a porous matrix in contact with an acid source is discussed by Cussler [19] and Kim and Cussler [20]. They present a simplified model which allows a qualitative prediction of dissolution. They also remark that a full treatment leads to a "moving boundary" problem, i.e., a differential equation wherein one of the boundary conditions is located at a point which moves in accordance with an additional subsidiary condition [11]. A full treatment of dissolution interfaces by moving boundary methods is extremely complex. More straightforward solutions are possible if simplifying assumptions are accepted [3].

Figures 3a and 3b show results from one of the simplified models, in this case with an analytical solution. The model assumes that the concentration of a contaminant remains constant at the waste - soil interface, and that the contaminant's solubility is controlled by reaction with a major element present in quantities much higher than that of the contaminant. The dissolution process therefore has little effect on the total

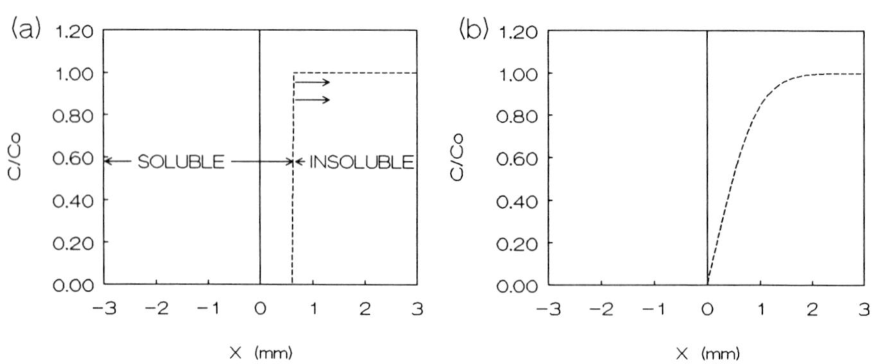

Figure 3. Dissolution interface. (a) Moving boundary dissolution.
(b) Diffusion-controlled release.

concentration of the major element and the solubility of the contaminant within the waste is nearly constant. Figure 3a shows a concentration profile for a contaminant which is slightly soluble at the interface. The profile clearly shows a moving boundary effect. The boundary is between a zone where only soluble contaminant exists and the zone where solid contaminant remains. Dissolution occurs only at the boundary and controls the overall rate of contaminant release. Figure 3b shows a concentration profile for a situation where the contaminant is strongly soluble at the interface. In this case, diffusion is the rate controlling step and the concentrations resembles a simple diffusion profile. It is interesting to note that, in both profiles, the same total mass of contaminant has crossed the interface. Conventional leach testing methods, which measure only contaminant flux out of a waste, would not allow the two processes to be distinguished from each other [21,3].

3.4 Precipitation interface

A precipitation interface is formed when a waste matrix containing one component of a precipitate is placed adjacent to a soil matrix containing the other component, and when the components are present in high enough concentration to cause precipitation at the interface. Precipitation interfaces are important when a contaminant from the waste matrix is involved in the precipitation reaction. The net release of contaminant from the waste is then controlled by its solubility at the interface.

Full treatment of precipitation interfaces leads to a "free boundary" problem which is similar to, but even more complex than, the moving boundary problem discussed above [11]. Again, simplifying assumptions lead to a more straightforward solution. Under the assumption that the precipitate is present at all points in the system, the concentrations of the two components are no longer independent and an analytical solution for the precipitation rate can be found. Numerical integration of the precipitation rate through time allows the concentration of precipitate to be predicted. [4]

Figures 4a and 4b show results from the model and from a diffusion tube experiment.

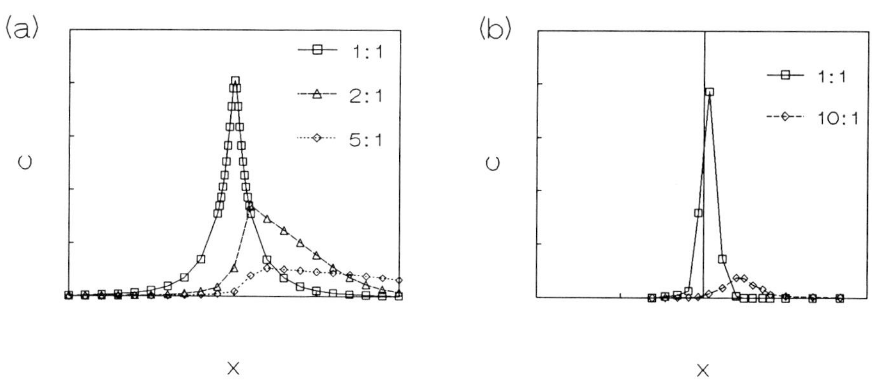

Figure 4. Precipitation interface. (a) Model predictions for different initial stoichiometries. (b) Experimental results.

The experimental results were obtained by placing a "waste" sand containing ^{45}Ca-tracer adjacent to a "soil" sand containing CO_3 ion. The solid curve in each figures show the distribution of Ca in the region of an interface where the counterion is present in stoichiometric proportions. The broken curve shows the Ca distribution when it is placed next to a matrix where the counterion is present in a concentration ten times that required by stoichiometry. The qualitative agreement between model and data is clear. Both show a localized $CaCO_3$ precipitation in the case of stoichiometric conditions and a precipitate shifted back into the "waste" when counter-ion is in excess. It is clear that this phenomenon controls the net release of Ca across the interface. Similar processes may control the release of precipitating contaminants from a waste phase.

3.5 Complex interface: Trace metal precipitation at a pH-interface

As mentioned above, pH interfaces are of interest because of their effect on contaminant mobility. It is well known that metals have a solubility minimum at moderately basic pH. Therefore, the combination of metal precipitation and a pH-interface leads to a complex interface of considerable practical importance.

A complex interface similar to that presented here is discussed by van der Sloot and Côté [5]. The earlier work modelled the main features of the data very successfully using a numerical solution. More recent attempts with a combination of a semi-analytical solution from a simple pH-interface and a numerical model of interface precipitation were more succesful in modelling some of the details more accurately [2].

Figures 5a and 5b show diffusion tube results obtained by placing a clay matrix containing ^{59}Fe adjacent to a basic fly ash. Although the clay matrix was extremely acidic, the strong buffering of the fly ash led to a relatively stable pH-interface similar to that shown in Figure 3b. The resulting interface pH was near the minimum for Fe solubility and the metal was precipitated. Continuing diffusion drew soluble, therefore "leachable", Fe from throughout the matrix and concentrated it in a less leachable form at the interface. Similar processes can be expected for other metals, including contaminants [2,5].

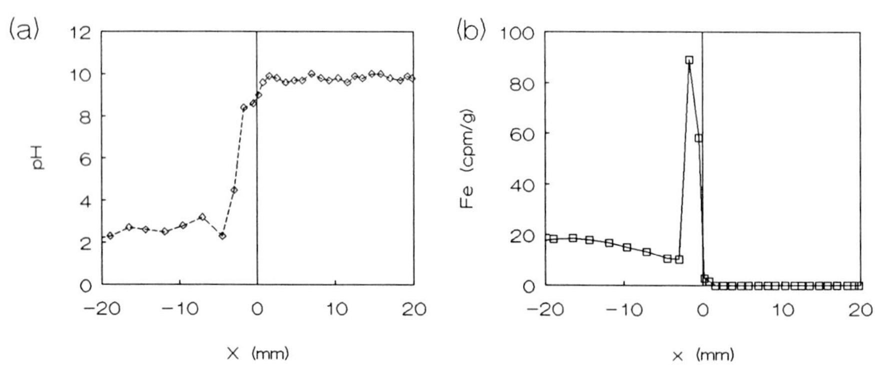

Figure 5. Complex interface experiment. (a) pH profile after 30 days. (b) ^{59}Fe profile showing precipitate peak.

3.6 Complex interface: Major element dissolution and precipitation

Major element dissolution and precipitation can involve sufficient quantities of material to cause physical changes at an interface. Hockley and van der Sloot have identified a system wherein major elements from a surrounding medium, in this case seawater, react with dissolution products of a stabilized waste to create pore-filling precipitates [22]. The beneficial effects of pore sealing are that both water permeability and contaminant diffusion are restricted [6,7].

Rigorous modelling of the dissolution and precipitation processes is an exceedingly complex problem, as the free and moving boundary problems become coupled. A numerical method such as that suggested by Yeh and Tripathi [9] provides as yet the only practical approach. Even this approach is not free from problems. The sharp gradients at the interface cause instability and excessive amounts of computer time are required to achieve a solution. Furthermore, no satisfactory method has yet been developed to include the physical effects of the precipitate on further diffusion.

Figures 6a and 6b show results of the numerical model for a system where $Ca(OH)_2$ is originally present in one matrix and the $MgCO_3$ is present in the other. After a unit dimensionless time, the two solids have dissolved away from the interface as moving boundaries. Two less soluble solids, $CaCO_3$ and $Mg(OH)_2$, have precipitated at the interface. A number of experiments have confirmed that the model predictions are qualitatively correct and that the precipitation eventually fills the interface pores sufficiently to create a seal [6,7].

4. Discussion: implications and opportunities

4.1 Waste testing and disposal regulations

Results from our studies of waste - soil interfaces show conclusively that the net release of contaminants is determined not only by the properties of the waste but also by interactions with surrounding matrices. However, current waste testing methods

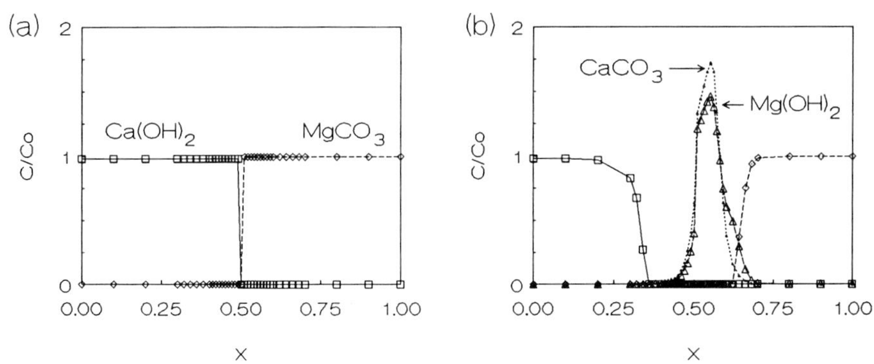

Figure 6. Complex interface. (a) Initial reagent profiles.
(b) Profiles at unit dimensionless time.

completely neglect interface effects. We know of no test wherein contaminant leaching from a waste or waste product is assessed in conjunction with a particular disposal or reuse environment. This deficiency is also reflected in current regulations. Any regulation which uses only the properties of the waste to govern disposal and reuse options implicitly assumes that the conditions into which the waste is placed are irrelevant.

One obvious implication of our results is that current testing and regulations are not always fair to particular disposal or reuse options. For example, some stabilized wastes leach high concentrations of metals. The precipitation of these metals at interfaces can substantially reduce the contamination hazard. Of course, there are practical difficulties with testing and regulating an infinite number of waste - soil combinations. Nonetheless we believe that some provision should be made for industries which wish to undertake an integrated assessment, including interface testing, to demonstrate that their waste - soil system poses no environmental hazard.

A second implication of our results is that the numbers obtained from current test methods may be much less important than the fundamental information they convey. It is the underlying parameters, diffusion coefficients, K_d-values, solubilities, etc., that are needed to predict interface phenomena and the net release of contaminants. In this respect, our work joins many other recent studies in indicating the need for a closer relationship between regulatory test methods and fundamental properties.

4.2 Macroscopic contamination models

Macroscopic soil and groundwater contamination models provide a necessary link between scientific results and public policy. The results of our study contribute to the development of macroscopic models by clarifying small-scale processes at the waste - soil interface. Unfortunately, the difference in length scales makes it difficult to directly combine the interface models with macroscopic models.

An alternative is to use simplified versions of the interface models as source terms for the macroscopic models. Examples of simple source terms are constant concentration sources arising from solubility control and effectively instantaneous inputs arising from a K_d-interface. A number of the individual manuscripts present simple source terms and define the conditions under which they are correct [1,2,4].

4.3 Engineering of interfaces

Once interface phenomena have been identified and understood, the logical next step is to attempt to optimize their beneficial effects. "Engineering" of waste - soil interface may range from choosing a disposal environment which minimizes the leaching of one contaminant to elaborate pretreatments designed to provide comprehensive protection.

The use of precipitation reactions to physically seal the waste - soil interface is the only example which has been fully explored in our work to date [6,7]. The addition of counterions to control precipitation reactions and the addition of sorbing phases are examples for future research. Continuing work on complex interfaces may lead to the identification of other processes amenable to engineering control.

5. Conclusions

Our work on waste - soil interfaces has demonstrated the importance of interactions between the two matrices in controlling the net release of contaminants from waste disposal and reuse activities. A number of simple interface types have been defined, modelled, and demonstrated in the laboratory. Consideration of simple interfaces leads to directly useful results and provides insight into the nature of processes occurring in more complex systems. A limited number of complex interfaces have also been modelled and demonstrated in the laboratory.

Interface phenomena are particularly interesting because they are not evident in any test where only one of the phases is present. This aspect implies that methods currently used to regulate waste disposal and reuse, based only on tests of the waste itself, can be inadequate. Furthermore, technological research focused on only one of the adjacent media may overlook processes which could be utilized to reduce the risk of environmental contamination resulting from waste disposal and reuse activities.

6. Acknowledgements

Jan Wijkstra performed most of the laboratory studies described herein. Funding for this project was obtained from NOVEM, the Netherlands Agency for Energy and the Environment.

7. References

[1] D.E. Hockley, J. Wijkstra, & H.A. van der Sloot. *Diffusion across K_d-interfaces: Model, experiment and implications for waste assessment.* Submitted to Environ. Sci. Technol. 1991.

[2] D.E. Hockley, J. Wijkstra, & H.A. van der Sloot. *Diffusion across simple and complex pH-interfaces*. In preparation.

[3] D.E. Hockley, J. Wijkstra, & H.A. van der Sloot. *Moving boundary dissolution of contaminant-controlling phases in wastes*. In preparation.

[4] D.E. Hockley, J. Wijkstra, & H.A. van der Sloot. *Diffusion and precipitation at an interface between adjacent porous media*. In preparation.

[5] H.A. van der Sloot and P.L. Côté. *Modelling chemical interactions at a waste/waste interface*. Environ. Technol. Letters, vol 10, pp 969-976, 1989.

[6] H.A. van der Sloot, D.E. Hockley, & J. Wijkstra. *Zelf-vormende en zelf-herstellende afdichtingen: Concept, modellering, en laboratoriumresultaten*. Energie & Milieu Technol., vol 1/2, pp 27-31, 1991.

[7] D.E. Hockley, J. Wijkstra, & H.A. van der Sloot. *Afdichtingen voor deponie en hergebruik van kolenreststoffen: Opsluiten van afval binnen 'natuurlijke' barrieres*. Energie Spectrum, vol 15, no 2, pp 50-55, 1991.

[8] J. Rubin. *Transport of reacting solutes in porous media: Relation between mathematical nature of problem formulation and chemical nature of reactions*. Water Resour. Res., vol 19, no 5, pp 1231-1252, 1983.

[9] G.T. Yeh, & V.S. Tripathi. *A critical evaluation of recent developments in hydrogeochemical transport models of reactive multichemical components.* Water Resour. Res., vol 25, no 1, pp 93-108, 1989.

[10] J. Crank. The Mathematics of Diffusion, Clarendon Press, Oxford, 1975.

[11] J. Crank. Free and Moving Boundary Problems, Clarendon Press, Oxford, 1984.

[12] H.S. Carslaw & J.C. Jaeger. Conduction of Heat in Solids, Clarendon Press, Oxford, 1959.

[13] P.H. Nye. *Diffusion of ions and uncharged solutes in soils and soil clays.* Advances in Agronomy, vol 31, pp 225-272, 1979.

[14] H.A. van der Sloot, O. Hjelmar, & G.J. de Groot. *Waste/soil interaction studies-The leaching of molybdenum from pulverized coal ash.* In: Flue Gas and Fly Ash. P.F. Sens & J.K. Wilkinson, eds. Commission of the European Communities, Elsevier Applied Science, London, 1989.

[15] H.A. van der Sloot. *Leaching behaviour of waste and stabilized waste materials; characterization for environmental assessment purposes.* Waste Management Res., vol 8, pp 215-228, 1990.

[16] D. Rai,& J.M. Zachhara. Chemical Attenuation Rates, Coefficients, and Constants in Leachate Migration Volume 1: A Critical Review, EA-3356; Electric Power Research Institute, Palo Alto, CA, 1984.

[17] S.W. Karickhoff. *Organic pollutant sorption in aquatic systems.* J. Hydraul. Eng., vol 110, pp 707-735, 1984.

[18] J.J. Lewnard, E.E. Peterson, & C.J. Radke. *Diffusion of H^+ and OH^- in porous solids.* J. Chem. Soc., Faraday Trans., vol 84, no 11, pp 3927-3939, 1988.

[19] E.L. Cussler. *Dissolution and reprecipitation in porous solids.* AIChE J., vol 28, no 3, pp 500-508, 1982.

[20] J.L. Kim & E.L. Cussler. *Dissolution and reprecipitation in model systems of porous hydroxyapatite.* AIChE. J., vol 33, no 5, pp 705-710, 1987.

[21] American Nuclear Society Standards Committee Working Group ANS16.1. American National Standard Measurement of the Leachability of Solidified Low-Level Radioactive Wastes by a Short-Term Test Procedure, American Nuclear Society: La Grange Park, IL 1986.

[22] D.E. Hockley & H.A. van der Sloot. *Long term processes in a stabilized coal-waste block exposed to seawater.* Accepted by Environ. Sci. Technol., 1991.

Waste Materials in Construction.
J.J.J.R. Goumans, H.A. van der Sloot and Th.G. Aalbers (Editors)

PROBABILISTIC MODELLING OF ENVIRONMENTAL IMPACT OF WASTE MATERIALS IN HYDRAULIC ENGINEERING

F.A. SWARTJES[1], G.J. MULDER[2], L. DE QUELERIJ[1] and G.A.M. VAN MEURS[2].
[1] Fugro. P.O. Box 1471 BL Nieuwegein (The Netherlands)
[2] Delft Geotechnics. P.O. Box 69. 2600 AB Delft (The Netherlands)

SUMMARY

A conceptual model for the evaluation of the environmental impact of waste materials in hydraulic engineering is presented. The emission of cadmium, originating from a dike enlargement of fosfogypsum, has been calculated by means of a probabilistic model. To this purpose, probability density functions for all input parameters have to be defined. As a result, confidence intervals are given, for the time at which critical cadmium concentration levels are exceeded in groundwater. The conceptual model proves to be fairly applicable and the probabilistic model is a useful tool to assess environmental impact.

1. INTRODUCTION

Since 1987 the Centre for Civil Engineering Research and Codes (CUR) in the Netherlands is commissioning investigation into the potential applications of waste materials in hydraulic engineering. Except for construction and financial aspects, research has been stressed into environmental implications of waste materials.

Because of the relatively high amount of potential contaminants, immission can be initiated from waste materials, leading to pollution of soil and groundwater by emission. Due to this environmental aspects can involve important restrictions for application of waste materials in hydraulic constructions.

A probabilistic model can be used, for assessing the environmental impact. In this paper, the principle of probabilistic modelling in environmental risk assessment related to the quality of soil and groundwater is evaluated. Some results are discussed, concerning the one-dimensional contaminant migration, due to the application of waste materials, i.e. a dike enlargement with fosfogypsum, in the Netherlands.

2. CONCEPTUAL MODEL

To analyse the migration of contaminants originating from waste materials a methodology has been devised that includes the following elements (1): a conceptual model, a data base of relevant parameters and a numerical model. In this paper a description of the elements mentioned above is given together with the methodology of interactions between these elements.

2.1 Conceptual model description

The first step to be made is to describe the conceptual model. It consists of two parts, the process identification and the geometrical information (2). Physical processes that are relevant to the model are groundwater flow (advection), hydrodynamical dispersion (molecular diffusion and mechanical dispersion, (3)) and other kinds of mixing processes. The second group of important processes are those of the chemical kind, i.e. retardation (including sorption and precipitation/solution) and production/decay. Description of these physical and chemical processes takes place in section three of this paper. The second part of the conceptual model, the geometrical information, concerns the physical circumstances and boundary conditions applicable (see section 2.2).

The next stage involves the collection and interpretation of appropriate values for the model parameters. After that construction of a numerical model can be started. Test calculations (benchmarking and application of the model without the possibility of using location specific data) are needed to validate the model before predictive calculations can be carried out.

2.2 Geometrical information

The geometry of a dike has to be described with suitable dimensions that are valid for a generally occurring situation in the Netherlands. The method of applying waste materials in such a construction has to be taken into account. Figure 1 presents a vertical cross-section perpendicular to the length-direction of a dike along a river in the west part of the Netherlands. In the figure an enlargement at the inner dike site is shown. At the foot of the inner dike site a ditch has to drain water flowing horizontally through the dike and leakage water rising upwards to the ditch (in the west part of the Netherlands).

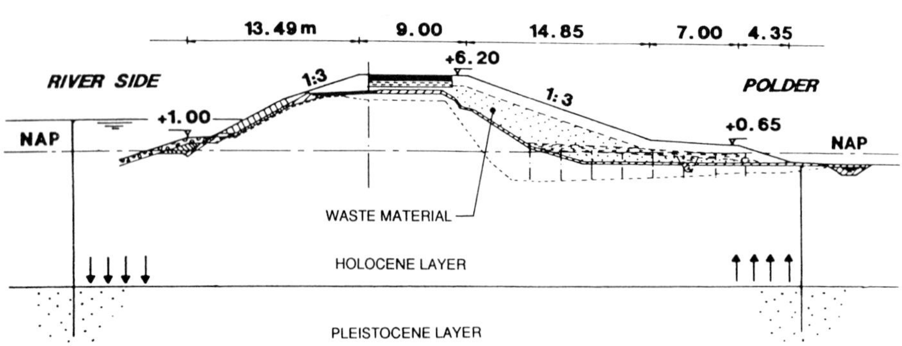

Fig. 1. Geometrical construction of a dike along a river in the west part of the Netherlands with an enlargement at the inner dike site, containing waste materials.

In this context the presence of subsoil layers is important in modelling groundwater flow and chemical interactions with the different soil compositions. The first aquifer (pleistocene) is confined at the bottom by a leaky layer with generally a high resistance to flow and at the top by a holocene surface layer.

Boundary conditions are needed at the top and bottom of the system. The amount of infiltrating rainfall provides the condition at the top and at the bottom the hydraulic head of the first aquifer can be taken.

3. PROCESS IDENTIFICATION

3.1 Geometry

The one-dimensional migration of contaminants under the dike enlargement has been considered. The mobility of contaminants varies considerately, mainly due to differences in the soil chemical features. Depending on the type of contaminant under consideration, the following subsystems can be influenced by the contaminants: a) the soil and groundwater in the dike core, partly saturated, partly unsaturated and in the Holocene top layer; b) the groundwater quality in the upper aquifer.

3.2 Contaminant migration

The migration of contaminants in soil can include the following components:

- The transport of mobile heavy metals via the liquid phase. The transport can be subdivided into advection, dispersion and diffusion.

- The interaction of contaminants between the solid and the liquid phase (retardation). The retardation, being the result of the combined effect of sorption and reversible chemical reactions, is assumed to be a reversible process. It determines the amount of contaminants available for transport in the liquid phase.

- Production/decay, i.e. the irreversible source or target of contaminants respectively.

The mathematical formulation of contaminant migration in soil follows from the combined description of the above processes (4):

$$\frac{\partial}{\partial t}(R\theta c) = \frac{\partial}{\partial z}\left(\theta D \frac{\partial c}{\partial z} - qc\right) - \mu\theta c \qquad [1]$$

θ = volumetric water content $[m^3.m^{-3}]$,
c = solute concentration $[\mu g.l^{-1}]$,
D = apparent dispersion coefficient $[m^2.s^{-1}]$,
q = volumetric velocity $[m^3.m^{-2}.s^{-1}]$,
μ = empiric production/decay coefficient $[s^{-1}]$,
t = time [s],
z = depth [m].

Here, R [-] is the retardation coefficient, which for linear retardation is expressed as:

$$R = 1 + \frac{\rho k_v}{\theta} \qquad [2]$$

ρ = bulk density soil [$kg.m^{-3}$]
k_v = linear retardation coefficient [$l.kg^{-3}$]

The linear retardation characteristic is expressed as:

$$Q = k_v c, \qquad [3]$$

where Q = total solid phase content [$mg.kg^{-1}$].

3.3 Boundary conditions

It has been assumed that the average effective yearly rainfall, $q_0(t)$, percolates through the dike enlargement. At the lower boundary of the dike enlargement, which is the upper boundary of the system under consideration, a third type of boundary condition has been formulated:

$$\left(-\theta D \frac{\partial c}{\partial z} + qc\right)_{z=0} = q_0(t)c_0(t), \qquad [4]$$

where $c_0(t)$ = concentration of the infiltrating water [$\mu g.l^{-1}$], being calculated from the leaching characteristic.

At the lower boundary of the system, such a pressure head has been calculated such that the water table is found at 3 m depth (see Fig. 1), given the value for $q_0(t)$.

4. SELECTION OF A CONTAMINANT

There are three important criteria which determine the potential for subsoil pollution (soil and groundwater) below a fosfogypsum construction and these are: availability of a contaminant due to leaching, its mobility in the subsoil and its toxicity. These criteria will be used to select a component of interest. Therefore it is essential to know the leaching characteristics of the components from the waste material fosfogypsum.

4.1 Availability

The proportion of the different components present in fosfogypsum limits their availability due to leaching. In this paper the immission into the subsoil will be presented through the leachate concentrations as a function of time. In figure 2 the leachate concentrations of cadmium (obtained by batch and column experiments) are used to construct a linearised leaching function setting concentrations against time and against percolating amount of rain/groundwater per unit mass of material (L/S). A step function is derived from this linearised leaching function by using linear interpolation between the measured concentration values. However due to the small amount of data used the application of the leaching function as an immission function into the subsoil has its limitations.

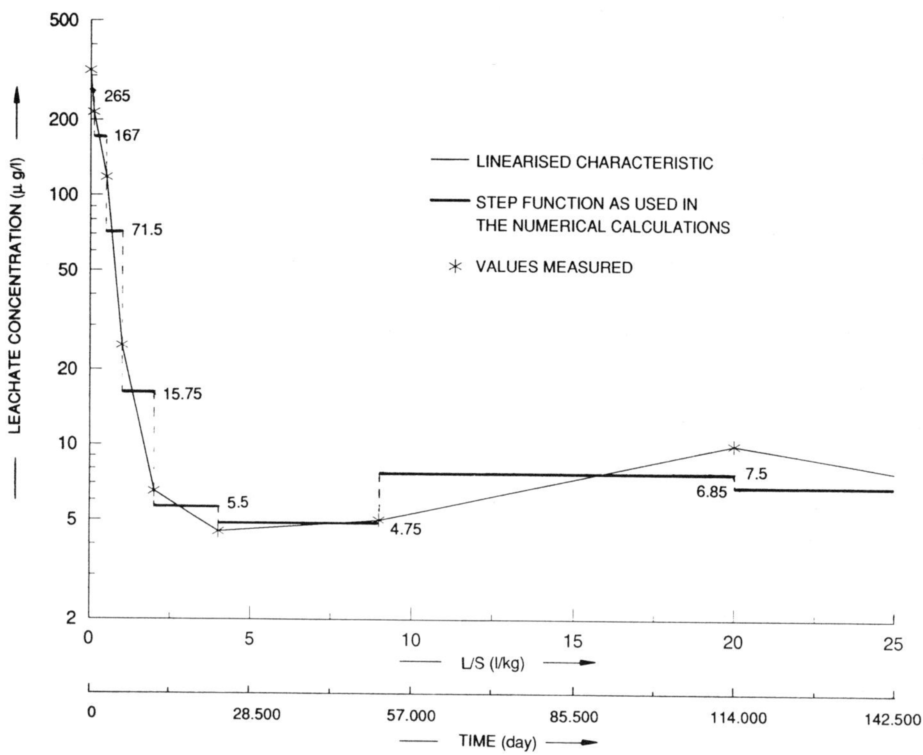

Fig. 2. Step function of leachate concentrations of cadmium set against time and against percolating amount of rain/groundwater per unit mass of material (L/S).

4.2 Mobility

Appropriate components to model are those that possess upon leaching a certain amount of mobility in the subsoil. This is dependant upon the groundwater flow (influenced by the restrictions of the geometrical system and the physical soil properties) and the retardation and physico/chemical behaviour of the component concerned (influenced by the chemical soil properties, including the presence of adsorbents).

4.3 Toxicity

If certain concentration levels are exceeded, chemical components can be harmful for the ecosystem. Therefore it is important to know the degree of toxicity of the components under investigation. It is difficult to find a good standard for measuring toxicity. Use of MAC-values or ADI-values is justified only for special cases (concentrations in the air or amount of oral intake). In this paper it is most convenient to make use of the Dutch reference (A), trigger (B) and action (C) values for concentrations in groundwater and soil.

4.4 Selection of most important component

A choice between the different components leaching out of fosfogypsum has to be made to limit the number of predictive calculations. Two components, cadmium and sulfate, do reply quite well to the mentioned criteria. In the following sections of this paper the leaching of cadmium from fosfogypsum is used to model the migration from the waste material in the subsoil under the dike enlargement.

5. PROBABILISTIC MODELLING

5.1 Probabilistic model

Because of uncertanties in relation to model assumptions and parameter identification, the reliability of contaminant profiles, as calculated with a conventional numerical model, is questionable. By probabilistic modelling, the uncertanties can be translated in terms of stochastics.

Probabilistic modelling has been achieved by coupling a numerical model for simulating migration of contaminants with a probabilistic module. In the numerical model, based on the Galerkin finite element method, the water dynamics are calculated for each time step, at forehand.

5.2 Input parameters

The soil geometry has to be defined. Furthermore, the following input parameters are required for each soil layer:

- soil hydraulic parameters (soil retention curve, saturated conductivity, unsaturated conductivity relation, residual water content and saturated water content);
- dispersion/diffusion characteristics (dispersivity, diffusion coefficient, tortuosity);
- retardation coefficient;
- production/decay coefficient.

Finally, the boundary conditions have to be defined. For some parameters, a detailed determination of the probabilitic density function is cumbersome, because of lack of data.

For all parameters, probability density functions of the input parameters are required, rather than deterministic parameter values. For most of the parameters, a normal distribution has been assumed. Based on numerous investigations reported in the literature, a log-normal distribution has been taken for the saturated hydraulic conductivity.

6. RESULTS

In Fig. 3, the initial cadmium profiles are presented and for a period of 20, 50 and 100 years after dike enlargement.

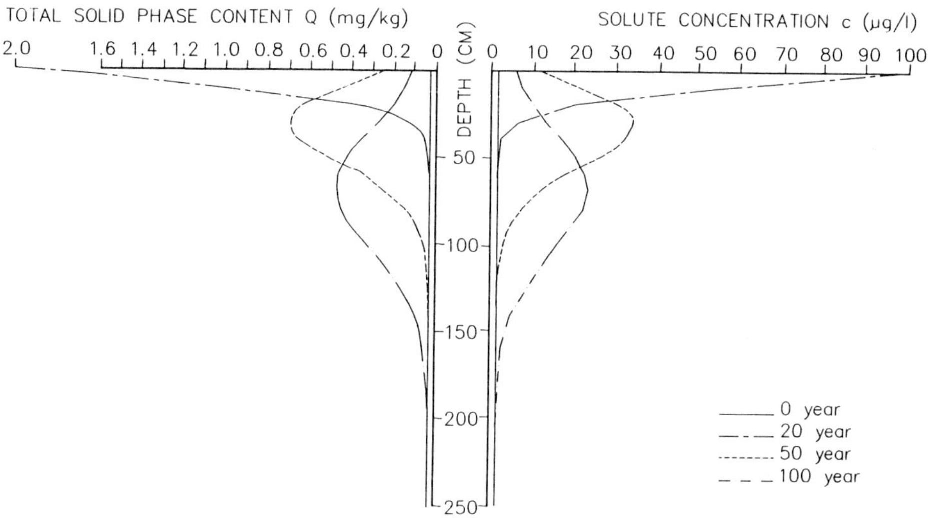

Fig. 3. Cadmium contents in the solid and liquid phase as a function of depth, at different time levels after dike enlargement.

From the Figure, the following can be concluded:

- Initially a homogeneous cadmium distribution has been assumed, where the solute concentration equals, and the solid phase content is lower than the Dutch A-value;
- After 20 years, an enormous cadmium increase is registrated in the upper soil in both phases, over a depth of about 0.4 m, because of cadmium leaching from the fosfogypsum;
- After 50 years, the cadmium contents in the upper layer are reduced because of a decrease in leached cadmium from the fosfogypsum and downwards cadmium migration; the peak is found at a depth of about 0.3 m, where the cadmium front reaches a depth of 1.2 m;
- After 100 years, the peak is found at a depth of about 0.7 m and the cadmium front reaches untill a depth of about 2 m; the profile has been elongated and the peak value has been reduced, due to dispersion anddiffusion.

With the probabilistic model, the probability of exceedance of some critical values has been calculated at several depths as a function of time. To illustrate the principle of probabilistic modelling, the results are presented for the groundwater quality at 0.2 m depth under the dike enlargement, referred to the Dutch C-value (= 10 μg/l) and at 1.0 m depth under the dike enlargement, referred to the Dutch B-value (= 2.5 μg/l) in Figs. 4 and 5, respectively. Furthermore, the contribution of the different input parameters to the uncertainty in the calculated solute concentrations and the respective probabilties of exceedance are shown in Figs. 4 and 5. The contribution of the input parameters which contribute less than 10 % has been combined (REST).

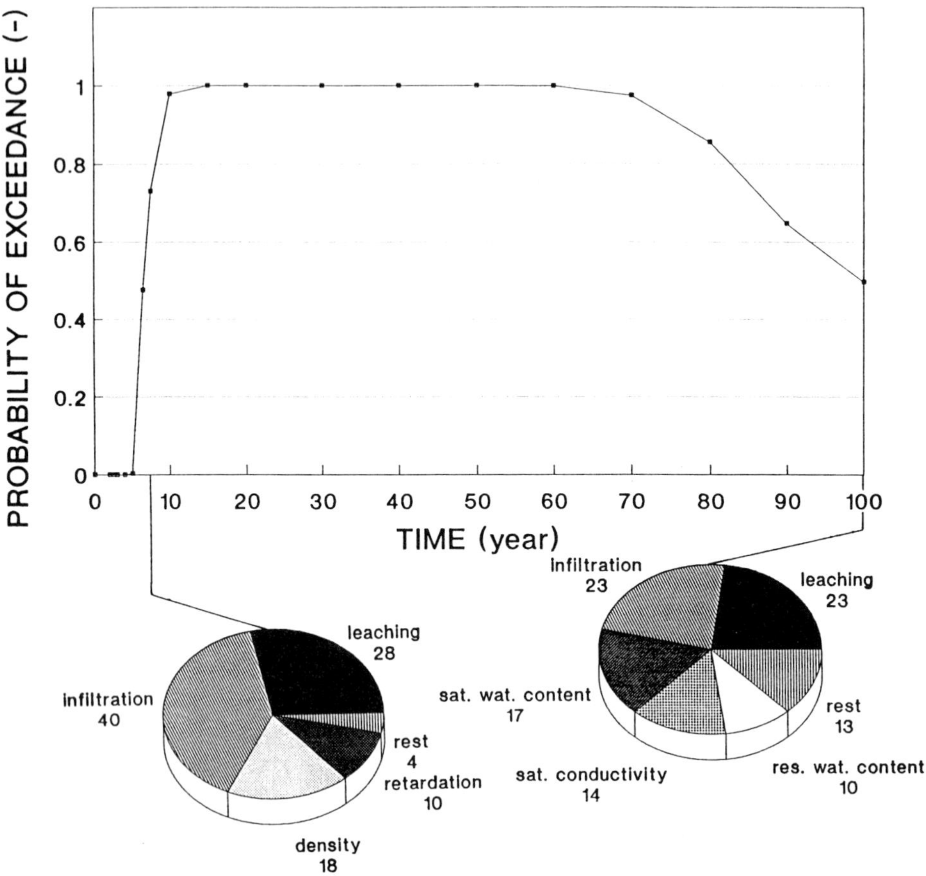

Fig. 4. Probability of cadmium solute concentration exceedance of the action value (dutch C-value) at 0.2 m under the dike enlargement, as a function of time; Contribution of the input parameters to the uncertainty in the respective probability of exceedance, as a function of time.

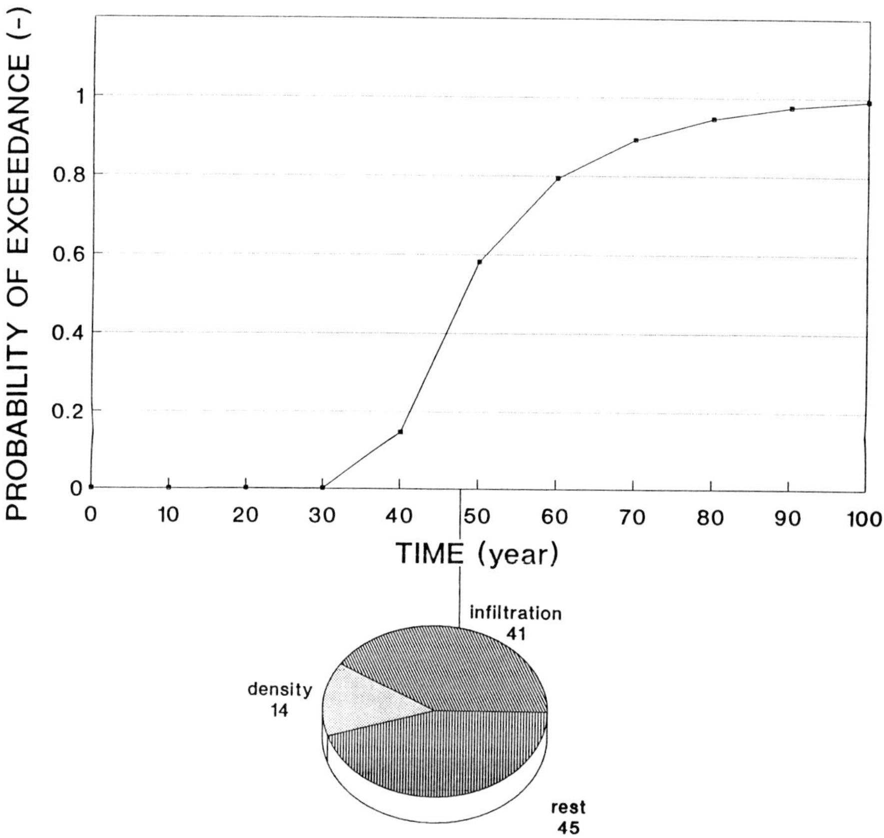

Fig. 5. Probability of **cadmium solute concentration exceedance** of the trigger value (dutch B-value) at 1.0 m under the dike enlargement, as a function of time; Contribution of the input parameters to the uncertainty in the respective probability of exceedance, at 48 years.

From the Figures, the following can be concluded:

- The time at which the solute concentration exceeds the C-value at 0.2 m depth is between 5 and 10 years (90% confidence interval), with an expectation of 7.5 years;

- After 7.5 years, being the time at which the solute concentration is expected to exceed the C-value at a depth of 0.2 m, the uncertainty in the calculated solute concentrations is mainly determined by the uncertainty in the infiltration rate (INFILTRATION, 40 %) and in the leaching characteristic (LEACHING, 28 %). Furthermore, the buffer capacity of the solid phase, i.e. the density (DENSITY) and the retardation (RETARDATION) play a role.

- The time at which the solute concentration drops under the C-value at 0.2 is expected to be about 100 year (90 % probability);

- After 100 years, being the time at which the solute concentration is expected to drop under the C-value at a depth of 0.2 m, the uncertainty in the calculated solute concentrations is mainly determined by the uncertainty in the infiltration rate (23 %) and in the leaching characteristic (23 %). Furthermore the uncertainties in the soil hydraulic input parameters, i.e. the saturated water content (17 %), the saturated hydraulic conductivity (14 %) and the residual water content (10 %) play a major role.

- The time at which the solute concentration exceeds the B-value at 1.0 m depth is between 35 and 85 years (90% confidence interval) with an expectation of 48 years;

- After 48 years, being the time at which the solute concentration is expected to exceed the B-value at a depth of 1.0 m, the uncertainty in the calculated solute concentrations is determined by the uncertainty in almost all parameters with a dominant role for the uncertainty in the infiltration rate (41 %).

7. CONCLUSIONS

The overall conclusion of this study is that the conceptual model is a useful tool in assessing the environmental impact of waste materials in constructions.

Application of the numerical model, including a probabilistic module, gives stochastic results in terms of the probability that concentration levels are exceeded. Futhermore, the following conclusions can be drawn:

- With 90 % probability, at 0.2 m depth the C value will be exceeded within a period of 5-10 years and at 1.0 m depth the B value will be exceeded within a period of 35-85 years.

- Further investigation is needed onto the form of the probability density functions, especially the leachate characteristic, being linearised from a small amount of data.

REFERENCES

1 L.F. Konikow. Role of numerical simulation in analysis of groundwater quality problems. The Science of the Total Environment 21:299-312, 1981.

2 Chin-Fu Tsang. Comments on model validation. Transport in porous media 2:623-629, 1987.

3 J. Bear. Hydraulics of Groundwater. McGraw-hill Inc., New York, 1979.

4 M.Th. VanGenuchten. Mass Transport in Saturated-Unsaturated Media: One-Dimensional Solutions. Water Resources Program, Princeton University, Research Report 78-WR-11. 118 p. 1978.

Waste Materials in Construction.
J.J.J.R. Goumans, H.A. van der Sloot and Th.G. Aalbers (Editors)

LONG TERM ENVIRONMENTAL IMPACT BY USE OF WASTE MATERIALS*: AN ASSESSMENT SYSTEM[1]

M. van Herwijnen**, P. C. Koppert*** and A.A. Olsthoorn**

* Commissioned by Rijkswaterstaat Dienst Weg- en Waterbouw, Hoofdafdeling Milieu, Delft

**Institute for Environmental Studies (IES), Free University, De Boelelaan 1115, 1081 HV Amsterdam

***Now at Erasmus Center for Environmental studies. Erasmus University, Rotterdam

1 INTRODUCTION

A current Dutch environmental problem is how to deal with the increasing production of coal-fly ash, caused by the growth in powder-coal based production of electricity. This prompted Rijkswaterstaat to commission the development of an assessment system, which main purpose is to calculate future emissions related to policy scenarios for future use of waste materials in building and construction and which may be used in life-cycle management of waste materials.

2 THE MODEL

Use of the system starts with the definition of a flow scheme wich shows current and conceived uses of the waste material. Figure 1 shows the scheme used to elaborate the case fly-ash use, the example used to demonstrate the system.

Figure 1. Flow scheme of current and conceived fly-ash use as used in example.

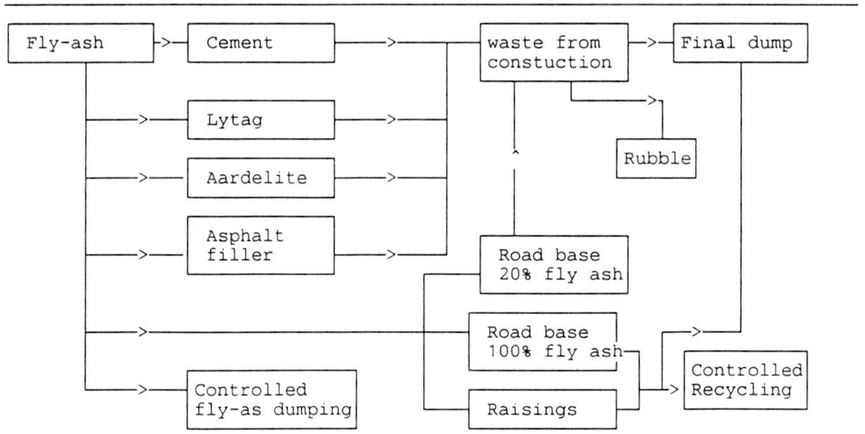

Future leaching is calculated in a two-step procedure. The first step is the calculation of future flows of fly ash and fly-ash products from: 1) scenarios for the developments of current and possible markets for (waste) products; 2) scenarios for the production of waste

materials; 3) data on life-cycles of waste-material applications; and 4) from an assumed **hierarchy in market preferences** to use waste materials. Such a hierarchy represents the simulated policy towards the use of waste materials. In the fly-ash example (figure 1) for instance, use in cement is thought as the most preferable, followed by use in lytag, while dumping is the least preferred. A fifth type of data needed for this calculation are preferences for materials from the point of view of the markets (e.g. controlled recycling in figure 1).
The second step is the calculation of the emissions. For each specific fly-ash use considered then so called emission functions are needed, which model the time dependent rates of leaching of the species considered from specific fly-ash constructions. To be able to demonstrate the system a small number of emission functions are estimated[1]. They are based on limited experimental information obtained in laboratory and practice[2] while accounting for the specific construction characteristics of the uses, such as the width of a road.
The system is open; data and assumptions on scenarios and emission functions can be added, changed and deleted. It can be implemented on a MS-DOS computer.

3 RESULTS AND CONCLUSIONS

The case as indicated in figure 1 is elaborated to demonstrate the system. Characteristics of the scenarios are: fly-ash production from powder coal units increases up to the year 2000, is constant until 2035 (change to coal gasification technology) and then starts decreasing. The main outlet for fly-ash is cement production. This market cannot, however, adsorb all future production. Figure 2 shows the calculated leaching of molybdenum in the period 1990-2070 from the two types of road bases for which fly-ash use is considered. Leaching starts when fly ash is "forced" to be used in these applications, preferable markets then being saturated. The subsequent course of the leaching is determined by changes in leaching rates (emission functions), by the life of road bases (assumption of 50 year), and the eventual producion decrease of the fly ash considered. The system described here is a tool to assess long-term environmental consequences of decisions on how and where to use waste materials, it is therefore of interest for life-cycle management of waste materials.

Figure 2. Leaching of Mo from road bases

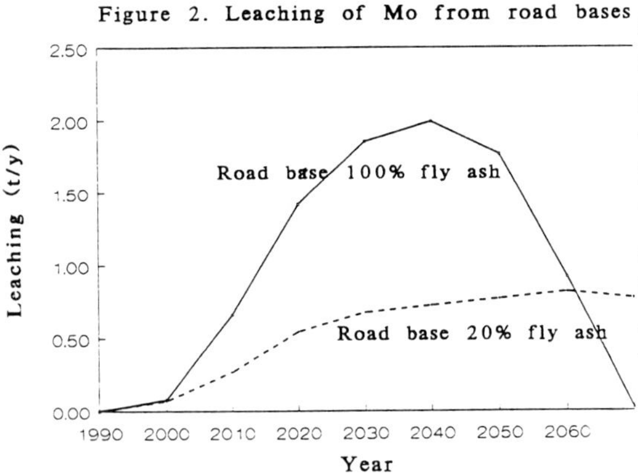

REFERENCES

1 Herwijnen, M. van, P.C. Koppert and A.A. Olsthoorn, Long Term Environmental Impact by use of Waste Materials (In Dutch), IES E-89/01 Amsterdam and RWW-DWW MI-OW-89-38 Delft.

2 N. Bolt en H.A. van der Sloot, Environmental implications of fly-ash use in road construction. Evaluation pilot projects and proposals for guidelines (in Dutch), KEMA-ECN 71911-SBA, Arnhem, 1988.

Waste Materials in Construction.
J.J.J.R. Goumans, H.A. van der Sloot and Th.G. Aalbers (Editors)

Leaching from Building Waste

Jan Folkenberg[1] and Benthe Rasmussen[2]

1. Danish Technological Institute, Department of Building Technology, P.O. Box 141, DK-2630 Taastrup (Denmark).

2. Danish Technological Institute, Department of Environmental Technology, P.O. Box 141, DK-2630 Taastrup (Denmark).

Summary

An estimated 2-4 mill. tons of building waste is deposited in Denmark every year. Part of this waste contains environmentally hazardous substances. On behalf of the Danish National Agency of Environmental Protection a number of leaching tests have been carried out on building waste sorted at source in order to establish which quantities of harzardous substances are released when deposited.

Preface and Background

An estimated 2-4 mill tons of building waste is deposited every year in Denmark. The depositing rises several questions on potential environmental risks, especially in connection with that part of the building waste that is not recycled fully, or for which options for recycling have not yet been made.
Building waste is used as a collective name covering a number of subcomponents in construction including concrete, brick, glass, plastics, metals, roofing paper, glazed bricks, etc. Various types of building waste contain environmentally hazardous substances in greater or smaller amounts. The aim of this survey is to examine if the types of waste considered problematic when deposited, in reality are.

Types of Building Waste

In recent years demolishing techniques based on selective house breaking have been developed in Denmark. Selective house breaking implies a current sorting at source of the building waste when a house is demolished. Faced with the problem of choosing the sections of building waste to be examined for the content of environmentally hazardous substances, it was relevant to ascertain how, in practice, a sorting of building waste in connection with a house breaking can take place.

Based on experience of selective house breaking combined with an intermediary knowledge of the theoretical content of environmentally hazardous substances in various types of building waste, we decided to pursue the following sections: Roofing felt, facade bricks with soot, chimney pipe with soot, insulating materials (gölasswool/rockwool), painted wood, pressure impregnated wood, window glass, glazed sanitary installations and plastics including pure PVC. The sections 1-4 include Polyaromatic hydrocarbons (PAH) compounds and the remaining sections 5-9 contain various complex heavy metals which may be mobilised when deposited.

Since there is a lack of sufficient knowledge about the amounts the various compounds will yield from the landfill during the leaching process, the results should be assessed as follows:

- What are the expected amounts of environmentally hazardous substances derived from various types of waste.

Based upon this information the following criteria will be set up:

- How detailed should a selective house breaking be in order to avoid spreading of environmentally hazardous substances in relation to recycling, and
- which precautionary measures should be taken in connection with a possible landfilling.

Method Chosen for Leaching Test

In 1986 The Environmental Protection Agency (EPA) in the USA released a directive on the establishment of the mobility of environmentally hazardous substances during the leaching of firm compounds/substances known as "Toxicity-Characteristic-Leaching-Procedure" (TCLP). This method has been applied to a comprehensive range of products including contaminated soil, flying ashes, slag, sludge, coal, etc. This method has been applied, although with minor adjustments to the machinery/appliances.

Analysing Results:

The results of the analyses in Tables 1 and 2 are shown as an average of the five tests on each type of building waste. The width of variation is shown for the individual products of the organic analyses. The analysing results for facade bricks are not yet available.

Table 1. Organic analyses

Components	Roofing felt	Rockwool	Glasswool	Chimney-pipe	Reference samples
Phenoles	4,7	0,02	0,06	nd	nd
Cresoles	9,1	0,02	0,11	nd	nd
Xylenoles	10,6	0,01	0,02	nd	nd
Naphtalene	1,3	0,003	0,002	nd	nd
Me-naphtalene	0,62	nd	0,003	nd	nd
Di-me-naphta-lene	1,22	nd	nd	nd	nd
Acenaphtylene	0,22	nd	nd	nd	nd
Fluorene	0,09	nd	nd	nd	nd
Phenanthrene	0,4	nd	nd	nd	nd
Antracene	0,1	nd	nd	nd	nd
Fluoranthene	0,09	nd	nd	nd	nd
Pyrene	0,05	nd	nd	nd	nd
Benz(a)an-tracene	0,01	nd	nd	nd	nd
Chrysene	0,005	nd	nd	nd	nd
Formaldehyde	-	20,7	38,1	-	-

All values are leaching quantity in mg/kg building waste.
nd - means: Not detectable

Table 2. Inorganic analyses

Compo-nents	Impregn wood, Type 1	Impregn wood Type 2	Painted wood	New PVC	Old plastics	Window glass	Glazed sanitary
Ba	-	-	8,4	-	-	10,44	-
Cr	14,76	16,4	1,24	nd	nd	-	-
Ni	-	-	-	-	0,3	-	-
Cu	146,8	153,6	-	-	-	-	-
Cd	-	-	-	0,0234	0,0168	-	-
Hg	-	-	nd	-	-	-	-
Pb	-	-	128	170	52	-	2.063
As	198	100	-	-	-	0,0104	-
Sb	-	-	-	-	-	nd	-

All values are leaching quantity in mg/kg building waste.
nd - means: Not detectable.

Waste Materials in Construction.
J.J.J.R. Goumans, H.A. van der Sloot and Th.G. Aalbers (Editors)

LEACHING TESTS AND THE INFLUENCE OF OXIDATION-REDUCTION PROCESSES

C. ZEVENBERGEN and W. F. HOPPE
IWACO B.V., P.O. Box 183, 3000 AD Rotterdam (The Netherlands)

INTRODUCTION

Many processes that regulate the chemistry of toxic metals from waste materials are influenced by the pH and the redox potential (Eh). In contrast however to the pH, the redox potential has received little attention in standardized leaching tests (e.g. TCLP, NVN 2508, DIN 38 414), that are used to assess the potential hazards of waste materials. Previous experiments carried out in our laboratory showed that the redox potential of shake and column extracts from organic waste (e.g. bottom sediments, contaminated soils) and to a lesser extent from fly ash and slag may drop drastic during the tests period after a few days to a few weeks depending on the conditions of mixing and flow rate and probably on the extend of weathering of the waste material. It was concluded that these redox changes may have a significant influence on the amounts of metals that are leached from waste materials in the laboratory.

To further study the influence of the redox changes on results of leaching tests on elemental mobilisation, saturated and unsaturated column tests were carried out. In order to obtain information to what extent equilibria are effected by redox-kinetics, aerobic and anaerobic batch tests were performed. Some results of these experiments are discussed in this paper.

METHODS AND MATERIALS

Waste material; the tests were carried out on a mixture of ash from a fertilizers plant and a loamy topsoil.

Batch test; in the batch tests glass bottles were used in which waste material and water were agitated in a 1 to 1 ratio with an extraction time of six weeks. Compressed air was used in open bottles to maintain oxidizing and mixed conditions. Capped bottles with no headspace was used to create a reducing environment.

Column test; saturated conditions within the column were obtained using a standard column test method (NVN 2508). To obtain unsaturated conditions within the column the normally adapted continious upflow was substituted with a sprinkler induced downflow. All the columns were percolated with demi water (flow rate 12 ml/h). The pH, redox potential and metal concentration of the extracts and percolates were determined.

RESULTS AND DISCUSSION

In the aerobic batches a fairly stable Eh (+100 mV) and As and Cd concentration is reached within 1 week of incubation. In the anaerobic batches a considerable increase of As and Cd concentration is observed during the first weeks, while the Eh remains relatively stable at -100 mV. Upon further incubation a stable (near) equilibrium As concentration and a continuous decrease of the Cd concentration is found. For both columns, Eh drops rapidly in the first days (L/S 0-1) after which a steady value of Eh is attained (unsaturated column: +100mV; saturated column: -150mV). The As concentration in the percolate of the columns is strongly effected by redox kinetics. With respect to the Eh, the observed leaching behaviour of As in the columns are in agreement with the findings in the batch tests. Probably due to the slow reaction kinetics, the leaching behaviour of Cd in the columns seems to be not influenced by redox conditions.

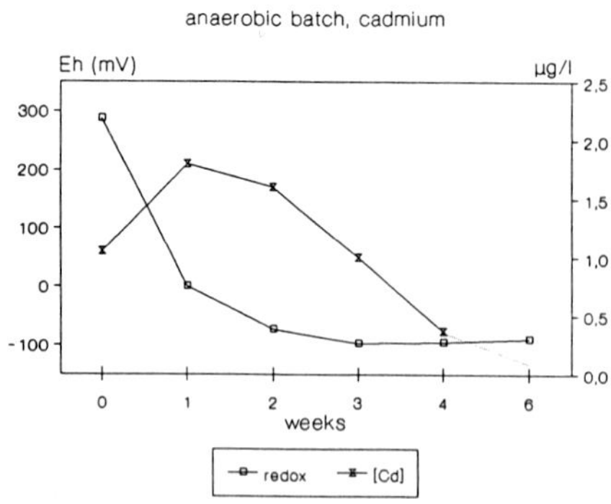

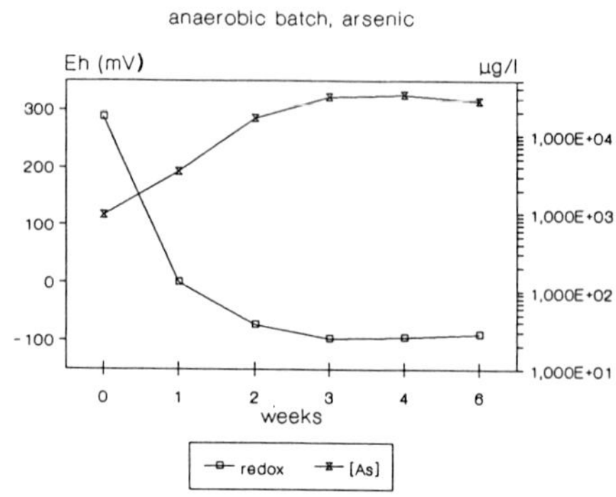

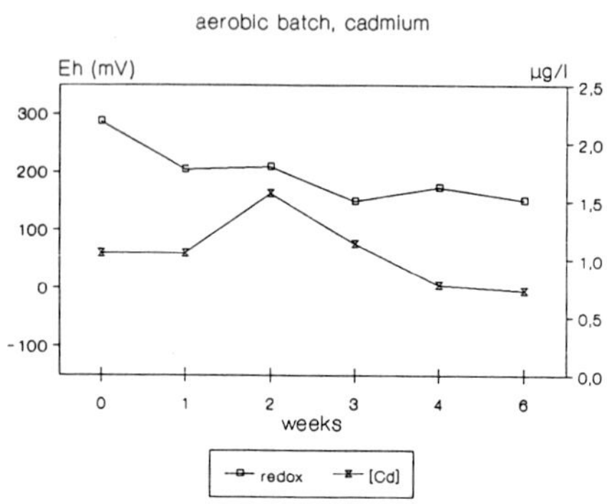

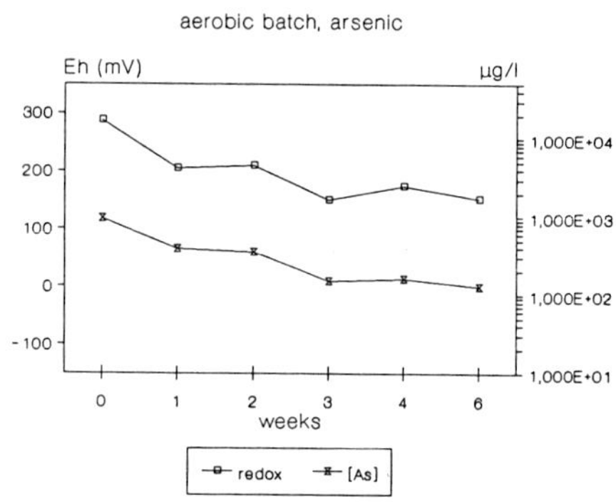

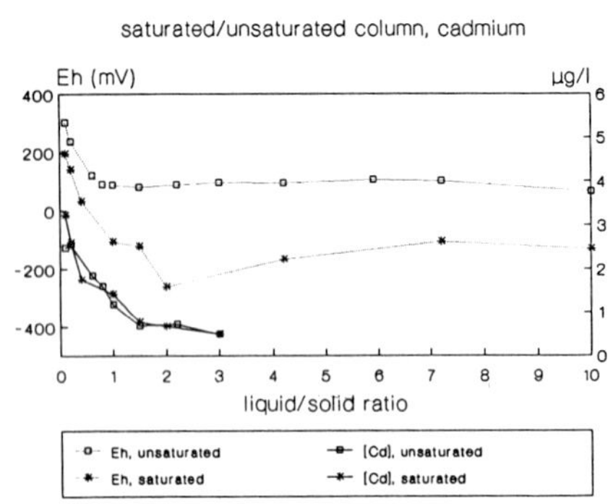

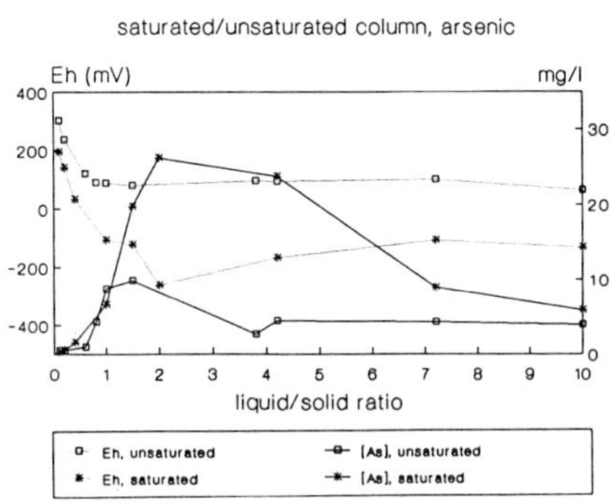

CONCLUSIONS

In these experiments, the importance of measuring the Eh in the percolate of the column test is demonstrated, as it can provide information about what processes are occurring within the column. Due to time dependant reactions or catalysis by microorganism, care must be taken in extrapolating column leaching data to field conditions.

Waste Materials in Construction.
J.J.J.R. Goumans, H.A. van der Sloot and Th.G. Aalbers (Editors)

CEMENT STABILIZATION/SOLIDIFICATION TECHNIQUES: pH PROFILE WITHIN ACID-ATTACKED WASTE FORM

K.Y. CHENG, P. BISHOP, and J. ISENBURG

Department of Civil and Environmental Engineering
University of Cincinnati
Cincinnati, Ohio, 45221 (U.S.A.)

SUMMARY

Leaching of cement-based waste form in acetic acid solutions has been investigated in this work. The examination of the pH profile along the acid penetration route by various pH colorimetic indicators is reported. A line of demarcation, believed to be the leaching boundary, was observed in every leached samples.

1. MATERIALS AND METHODS

Cement-based waste samples were made in the laboratory by mixing metal sludges with type I portland cement. The metal sludges were prepared from cadmium nitrate, lead nitrate, and sodium arsenite of 0.01 mole each. Samples with 0.6 sludge-cement weight ratio were cast as 3.72 cm diameter spheres and cured for two months before the leach test.

Table 1. The properties of the pH colorimetic indicators.

Indicator	pH range	Color change
Tetrabromophenolphthalein-ethyl ester, K-salt	3.0 - 4.2	Yellow to Blue
Congo Red	3.0 - 5.0	Blue to Red
Ethyl Red	4.0 - 5.8	Colorless to Red
Alizarin	5.6 - 7.2	Yellow to Red
Bromothymol Blue	6.0 - 7.6	Yellow to Blue
Phenolphthalein	8.2 - 10.0	Colorless to Pink
Alizarin Yellow R	10.1 - 12.0	Yellow to Red

The modified ANSI/ANS-16.1 leach test procedures were followed. Seven samples were leached in 0.2 N acetic acid solutions. The solutions were renewed at 1, 8, 15, 22, 29, 36, and 43 days. During each solution renewal period, one sample was removed and fractured. Seven pH colorimetic indicators were applied to the surface of the damp fractured samples in order to identify the pH profile along the radius of the samples. The properties of these indicators are listed in Table 1. The solutions of the indicators were prepared by the methods described in the CRC Handbook of Chemistry and Physics, 67th edition, p. D.147-D.148.

2. RESULTS AND DISCUSSION

On every fractured sample surface, a light grey kernel surrounded by three different color layers were observed. Starting from the solid/solution interface of the sample, these layers are the orange layer, the dark grey layer, and the medium grey layer (Figure 1). The application of pH indicators indicated that the pH of the medium grey layer and the light grey kernel was above 12, whereas the pH of the orange layer and the dark grey layer was below 6. A white demarcation line appeared between the dark grey layer and the medium grey layer. This demarcation line is believed to be the leaching boundary where the pH values changes from below 6.0 in the leached layer to above 12 in the unleached kernel. The white color of the demarcation line was caused by the reprecipitation of calcium hydroxide, which is the dominant alkaline species in the cement-based waste form.

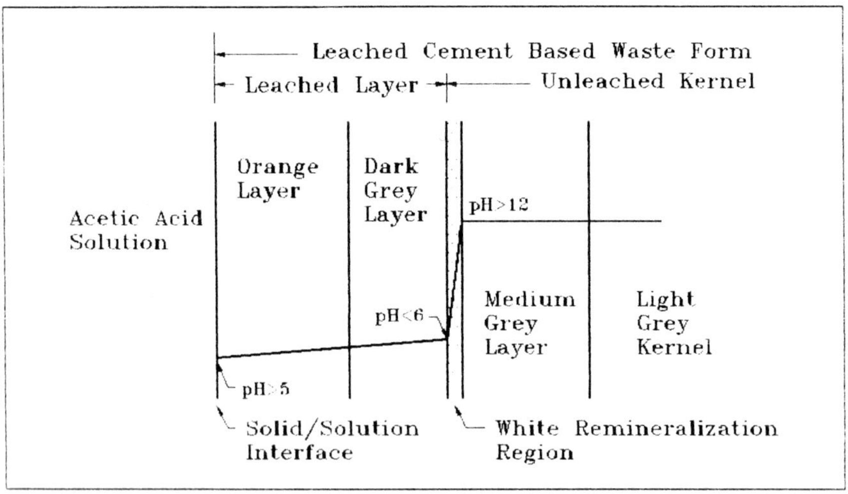

Figure 1. Schematic profile of a leached cement-based waste form.

ACKNOWLEDGEMENTS

This work was begun under U.S. EPA contract No. 68-03-3379, work assignment 2-7 and continued under Project Activity No. SF5, COE-UC/RREL Cooperative Agreement CR-816700. The work was done under the sponsorship of the U.S. EPA Risk Reduction Engineering Laboratory, Cincinnati, Ohio.

Waste Materials in Construction.
J.J.J.R. Goumans, H.A. van der Sloot and Th.G. Aalbers (Editors)

POTENTIAL FOR REUSE OF LEAD-CONTAMINATED URBAN SOILS

H.A. van der Sloot, J. Wijkstra and J. van Leeuwen

Netherlands Energy Research Foundation, ECN, Petten.
Geotechnics and Environment, Public Works, Rotterdam.

Introduction - As a result of industrial activities, in particular paint production and traffic emissions, soils of the center of major cities, such as Rotterdam, are contaminated with lead. These soils are uncovered in reconstruction and new building activities. Based on the guidance values in the Soil Protection Act [1], these soils should be treated according to regulations for contaminated soil. For construction in the urban environment, this results in high cost and serious delay. Based on earlier experience, the mobility of lead was expected to be low and, consequently, the risk of exposure was also considered small.

In this study [2], the potential for lead mobility in the long-term has been evaluated and recommendations for utilization lead-contaminated soils are given.

Experimental - Leaching tests according to NVN 2508 [3], diffusion measurements and experiments of lead contaminated soil with natural humic and fulvic acids were carried out.

Results - The concentrations of lead in urban soil varied from 3 to 3250 mg/kg. The availability for leaching based on NVN 2508 was higher in sandy soils (4-13 %) than in clay soils (1-4%). In figure 1 the correlation between available lead and total Pb concentration is given for sandy soil and for clay soil. A direct correlation between leachable quantity and organic content was observed (figure 2). In the soil samples studied, no correlation between the clay fraction (mass < 2 μm) and the leachable lead was noted. This indicates that the lead contamination seems primarily associated with the organic fraction of the soil. The leaching percentages of lead after extraction in a serial batch procedure without pH adjustment was less than 0.2% at LS 100 in the samples.
The effective diffusion coefficient of lead was found to be 2.5×10^{-14} m^2/s after a contact time of 327 days. In figure 3 the profile obtained for a mobility measurement of Pb in a highly Pb contaminated clay soil is shown.
Measurement of the mobility of Pb in soil samples spiked with Pb-210 has shown a very limited mobility of Pb under low flow conditions, even after percolating the soil with humic and fulvic substances. In figure 4, the concentration profile in a sliced column is presented for a percolation experiment of 132 days.

Conclusions - In view of the low mobility of lead, reuse of urban soils moderately contaminated with lead is recommended without further isolation in civil works in the urban environment. Overlaying the soil with a layer of clean soil is considered to lead to sufficient protection from direct (ingestion) or indirect (skin contact, crops) contact.

Acknowledgment - The study was funded by the Public Works, Rotterdam.

References

1. Soil Protection Act, Dutch Government, Staatscourant, 1986, 374.
2. H.A. van der Sloot, J. Wijkstra and P. Bonouvrie, ECN-C-90-037, 1990.
3. NVN 2508 Dutch standard for leaching of granular waste material, 1988.

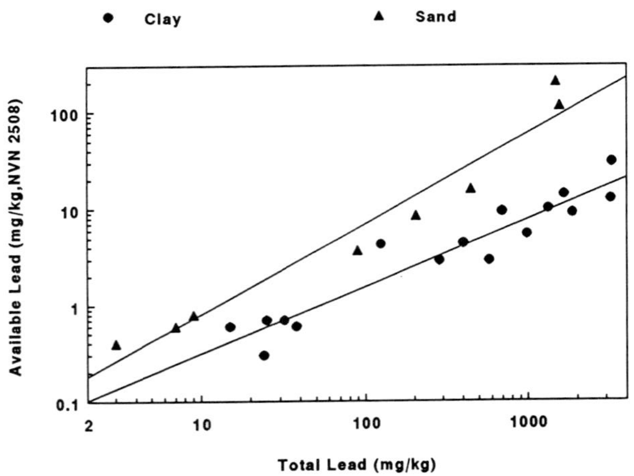

Fig 1. Relation between availability of Pb for leaching from contaminated soil and the total content

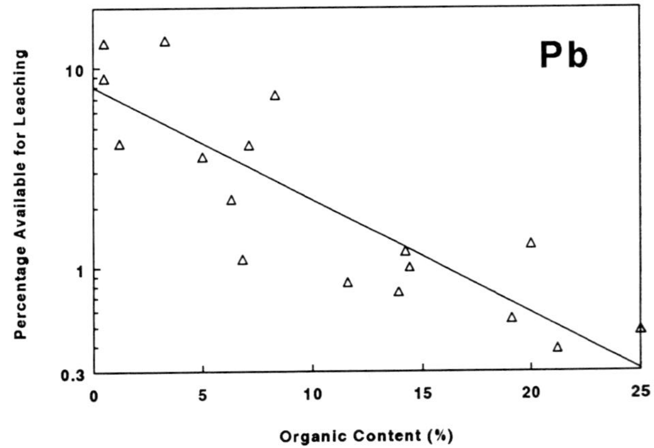

Fig 2. Correlation between the availability of Pb from contaminated soil with the organic carbon content

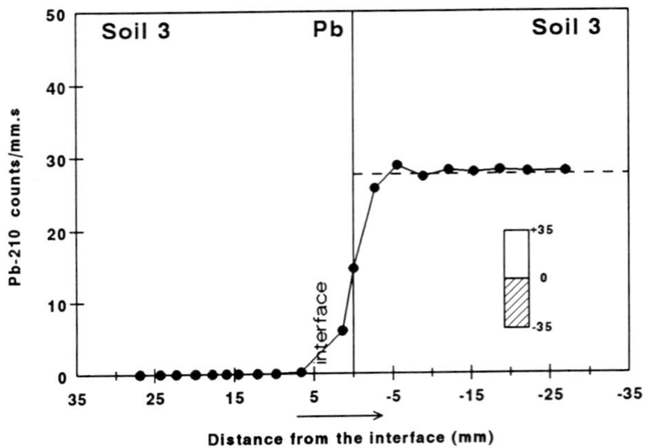

Fig 3. Mobility of Pb in contaminated soil (contact-time: 327 days)

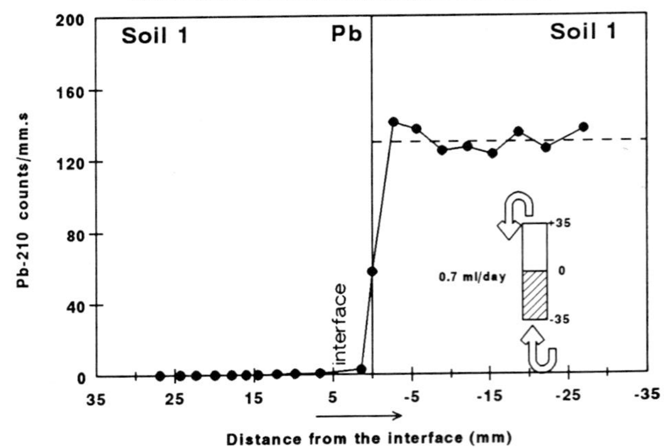

Fig 4. Mobility of Pb in contaminated soil under low flow conditions; percolation time: 132 days

Waste Materials in Construction.
J.J.J.R. Goumans, H.A. van der Sloot and Th.G. Aalbers (Editors)

STANDARD SAMPLE PREPARATION AND REFERENCE SAMPLES AS A TOOL FOR DETERMINATION OF THE ENVIRONMENTAL QUALITY OF BUILDING MATERIALS

F.J.M. Lamers[1], G.J. de Groot[2]

[1] MBN, Centre of Materials Samples of the Netherlands, P.O. Box 151, 6470 ED Eygelshoven (The Netherlands)

[2] ECN, Energy Research Centre Netherlands, P.O. Box 1, 1755 ZG Petten (The Netherlands)

SUMMARY

The determination of the environmental quality of building materials can be facilitated with the aid of reference samples in the field of chemical composition and leaching behaviour and with the aid of a leaching database. The Centre for Materials Samples of the Netherlands is specialised in the preparation of standard samples from bulk materials. An overview of its sample preparation possibilities is presented. In addition a strategy for environmental certification of building materials is presented, in close cooperation with ECN, the developer of the leaching database.

1. INTRODUCTION

In the near future building materials in the Netherlands will have to comply with strict limits in the field of chemical composition and leaching behaviour, as a consequence of the Building Materials Decree. It is important to be certain of a correkt ranking of a building material, regarding environmental quality. The analysis of composition and leaching behaviour of bulk building materials will be facilitated by the parallel analysis of known reference samples from the same type of building material. Furthermore, measures to reduce the amount of chemical analyses are welcome.

2. THE CENTRE FOR MATERIALS SAMPLES OF THE NETHERLANDS

MBN, the Centre of Materials Samples of the Netherlands, has been founded by several scientific institutes in the Netherlands, with the following tasks: a. The preparation of a collection standard and reference samples of a range of building materials. b. The distribution to the industry of knowledge regarding environmental classification of building materials.

A general outline to the role of MBN is presented, viz. a) as a centre for the preparation of standard samples, and b) as a consultant in the field of environmental certification.

3. STANDARD AND REFERENCE SAMPLE PREPARATION FROM BULK MATERIALS

3.1 Sample preparation equipment

The preparation of standard samples from bulk raw materials is only possible if one can reduce a bulk sample into identical sub-samples of 1 kg or less, so that one can be certain of reliable analysis of the total lot. MBN disposes of the following equipment to perform this operations:

a. A sample crushing and sample splitting apparatus with a capacity of 2 tons of material per charge (fig. 1 and 2). In the apparatus, all necessary precautions are taken to prevent contamination of the samples. It is known that a rotating sample splitter (spinning riffler) results in subsamples with the smallest possible standard deviation (fig. 3).

b. 2 Sizes small spinning rifflers (10 or 20 subsamples) with a capacity of max. 30 kg, and max. 200g.

c. Several crushing and milling machines for preparation of small size samples.

d. Sample preparation under cryogenic conditions.

d. Controlled climate cells, for preparation and testing in several gas environments like nitrogen and carbon dioxide, under controlled temperature and moist conditions etc.

e. Storage in the constant climate of one of the South Limburg caves.

Fig 1 Sample crusher and fig 2 Spinning riffler in the large scale sample preparation equipment

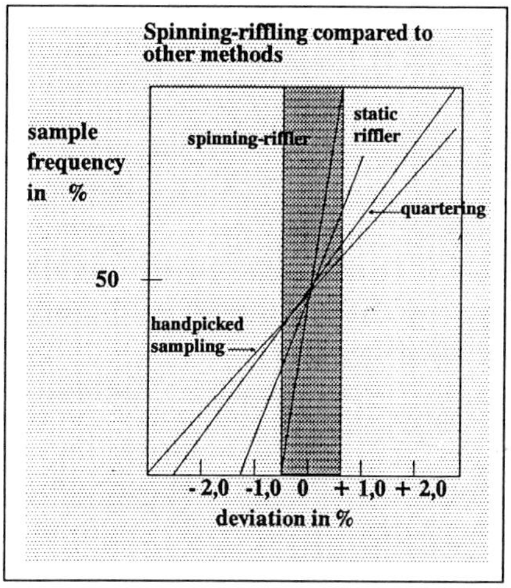

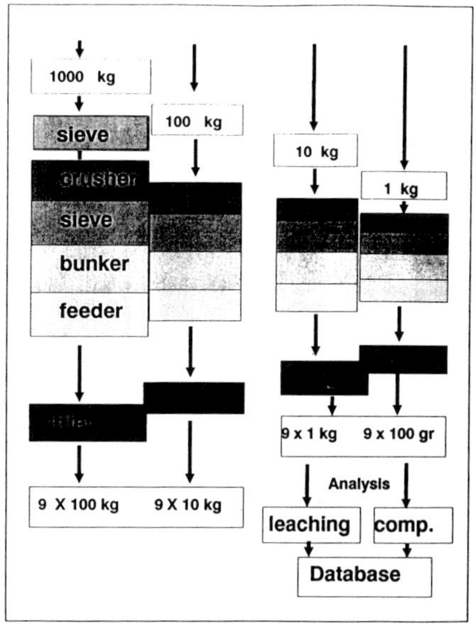

Fig 3 Standard deviation for several sample splitting techniques

Fig 4 MBN standard sample preparation scheme

3.2 Standard sample preparation

The total scheme for the preparation of a standard sample collection is presented in fig. 5. MBN aims to a proces certification of its sample preparation procedures according to ISO 9002. The characterization of the standard samples proceeds in round robins. The fully characterized standard samples will then be offered to chemical laboratories as known standard materials.

3.3 Sample collection

At this moment a sample collection is present with a range of primary and secondary building materials like: gravel, sand, cement, clay, coal fly ash, municipal waste incineration ashes, blastfurnace slags, phosphorous slags, concrete granulate etc. At present no results of round robins are available yet.

4. MBN AND ENVIRONMENTAL QUALITY DETERMINATION

In the field of environmental quality determination, MBN works closely together with ECN. MBN and ECN have the following approach to environmental quality determination of building materials:

A. In cooperation with the building materials producer and a technical certifying agency (ref. 2), a strategy for testing is described (BRL), based on *) requirements of the Building Materials Decree and **) existing systematical data on the environmental quality of said product. In the Building Materials Decree, determination of the chemical composition and leaching behaviour are demanded.

B. Existing and incoming data on the environmental quality are stored in a database allowing the grouping and statistical processing of data for gaining insight in leaching systematics of the product in question (Ref 1). Based on growing insight, short testing procedures for environmental certification can be developed and - probably repeatedly - be adjusted.

C. Standard and reference samples from MBN can be used as inner standard for the leaching analyses.

D. For environmental certification, the client can only get acces to the leaching database through MBN. Both MBN and ECN can act as a consultant to the client in this cooperation, each on its specific know how.

The advantage of this approach is that through the insight from the database, a lot of unnecessary (for the product in question) analyses, that are obliged in the Building materials Decree can be prevented.

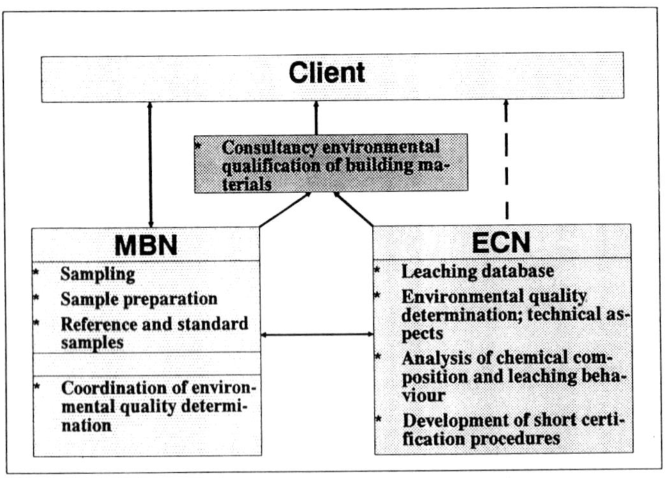

Fig 5 Cooperation MBN - ECN for environmental qualification of building materials

REFERENCES

1. G.J. de Groot, this Proceedings
2. P. Rademaker, this Proceedings

Waste Materials in Construction.
J.J.J.R. Goumans, H.A. van der Sloot and Th.G. Aalbers (Editors)

CERTIFICATION OF MSW SLAGS AS A ROAD CONSTRUCTION MATERIAL

J.J. STEKETEE and J.H. DE ZEEUW

TAUW Infra Consult B.V., P.O. Box 479, 7400 AL Deventer (The Netherlands)

SUMMARY

In the framework of the certification of MSW slags, procedures for sampling and sample preparation were developed. Data on the leaching behaviour of the slags of eight Dutch incinerator facilities are presented.

INTRODUCTION

In order to assure the market for Municipal Solid Waste (MSW) slags, the VEABRIN (Dutch Organization of Waste Incinerators) started a procedure for certification of the slags in 1987. The first certificate was granted to AVR (Afval Verwerking Rijnmond) in 1990.

In this certificate certain requirements have to be met regarding the environmental and mechanical properties of the slags. Procedures for sampling, sample preparation and the frequency of the investigations, which are prescribed in the certificate, are based upon preliminary research carried out by TAUW Infra Consult B.V.

In the first period of a year (1987/1988), samples were taken daily from four Dutch incinerator plants. This research produced much information regarding (seasonal) fluctuations of the quality. The following year (1988/1989), the quality control was continued and extended to nine Dutch incinerator plants.

MATERIALS AND METHODS

MSW slags are defined as the solid residue of municipal solid waste incineration, screened about 40 cm mesh, from which the iron is eliminated, without fly-ash. In the first research period (one year) samples were taken daily. These samples were mixed over a two week period. The mixed samples were divided by TAUW Infra Consult B.V. into two sub-samples, one for investigating the mechanical properties (carried out by Zuidelijk Wegenbouw Laboratorium,
Vught) and the other for investigating environmental properties (carried out by TAUW Infra Consult B.V.). For the time being, the Dutch government has ordered to establish the environmental quality of the MSW slags by

means of a cascade test (new regulations will come within a year). A cascade test implies a fivefold sequential extraction with demineralized water, acidified to pH 4. In each step the liquid/solid ratio is 20, so the cumulative ratio of the test is 100. The leaching fluids were analyzed on arsenic, cadmium, chromium, copper, molybdenum, nickel, lead and zinc.

Due to the rather limited fluctuation of the quality, it was concluded, after a year of research, that the frequencies of investigations could be reduced for most plants. Depending upon the production of the plant, a mixed sample was taken over a period of two, four and six weeks respectively.

ENVIRONMENTAL QUALITY RESULTS

Some of the results for the period 1988/1989 are summarized in table 1. This table includes the mean and standard deviation of the quantities leached in the complete cascade test (L/S100). Only the data for copper, molybdenum, lead and zinc are given. The leaching of the other elements is generally speaking, low.

TABLE 1

Environmental quality data of Dutch MSW slags. Mean and standard deviation of quantities of heavy metals leached (mg/kg d.m.) in the cascade test (L/S 100) N = 9-13.

	Incinerator Facility															
	1		2		3		4		5		6		8		9	
	m	s	m	s	m	s	m	s	m	s	m	s	m	s	m	s
Copper	12.2	3.7	4.3	1.7	8.5	1.5	25.8	4.7	27.8	5.7	11.5	4.4	15.3	5.4	8.1	7.0
Molybdenum	1.0	0.5	0.8	0.3	1.8	0.8	1.2	0.5	1.5	0.7	1.0	0.3	5.9	6.7	1.2	0.0
Lead	0.5	0.3	1.3	0.8	1.8	2.2	2.2	1.5	4.6	2.8	1.1	1.1	4.0	2.8	3.0	5.0
Zinc	1.3	1.4	1.2	1.1	1.9	1.2	0.7	0.8	1.7	1.4	1.5	1.2	2.4	1.6	1.4	1.3

CONCLUSIONS

1. Considering the inhomogeneous nature of waste and slags, the seasonal fluctuations in the compositions of waste and the accuracy of the method, the standard deviation of the quantities leached is fairly low (generally speaking < 100%, for copper and molybdenum < 50% of the mean).
2. There are striking differences in the leaching behaviour of slags from different incinerator plants. The input composition is probably largely responsible for this difference.
3. The quality of all slag samples meet with the (interim) approval of the Dutch government. In the foreseeable future, new quality requirements will probably result in a need for improving the leaching behaviour.

Waste Materials in Construction.
J.J.J.R. Goumans, H.A. van der Sloot and Th.G. Aalbers (Editors)

A REFERENCE STUDY ON LEACHABILITY OF METALS FROM NATURAL SOILS

J. Keijzer[1], C. Zevenbergen[1], P.G.M. de Wilde[2] and Th.G. Aalbers[2]

[1]IWACO b.v., P.O. Box 8520, 3009 AM Rotterdam (The Netherlands).
[2]National Institute of Public Health and Environmental Protection, P.O. Box 1, 3720 BA Bilthoven (The Netherlands).

SUMMARY

In this study the leachability of metals from natural topsoils has been investigated. For the determination of the leachability of metals a column test and an availability test was carried out with a wide variety of soil types. Within the investigated soils no uniform leaching behaviour of metals was observed. In general a significant relation was found between the emission on one hand and the content of organic carbon and cation exchange capacity (CEC) on the other hand.

INTRODUCTION

Dutch governmental regulations dealing with re-use of waste materials as building materials have been based on total elemental concentration and leachability. For the determination of the leachability of inorganic pollutants Dutch experts have develloped a Standard Leaching Test, consisting of a column test, a serial shake test and an availability test. Research has been carried out to establish reference levels of primary materials (natural materials) and secundary materials (waste materials) [1]. In contrast with most bulk waste materials (e.g. fly ash, incineration slag, granular waste), few information is available of the leachability of metals from soils. The soil clean-up operations in the Netherlands produce however more than 600.000 ton contaminated soil a year.

In this study the leachability of metals from natural topsoils has been investigated. Apart from metal speciation and metal content, leachability of metals from soils may vary to a large extend depending on typical soil parameters like organic carbon and clay content. In order to asses the soil paramaters which are controlling leaching, leaching tests have been carried out with a wide variety of soil types. Correlations of background concentrations of trace metals with organic and clay content have been reported earlier by Edelman [2].

This research has been financially supported by the Ministry of Housing , Physical Planning and Environment (VROM) and the Netherlands Integrated Soil Research Program (PCTB).

METHODS AND MATERIALS

The soil samples had been obtained from 19 natural locations in the Netherlands. They had been taken from the surface (0-10 cm). Each sample was dried and sieved prior to chemical characterization and determination of the leaching behaviour. A columntest [3] and an availability test [4] were carried out with each sample. In the column test a column of 5 cm diameter and a thickness of 20 cm was used. The columns were percolated (up flow) with acidified demineralized water (pH=4) untill a liquid/solid ratio of 10 was reached (after three weeks). The availability test was carried out with a 1:100 solid/liquid ratio. The suspension was stirred for 3 h, while the pH of the suspension maintained neutral (pH=7). After filtration, the residue was stirred again with the same amount of water (L/S=100) during 3 h, while the pH of the suspension was maintained acid (pH=4). The extract was filtered. The percolates from the columntest and the extracts from the availability test were analyzed on trace metals content by graphite furnace or flame AAS depending on trace element. Correlation coefficients and regressions equations were calculated.

RESULTS AND DISCUSSION

The following features had been derived from the results:

Elemental content

Significant relationship ($p < 0.05$) was found between the content of clay and the elements Ba, Cr, K, Co, Cu, Mg, Na, Ni, Sn, V and Zn and between the content of organic carbon and Cd, Pb and Hg. These relationships corresponded with earlier reported results by Edelman [2].

Emission

In table 1 the minimum and maximum emission of metals from natural soils measured by means of the column test had been shown.

TABLE 1

Emission (L/S=10) of metals from natural soils.

Element	Emission (mg/kg)	Element	Emission (mg/kg)
As	0.03 - 1.16	Mo	0 - 0.58
Ba	0.02 - 1.36	Na	11 - 203
Ca	6 - 500	Ni	0 - 0.93
Cd	0 - 0.034	Pb	0.01 - 1.74
Co	0 - 0.37	Sb	0 - 0.07
Cr	0 - 0.15	Se	0 - 0.09
Cu	0.07 - 0.39	Sn	0 - 0.08
Hg	0 - 0.029	V	0 - 1.03
K	14 - 232	Zn	0.10 - 4.74
Mg	2 - 88		

Within the investigated soils no uniform leaching behaviour of metals was observed. This could be attributed to differences in the kinetics of redox reactions and dissolution reactions occuring within the column. Most soils demonstrated elevated levels of trace metals in the first percolates of the columntest (rapid initial decline). In these percolates extreme low pH (3-5) were observed. However some leaching curves of soils have showed a different patern: delayed release or constant release.

The maximum available quantity of most elements was low related to the total elemental composition (< 10%).

In table 2 the significant ($p < 0.1$) correlations between the emission (L/S=10) and the soil parameters or the emission (L/S=200) measured by means of the availability test had been shown.

TABLE 2

Significant (p < 0.1) correlations between the emission (L/S=10) and the soil parameters.

Element	Elemental content	pH	Clay	Org C	CEC	Emission L/S=200
As	0	0	0	++	++	
Ba	0	0	0	+++	+++	0
Ca	+++	+++	++	+++	+++	+++
Cd	++	0	0	+++	+++	0
Co	++	0	+	0	0	
Cr	0	0	0	+++	+++	
Cu	0	0	0	0	0	
Hg	0	0	0	0	0	
K	0	0	0	+++	+++	0
Mg	+++	+++	+++	+++	+++	+++
Na	++	0	++	+++	+++	+++
Ni	+	0	++	+++	+++	0
Pb	+++	-	0	+++	++	
V	0	0	0	+++	+++	
Zn	0	0	0	+++	+++	0

0 not significant
+,- significance 0.05<p<0.1
++,-- significance 0.01<p<0.05
+++,--- significance p<0.01

The elemental content of Ca, Cd, Co, Pb, Mg and Na showed a statistically significant correlation ($p < 0.05$) with the emission measured by means of the column test (L/S=10). With the exception of Co, Cu and Hg, a significant relation was found between the emission (L/S=10) on one hand and the organic content and cation exchange capacity (CEC) on the other hand. In general there was no significant correlation ($p < 0.05$) found between emission and the clay content or pH. The emission of Ca, Mg and Na measured by means of the column test correlated well with emission measured by means of the availability test.

REFERENCES

1. Aalbers, Th.G., Keijzer J. and Gerritsen R., Juli 1990, Uitloog gedrag van primaire en secundaire grondstoffen. Mammoet deelrapport 7. RIVM/TNO rap.nr. 738504010.
2. Edelman, Th. and M. de Bruin, 1966. Background values of 32 elements in Dutch topsoils, determined with non-destructive neutron activation analysis. In: J.W. Assink and W.J. Van den Brink (Eds), Contaminated Soil. Nijhoff, Dordrecht, pp 89-99.
3. NVN 2508. Standard leaching test for combustion residues, Netherlands Normalization Institute (NNI), Delft, 1988.
4. Draft NVN 5432. Determination of the maximum leachable quantity and the emission of potentially hazardous components from construction materials, monolithic waste and stabilized waste products of mainly inorganic character, NNI, Delft, 1989

Section 3: Technology for the Re-use of Waste Materials

Waste Materials in Construction.
J.J.J.R. Goumans, H.A. van der Sloot and Th.G. Aalbers (Editors)

CONTRIBUTION OF POWDER COAL FLY ASH TO CONCRETE PROPERTIES

J. BIJEN[1&2], R. VAN SELST[2] and A.L.A. FRAAY[1]

[1]Delft Technical University, P.O. Box 5084, 2600 AA Delft (The Netherlands)

[2]Intron, P.O. Box 5187, 6130 PD Sittard (The Netherlands)

SUMMARY

The mechanisms through which powder coal fly ash contributes to cementitious binder systems are discussed. Special attention is given to the so-called pozzolanity of fly ash and the ability to densify the transition zone between the cementitious matrix and aggregate grains.
These various phenomena can lead to an improvement of the features of concrete of which part of the portland cement has been replaced in comparison with reference composites without cement replacement. The nature of these phenomena is discussed. It is shown that the fly ash performance in concrete not only depends on the features of the fly ash but also on other parameters, notably cement type and composition, water cement ratio and temperature.
A model for fly ash reaction in concrete is presented, explaining the various effects of fly ash on concrete properties.

1. INTRODUCTION

The most important application of powder coal fly ash in the world is its use in composite cements and as an addition to concrete. In both applications fly ash contributes to the development of concrete properties and can be regarded as a part of the binder, together with cement.
A number of phenomena appear to be involved. Those are:
* Fly ash acts as a nucleation site for lime. This accelerates the cement hydration.
* Fly ash reacts with lime generated by the cement hydration, giving cementitious products. This contributes to the so-called pozzolanity of fly ash.
* Fly ash improves the particle packing. Consequently it decreases the water required for a certain workability and it densifies the transition zone between cement matrix and aggregates and it reduces segregation.
* Fly ash with its round shaped particles has a ball bearing effect for the irregular cement particles and decreases water demand.

With respect to the contribution to strength development of these mechanisms two main effects can be distinguished. Fly ash has a filler effect and is a pozzolan. The former means that inert particles having the size and shape of fly ash contribute to strength. Pozzolanity is the ability to react with lime and water to cementitious compounds like calciumsilicate hydrate gel. It is known that this pozzolanic activity is only noticeable under ambient conditions after a few weeks of hardening. During this dormant period the fly ash is more or less inert.
Due to this 'delayed' reactivity of fly ash the cementitious structure of concrete of which part of the cement is replaced in these first weeks is relatively weak. The young concrete is more permeable and consequently more vulnerable to drying out and carbonation. When the latter processes are prevented concrete will in general become more dense after the pozzolanic reaction has developed.

The paper concentrates on the pozzolanity and also pays attention to the effects of fly ash on the interfacial zone between the cementitious matrix and aggregate grains.
Most of the work presented has been financed by the Dutch Organization for Energy and Environment in the framework of the (Dutch) National Coal Research Programme and has been coordinated by the Centre for Civil Engineering Research and Regulations, CUR.

For details about test procedures, see references (1-6).

2. SOLUBILITY OF FLY ASH GLASS

Fly ash of bituminous coal consists for the main part of glass. In most standards a minimum glass content is required, because this glass is the reactive part of fly ash. Silicon oxides and aluminumoxides are the major constituents of this glass.

Before formation of calciumsilicathydrates and calciumaluminatehydrates from fly ash can occur, the fly ash glass has to solve in the pore water first. The solubility of fly ash strongly depends on the alkalinity of the solution. Figures 1 and 2(2) show test results of the solubility of a fly ash in a solution of sodiumhydroxide (NaOH) with or without lime (CaOH) saturation. The composition of the fly ash is given in table 1. The concentrations in solution were determined after one week at respectively 2, 20 and 40°C.
Optional experiments at 65°C have been done.

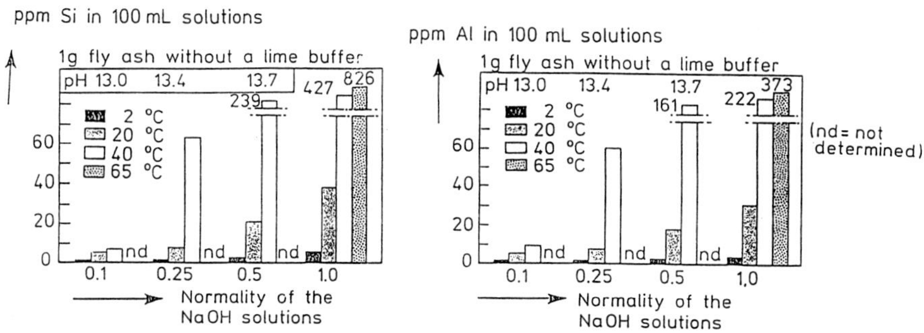

Fig. 1. Si and Al concentration in filtrates of samples after one week of immersion in sodiumhydroxide (NaOH) solutions of various normalities (2).

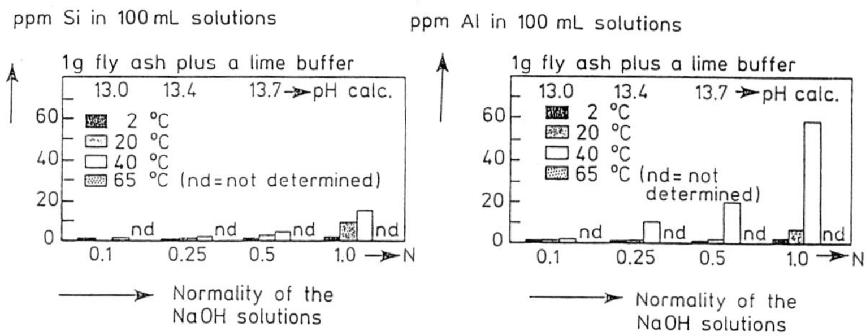

Fig. 2. Si and Al concentration in filtrates of samples after one week of immersion in sodiumhydroxide (NaOH) solutions of various normalities in the presence of lime (2).

TABLE 1

Components, specific surface and density of fly ash, portland cement and portland blast furnace slag cement

		fly ash	portland cement	portland blast furnace slag cement
Na_2O	% m/m	0.6	0.2	0.2
K_2O	% m/m	2.4	0.6	0.6
SO_3	% m/m	0.4	3.4	2.9
CaO	% m/m	2.9	63.0	49.0
SiO_2	% m/m	54.6	20.4	29.2
Al_2O_3	% m/m	27.0	5.5	7.5
Fe_2O_3	% m/m	7.6	3.3	1.7
L.O.I.	% m/m	2.1	0.6	0.3
HCL-solubles	% m/m	18.2	97.3	98.7
Blaine	m^2/kg	220	310	370
Density	kg/m^3	2190	3100	2950

At 20°C the pH required to trigger off significant solution of the fly ash appears to be at least 13. With lime buffer this 'threshold value' appears to be even higher: about 13.2. This may be explained from the precipitation of calciumsilicate hydrates and calciumaluminate hydrate in the vicinity of the fly ash directly after solving due to the very low solubility of there hydrates. In that case solving of the fly ash is not measured by determination of silicon and aluminum concentrations in the liquid phase.

3. COMPOSITION OF PORE WATER IN PORTLAND CEMENT PASTE

In view of the above results the composition of the pore water in cement paste must be of decisive importance for the solution of the fly ash glass and consequently for the pozzolanity. The pore water is the medium in which fly ash glass has to solve and through which cementitious compounds are formed. The composition of the pore water of cement paste has been analysed at various ages. The pore water was squeezed out of the paste by using a high-pressure device. The pore water was analysed for sodium-, potassium-, calcium-, sulfate-, hydroxyl-, silicate- and aluminate ions.

The chemical composition, specific surface and density of the ordinary portland cement and fly ash investigated are also given in table 1(1). In the case of cement replacement 20% was substituted for fly ash. The water/cement plus fly ash ratio (w/c+f) was 0.45. For comparison also pastes were made with quartz flour (<250 μm) as cement replacement instead of fly ash.
The experiments are carried out at 20°C and optionally at 2 and 40°C.

Figure 3 shows the results for 20°C. After mixing the pore water consists approximately of a saturated lime solution in the presence of sulphate. After about one week the hydroxyl ion concentration increases sharply together with the potassium- and sodium concentration. At the same time the lime and sulphate ion concentration decreases to a very low level in the end.

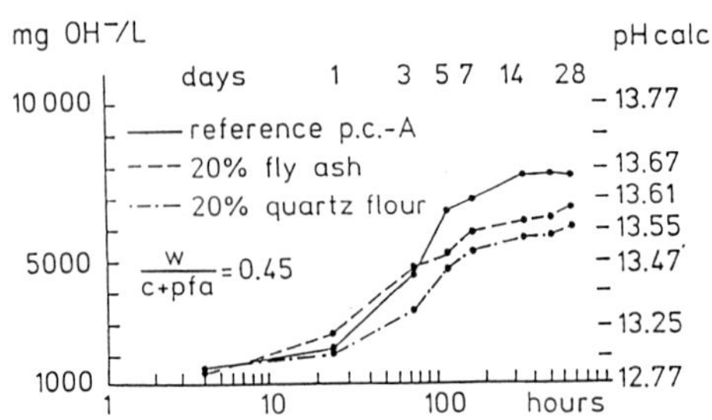

Fig. 3a) The development of the OH^- concentration in the pore water of portland cement paste with fly ash and quartz flour. Temperature 20°C, w/c+f=0.45.

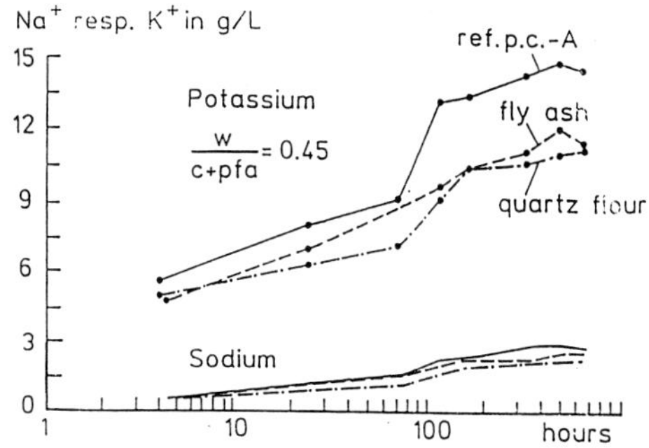

b) The development of the Na^+ + K^+ concentration (2) in the pore water of portland cement paste with fly ash and quartz flour. Temperature 20°C, w/c+f=0.45.

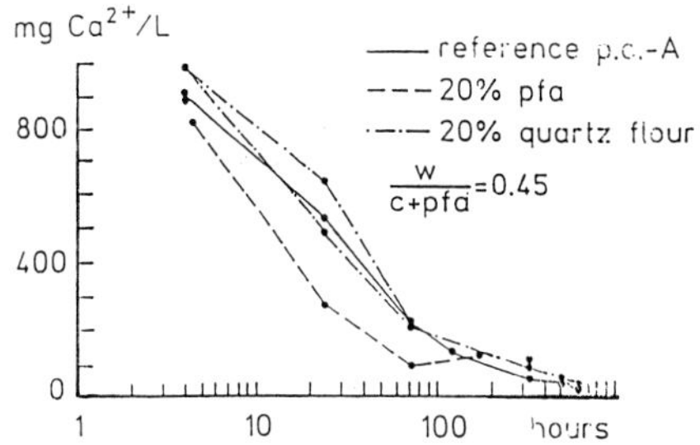

c) The development of the Ca^{2+} concentration (2) in the pore water of portland cement paste with fly ash and quartz flour. Temperature 20°C, w/c+f=0.45.

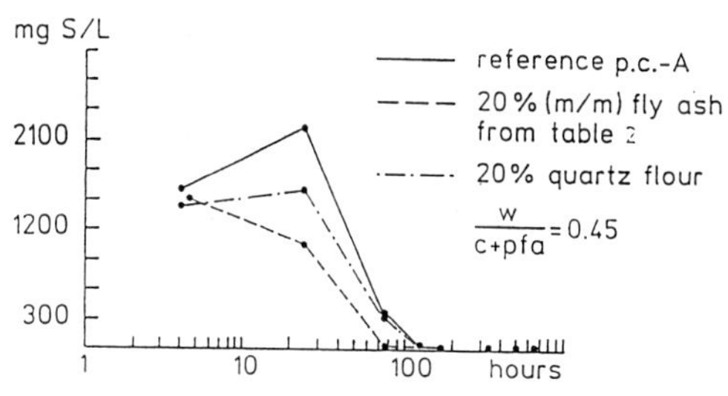

d) The development of the sulphate as S (2) in the pore water of portland cement paste with fly ash and quartz flour. Temperature 20°C, w/c+f=0.45.

In the case of cement replacement either by fly ash or quartz flour the development of alkalinity of the pore water is lower. This is due to the higher water/cement ratio (w/c). For the fly ash some contribution to the alkalinity of the pore water is observed. This is likely to be due to the presence of some sodium and potassium salts at the surface of the fly ash particles, notably sulphates. After some time the difference in alkalinity between the cement paste with inert quartz flour and the fly ash paste decreases again. In general the solution of fly ash decreases the hydroxyl ion concentration due to the reaction

$$SiO_2 + 2H_2O \leftrightarrows Si(OH)_4$$

$$Si(OH)_4 + 2Ca^{2+} + 4OH^- \leftrightarrows 2CaO \cdot SiO_2 \cdot 2H_2O + aq$$

and due to the incorporation of sodium- and to a lesser extent potassium-ions in the CSH-gel*.

The calcium and sulphate ion concentration during the first week of hardening appears to be lower for the fly ash pastes than for the cement and quartz flour pastes. This is likely to be bound to the ability of fly ash to act as a nucleus for lime precipitation.

4. CEMENT TYPE

Besides ordinary portland cement also pastes with portland blast furnace slag cement (about 70% m/m of ground granulated slag) and with a rapid hardening portland cement have been investigated. The composition of the slag cement is given in table 1.

In figure 4(2) the development of the hydroxyl ion concentration in the pore water is shown for the portland blast furnace slag cement.

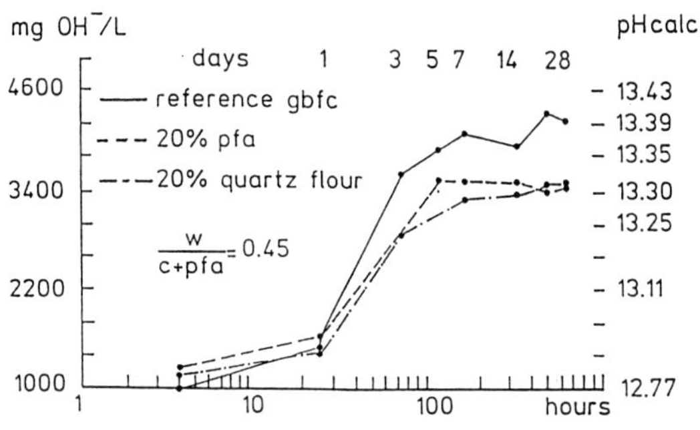

Fig. 4. The OH^- concentration development in paste of portland blast furnace slag cement (gbfc) with and without 20% m/m replacement by fly ash or quartz flour at 20°C and w/(c+f) =0.45 (2).

The alkalinity development is substantially lower than for the ordinary portland cement and hardly exceeds the level of pH 13.2. Therefore a lower pozzolanic behaviour of fly ash if combined with this type of cement in comparison with portland cement can be expected. Another cause for low pozzolanic activity of fly ash with this cement is the reduced availability of free lime.

* CSH is abbreviation of calcium silicate hydrate gel.

The rapid-hardening portland cement applied has the same chemical and mineral composition as the ordinary portland cement but has been ground finer. Figure 5(2) shows that the alkalinity of this cement develops much faster than that of the ordinary portland cement. The pozzolanity of the fly ash will therefore start earlier.

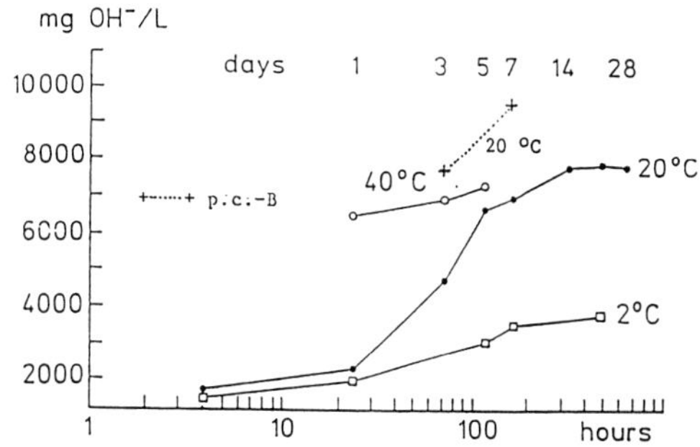

Fig. 5. The OH^- concentration development in paste of ordinary portland slag cement for 2, 20 and 40°C and rapid-hardening portland cement (pcB) at 20°C and w/c=0.45 (2).

The influence of the cement type on the development of the pozzolanity is reflected in the contribution of the fly ash to strength development of the fly ash. In figure 6 (3) the compressive strength based cement equivalence factor (k-value) is shown as a function of time for the three types of cement discussed.
For the portland blast furnace slag cement the k-value of the fly ash decreases in time, indicating that the fly ash acts mainly as an inert filler. The k-values of both ordinary and rapid-hardening portland cement increase in time, proving the pozzolanic behaviour of the fly ash.

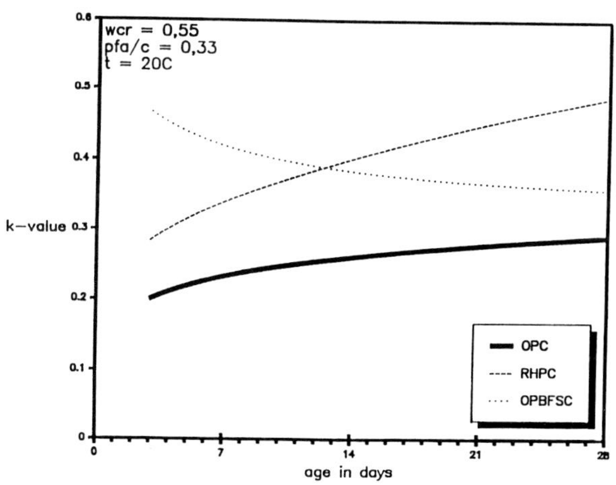

Fig. 6. Development of compressive strength based cement equivalence factor of fly ash (k-value) in time for respectively ordinary portland cement, rapid-hardening portland cement and portland blast furnace slag cement at 20°C and w/c 0.55 and 25% cement replacement (3).

5. WATER/CEMENT RATIO AND TEMPERATURE

A higher water/cement ratio means more dilution of solving cement constituents. This is reflected in the development of alkalinity for pastes with various water/cement ratios as shown in figure 7(1).

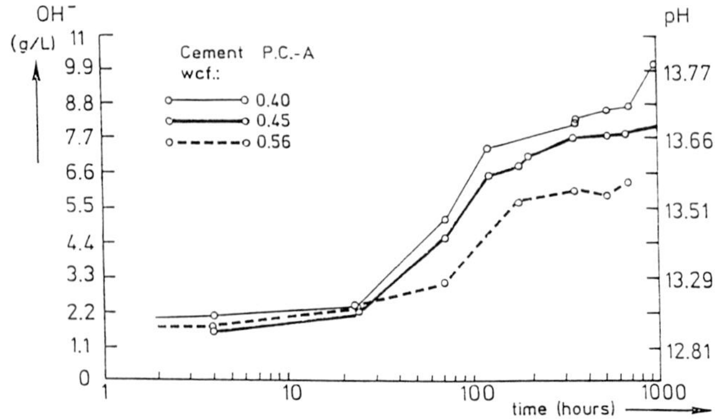

Fig. 7. Influence of water/cement ratio on the alkalinity development of ordinary portland cement with w/c=0.45 (2).

Consequently it can be expected that with increasing water/cement ratio the contribution of fly ash to strength development will decrease. Figure 8(3) shows the k-value for the fly ash after 28 days as a function of the water/cement ratio. It appears that for cement fly ash compositions with increasing pozzolanic behaviour of the fly ash the dependency on the water/cement ratio increases.

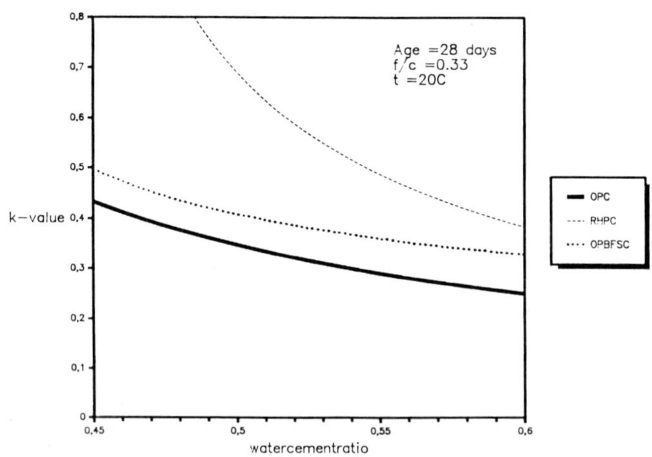

Fig. 8. 28 Days compressive strength based cement equivalence factor for fly ash (k-value) as a function of the water/cement ratio for respectively ordinary portland cement, rapid hardening portland cement and portland blast furnace slag cement at 20°C.

In figure 5(2) the effects of temperature on the alkalinity development is shown. The alkalinity at 2°C is substantially lower than at 20°C. At 40°C a high alkalinity is already present within one day. At increasing temperature not only the alkalinity development of the pore water will increase but also the activity of fly ash.
This means e.g. that in summer the activity of fly ash will be higher than in winter, that rapid-hardening cements will show a higher activity also because of higher temperature development, etc.

6. REACTION PRODUCTS FORMATION

When the silicate and aluminate ions solve they will react with lime to calcium hydrates. If the alkalinity develops the first solving silicates and aluminates will react and precipitate in the near vicinity of the particles because they meet a still highly lime concentrated solution. Later when the alkalinity increases further and the calcium-ion concentration diminishes precipitation of hydrates further away from the fly particles is possible.
This model corresponds to the model for the reaction of blast furnace slag with portland clinker proposed by Bakker [see fig. 9(6)].

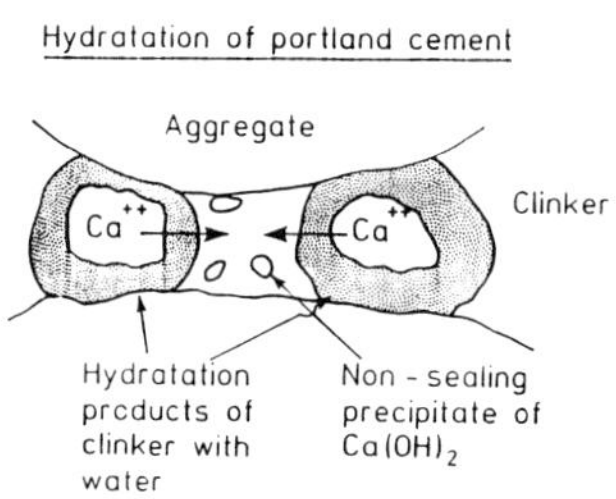

a) cement particle

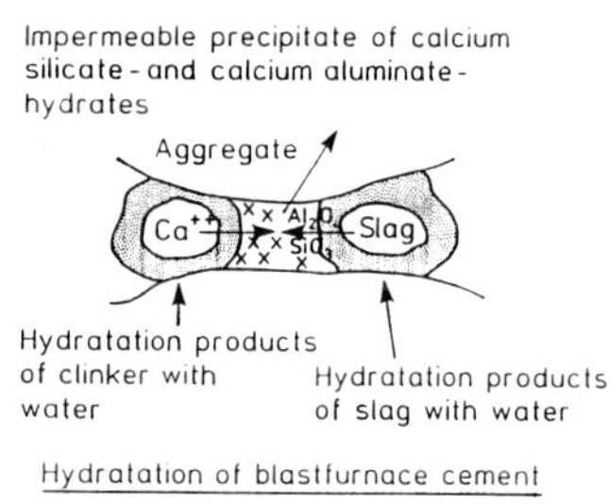

b) cement and slag particle

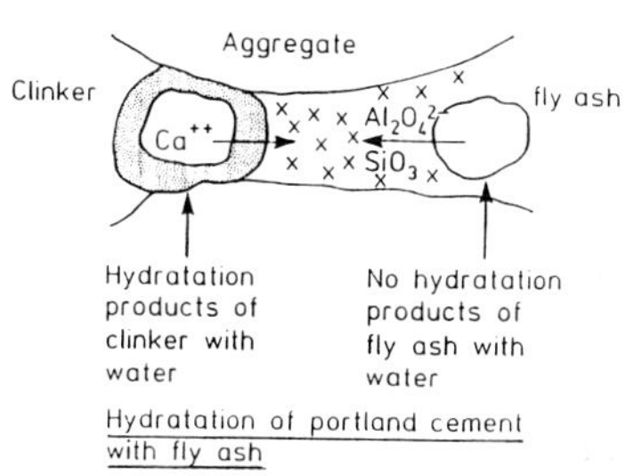

c) cement and fly ash particle.

Fig. 9. Hydration model according to Bakker (6)

Bakker explains the lower permeability of portland blast furnace slag cement in comparison with portland cement by the location where hydration products precipitate. For portland cement precipitation occurs close to cement particles. The hydrating cement particles grow to each other, leaving capillary pores in the larger spaces between them (see figure 10).
In the case of the presence of blast furnace slag or fly ash the precipitation will also occur in the open space, thus rendering the capillary pores less coarse.

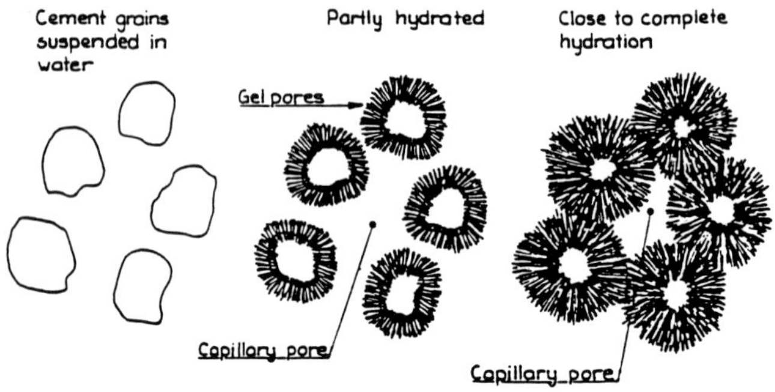

Fig. 10. Model of growing portland cement particles.

The reaction with free lime decreases the free lime content in cement paste as shown in figure 11(1). From this figure it is also clear that there is no significant lime consumption within one week.

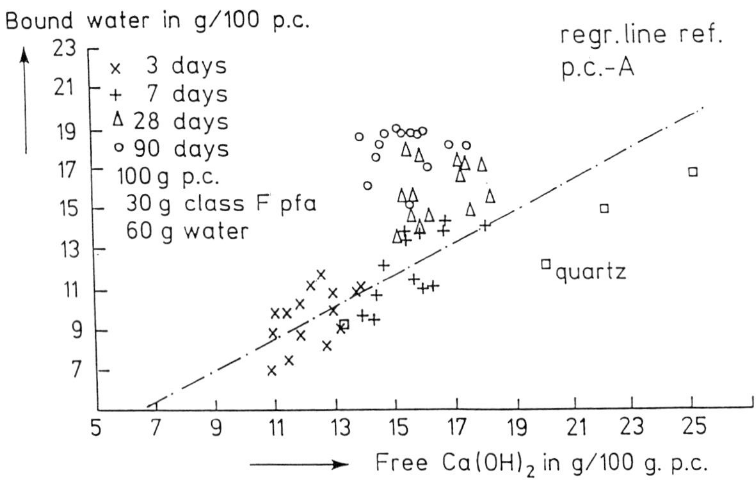

Fig. 11. Bound water and free lime in cement stone (1).

The formation of reaction particles is noticeable from the development of the pore structure. Figure 12 shows the pore size distribution for pastes measured with mercury intrusion porosimetry.

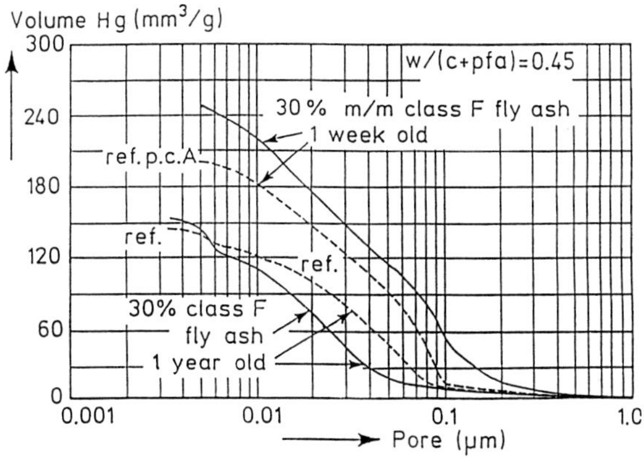

Fig. 12. Intruded volume of mercury as a function of pore size in portland cement paste with and without fly ash after respectively a week and a year hardening (1).

The pozzolanic reaction renders the pore structure finer in time. However, after one week both the porosity is larger and the pore structure more coarse when cement is replaced by fly ash. The effects of fly ash on the pore structure are reflected in the electrical (ohmic) resistance of concrete as shown in figure 13(3). The ohmic resistance starts to increase significantly with portland cement and fly ash after about two weeks. No substantial effect is observed for fly ash in combination with portland blast furnace slag cement.

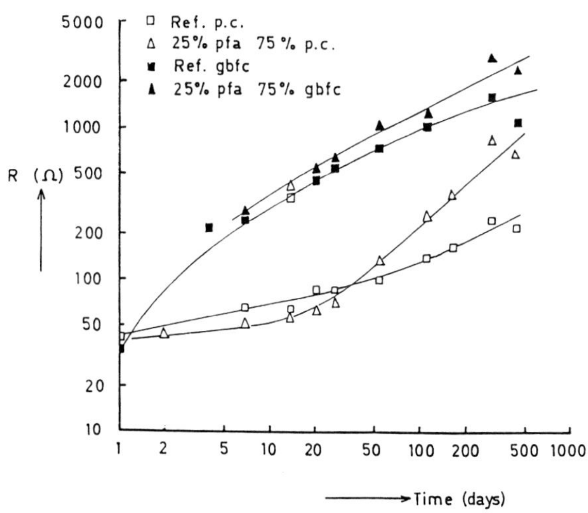

Fig. 13. Development of ohmic resistance of concrete with respectively ordinary portland cement and portland blast furnace slag cement with and without fly ash.

7. TRANSITION ZONE

It is known that fine fillers and pozzolans influence the transition zone between the cementitious matrix and the aggregate grain. For concrete made with ordinary portland cement and river gravel this zone extends from 30 to 70 μm. The zone is porous and the solids consist mainly of calciumhydroxide, which has no bonding capabilities. The volume percentage relative to the total volume of the cementitious matrix can amount up to 50%. It is the weak link in concrete. Investigations (4) have shown that fly ash substantially decreases the thickness of this interfacial zone. Figure 14 shows the so-called lime orientation index after 7 days of hardening for ordinary portland cement and for samples in which 20% has been replaced by respectively quartz flour, fly ash (LM) and fly ash originating from a slag tap boiler (EFA). In figure 15 the orientation index after 28 days is shown.

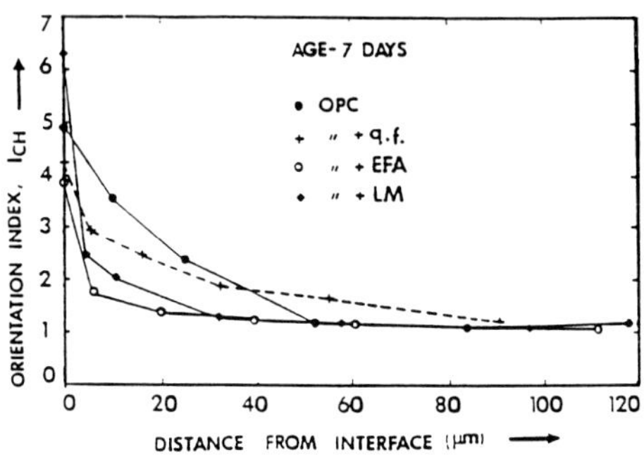

Fig. 14. Orientation index of lime crystals at the transition zone of paste and aggregate grain after 7 days (4)
OPC = ordinary portland cement q.f. = quartz flour
EFA = fly ash of wet bottom boiler
LM = fly ash of dry bottom boiler

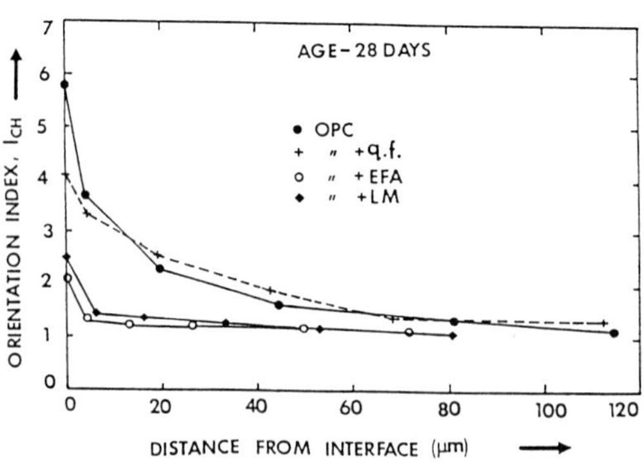

Fig. 15. Orientation index of lime crystals at the transition zone of paste and aggregate grain after 28 days (4)
OPC = ordinary portland cement q.f. = quartz flour
EFA = fly ash of wet bottom boiler
LM = fly ash of dry bottom boiler

The quartz flour has a small effect on the orientation index, which does noet change very much in time. The fly ashes reduce the thickness of the interfacial zone, where orientation occurs more substantially, which effect increases in time. The improvements in the interfacial zone are likely to be due to the improved particle packing. The fly ash particles are smaller than the portland cement particles and can fill up the interstices between them. Therefore at the aggregate grain surface more solid material will be present than without fly ash.

8. CONTROL OF FLY ASH REACTION

The above results prove the importance of the alkalinity of the pore water for the contribution of fly ash to the development of various properties of concrete. This knowledge can be used to control the pozzolanity of fly ashes in concrete to a certain extent. Cements which develop high alkalinities and/or will develop their maximum alkalinity fast will show a favourable fly ash contribution.

For systems with an alkalinity development too low for a substantial fly ash activity measures can be taken to increase the alkalinity artificially. This principle has been used in an investigation into fly ash cement stabilization for road foundation. These stabilizations consist of more than 90% m/m fly ash and up to 10% cement. It has been found that in many cases the pH does not exceed the level of 13. The alkalinity developed depends on the fly ash composition in this high volume fly ash application. By increasing the pH by adding sodiumhydroxide strength gains up to 300% were achieved. Figure 16(5) shows the strength development for a composition with 6% m/m ordinary portland cement and 94% m/m of a fly ash with and without sodiumhydroxide addition at 20°C.

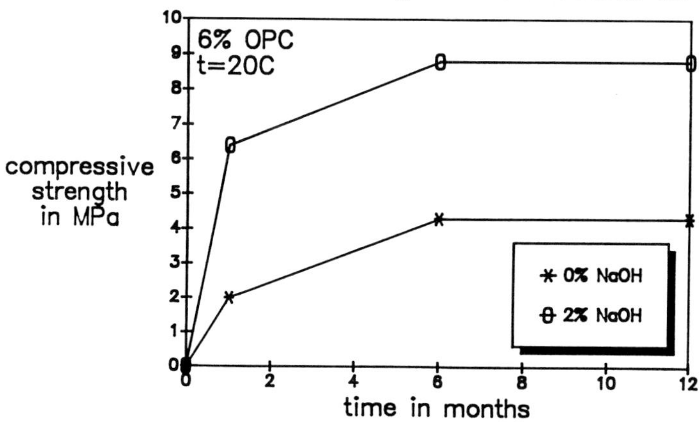

Fig. 16. Compressive strength development of fly ash cement stabilization with 94% m/m fly ash and 6% m/m ordinary portland cement with and without 2% NaOH addition as a function of time at 20°C.

Another measure to control fly ash pozzolanity is to adjust the temperature. However, for many applications this is not practicable.

9. CONCLUSIONS

The fly ash pozzolanic activity strongly depends on the development of the alkalinity of the surrounding pore water. It has been shown that for ordinary portland cement at ambient temperatures the alkalinity is only high enough to activate fly ash after a period of some weeks. This explains the dormant period observed in the contribution of fly ash to concrete properties.

Cements with a relatively low pH development will show less pozzolanic behaviour than cements with a high pH. The dormant period can be decreased in time if rapid-hardening cements are used. Also an increase in temperature decreases the dormant period. Not only because the alkalinity develops faster but also because fly ash itself is more active.

An increase in water/cement ratio decreases the development of the alkalinity of the pore water and consequently the contribution of fly ash to properties of concrete, such as strength.

The use of fly ash substantially reduces the thickness of the interfacial zone and therefore substantially improves this weakest link in concrete.

It is shown that for systems of which the alkalinity is too low for a substantial pozzolanic activity an artificial increase in alkalinity by adding sodiumhydroxide can bring about large improvements.

REFERENCES

1 A.L.A. Fraay, Fly ash as a pozzolan in concrete, PHD-thesis TU Delft (1990).
2 A.L.A. Fraay, J. Bijen and Y.M. de Haan, The reactons of fly ash in concrete, a critical examination, Cement and Concrete Research, Vol. 14 (1989), 235-246.
3 J. Bijen and R. v. Selst, Fly ash as an addition in concrete (Compilation of Dutch research on fly ash in concrete) to be published (in English) by CUR in 1991.
4 J.A. Larbi and J. Bijen, Evolution of lime and microstructural development of the paste-aggregate interfacial zone in fly ash portland cement systems, in press, Materials Research Society, Boston, U.S.A. (1990).
5 Fly ash cement stabilizations, Intron-report to be published (1991) on behalf of Novem.
6 R.F.M. Bakker, Permeability of blended cement concretes, paper SP 79-30 (1983), 586-605, ACI SP 79.

Waste Materials in Construction.
J.J.J.R. Goumans, H.A. van der Sloot and Th.G. Aalbers (Editors)

EFFECTIVENESS OF FLY ASH PROCESSING METHODS IN IMPROVING CONCRETE QUALITY

R. HÄRDTL

Institute for Building Research (ibac),
Aachen University of Technology,
Schinkelstr. 3, D-5100 Aachen (Germany)

SUMMARY

Fly ash is used as a component of blended cements or as addition in concrete production for many years. Besides other properties the fineness of fly ash has a large influence on the quality of concrete. The effectiveness in improving mortar properties of two processing methods (air classification and grinding), which increase the fineness of fly ash, is compared. For the combination of cement and fly ash investigated in this paper, air separation has a more favourable effect on workability, whereas grinding shows a higher effectiveness in early strength and pozzolanic reaction.

1. INTRODUCTION

The utilization of fly ash as component of blended cements or as addition in concrete production is well established worldwide for many years. The use of this by-product in concrete avoids the disposal of large amounts of fly ash and therefore contributes actively to reduce enviremental pollution. A second positive effect is the energy saving as the result of partial replacement of energy-intensive produced cement by certain amounts of fly ash. In future intensified efforts have to be made to reach a further reduction of waste disposal and the interrelated environmental implications.

The general basis for the use of fly ash in concrete is, that the quality of concrete, especially properties like workability of fresh concrete, strength development, permeability or durability, will not be influenced negatively. Preferably the use of fly ash should influence concrete properties positively. Therefore a special quality of fly ash is necessary. Results of previous research indicate that, besides the chemical and mineralogical composition, the granulometric parameters like fineness, particle size distribution or particle morphology of fly ash characterize their quality and effectiveness in concrete (1)(2).

This paper reports on results of investigations to compare the effectiveness of two processing procedures with respect to the influence on mortar properties. The influence of processed fly ash fractions on workability of fresh mortar, compressive strength development and porosity will be discussed.

2. FLY ASH PROCESSING METHODS

To improve fly ash properties, particularly the granulometry, various processing methods have been developed (3). The most frequently-used procedures in practice are classification and crushing by grinding.

Classification methods are sieving, flotation or air separation. Sieving and flotation are only used for specific applications, i. e. the separation of very coarse particles and the production of hollow fly ash spheres (cenosheres), respectively. The commonly used method to divide fly ash in different fractions is the classification by air separation. The advantage of this procedure is the flexibility of cut size and the high efficiency (4). Coarse fly ash fractions frequently containing higher amounts of less-reactive and unregular shaped particles are separated. In the most cases these residues have to be dumped, unless another utilization is practicable.

By grinding the chemical-mineralogical composition of fly ash remains unchanged. Depending on the grinding system and the grinding time particularly hollow spheres are crushed whereas a large proportion of fly ash particles seems to stay intact (5). In fly ash cement production grinding and air classification are often combined in one processing system.

3. EXPERIMENTAL

In a test program a fly ash possessing a mark of conformity as concrete additive according to the German concrete standard DIN 1045 received from a dry bottom furnace was used. It was processed applying two different processing methods:

- Classification of the fly ash with a laboratory air separator. Fractions containing particles < 40 µm, < 20 µm, and < 10 µm were produced. The remaining fraction > 10 µm also was used in the tests.

- Grinding the as received ash using a laboratory mill.

The granulometric parameters of the processed fly ash fractions are summerized in Table 1. The resulting particle size distributions of the ashes are shown in Fig. 1.

TABLE 1

Granulometric parameters of the processed fly ash fractions

Fraction	Density	Specific Surface	Particle fraction < 40 µm	< 20 µm	< 10 µm	Median d_{50}
	g/cm³	cm²/g	M.-%			µm
> 10 µm	2.24	1660	68.6	34.6	7.1	27.1
as received	2.32	3090	82.1	60.2	36.4	14.7
< 40 µm	2.38	3280	98.4	80.8	47.8	10.6
< 20 µm	2.40	4110	100	91.5	55.8	9.2
< 10 µm	2.42	5370	100	100	85.8	6.0
ground	2.67	6130	94.1	76.7	55.7	8.3

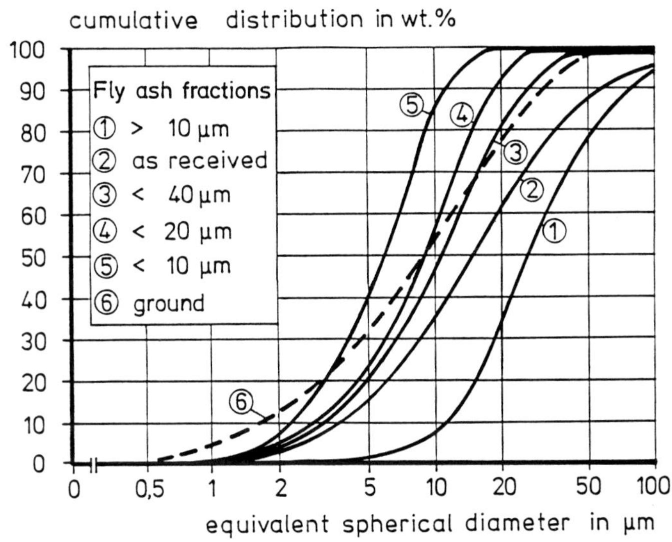

Fig. 1. Particle size distribution of the processed fly ash fractions (SEDIGRAPH)

Only small amounts of fly ash could be processed using the laboratory equipment. Therefore the tests were performed with mortars. Equal volumes of fly ash were substituted for corresponding quantities of a portland cement class 35 (volume fly ash/volume cement = 0.4). The volume ratio water/binder was kept constant. Aggregate and mixing procedure corresponded to the German cement standard DIN 1164, Part 7. The specimens were stored in water until testing. More experimental details are published in (5).

4. TEST RESULTS

4.1 Workability

As a parameter for the mortar workability the spread at flow table was tested. Fig. 2 illustrates the influence of the different fly ash fractions on the relative mortar spread, that means the spread of the cement+fly ash combination related to the spread of the reference mix (pure cement).

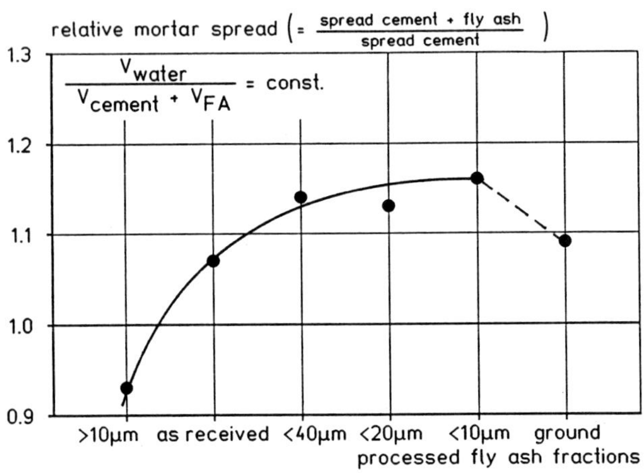

Fig. 2. Influence of the processed fly ash fractions on the workability expressed by the relative mortar spread

The "as received" fly ash shows an improvement of the spread of arround 8 % compared to the spread of the mortar without fly ash. With increasing fineness as result of a higher separation grade there is an additional improvement of workability. One general reason for this is that the total particle size distribution of the cement + fly ash mix becomes broader. The increased percentage of fine particles releases water from the interstitial volumes of the bulked particles and thus the water demand decreases.

The increase in spread is markedly expressed for the fraction < 40 µm whereas the additional effect for the finer fractions

< 20 µm and < 10 µm formed out to be negligible. This can be explained by the fact that the coarsest fly ash fractions contain higher amounts of particles with an unregular, partly porous particle shape having a higher water demand. The removal of these particles affects the workability of the mortar most positively.

Grinding the fly ash shows only a little change of mortar spread compared to the unprocessed ash, although the ground ash has the highest amount of fine particles. The positive effect of increased fineness is reduced by producing particles with unregular formed, cubic or splintered particle shape.

4.2 Compressive strength

Fig. 3 shows the influence of the different fly ash fractions on relative compressive strength (strength of cement + fly ash mortar related to strength of cement mortar) for different hydration ages. At all ages there is an increase in relative strength with increasing fly ash fineness. Up to 28 days the ground fly ash with the highest amount of very fine particles (see Fig. 1) shows the highest strength contribution. In this first phase of hardening no notable strength contribution is to be expected from pozzolanic reaction of the fly ash (5). An increase in strength may therefore be ascribed to a physical filler effect in the hardening mortar. Small fly ash particles are filling voids between the coarser cement and aggregate particles which results in a denser mortar structure and a higher early strength. If there are any other influences, i. e. an acceleration of the cement hydration or near-surface reactions of the processed fly ash, especially of the ground ash, cannot be answered at the moment.

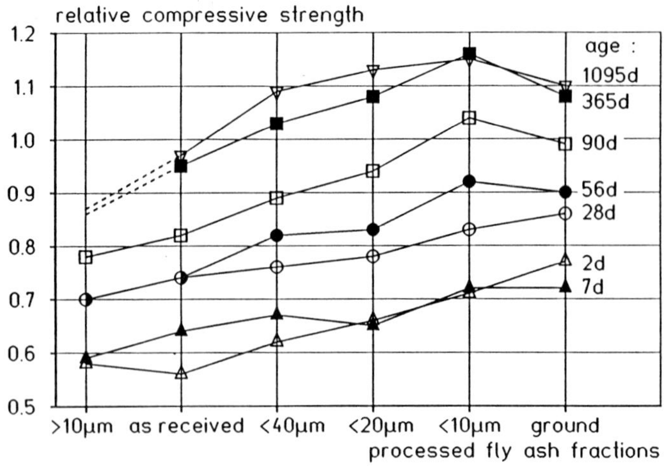

Fig. 3. Influence of the processed fly ash fractions on compressive strength for different hydration ages

At more advanced ages (including 28 days) the pozzolanic reactivity of the fly ash affects strength behaviour. Above 28 days of hydration the classified fraction < 10 µm shows higher strengths than the ground ash. At an age of 1 year the fraction < 20 µm also reaches the strength level of the ground ash. After 3 years even the fraction < 40 µm reaches the strength contribution of the ground ash.

With increasing fineness by excluding coarser particles the fly ash will contain higher proportions of reactive glassy constituents. The ground fly ash, however, has the same unchanged, chemical-mineralogical compostion as the "as received", unprocessed ash and therefore a higher amount of partially crystalline particles compared to all classified fractions investigated. As a result the effect on the pozzolanic reaction caused by the higher fineness of the ground ash with a larger reaction surface is deminished with time by the smaller content of reactive particles. Nevertheless the compressive strength is significantly improved by the ground fraction compared to the unprocessed fly ash.

4.3 Porosity

Most of the durability properties of concrete are connected with transport phenomenons. Therefore the porosity, especially the pore size distribution is one of the most important parameters to characterize the resistance of concrete against possible attacks.

Pore size distribution of the mortars at an age of 3 years has been measured by mercury intrusion porosimetry. Fig. 4 compares the influence of the fly ashes on the median pore radius representing the pore radius which corresponds to a total pore volume of 50 % of the cumulative distribution.

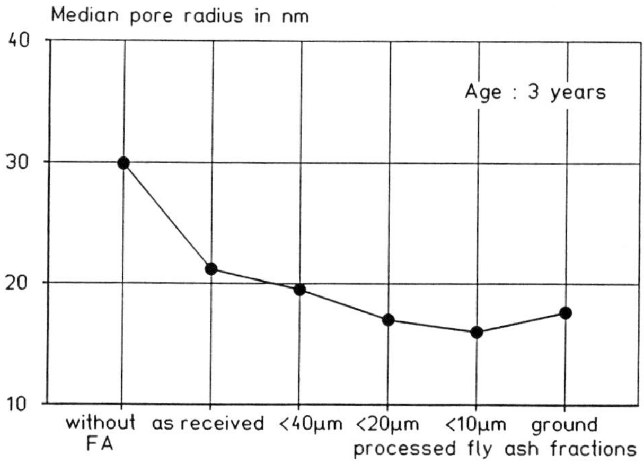

Fig. 4. Influence of the processed fly ash fractions on pore structure represented by median pore radius.

The as received fly ash shows a clearly lower value for median pore radius compared to the mortar without any replacement of cement by fly ash. That means, that the use of fly ash results in a denser pore structure. The processed fractions lead to an additional reduction of median pore radius, which is closely related to the compressive strength results illustrated in Fig. 3. However, the increase in densification by processing the fly ash is lower than the increase caused by replacing a certain volume of portland cement by as received fly ash.

5. CONCLUSIONS

The test results indicate that both investigated processing methods, air classification and grinding, show positive effects on mortar properties, which are transferable to concrete in a similar way.

The use of air classified fly ash improves the workability of cement mortars compared to unprocessd fly ash. The increase is markedly expressed when only the coarsest fraction (< 40 μm) is separated. With higher separation grade the pozzolanic reactivity increases related to a higher specific particle surface and glass content. Nevertheless it must be noted that simultaniously the amount of removed, oversized particles is increased. If no other utilization is practicable, this material must be disposed.

Grinding the fly ash results only in a little change of workability. The effect of higher fineness is reduced by a more unregular particle shape. On the other hand there is a significant increase in early strength and in pozzolanic reaction.

The test results elucidate, that with both processing methods there is of course an improvement of properties. But there seems not to be one single procedure, which generally can be defined as the optimal one. For the combination of cement and fly ash investigated in this paper air separation has a more favourable effect on workability, whereas grinding shows a higher effectiveness in early strength an pozzolanic reaction. Finally the application of any of these processing procedures depends on the commercial feasibility. A combination of the two processes - separation at first and then grinding of the coarser particles - may lead to an optimization.

REFERENCES

1. F. Sybertz, Betonwerk + Fertigteil-Technik, 45 (1988), No. 1, pp. 42-47, No. 2, pp. 80-88.
2. B.P. Hughes and M.N.A. Al-Ani, Magazine of Concrete Research, 41 (1989), No. 147, pp. 99-105.
3. W.B. Butler and M.C. Mearing, MRS Symposia Proceedings: Fly Ash and Coal Conversion By-products; Characterization, Utilization and Disposal II, 65 (1986) pp. 11-17.
4. J.O. Cleemann, Zement-Kalk-Gips International, 39 (1986), No. 6, pp. 295-304.
5. P. Schießl and R. Härdtl, in: V.M. Malhotra (Ed.), Proceedings of the 3rd International Conference on Fly Ash, Silica Fume, Slag and Natural Pozzolans in Concrete, Trondheim 1989, Supplementary Papers, pp. 277-294.

Waste Materials in Construction.
J.J.J.R. Goumans, H.A. van der Sloot and Th.G. Aalbers (Editors)

DEVELOPING A NEW FIELD OF UTILIZATION OF CONCRETE WITH WASTE MATERIALS

PAO YING.
P . Y. New Materials Department, The Institute of Hubei Province Building Materials Research and Design, Zhong Bei Road, Wuchang, P.R. of China. P.O. Box 430071.

SUMMARY

At present the protection of environment has become a severe problem which every country in the world must consider now. But how to treat properly about the solid waste including toxic and low radiation waste embarrasses scientists and Government.
The utilization of concrete must break through the traditional idea that only used in industrial reconstruction. It should take part in the subject of purifying the earth.
The authors of this paper provided a new thought, We can solidify together the solid waste and poisonous solid waste as a strengthened construction materials. It can be used as the base civil engineerings or highway basement.

1. INTRODUCTION

1.1 The fly ash contains radioactive matter. Many researches pointed out, The radioactivity of fly ash from power plant (burning coal) to the environment is not lower the same power of operating nuclear power station. When we use the fly ash as a renewable resource, we must think about its influence on the environment pollution. The radioelement contents of fly ash is 2-5 times higher than that of original coal, because of rich concentration of radioelement in fly ash.
The influence of the radioactivity of fly ash on the agriculture crops. When the fly ash as a kind of soil is used to cultivate, its radioelements are concentrated in roots. The contents order, root-stalk-leaf-seed (or pod). Tests indicate,

1.1.1. The quantity of U and TH absorbed by wheat approximates to that of rice (or rape), it not increases with the amount of fly ash in soil.

1.1.2. The amount of TH absorbed by soybean pod increase multiply with the fly ash increase in soil.

When we think that the waste stock of fly ash can be used to cultivate the agriculture crops (except beans) or vegetables.

1.2 The radioactivity of fly ash products as building materials. When the fly ash products are used as a wall body materials, the Radon content in closed room is increased with 150%—200% compared with that of traditional building materials. If the door and window are opened and keep the circulation of air in room, the toxic gas content is not harm to the people.

Therefore we consider that the fly ash used as feeling materials is better than other.

2. Extra quantity fly ash instead of cement and small sand in concrete and its properties

After 1980 decade, the fly ash content added in concrete increases continuouly from equal same instead of cement to extra quantity.

This is due to use further fly ash instead of part sand.

The results of tests see table 1. and Fig 1.

TABLE 1

The proportioning ratio and strength of concrete

No	fly ash content	sand ratio (%)	plasticizer (%)	Compressive strength Mpa. 7d.	28d.	120d.	180d.
E1	0	37	0.2	17.9	24.8	26.0	28.6
E2	50%	33	0.2	9.1	13.7	23.5	29.2
E3	54%	32	0.2	10.6	18.3	28.2	32.9

Table 1. shows the strength developing rule of extra quantity fly ash concrete in all ages are larger than equal same fly ash concrete. The early strengths of concrete before 28 days.

no addition > extra q. > equal q.

The later strengths after 180 days developing tendency

extra q. > equal q. > no addition.

test used ordinary portland cement.

2.1 The drying shrinkage is the main cause to present craks in the surface layer of concrete, Fig 1. shows that the drying shrinkage rate is decreased with increase of fly ash added, according to the order

extra q. < equal q. < no addition

Notes. E1—E3 in the Fig 1. and table 1. they are same

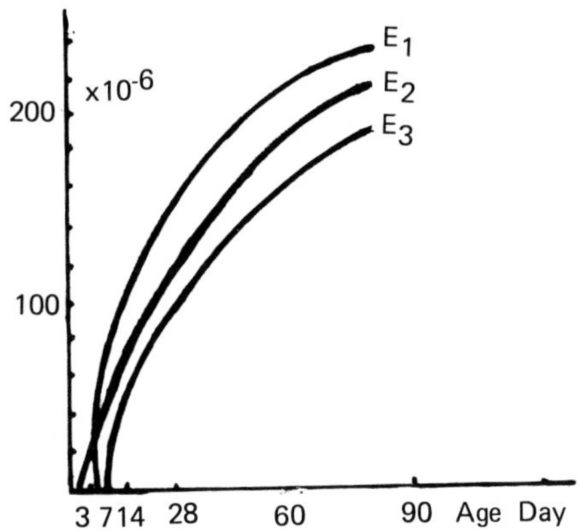

Fig 1. dry shrinkage of concrete and dry shrinkage rate

2.2 The crack resistance The crack of concrete often present in the early period, because when the strength of concrete is very low, but the temperature in the inner part is higher. In this condition the cracks present due to surface shrinkage when especially the mass body are attacked incidentally by the strong wind and cold current.

Data from table 2. indicate that the ultimate elongation value of extra quantity fly ash concrete is the lowest at 7d. but at 120 days is the highest. We can say that in the early period it is disadvantage to crack resistance, but at later age it will be improved gradually.

TABLE 2
The ultimate elongation deformation of extra quantity fly ash instead concrete.

No	axis tensile strength (Mpa)			ultimate elongation ($\times 10^{-4}$ m/m)		
	7d	28d	120d	7d	28d	120d
E1	1.73	2.98	2.60	0.58	0.74	0.80
E2	0.71	1.37	2.68	0.41	0.49	0.77
E3	0.75	1.50	3.0	0.39	0.54	0.84

	modulus of elasticity (Mpa)		
	7d	28d	120d
	32242	34104	35966
	21364	37142	40866
	26068	32928	40964

E1—E3 also as the same as that in TABLE 1.

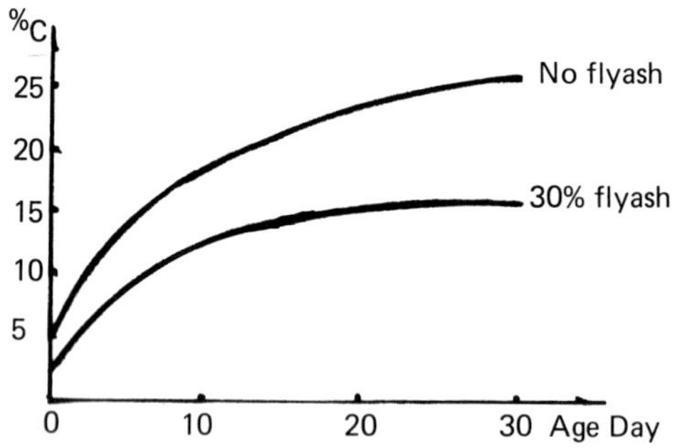

Fig 2. temperature rise curve of concrete

On the otherhand, the absolute temperature rise and temp. difference of the concrete body will decrease obviously. Therefore we also can say that it is advantage to crack resistance. In Fig 2. these two temp. rise curves represent at same condition in different part region. According to the calculation decreasing 6.2°c of the absolute temp. rise of concrete corresponds to ultimate elongation value 0.19×10^{-4} (m/m) in 7 days.

Fig 2. shows that the absolute temp. rise of the fly ash 30% concrete decreases over 6°c than that of no fly ash concrete. If the more fly is added to the concrete, the more absolute temp. rise decreased with extra quantity fly ash concrete. It possesses more advantage .

2.3 Frost resistance There are different points of view on the frost resistance of fly ash concrete. Japanese researchers consider that comparing the properties of the concrete added or not fly ash, their durabilities under freezing and thawing are same, when their strength and sum voids are same. This is due to that the properties of Japanese fly ash is better.

Chinese researchers think that the ignition loss at 1000°c > 4.0% or residue on 80u aperture sieve > 6.0% influences the frost resistance of concrete with fly ash more intensive than the mechanical properties. In general, the fineness of chinese fly ash is coarse due to water collection, the frost resistance of concrete is decreased with the fly ash content increase. Tests are tabulated in table 3.

TABLE 3
The Frost resistance Tests

No	fly ash instead of cement(%)	fly ash instead of sand (%)	plasticizer (%)	voids content (%)
E4	0	0	0.2	2.31
E5	30	4	0.2	2.30
E6	50	4	0.2	2.34

workability slump(cm)	28d Mpa.	times of freezing & thawing resistance 28d.	times of freezing &thawing resistance 120d
4.5	31.6	13	> 140
4.6	25.6	11	110
5.3	22.6	/	6

Notes, sample dimension is 10×10×10 cm, the center of sample freezed to −17±2℃ in center through 100-120 min. temperature drop thawed to

7±2℃ in center through 45-60 min.
The frost resistance of concrete added water reducer can improve resistance distinctly.

2.4 Carbonation Fly ash instead of cement according to ratio 0.10.20.30.40.% prepare 5 kinds of concrete briquettes 10×10×10 (cm), the aggregate and sand are the same. Each of them has two sections of specimen. After curing 28 days, one is placed continueously in the atomizer room, another in the air. In the center of each briquette a stainless steel rod is placed in order to measure the stain erosion depth.
Tests indicate, curing at atomizer room, the strengths of addition fly ash or not concrete grow continueously till 15 years, but the speed of growth of fly ash concrete is larger than that of no addition. In air the increase of strength by adding fly ash concrete is smaller than that of no addition and the rate is lower.
The all specimens, cured in atomizer room are carbonated only at the surface layer and the each steel rod is not stained after 20 years. Therefore we can infer that at water saturated condition and in the inner part of mass concrete, the fly ash contents in concrete is not influenced by carbonation because of air isolation. Contrarily the specimens placed in air the carbonation of the concrete are much influenced by the fly ash contents, such as, after 20 years the depth of carbonation of 20% fly ash concrete is 25(m/m), while 30% fly ash increases to 33(m/m), the more fly ash addition, the more depth of carbonation, a few steel rod also are stained. If the engineering exposed in air the fly ash addition must be controled below 10%.

3. HIGH content fly ash rolled concrete

In China the high content fly ash rolled concrete had applied to the hydraulic engineering and highway base successfully. It can not only shorten the time of construction but also spare the investment 15-30%. The cost of one m^3 concrete is about 50% of the ordinary.

3.1 The cementing materials used one m^3 concrete containing 65% fly ash, 16% slag. Design compressive strength of concrete on 90 days is 14.7(Mpa). The quality of rolled concrete is better since1985 see table4

TABLE 4
The core drill specimen measure results

age day	Comp. strength (Mpa)	tensile strength (Mpa)	shearing resistance in contact suture (Mpa)	impermeability
2	2	/	/	/
7	6.5	0.35	/	/
28	11.5	0.99	/	S>6
90	18	1.96	1.33	S>6
180	23	/	1.76-1.96	/

(S > 6 = 0.419 × 10^{-8} cm/s)

3.2 The cementing materials used one m^3 concrete containing 57% fly ash from electric dry collecter. The rolled concrete engineering quality is better since 1986 year. The properties test of rolled concrete, please see table 5.

TABLE 5

one m^3 concrete fly ash(kg)	plastizer (%)	stiffness sec.	comp. strength (Mpa) 3d	7d	28d	90d	180d	365d
80	0.2	13	4.1	6.4	9.8	20.5	23.5	24.4

Tensile strength (Mpa) 7d	28d	90d	180d	365d	ultimate elongation 28d	90d
0.33	o.67	1.93	2.58	2.77	0.66 × 10^{-4}(m/m)	0.98 × 10^{-4}(m/m)

Static modulus of elasticity (Mpa) 28d	90d	Absolute temp. rise (℃) 28d	final
14710	22653	13.2	14.2

Impermeability (90 days)	Density (kg/m^3)
S > 12 = 0.129 × 10^{-8} (cm/s)	2324

The fly ash content is extra instead and the sand rate is 36.8%.

3.3 The reacting mechanism of high contents flyash in rolled concrete

3.3.1 Water reducing and plasticity reinforcing effect

3.3.2 The improvement of porous structure of concrete

3.3.3 The concrete has got a good dense and Active Network

4. The solid waste can be treated as artificial sand and aggregate in high fly ash content rolled concrete

4.1 According to the above research results, we can develop a new field of utilization of concrete for stabilizing the waste materials in the future. Many developing countries meets a trouble problem the waste materials pollution to the soil and ground water. This method is submitted as a theoretical and practical example to solve the above question. Under wet condition or inner part in large body, the every composition of stabilizer is a important role in stimulating activity each other for a long time, therefore it improves the properties of concrete such as dry shrinkage, crack resistance, porous structure, impermeability and low heat of hydration. On the other than an ordinary base engineerign, it especially appropriates to mass volume concrete under ground and harbour engineering that is prestressed or rolled. This mass concrete can be used as fillings underground, or in seashore, the highway base and soft base for mashland. It is not only a waste stack storage, but also a utilization base.

4.2 If the bonded object is poison waste, the concrete with waste materials can protect soil and ground water, because of it has good impermeability property.The authors of this article considers that the nuclear waste stockpile if not be established under ground concentrical -ly. They may be used the concrete with waste materials to bond nuclear waste at appropriate wasteland or uncultivated island separately E.E.C.decided recently that the poisonous waste must be treated in the that place where it produces. They may use it to treat.

Waste Materials in Construction.
J.J.J.R. Goumans, H.A. van der Sloot and Th.G. Aalbers (Editors)

POWER CONCRETE

R.W.M. Faase , Bruil Wegenbouw B.V., Ede
J.H.J. Manhoudt, BVN Raadgevend Ingenieursburo B.V., Rijswijk
E. Kwint , B.V. Vasim, Nijmegen

Summary

Power concrete is a concrete made of the waste products from energy production. The waste used is fly ash. Lytag is sintered fly ash. Lytag, mixed with cement, sand and water, is Lytag-concrete, in other words, power concrete.

Electricity generation in the Netherlands, provided by five producers, creates a large volume of waste products every year.
One of the waste products, produced particularly by coal-fired power stations, is fly ash. In 1990, approximately 850,000 tonnes of fly ash were produced in the Netherlands.
Stimulated by government policy, the Province of Gelderland began in the 1980s, to look for a solution to the fly ash problem. This solution was found in the form of a factory where fly ash is converted to Lytag, via a sintering process.
Lytag is a lightweight aggregate material, which can be used to replace gravel in concrete. This means that the solution is also helping to limit the volume of minerals (gravel) which must be quarried.
Thanks to the forward-looking attitude of the Province of Gelderland, the product was used for the first time in the construction of a road bridge over the Rhine in Arnhem. This project was carried out entirely in Lytag concrete. The construction and design of the bridge were implemented by the BVN Engineering Consultancy from Rijswijk, the Netherlands.

1. The production of waste materials

It is impossible to imagine today's society without electricity. Electricity is used for countless purposes, and is taken very much for granted.
Electricity generation in the Netherlands is provided by five electricity producers (see Fig. 1).

For electricity generation, you must have fuel. This may be in the form of coal, natural gas, oil or uranium (atomic energy).
The majority of the production costs arise from the fuel costs. As a result of objections to atomic energy, and after the oil crisis of 1973, the Dutch

Rechtsboven: Electriciteits Produktiebedrijven in Nederland.
1 N.V. Energieproduktiebedrijf UNA (Utrecht, Noord-Holland, Amsterdam) (UNA)
2 N.V. Electriciteits-Produktiemaatschappij Oost- en Noord-Nederland (EPON)
3 N.V. Electriciteits-Produktiemaatschappij Zuid-Nederland (EPZ)
4 N.V. Provinciale Zeeuwse Energie-Maatschappij (PZEM)
5 N.V. Electriciteitsbedrijf Zuid-Holland (EZH)

Geheel rechts:
De centrale Maasvlakte met op de voorgrond de hoogspanningsverbindingen

Lijnenspel aan de achterzijde van de centrale Maasvlakte

Figure 1

verdeling brandstoffen pakket openbare centrales

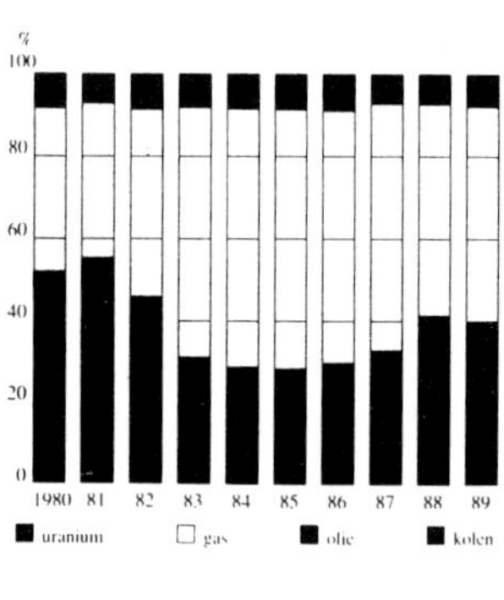

Figure 2

government decided to increase the use of coal-fired electricity power stations. Coal is cheap, in plentiful supply, and with today's environmental technology, can be burnt safely.

Figure 2 shows that approximately 40% of electricity produced is generated through burning coal. The disadvantage of coal, as compared to oil and gas, is the large volume of waste products left after burning.

- The sulphur in the coal is converted, during burning, to sulphur dioxide (SO_2). This environmentally-damaging material is subsequently converted to plaster by means of a flue gas desulphurisation plant. This plaster is an excellent substitute for natural plaster, and is taken up in its entirety by the plaster processing industry.
- The lighter ash particles (fly ash) are carried up with the flue gases. Without certain measures being taken, this would create an unacceptable dust pollution level. Electrostatic filters absorb practically all of the fly ash. Periodically, the filters are "beaten", whereby the fly ash falls to the bottom of the filters. A conveyor system transports the fly ash to silos, from where the fly ash is removed in dry, or slightly damp form, for further use.
- The heavy ash particles are captured in the boilers and removed. This so-called bottom ash is directly usable as a foundation material for road construction, and as a raw material for the production of, for example, building blocks.

2. The policy on waste products

The policy of the Dutch government is aimed at serving the environment as well as possible, and at stimulating the re-use of waste products. This fits in perfectly with another policy area, the reduced use of natural raw materials. If a natural raw material can theoretically be replaced by a waste product, then this theory must be put into practice. The Dutch government is attempting to encourage this type of activity, for example, by increasing the cost of dumping waste products.

Provincial government policy

Obviously, the policy chosen by the national government is also carried out by the various provincial governments in the Netherlands.

In the beginning of the 1980s, with the construction of a new coal-fired production unit in Nijmegen, the Province of Gelderland was faced with a considerable increase in the production of waste products, including "fly ash". An area of application for this fly ash was looked for, and eventually, the Provinciale Gelderse Energie Maatschappij (PGEM) - the Province of Gelderland's energy company - responsible for the production and distribution of electricity in Gelderland, decided in 1981 to turn to the construction and running of a factory for the manufacture of sintered fly ash, the B.V. Vasim in Nijmegen. The factory went into service in 1985. The end product is a lightweight aggregate material, which can be used as a substitute for gravel in concrete. The product is named Lytag, which comes from the English term "Lightweight aggregate". The factory's annual production capacity is between 160-200,000 tonnes.

Figure 3 is a summary of the total production and application of fly ash in the Netherlands. In 1990, fly ash production was at a level of approximately 850,000 tonnes, and it is expected that this figure will further increase over the coming years.

The choice of Lytag represents a double-edged sword for the government.

- It helps to alleviate the waste product problem (fly ash);
- It helps to limit the volume of minerals (gravel) quarried.

The Province of Gelderland chose to implement its policy immediately, and initially decided to construct a number of large concrete structures in Lytag concrete, instead of gravel concrete.
One of the largest projects is the Pley Bridge over the River Rhine. The bridge is 760 metres long, and is the first bridge in the world to be constructed entirely in Lytag (lightweight concrete). The project was completed in 1988. At the beginning of this year, a viaduct was built over the Oude IJssel River in Ulft, also in Lytag concrete, on behalf of the province.

3. Lytag production (Figure 4)
The Lytag process was developed in England, in the 1950s. Partly as a result of a shortage of light aggregate material for the concrete industry, the company Laing began, in 1960, to produce Lytag. Currently, there are 3 factories in England, which collectively produce approximately 600,000 tonnes of Lytag per year.

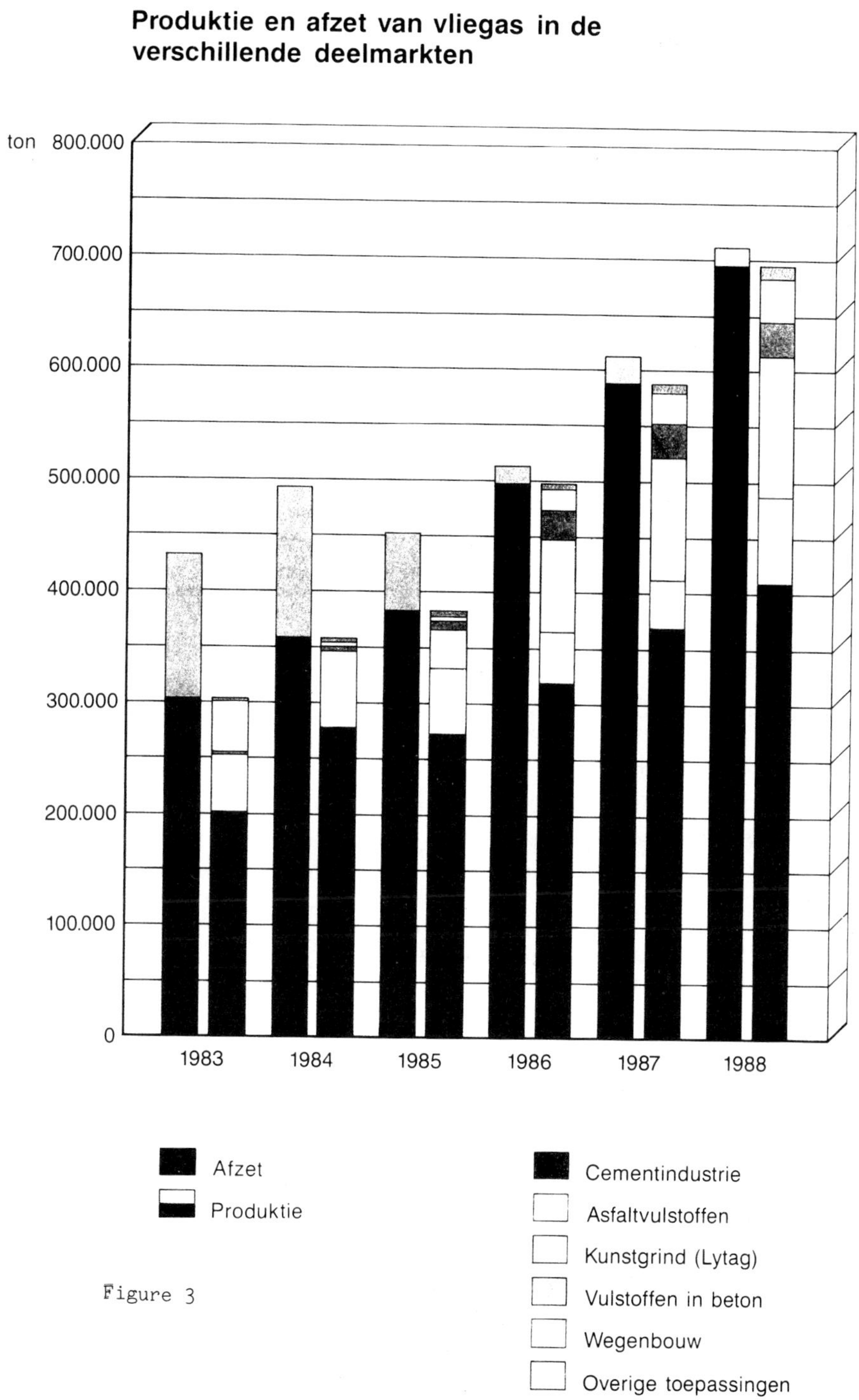
Produktie en afzet van vliegas in de verschillende deelmarkten
ton
800.000
700.000
600.000
500.000
400.000
300.000
200.000
100.000
0
1983
1984
1985
1986
1987
1988
Afzet
Produktie
Cementindustrie
Asfaltvulstoffen
Kunstgrind (Lytag)
Vulstoffen in beton
Wegenbouw
Overige toepassingen

Figure 3

The production process is divided into four steps.

a. Mixing

In order to sufficiently fuel the subsequent sintering process, the fly ash must contain between 3 and 5% carbon. If this level is not reached, powdered coal is added. A small quantity of water ensures that the mixture binds well.

b. Pelletising

The dampened powder coal/fly ash mix is fed into a pelletising machine. This machine is made up of rotating pans, which can be set at an angle. The mixture poured into the pans bonds together in pellet form.

c. Sintering

The "green" pellets pass along a conveyor belt and join the sinter belt via a hopper. The pellets are spread on the belt in a layer 30 cm thick and 2 m wide. The sinter belt moves horizontally and passes under an ignition hood, where the upper side of the bed is ignited. Subsequently, air is drawn from top to bottom through the sinter belt. Gradually the burning zone spreads through the whole pellet bed, until all of the carbon is burned. Sintering takes place at a temperature of approximately 1100 degrees Celsius.

d. Final processing

The last production phase is final processing. Any pellets which have become baked together are separated in the crusher. The screen system divides the pellets into three grades: 0.5-4 mm, 4-8 mm and 6-12 mm. The 4-12 mm grade, often used in the concrete industry, is a combination of the last two sizes. Pellets larger than 12 mm are recrushed, and anything smaller than 0.5 mm is returned to the process.

Lytag pellets

The pellets have a glass-like, sealed surface with a closed pore structure. In other words, the pores are themselves linked together, and have the same glass-like surface as the pellet. Any environmentally damaging characteristics of fly ash are completely negated by sintering.

The dry loose bulk density is approximately 800 kg/m3; pellet density is approximately 1420 kg/m3, and pellet strength is somewhere between 5 and 9 Mpa. Processing and internal quality control at B.V. Vasim are monitored by the

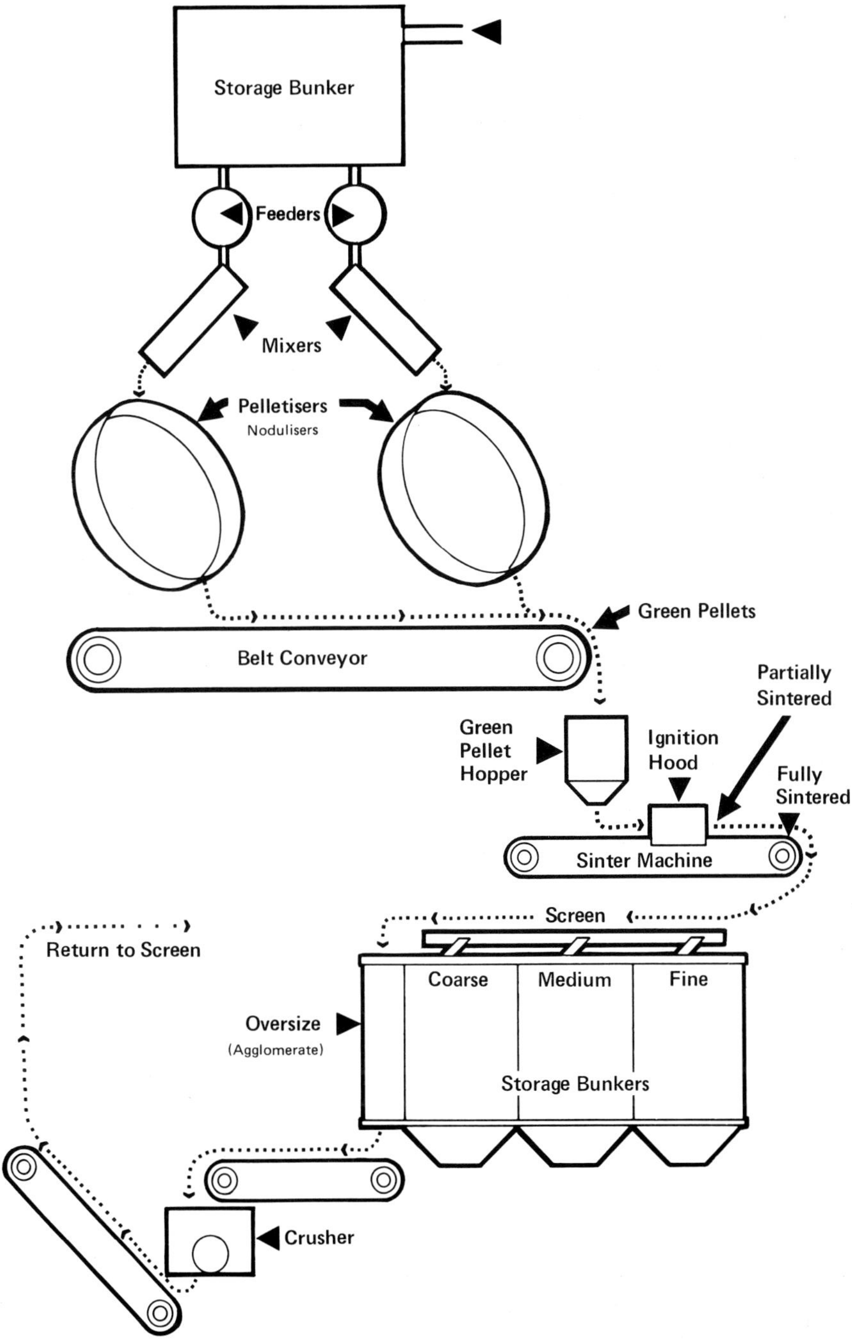
Storage Bunker
Feeders
Mixers
Pelletisers
Nodulisers
Green Pellets
Belt Conveyor
Partially Sintered
Green Pellet Hopper
Ignition Hood
Fully Sintered
Sinter Machine
Screen
Return to Screen
Coarse
Medium
Fine
Oversize
(Agglomerate)
Storage Bunkers
Crusher

FIGURE 4

Stichting BMC in Gouda, the Netherlands, and Baustoffüberwachung, in Aachen (FRG). On this basis, Lytag is supplied with a KOMO-Betonvereniging certificate, which means that it fulfils the specifications laid down in the Dutch standard NEN 3543 "Coarse, light aggregates for concrete".

4. Lytag applications

The steel industrie

10% of all Lytag producted is employed in the steel industry. The pellets are used as an insulating material for molten steel. Its use prevents the cooling of the hot mass during transportation from the furnace to the foundry.

Ready-mixed concrete

The remaining 90% of the Lytag producted is used as an aggregate in ready-mixed concrete in the concrete industry. It is generally used in circumstances where its specific characteristics are best employed.

- in lightweight, high-quality construction concrete;
- as a substitute for fine gravel (nominal value 16).

Volumes of the various types of concrete used on the market, are as follows:

A. prefabricated concrete elements		70%
- paving brick, tiles and concrete strips	: 5%	
- building blocks	: 25%	
- floors	: 20%	
- walls	: 10%	
- general prefab elements (piles, girders, etc.)	: 10%	
B. concrete poured on site		30%

Currently, in the Netherlands there are a number of examples of Lytag applications, in various sectors:

a. bridges and viaducts.
The viaducts on the Koningspley route, the largest example being the bridge over the Rhine, with a total length of 760 m, and a largest single span of 133 m.
The bridge over the Oude IJssel at Ulft, with a total length of 96 m.

b. public, commercial and industrial building.
Offices and shops in Arnhem and Tiel.
The hydroelectric power stations in Maurik, and Alphen aan de Maas.
Cooling water channels for the AMER power station at Geertruidenberg.

c. prefab construction.
Channel plates.
Floor plates.
Driven piles.

5. Characteristics of Lytag comcrete

a. Lytag concrete is almost as strong as gravel concrete, but is more sensitive to impact forces.

b. Workability is good; in general terms Lytag concrete can be used as a substitute for gravel concrete.

c. Work on Lytag concrete, such as drilling and sawing, is simple.

d. Crack forming during hardening, and temperature influences are less than with gravel concrete.

e. Lytag concrete weighs approximately 20% less than gravel concrete, which can be advantageous for transportation and lifting.

f. At a moisture content level of 3%, the coefficient of heat conductivity of a volume mass of 1750 to 1850 kg/m3 of Lytag concrete is equal to between 0.81 and 0.88 W/m C.

g. Generally, the sound-insulating quality of a unit in a structure is heavily dependent on its volume mass. This means, that if no other provision is made, using Lytag concrete in separating floors, for example, can lead to greater construction thicknesses.

6. Price comparison with gravel

One cubic metre of gravel concrete contains approximately 1100-1200 kg of gravel. If gravel concrete is substituted by Lytag concrete of the same quality, then this concrete contains approximately 600-660 kg of Lytag and approximately 10 % extra cement. If we assume that on delivery the gravel contains 2% moisture and the Lytag 10%, this means that for the manufacture of 1 m3 of concrete of the same quality and cost price, the cost of the Lytag aggregate material per tonne could be approximately 1.5 times higher than the cost per tonne of the gravel to be substituted.

Photograph : Pley Brigde under construction

References

1. Stichting CUR : Preadviescommissie PD 9, Lytag als Toeslagmateriaal in beton, rapport 89-3.

2. TU-Delft : Eigenschappen van lichtbeton met gesinterde vliegas als toeslagmateriaal, Deel 1 t/m 7.

Deel 1 (05-86-04): Ir. J.W. Frenay / Dr.ir. H.A.W. Cornelissen.

Deel 2 (05-86-07): " "

Deel 3 (05-86-10): " "

Deel 4 (05-86-17): Ir. J.A. den Uil.

Deel 5 (25-88-32): Ir. J.J.W. Gulikers / Ir. J.W. Frenay.

Deel 6 (25-88-10): Ing. A.P. van der Marel.

Deel 7 (25.5-89-08/C: Ir. J.A. den Uil.

3. Tauw B.v. : Onderzoek naar het uitlooggedrag van Lytag, rapportnr. 51346.07

4. Tauw B.V. : Onderzoek naar het uitlooggedrag van beton waarin Lytag is verwerkt, rapportnr. 51346.09

5. Technisch informatiebulletin Lytag : B.V. Vasim, Raadgevend Ingenieursbureau BVN, Dienst Wegen, Verkeer en Grondzaken van Provincie Gelderland.

Waste Materials in Construction.
J.J.J.R. Goumans, H.A. van der Sloot and Th.G. Aalbers (Editors)

PRODUCTION AND PROPERTIES OF SINTERED INCINERATOR RESIDUES AS AGGREGATE FOR CONCRETE

P J WAINWRIGHT[1] P ROBERY[2]

1. Department of Civil Engineering, University of Leeds, Leeds LS2 9JT, England.

2. Stangers Consultants Ltd, Fortune Lane, Elstree, Herts, WD6 3HQ, England.

SUMMARY

A production process is described which is designed to remove all the ferrous and non-ferrous metals from the residues of incinerators used to burn domestic refuse. The material remaining is further treated by crushing, blending with clay and firing in a kiln to produce an artificial aggregate for use in concrete. Details of the resulting concrete properties are given together with information on the composition and variability of the incinerator residues from a number of different sources and on the properties of the aggregate.

1. INTRODUCTION

Today people are being made ever more aware of the need to protect and conserve the environment in which they live. Two of the many environmental issues which give cause for concern are those of waste disposal and the depletion of the world's supply of natural resources. Recycling of any waste material/industrial by-product is obviously an attractive proposition as it goes someway towards solving both of the problems referred to above.

The work presented in this paper describes a process which is designed to remove all the ferrous and non-ferrous metals from the residues of incinerators used to burn domestic refuse. The material remaining is then converted into an artificial aggregate, suitable for use in concrete, by blending it with clay and firing it in a kiln. Potentially the process is very attractive because it results in complete utilisation of a waste material, however the energy requirements resulting from incineration and from firing the aggregate may be a major drawback.

In addition to the development of the production process information is also given on:- the composition and variability of incinerator residues from different sources, the properties of the artificial aggregate and of the concrete made from it.

2. PRODUCTION PROCESS

Details of the pilot scale production process have been given in previous publications (1,2). Briefly the process involves passing the residue through a series of crushers, screens and magnetic separators to remove the ferrous and non ferrous metals. Initial crushing is by jaw crusher and then by a series of roller crushers; the roller crusher tends to flatten the relatively soft non-ferrous metals into platelets which can then be removed by screening on a 900μm sieve. The resulting residue resembled a coarse sand and this was further crushed, to pass a 300μm sieve, blended with clay, pelletised and fired in a rotary kiln to produce a smooth spherical aggregate. Tests were performed to obtain the optimum firing time, temperature and the percentage clay to give the strongest aggregate as determined by a splitting tensile strength test. A 15/85% blend of clay/incinerator reside fired at 975°C for 13 $^1/_2$ minutes proved to be the best combination.

The major oxide composition of the treated residues (before blending and firing) and of the aggregate are shown in Table 1 and some of the physical properties of the aggregate are shown in Table 2.

Table 1
Major oxide composition of treated residues (before blending and firing) and of the resulting aggregates

Oxide %	Composition (%)			
	Edmonton		Rotterdam	
	Residue	Aggregate	Residue	Aggregate
SiO_2	50.25	54.14	57.50	63.42
Fe_2O_3	15.18	10.20	10.17	8.01
CaO	9.77	12.88	9.43	10.32
Al_2O_3	6.75	8.90	6.55	7.36
Na_2O	5.67	5.04	5.32	4.26
SO_3	1.14	1.52	0.88	0.76
K_2O	1.04	1.47	1.19	1.56

Table 2
Physical properties of the aggregates used

Source	Cylinder-split tensile strength (N/mm^2)	Bulk density (kg/m^3)		Relative density	Water absorption (% SSD)	Porosity (%)
		Loose	Rodded			
Edmonton	3.65	909	955	2.18	15.10	38.04
Rotterdam	7.27	1059	1121	2.29	10.11	24.96
10mm gravel	not measured	1525	1626	2.61	0.6	not measured

During the course of this project a full-scale plant was built in Holland, by a company in the UK, to recover all ferrous and non-ferrous metals from the incinerated domestic refuse. The author was able to obtain large quantities of the resulting residue for aggregate production and was also able to carry out a detailed study into the long term seasonal variations in the composition of the material. Unfortunately after about one years operation mechanical problems led to closure of the plant.

3. CONCRETE PROPERTIES

3.1 Material Properties - Mechanical

Some of the aggregates produced from the above process were used to make concrete and tests were performed on this material up to a period of 4 $^1/_2$ years[2,3]. The properties of the aggregates used are shown in Table 2, as can be seen aggregates were produced from incinerator residues obtained from two different sources, namely:-

(i) Edmonton London - the residue being treated in the pilot scale plant described above.

(ii) Rotterdam Holland - the residue being treated in the full-scale commercial plant.

Mechanical properties of the concrete monitored included - compressive and tensile strength, modulus of elasticity, shrinkage and creep. Some typical results of compressive strength development are shown in Figure 1. As expected the artificial aggregate yielded lower strengths than the natural aggregate but more importantly there was no reduction in strength during the 4 $^1/_2$ year test period. The artificial aggregate concrete also had lower elastic modulus (Fig 1), higher shrinkage and creep than that made with natural aggregates.

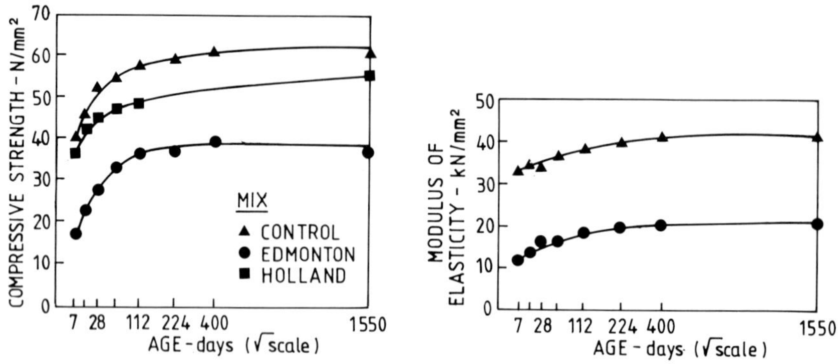

Fig. 1. Some properties of hardened concrete made with natural and with incinerator residue aggregates.

3.2 Material Properties - Durability

A number of tests were performed to study certain long term durabilty related properties of the concrete made with the artificial aggregate. Using a technique based on linear polarization tests were performed to measure corrosion rates of mild steel bars embedded in different concrete samples made with the artificial aggregate and with natural aggregates. After 12 months immersion in a 2.5% NaCl solution it was found that when comparisons were made on the basis of equal w/c and cement content, the corrosion rates were greater in those concretes containing the artificial aggregate. It was concluded though that similar corrosion rates could be achieved by reducing the w/c of the incinerator aggregate concrete by about 20% compared with that of the natural aggregate concrete.

In a separate study[4] carried out using similar materials, tests were conducted to determine the potential alkali-aggregate reactivity of the aggregates using both the ASTM chemical and mortar bar tests. In general no reactivity was indicated although certain inconsistencies between the two tests were observed. In addition scanning electron microscopy was used to examine the interface between the aggregate and the cement paste in concrete samples made with an exceptionally high water/cement ratio of 1.0. The high water/cement ratio of 1.0 was chosen to accelerate any possible reaction; after

6 months storage in a mist room at 20°C no abnormal reaction products were identified.

3.3 Structural Properties

Robery[5] studied the structural behaviour of reinforced concrete beams made with natural dense aggregate, a commercially available light weight aggregate (Lytag-sintered pfa) and the pelletised incinerator ash aggregate. Amongst other things, he looked at load/deflection characteristics of under and over reinforced beams, crack widths, and bond stress characteristics. In all tests he found that the concrete, made with the incinerator residue aggregate, performed satisfactorily when compared with the other two.

4. COMPOSITION OF INCINERATOR RESIDUES

In parallel with the work described above Hadzinakos[4] carried out a detailed X-ray fluorescence analysis of the incinerator residues obtained from Edmonton and Rotterdam and also from a number of other incinerators within the UK. The data collected from Holland enabled measurements to be made on:-

(i) short-term variations from daily samples collected over a one month period.

(ii) long-term variations from monthly samples collected over a 10 month period.

The frequency of sampling from the other incinerators in the UK was not as great but it was sufficient to enable comments to be made on variations between the different incinerators. Analysis was carried out on 15 oxides and a summary of the results, showing only the major oxides, is shown in Table 3. The main conclusions to be drawn from these results are:-

(1) The main compositions of the ashes from the various sources are significantly different.

(2) The long term variations over a 10 month period from the one incinerator in Holland were not significant.

(3) The somewhat larger variations over the shorter term from Blackburn are probably due in some part to problems with sampling.

Apart from analysing the residue Hadzinakas[4] also carried out a detailed analysis of the material at different stages of the aggregate production process (see Section 2) and also of the aggregate itself after blending with clay and firing. He concluded that successive rolling and screening through a 900μm sieve did remove the majority of the non-ferrous metals. Also any very fine ash could be removed by screening the material successively through 3.4mm and 300μm sieves having first been passed through the jaw crusher and before any additional crushing by roller has taken place. This fraction was found to contain relatively high proportions of oxides such as SO_3, ZnO, and PbO the removal of which would be beneficial to the long term stability of the concrete. Removal of this fraction was also shown to increase the strength of the resulting aggregates.

Table 3

Variation in Composition of Incinerator Residues From Different Sources after Extraction of Ferrous and Non-Ferrous Metals

Oxide %	Country			
	Holland(1)	Blackburn(2) (UK)	Sheffield(3) (UK)	Edmonton(4) (UK)
SiO_2	57.5 - 61.66	45.9 - 56.7	49.7	50.3
Fe_2O_3	7.6 - 10.3	8.3 - 16.7	12.8	15.2
Al_2O_3	6.0 - 9.6	5.5 - 9.9	7.2	6.8
CaO	7.9 - 9.4	9.6 - 13.8	10.8	9.8
Na_2O	4.3 - 7.0	3.6 - 6.8	6.0	5.7
MgO	1.3 - 1.9	1.0 - 1.9	1.2	1.0
SO_3	0.4 - 0.9	0.6 - 2.0	0.8	1.1

(1) Range of 45 samples taken over a 10 month period
(2) Range of 24 samples taken over a 4 month period
(3) Average of 4 samples taken on one day
(4) Average of 2 samples taken on one day

When examining the influence of firing temperature and kiln retention time on the aggregate properties no significant changes were observed except with some aggregates fired at the lowest temperature of 850°C. In one particular concrete made with Holland residue fired at 850°C for 2 $^1/_2$ - 3 minutes a reaction was observed almost immediately after mixing the aggregate with the cement and water. The reaction was assumed to be one between aluminium and cement resulting in the evolution of hydrogen gas. However, no reaction took place with any of the aggregates fired at the higher temperatures or at the longer retention times so it is to be assumed that most, if not all, of the aluminium metal is converted to aluminium oxide which is unreactive.

Apart from the aluminium the only other material found in significant quantities in the aggregates that might give cause for concern was sulphates. SO_3 concentrations of between 1.5 to 2.5% were found in the aggregates made from the Edmonton incinerator residue (see Table 1). Since the limit imposed by British Standards[6] is only 1.0% this may be the source of a potential problem. It should be remembered however that no strength reductions were observed over a 4 $^1/_2$ year period in concrete cubes made with this material. Also as discussed earlier removal of the fine ashes by initial screening through a 300μm sieve should significantly reduce the level of SO_3 remaining in the residue.

5. USE OF UNTREATED RESIDUE AS A FINES REPLACEMENT IN CONCRETE

Robery[5] looked into the feasibility of using the untreated residue (i.e. after metals separation, but not blended with clay and fired) as a partial replacement to the fines in concrete. He used incinerator ashes from two of the sources referred to before namely Sheffield and Rotterdam and processed them in the way described to remove all ferrous and non-ferrous metals. Like previous investigators[7] Robery found the Sheffield ash to be unsuitable due to an expansive reaction taking place within 24 hours of casting and also retardation of strength development. The ash from Holland produced markedly different results; there was no expansive reaction and strengths continued to increase over a period of two years. The composition of the two residues was very similar and it was concluded that the main reason for the difference in behaviour was that the temperature in the Sheffield incinerator was lower than the Dutch and was probably not high enough to cause complete "burn out".

CONCLUSION

1. A production process has been developed which can successfully remove the majority of the ferrous and non-ferrous metals from the incinerator residues.

2. The resulting residue from this process can itself be further processed by blending with clay and firing in a rotary kiln to produce an artificial aggregate. This aggregate has been shown to perform satisfactorily in concrete for a period of up to 4 $^1/_2$ years.

3. Although tests showed that the artificial aggregate performed satisfactory in structural concrete it is suggested that its use should be restricted to non-structural concrete until further long term tests can be undertaken.

4. Conflicting results were obtained as to the suitability of using the residue remaining after metals extraction (but before blending and firing) as a fines replacement in concrete.

5. The composition of incinerator resides was shown to vary from source to source but the variation from one source over a 10 month period was not considered to be significant.

6. No data was obtained on energy consumptions but because the process involves incineration and kiln firing the energy requirements are likely to be quite high.

REFERENCES

1. Wainwright P.J and Boni S.P.K. Artificial Aggregate from Domestic Refuse. Concrete Vol.15, No.5 May 1981, pp 25-29.

2. Boni, S.P.K. The Use of Sintered Pelletised Domestic Refuse as an Aggregate in Concrete. PhD Thesis, Dept of Civil Engineering, Leeds University, England. 1980, pp 285.

3. Wainwright P.J and Boni, S.P.K. Some Properties of Concrete Containing Sintered Domestic Refuse as a Coarse Aggregate. Magazine Concrete Research Vol.35, No.123. June 1983, pp 75 - 85.

4. Hadzinakos, I. "The Chemical and Other Properties of Sintered Refuse Slag as an Aggregate in Concrete". PhD Thesis, Dept of Civil Engineering, Leeds University, England. 1980, pp 243.

5. Robery, P.C. The Use of Domestic Incinerator Ash as an Aggregate in Concrete. PhD Thesis. Dept of Civil Engineering, Leeds University, England. 1981, pp 393.

6. British Standards Institution. Specification for Lightweight Aggregates for Concrete. London BS 3793. 1971.

7. Wainwright, P.J. et al. A Review of the Methods of Utilisation of Incinerator Residues as a Construction Material. Proc. Int. Conf. Low-Cost and Energy Saving Construction Materials. Rio de Janeiro, Brazil, July 1984.

Waste Materials in Construction.
J.J.J.R. Goumans, H.A. van der Sloot and Th.G. Aalbers (Editors)

UTILIZATION OF ASH AND GYPSUM PRODUCED BY COAL BURNING POWER PLANTS

F. Gera, O. Mancini, M. Mecchia, S. Sarrocco, A. Schneider

ISMES S.p.A., Via dei Crociferi 44, 00187 Roma (Italy)

SUMMARY

By-products of coal burning in power plants, mainly fly ash and gypsum, have a number of possible applications in the building industry and in road construction. However, due to the very large production, complete utilization of the by-products is not always possible. Considering that the production of by-products will increase in the future, it is important to explore new useful applications of fly ash and gypsum in order to limit the amounts disposed of in landfills.

The use at fly ash and gypsum to make blocks utilized in construction of artificial reefs is an interesting concept. Experiments carried out in USA and Japan have given promising results. ISMES has started the investigation of alternative procedures to produce blocks suitable for artificial reef construction.

1. INTRODUCTION

Coal is by far the most abundant fossil fuel on earth. In addition, the geographic distribution of coal reserves is sufficiently different from the distribution at hydrocarbon deposits to ensure coal utilization for strategic reasons.

Coal burning has negative environmental impacts both on the local and global scales. In order to mitigate the release of sulfur dioxide, which is the main responsible of acid rain, modern power stations are equipped with flue gas desulfurization (FGD) systems. Consequently coal burning produces large amounts of solid by-products: fly ash and FGD gypsum.

A coal burning power station with an installed capacity of 2.600 MWe will produce annually a million tons of fly ash and about 270.000 tons of gypsum (if the sulfur content of coal is 1%). The fraction of by-products that cannot be utilized will require landfilling, combining two undesirable effects: the waste of potential resources and the risk of environmental pollution. It is obvious that the most favourable strategy is to utilize entirely the by-products, preventing them from becoming wastes;

this outcome requires the development of innovative uses for fly ash and FGD gypsum.

The following pages will review the current uses of these by-products and discuss some innovative concepts that might succeed in expanding their utilization.

2. BY-PRODUCTS

2.1 Fly ash

The pulverized coal burnt in power stations contains a significant fraction of unburnable minerals that melt in the combustion chamber and form the ash; about 85% of the ash consists of glassy microspheres that are carried by the exhaust fumes. In order to prevent the excessive release of particulate matter, the fumes are treated by some kind of filtration system that intercepts the majority of fly ash. The remaining 15% of ash is formed by coarser material and accumulates on the bottom of the combustion chamber (bottom ash).

Fly ash is a very fine dust formed by silico-aluminous compounds in the form of subspherical particles ranging in diameter between 1 and 150 μm, but mostly less than 45 μm. Fly ash has a chemical composition quite similar to pozzolan and therefore can be mixed, as a substitute for the natural material, with portland cement. A proportion of 15 to 40% by weight of pozzolan or fly ash has beneficial effects on concrete, making it very resistent to penetration and to corrosion by salt water. The siliceous material in both pozzolan and fly ash reacts with calcium hydroxide, which is liberated as concrete hardens, forming compounds with cementitious properties.

2.2 Desulfurization residues

There are several desulfurization processes and each one produces a certain kind of residue, for example ammonium sulfate, calcium sulfite and calcium sulfate (gypsum). The most common process uses a humid reaction between the exhaust fumes and lime (CaO) or limestone powder ($CaCO_3$); the reaction products are a sludge of calcium sulfite and sulfate or granular gypsum ($CaSO_4.2H_2O$). Desulfurization in the United States often produces sulfite-sulfate sludges, while the most common by-product in Europe is gypsum.

FGD gypsum is composed of crystalline particles with dimensions usually less than 0.2 mm. The purity of FGD gypsum is high, usually higher than natural gypsum.

3. UTILIZATION EXPERIENCE

Fly ash and gypsum are by-products of the same plants, but the experience about their use and disposal is quite different. Fly ash is an unavoidable end product of coal burning and as such has been a disposal problem for many decades; on the other hand, FGD gypsum is a relatively recent addition to the environmental questions related to electricity generation by coal burning plants. It must be pointed out that flue gas desulfurization has been introduced as a mitigation measure; it is therefore essential that the unavoidable environmental impacts of desulfurization be significantly less than the impacts that would be incurred by direct release of the combustion products.

Possible utilizations of fly ash have been the subject of many studies. The Electric Power Research Institute (EPRI), among other organizations, has promoted a specific project aimed at assessing numerous uses of fly ash, with particular attention to those applications that make use of large volumes of material.

In a progress report of the project (1) several examples of fly ash utilization are described, such as filling up of excavations, construction of embankments and dykes, as daily cover of municipal waste in landfills, as soil conditioner, as foundation layer and additive to asphalt in road construction. However one of the most interesting applications of fly ash is as an additive to cement. It is known, for example, that the addition of fly ash can improve the long-term performance of concrete in dam construction (2). Fly ash in concrete can replace both cement (partially) and sand. As a substitute for cement as much as 30% of fly ash can be used; if sand is replaced also, the fly ash content can be even higher. Fly ash containing concretes can be used for dam construction in the traditional way or in a new and more convenient way, which implies, in gravity dams, the spread and compaction of concrete with methods normally used for loose materials.

The production of roller compacted concrete (RCC) implies the use of a material having particular characteristics. In the first place, it must be dry enough to make possible the movement of vehicles and earth moving equipment, and at the same time, fluid enough to reach the maximum density during compaction; to meet these requirements, a high content of fine material can be used since it provides a lubrification action between the particles, minimizing the effects of segregation. Fly ash has favourable characteristics when used as fine f-action in concrete.

In most of the applications of the RCC method, a ratio ash to cement+ash close to 0.30 was used, obtaining a material similar to concrete. Nevertheless, a higher ratio, reaching 0.80, was used in some projects where innovative criteria were applied; an example is the Upper Stillwater dam in the USA.

4. ARTIFICIAL REEFS

4.1 Previous experience

The construction of marine artificial reefs with blocks made of mixtures of FGD gypsum and fly ash represents an interesting and innovative way to utilize these by-products.

Artificial structures, either floating or submersed in sea water, have been constructed and utilized widely in the past to improve the quantity and/or quality of the fish catch (3); the phenomenon of attraction of marine organisms towards hard substrates is known as "tigmotrophism" and results in the colonization of the reef by various species of fixed and moving organisms with a marked increase of biomass and the formation of new local ecosystems.

Various kinds of raw materials have been utilized for building these structures, including dismantled ships, junked cars, tyres, concrete blocks, etc.. The first experiences in the utilization of FGD gypsum and fly ash in this field have been carried out in Japan and in the USA.

In Japan, the Electric Power Development Company (EPDC) prepared a mixture called "FGC concrete"; several years of research showed its suitability for the construction of submersed barrieres. The FGC concrete is made up of fly ash (F), FGD gypsum (G), portland cement (C), and marine or fresh water. Tests carried out were oriented to identify the best composition of the mixture, the influence played by marine or fresh water, the effects of treatment methods on the material characteristics, and its suitability for the construction of artificial reefs. The results can be synthesized as follows (4):

- the highest compressive strength of the material is obtained with a water content equal to 30%;
- the optimum gypsum content, that is the amount that gives a material with the highest compressive

strength, is inversely proportional to the quantity of cement in the mixture;

- the strength of the material improves with age;
- the compressive and bending strength of the FGC concrete is higher with marine than with fresh water;
- the material develops a high compressive strength if treated in autoclave, both with steam or immersed in hot water.

The results showed that the FGC concrete has a mechanical resistance equal or even higher than traditional concrete. Leaching and biological compatibility tests gave favourable results as well.

In the United States, an experimental study on the use of coal by-products for artificial reefs was carried out by EPRI in the period 1978 - 1984 (5). The research, called "Coal Waste Artificial Program", was aimed at finding a mixture suitable for the construction of barriers, and at defining its environmental compatibility. The 20×20×40 cm blocks eventually used were the result of a series of experiments aimed at identifying the optimum composition of the mixture and the more convenient form of the blocks themselves. Initially, 0.8 m^3 blocks were produced, but, because of their great weight, handling caused significant damage; it was then decided in favour of smaller blocks which could be handled easily and produced with machines already existing in the concrete industry.

The procedure for blocks production was at first defined experimentally and then applied on a semi-industrial scale. The procedure was based on the following phases:

- fly ash and dewatered FGD sludges were blended together in the ratio 3.0 to 1 or 1.5 to 1 (dry weight);

- weighed batches of the blended material were transported to a mixer where 6% hydrated lime and 3% portland cement were added; water was then added to reach the optimum water to solid ratio;

- the mixtures were discharged in the moulds and blocks were produced;

- the blocks were cured by keeping them in an underground rotary "wet" steam kiln for 1 or 2 days at a temperature of 70°C.

The in-situ testing phase consisted of placing 15.000 blocks on a sandy sea floor, at a depth of about 20 meters, south of Fire Island, Long Island, in the New York Bight; a monitoring program was carried out for three years for the collection of physical, chemical and biological data, paying particular attention to the structural longevity of the blocks, to the colonization of the new reef, and to the environmental toxicity of substances leached out of the blocks.

The research demonstrated the feasibility of using coal by-products in the marine environment; in particular, the following results were obtained:

- the leaching of trace elements was minor;
- the structural integrity of the barrier can be garanteed for relatively long periods since the physical properties of the blocks were stable and the leaching of calcium, main component of the mixture, decreased exponentially after the initial peak;
- negative effects on marine organisms were not detected in normal conditions; some inhibitory effects were found only in tests performed in closed basins with particularly high concentrations of fly ash and gypsum.

4.2 ISMES approach

Studies on alternative procedures and mixtures for the production of blocks usable in the costruction of artificial reefs have been initiated also by ISMES. In particular, the ISMES research is aimed at defining the characteristics of blocks produced with simpler procedures than those applied in Japan and in the United States.

In the first phase of the research (6), carried out in the period 1989 - 1990, cylindrical blocks of small dimensions were produced by Calcestruzzi S.p.A. with fly ash and gypsum in the ratio 3 to 1 or 2 to 1. Since FGD gypsum has not yet been produced in Italy, natural gypsum was used in the mixture.

The procedure utilized for producing the blocks was based on cold pressing of the dry mixture without the addition of binders.

Given the small number of blocks available, mechanical tests were not carried out and the determinations focused on the leachability characteristics of the samples. On four of these blocks, for which the main characteristics are shown in Table 1, leachability tests were performed following the procedure defined in ISO testing standard N. 6961.

Each block was suspended in a beaker, immersed in a volume of leaching solution ranging between 370 and 471 mL. Leaching by artificial sea water lasted for a total period of 9 weeks; substitutions and analyses of the solution were carried out after 1, 3, 4, 7, 10, 14, 21, 28, 35, 42, 49, and 59 days; changes of block volume and weight losses were measured as well.

As the experiment began, it was clear that the mechanical strength of the blocks was very low since the four samples started to develop obvious signs of physical damage. The mechanical characteristics of the blocks appeared to be related to the pressure applied for their construction and to the ratio fly ash to gypsum in the mixture; the four samples could be ordered, by resistance to damage, according to the sequence: 1 > 3 > 4 > 2.

The blocks resulted to be unsuitable for the construction of a marine reef. The extrapolation of the calcium leaching curve showed that the complete solubilization of gypsum in the samples would occur in a time span ranging from 2.5 to 6 months.

As regards trace elements, their content in the solution was compared with the limits of concentration fixed by EPA for toxic elements extracted from materials that can be disposed of in the environment; the EPRI research referred to the same limits for the assessment of the toxicity of the blocks (7).

TABLE 1

Samples description

n.	ash/gypsum ratio by weigth	press.	dimensions height mm	diam. mm	area cm^2	weight g	vol. mL	dens. g/mL
1	3 / 1	high	10.0	40.0	37.7	21.1	12.56	1.680
2	3 / 1	low	14.0	40.0	42.7	19.8	17.58	1.126
3	2 / 1	high	11.5	40.0	39.6	25.8	14.44	1.787
4	2 / 1	low	17.5	40.0	47.1	25.9	21.98	1.178

The concentrations of arsenic, chromium, lead and copper in the solution, normalized to keep into account the different procedures followed in the leaching tests, resulted to be lower than the EPA limits. This result is probably due to the small solubility of ash and to the low content of toxic elements in the mixture.

5. CONCLUSIONS

The initial tests on leaching materials formed by cold pressing of fly ash and gypsum have given negative results. However the development of simplified procedures for producing blocks suitable for making artificial reefs is a worthwhile objective. In particular, avoiding curing at high temperature would allow significant savings.

Depending on the availability of funds the logical continuation of the study should include testing blocks produced applying much greater pressures and with the addition of small amounts of binders. Many types of binders should be tested including a variety of polymeric compounds.

REFERENCES

1 EPRI, Fly Ash Design Manual for Road and Site Applications, Volume 1: Dry or Conditioned Placement, Electric Power Research Institute, CS-4491, Palo Alto, California, 1986.

2 ICOLD, L'impiego delle ceneri di carbone nei calcestruzzi per dighe, Comitato Nazionale Italiano delle Grandi Dighe, Rome, 1990.

3 G. Bombace, Artificial Reefs in the Mediterranean Sea, Bullettin of Marine Science, 44 (r), 1989, pp. 1023-1032.

4 M. Yasuda e T. Sawada, Study of FGC Concrete and its Use for Artificial Fishing Reefs, Proceedings of the Seventh International Ash Utilization Symposium and Exposition, DOE/METC - 85/60, 2, 1985, pp. 763-775.

5 EPRI, Coal-Waste Artificial Reef Program. Final Report Electric Power Research Institute, CS-3936, Palo Alto, California, 1985.

6 ISMES, Utilizzo dei manufatti ceneri/gesso per la costruzione di scogliere artificiali sottomarine in connessione con la costruzione di una centrale offshore, Unpublished report, Rome, 1990.

7 EPRI, Coal-Waste Artificial Reef Program. Phase 4A. Interim Report, Electric Power Research Institute, CS-2574, Palo Alto, California, 1982.

Waste Materials in Construction.
J.J.J.R. Goumans, H.A. van der Sloot and Th.G. Aalbers (Editors)

Quality and environmental aspects in relation to the application of Pulverized Fuel Ash.

J.W. van den Berg,
VLIEGASUNIE b.v. (Dutch Fly Ash Corporation),
P.O. Box 3254, 5203 DG 's-Hertogenbosch, The Netherlands

1. Summary

In the Netherlands pulverised coal from different parts of the world is used in 5 coalfired power stations which each have 1 to 3 boilers, which in turn are equipped with various types of burners. This causes a large variation in the composition of the pulverized fuel ash (P.F.A.). The P.F.A. is marketed in several areas, each of which has its own specific quality requirements. These requirements are partially dictated in standard specifications.
In order to obtain insight into the quality of the P.F.A., samples are taken and analysed daily.
Rules for certification, concerning the quality and the quality controle of P.F.A. for use in concrete in The Netherlands have been agreed upon.
The environmental aspects in relation to the application of P.F.A. concern the working conditions and health aspects during processing and the environmental impact when P.F.A. is used as a building material. The Dutch legislation concerning the environmental consequences of the application of P.F.A. and other secondary materials is currently under review.

2. Introduction

In 1990 the 10 electricity production units, divided over 5 locations in The Netherlands used approximately 9 million tons of coal fuel.

Locations:

Nijmegen	Unit 13	603 MW
Haelen	Unit 4	120 MW
	Unit 5	177 MW
	Unit 6	223 MW
Geertruidenberg	Unit 4	223 MW
	Unit 5	223 MW
	Unit 8	644 MW
Borssele	Unit 12	402 MW
Rotterdam	Unit 1	518 MW
	Unit 2	518 MW

The total coal fueled production capacity of electricity in operation at this moment is 3651 MW.

In the comming decade a number of the small old units will be phased out of production, whilst a total of four new coal fueled units, each with an output of 600 MW will be started up in 1992, 1994, 1997 and 2000. The first three of these units will be fueled by pulverised coal and the unit planned in the year 2000 will probably be fueled by coal gas.

Coal was re-introduced as fuel for the generation of electricity in The Netherlands as a result of the oil crisis in 1973 and 1978 and consequently the problem of coping with the by-products had to be tackled.

The capacity of the few available dumping sites was limited, so that suitable applications for these by-products had to be sought.

The Dutch Fly Ash Corporation is a 100 % daughter company of the Association of Dutch Electricity Boards and was founded in 1982. The objective of the Dutch Fly Ash Corporation was to centrally explore potential markets and to co-ordinate and stimulate the developement of new applications of the by-products, produced by all the coal fueled power stations in The Netherlands.

These by-products are:

Pulverized fuel ash (P.F.A.)
Bottom ash
Fluegas desulfurisation gypsum (F.D.G. gypsum).

3. Origins

P.F.A. is formed during the process of combustion of pulverised coal in the furnace of the boiler of a power station. The coal is ground to powder-coal in coal mills, before being blown by compressed air into the furnace and burned at temperatures ranging between 1300 and 1600 °C, depending on the type of boiler in question.
Coal contains 10 to 15% of non-combustible mineral material. Most of this material melts in the furnace. The majority (approximately 85%) of the non-combustible material is transported into the stack by means of the fluegases, where it solidifies into fine grained ash particles (P.F.A.).
These ash particles are extracted from the fluegases by electro-static precipitators which have an efficiency of almost 100%. The P.F.A. is transported from the precipitators to silos for dry storage.
The P.F.A. which cannot be marketed immediatly is moistened and stored in open air depots.

Bottom ash is formed by the slagging of ash particles in the lower parts of the furnace. The amalgamated ash falls through open grids in the furnace floor where it is collected in hoppers. The bottom ash is then moistened and stored in open air depots.
Bottom ash is used in road construction as a light weight base material and as raw material in the building block industry.

F.D.G. gypsum is formed during the removal process of SO_2 from the fluegases. The fluegases are led through a scrubber and oxydation tower, where they are brought into contact with a lime or limestone suspension and, after an interaction with an excess of air, F.D.G gypsum ($CaSO_4$ $2H_2O$) is formed. This gypsum is primarily used in the fabrication of plaster board in The Netherlands and Belgium.

4. Production and Applications

P.F.A. is primarily used in the building industry. In table 1 the production and the application in the various market areas are listed for the period between 1986 and 1989.

YEAR	**1986**	**1987**	**1988**	**1989**
P.F.A production (tons)	513.500	613.700	712.400	766.535
P.F.A utilisation (tons)				
Cementindustry	319.900	369.300	409.300	608.125
Asfaltfiller	46.000	43.200	78.500	72.700
Artifical gravel (lytag)	84.000	110.400	126.900	121.465
Filler in concrete	27.400	30.400	31.800	28.860
Road constructions/ embankments	18.300	29.000	38.200	150.430
Other applications	3.400	7.500	11.700	1.610
	499.000	589.800	696.400	983.190

Table 1: Production and application of P.F.A.

The applications of P.F.A. in the various market areas are:

* Cement industry
- As raw material for Portland Fly ash cement.
- As raw material for Portland clinker.
* Concrete mix and concrete commodities industries
- P.F.A. is used as a partial replacement of cement or as filler.
* Mineral aggregates
- Lytag
- Aardelite

Both Lytag and Aardelite can be used as light weight replacement material for conventional mineral aggregate (gravel and/or crushed rock).

* Roadconstruction
  - as filler for the asphalt industry
  - as (partial) replacement of sand in cement stabilized road base material.
  - in embankments

5. Composition

The Dutch power stations are fueled with bituminous coal, which is imported from various coal mines in different countries throughout the world.
The geographic origin of the coal used in 1989 was as follows:

Australia	46 %
U.S.A.	27 %
Columbia	16 %
Poland	9 %
China	1 %
Indonesia	1 %

As already stated, there are 5 coal fueled power stations with a total of 10 boilers operational in The Netherlands at this moment. A number of these boilers still have conventional burners, the remainder have so called low-NOx burners.
The diversity in geographical origin of the coal in combination with the variety of boiler types, leads to a great variation in the composition of the P.F.A.
The average value, the standard deviation and the range of a number of chemical compounds in Dutch P.F.A., produced in 1990, are given in table 2.

Compound	Average	Minimum	Maximum	St. dev.	Unit
SiO_2	59.1	37.7	78.0	5.6	%
Al_2O_3	26.0	11.7	37.1	4.0	%
Fe_2O_3	6.2	0.9	11.6	1.5	%
CaO	2.3	0.2	14.0	1.6	%
MgO	1.0	0.2	6.4	0.7	%
Na_2O	0.4	0.0	3.1	0.3	%
K_2O	2.0	0.1	8.1	0.8	%
pH	8.9	3.5	12.8	2.0	-
C*	5.4	0.5	16.5	2.3	%
< 32 mu	73.6	42.3	97.3	7.1	%
< 45 mu	81.7	50.0	99.0	6.2	%

table 2 : Composition of P.F.A. (1990)

This table shows that the diversity in geographical origin of the coal in combination with the variety of boiler types, can lead to a great variation in the composition of the P.F.A.

6. Quality control

P.F.A. is marketed in several areas, each of which has its own specific quality requirements. These requirements are partially dictated in standard specifications, which are being drafted at this moment by CEN as part of the European Regulations.

An example of such specifications is given below as the proposed requirements for fly ash in concrete (Draft ENV - 197).

Fly ash: Fine powder of mainly spherical, glassy particles, derived from pulverized fuel, which has puzzolanic properties and consists essentially of SiO_2 and Al_2O_3. Reactive CaO should generally be less than 5% by weight. At least 25% by weight should be reactive SiO_2. The loss on ignition should not exceed 7% by weight.

Some of the quality requirements are dictated by the relevant industries themselves, for example the content of Al_2O_3 for certain applications in the cement industry.
In The Netherlands the requirements for P.F.A., for utilisation in concrete are dictated in the CUR Recommendation nr. 12. Based on these requirements a procedure for the certification of P.F.A. has been agreed upon with the certification institute KIWA. The procedure consists of a set of guide lines for the assessment of P.F.A. (Beoordelingrichtlijn BRL K223/02) and incorporates an internal quality monitoring scheme (Intern Kwaliteits-
bewakingsschema IKB), which is performed under the auspice of the certification institute KIWA.
The quality controle of the P.F.A. production is part of the IKB scheme.

On delivery at the power station the coal is analysed. Based on the results of this analysis, and taking into account the unit and the burner in which the coal will be fired, it is possible to predict the composition and the quality of the P.F.A. that will ultimately be produced. This gives the great advantage of being able to select the possible applications and thus the potential markets for the P.F.A. before it has been produced.
The quality controle is performed at the power stations, which all have a modern and adequately equipped laboratory at their disposal.
Each day samples are taken for chemical and physical analysis. The test results are forwarded to the Fly Ash Corporation where they are compared with the results of the afore mentioned prognosis. If necessary the logistics concerning the destination of the P.F.A. can be adjusted.

Depending on the client's quality requirements, extra samples can be taken and analysed before and during the loading of specific shipments. In this way the client can be given the assurance that the particular shipment of P.F.A. meets his specific requirements.

7. Environmental aspects

The following aspects should be considered when working with P.F.A.:

- Working conditions and health issues during production and application.
- The chemical composition and leaching behaviour of the products.
- Radioactivity.

7.1. Working conditions and health issues

During the production and the application of P.F.A one should consider the following aspects in relation to labour and health:

- Particle size in connection with the dust yield.
- The content of quarts in connection with silicosis.
- The content of toxic inorganic compounds.
- The content of organic compounds in relation to possible mutagenic and carcinoganic effects.
- The content of natural radioactive nuclides in relation to a possible increased dose of radiation.

After a 10 year research and study period on the possible health effects of working with P.F.A., a booklet was published in 1989 by the Labour inspection of the Ministery of Public Health, entitled: "Working with P.F.A. fly ash".
In the booklet several stages of the P.F.A. production and application were analysed and the possible health effects evaluated.
The booklet recommends that when working with P.F.A. no more than the normal precautions for working with fine powders need be taken.

7.2. <u>Chemical composition and leaching</u>

The Dutch environmental legislation on this topic is currently under review. Part of this legislation is the green paper on the use of building materials (Bouwstoffenbesluit). The paper specifies the range of acceptance of the potential environmental impacts from the use of building materials, and reviews both building materials manufactured from natural raw materials and building materials manufactured from the so called secundary raw materials (by-products).
The paper indicates in which cases specific environmental protective measures will be necessary in and which cases not.
It seems that the concentration and the leachability of pollutive compounds in building materials will determine the environmental acceptability.

Much is known about the composition and the leaching behaviour of P.F.A. The concentrations of trace elements and their availability for leaching is indicated in table 3.

Component	Concentration mg/kg	Leachability mg/kg
As	58	1
Ba	1882	
Cd	1	< 0.01
Cr	161	4.6
Cu	284	0.09
Hg	0.2	
Mo	17.5	16
Ni	145	< 0.7
Pb	97	0.11
Sb	14	
Se	20	
Sn	10	
V	360	17.6
Zn	192	0.4

Table 3: Concentration and leachability of trace elements

7.3. Radioactivity

Compared to the natural concentration in the ground, P.F.A. contains a somewhat increased concentration of radioactive elements (i.e. Uranium and Thorium and the decay products). Therefore there was public concern about a possible increased dose of radiation from building materials in which P.F.A is utilised. One of the daughter products from Uranium and Thorium is the radioactive noble gas Radon.
Since the general acceptance of the fact that the indoor radiation dose received by the inhabitants of buildings and houses is not only determined by the direct radiation from the structure, but to a great extent by the indoor Radon concentration in the air, the public opinion on this matter has changed.

The indoor concentration of Radon is (among other sources) determined by the degree of exhalation from the structures containing the building material. Due to the glassy nature of P.F.A. this exhalation is far less than in most common building materials and compensates for a possible increase direct radiation.

Extensive studies by the Ministery of Housing and Environmental Affairs have shown that in comparison to conventional building material components, the utilisation of P.F.A. will not increase the total radiation dose of the inhabitants.

For the time being the Ministery has decreed the preservation of the current guide lines for indoor radiation doses. In this respect no selection of P.F.A. on its radioactive behaviour is necessary.

Waste Materials in Construction.
J.J.J.R. Goumans, H.A. van der Sloot and Th.G. Aalbers (Editors)

THE USE OF FLY-ASH IN THE CLAY PRODUCTS STABILIZED WITH CEMENT AND LIME, OBTAINED THROUGH EXTRUSION

M. TEMIMI, A. AIT-MOKHTAR, J.P. CAMPS, M. LAQUERBE.

Laboratoire G.T.Ma., Département Génie Civil, I.N.S.A., 20 Avenue des Buttes de Coësmes, 35043 RENNES, FRANCE.

SUMMARY

The fly-ash valorization by their use in cold stabilized clay products has a double advantage :
- it is an economical solution to elaborate a competitive building material,
- it contributes to reduce pollution problems.

The properties of elaborated products have been greatly improved thanks to the grain shape and size and to the pouzzolanic effect of silico-aluminous fly-ash.
The grain shape, indeed, reduces the shrinkage and improves the mechanical resistances at early ages by allowing a better hydration of binding material and by acting as a filler. The pouzzolanic effect increases the long-term mechanical resistances.

1. INTRODUCTION

The fly-ash are a waste material, residue of the coal combustion in steam generating station. In the field of Civil Engineering, they may be eliminated by incorporating them into building materials (cements, concrete, soil stabilization, embankments, grouts...). Their satisfactory use in this field (1) lead to study their ability to be incorporated into the clay products stabilized with binding materials and shaped by extrusion. For some years, the G.T.Ma. laboratory has undertaken some research works in order to check out new construction elements. In this field, the study of clay cold stabilization, for example the cold stabilized bricks (B.S.F. : Briques Stabilisées à Froid)(2),(3) was particularly developped.

The aim of the fly ash use in such materials is (4):
- to improve the performances and the qualities of this product,
- to contribute to reduce the pollution they might cause.

2. THE FLY-ASH IN THE STABILIZED CLAY PRODUCTS

2.1 Used materials

2.1.1 The fly ash. The used fly-ash have been supplied by the steam generating station of Cordemais (Loire Atlantique, France). They are silico-aluminous fly-ash which have pouzzolanic properties.

2.1.2 The clay. It is a monomineral China Clay coming from the Ploemeur deposit (Southern Britanny). It has been selected because it allows a good, reliable and reproductible comparison between various stabilization methods.

2.1.3 The stabilizing substances. Two different binding materials were used :

- an Ordinary Portland Cement, C.P.J. 45 whose compressive strength nominal value is 45 MPa (N/mm^2) at 28 days,
- a Hydraulic Lime "LAFARGE 100" whose compressive strength nominal value is 10 MPa, according to French Standards (5).

2.2 Mixtures

2.2.1 Manufacturing process. The samples are made through extrusion with a laboratory machine operating under vacuum like an industrial one used in brick field. After mixing, the mixture is introduced into the extruder. An endless screw pushes the material through an extrusion die, the shape and the size of which have been choosen to make a prismatic sample.

2.2.2 Samples. Clayless mixtures have not a sufficient cohesion to be extruded. Therefore, various compositions were tested in order to obtain the lowest clay proportion able to give extrudable products.

The experiments show that the extrusion water amount is close to the plasticity limit of the mixture (3). The composition of studied mixtures are given in table 1.

The prismatic samples (4x4x16 cm3) are stocked at 20°C and 50% relative humidity.

2.3 Testing procedures

The dimensions, the weight variations and the mechanical resistances were measured at different ages. The X-Ray diffraction analysis was carried upon powder for samples aged, respectively, 1 day and 90 days.

TABLE 1
Composition of mixtures

Mixtures	Clay %	Fly-Ash %	Cement %	Lime %	Water %
MT1	94	0	6	0	28
MT2	88	0	12	0	28
MT3	82	0	18	0	28
M1	40	54	6	0	28
M2	40	48	12	0	28
M3	40	42	18	0	28
MT4	94	0	0	6	28
MT5	88	0	0	12	28
MT6	82	0	0	18	28
M4	40	54	0	6	28
M5	40	48	0	12	28
M6	40	42	0	18	28

2.4 Results and discussion

2.4.1 Shrinkage and weight loss. The figures 1 and 2 show that the greater part of the shrinkage and the weight loss of all the mixtures occur at the early ages. After a while, about 14 days, these parameters are stabilized to a maximum value. This value decreases with fly-ash addition. An explanation may be put forward : because of its leaf structure and particle low size, the clay is able to retain more water than fly ash.

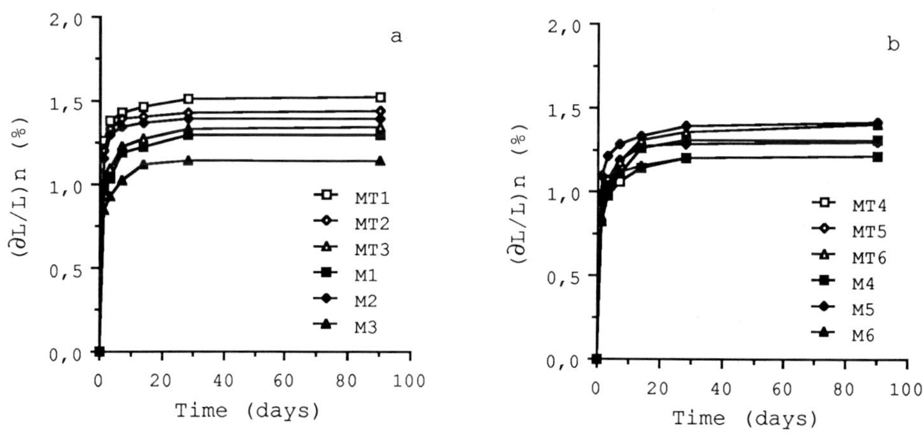

Fig. 1. Diagram showing the variations of shrinkage $(\delta L/L)_n$ v. time for : (a) cement, (b) lime.

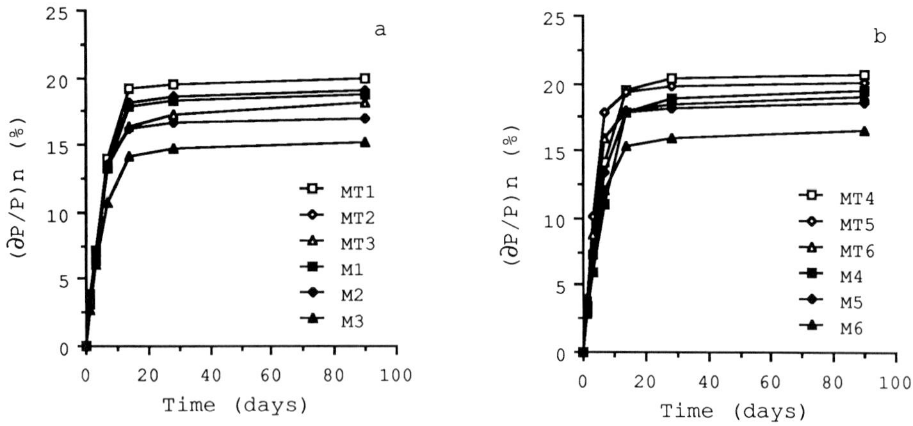

Fig. 2. Diagram showing the variation of weight loss $(\delta P/P)_n$ v. time for : (a) cement, (b) lime.

Therefore, a part of the clay being replaced with the fly-ash, more free water will be present for a better hydration of binding material.

2.4.2 Mechanical resistances. In figure 3 and figure 4 a clearer difference between the mixtures containing fly-ash and

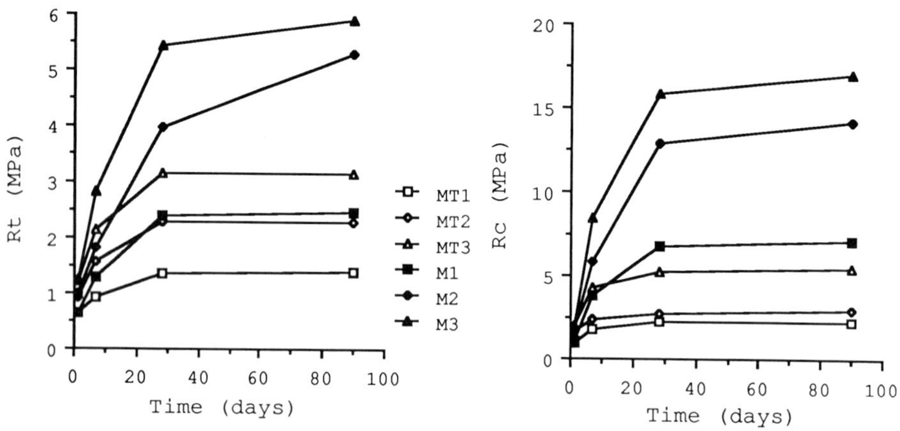

Fig. 3. Diagram showing the variations of tensile (Rt) and compressive (Rc) strength v. time for cement samples.

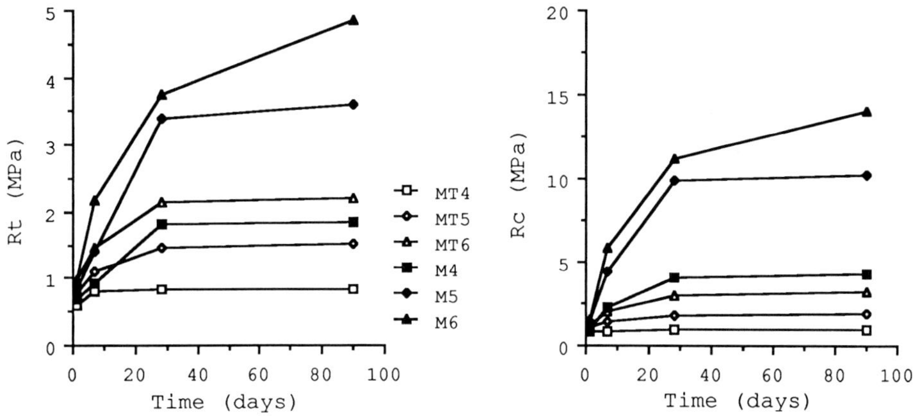

Fig. 4. Diagram showing the variations of tensile (Rt) and compressive (Rc) strength v. time for lime samples.

their respective reference samples may be observed. The resistance increase in the fly-ash mixtures is always the greatest. This can be explained :

- at early ages, by a better binding material hydration as shown above,
- after about 28 days, by the pouzzolanic effect of the silico aluminous fly-ash which allows the resistances to increase instead of being stabilized.

2.4.3 X-Ray diffraction analysis. In order to detect hydrated phases which might appear throughout the period, a X-Ray analysis was carried out for 4 samples, M3 and M6, and for their respective reference samples, MT3 and MT6 (Figure 5 and figure 6). Whatever the date, it is observed for all the mixtures :

- the Calcite, $CaCO_3$,
- the Kaolinite, $Al_2Si_2O_5(OH)_4$,
- the Quartz, SiO_2,
- the Muscovite, $KAl_2(Si_3Al)O_{10}(OH,F)_2$

1 day old MT3 and MT6 do not contain any hydrates when the following hydrates are found in M3 and M6 :

- Ettringit, $Ca_6Al_2(SO_4,SiO_4,CO_3)_3(OH)_{12},26H_2O$
- Gismondite, $CaAl_2Si_2O_8,4H_2O$
- Hydrated Calcium Aluminate, $2CaO,Al_2O_3,8H_2O$

M6 also contains a Hydrated Calcium Silicate, $2CaO,SiO_2,H_2O$,

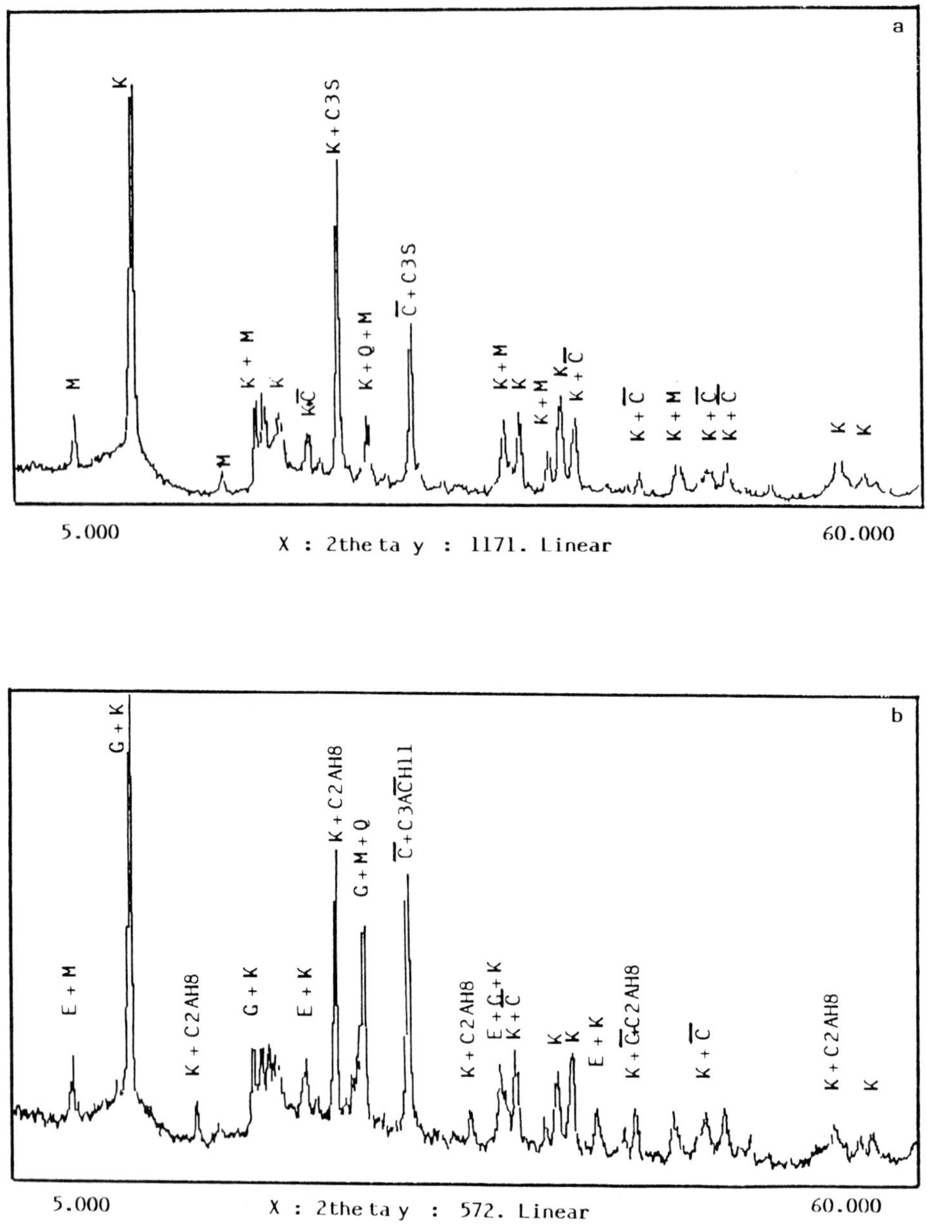

Fig. 5. X-Ray diffraction diagram at 90 days for : (a) MT3 reference sample, (b) M3 sample

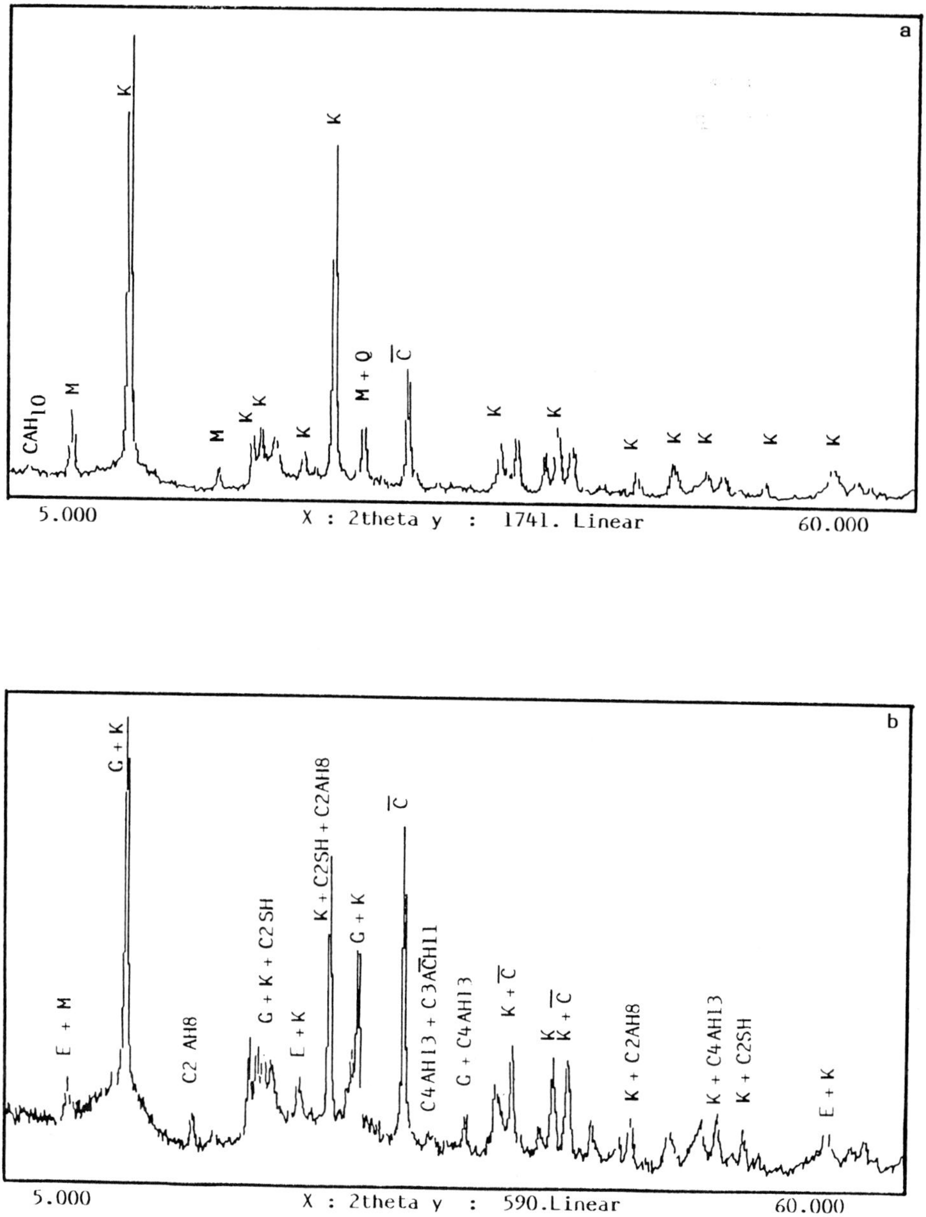

Fig. 6. X-Ray diffraction diagram at 90 days for : (a) MT6 reference sample, (b) M6 sample

90 day old MT6 contains the only observed hydrate : an hydrated calcium aluminate $CaO,Al_2O_3,10H_2O$. No hydrate is detected in MT3 at the same age.

Added to the precedent hydrates present at 1 day, M3 and M6 contain $3CaO,Al_2O_3,CaCO_3,11H_2O$, an hydro-carbonated calcium aluminate. At last, M6 also contains another hydrated calcium aluminate $4CaO,Al_2O_3,13H_2O$.

3. CONCLUSION

The addition of fly-ash to cold stabilized clay mixtures greatly improves the various properties of the final products. The shrinkage and the weight loss are lowered in comparison with the reference samples. Better mechanical resistances of fly-ash added samples are recorded. The values of tensile strength of these samples are twice the value of reference samples, the increase is three to four times for the compressive strength. It seems that the better hydration of binding materials and the pouzzolanic effect of these fly-ash are responsible for this. Besides, the pouzzolanic property was demonstrated by the X-Ray diffraction analysis.

To complete this study a water resistance test was carried out. 28 day samples were immersed into water for several months. No degradation (cracks for instance) was observed for fly-ash added samples, showing their qualitative superiority.

The use of fly-ash as an additive in cold stabilized clay, seems to be greatly beneficial to the binding materials used during this work. After this first step, it is necessary to continue such a resarch in order to explain some mechanisms which were observed.

REFERENCES

1 A. Jarrige, Les cendres volantes, Eyrolles Edition, Paris, 1971.

2 A. Aït-Mokhtar and M. Laquerbe, Contribution à l'étude de la stabilisation à froid des argiles, Rapport de Laboratoire, I.N.S.A., Rennes, 1988.

3 J.P. Molard; J.P. Camps and M. Laquerbe, Etude de l'extrusion et de la stabilisation par le ciment d'argiles monominérales, Materials and Structures, 1987, pp.44-50.

4 M. Temimi, Utilisation des cendres volantes dans des mélanges à base d'argile stabilisés à froid et mis en forme par extrusion, Rapport de D.E.A., I.N.S.A., Rennes, 1988.

5 Normes A.F.N.O.R., NFP 15 300 to 15 312, Paris.

Waste Materials in Construction.
J.J.J.R. Goumans, H.A. van der Sloot and Th.G. Aalbers (Editors)

PRODUCTION OF LIGHTWEIGHT AGGREGATE FROM WASTES: THE NEUTRALYSIS PROCESS

Andre Krol[1], Ken White[2] and Bill Hodgson[3]

[1]Department of Chemical Engineering, The University of Queensland, Q4072, Australia
[2]Neutralysis Industries Ltd, PO Box 243, Brisbane Markets, Q4106, Australia
[3]Department of Civil Engineering, Queensland University of Technology, Q4000, Australia

SUMMARY

The Neutralysis process fires a pelletised mix of refuse derived fuel, liquid waste and clay in a rotating kiln system to produce a lightweight aggregate for use in the construction industry. The produced aggregate has been subject to a testing programme designed to assess its marketability. A variety of leachate tests have been used to establish that the leachability of metals is very low, and that the total concentration of environmentally-significant organics is below detection limits. Engineering tests on the aggregate have shown that it has similar characteristics to lightweight aggregate manufactured by traditional routes, and that it shows great potential for use in masonry, structural concrete and road base material.

1. INTRODUCTION

The Neutralysis process utilises municipal solid waste (MSW), non-hazardous liquid wastes (such as sewage sludge and grease trap wastes) and clay in the production of lightweight aggregate. In August 1988 a pilot plant capable of processing 25 tonne/day of MSW was completed in Brisbane, Queensland with partial funding of A$1.5 million provided by the Australian Government. This facility has been used to prove and tune the process, and to provide aggregate for a testing programme designed to assess its environmental acceptability and marketability.

The Department of Chemical Engineering, The University of Queensland and the Australian School of Environmental Studies, Griffith University provide advice and testing services on environmental aspects connected with the process and its commercialisation. The Department of Civil Engineering, Queensland University of Technology has a similar role on aspects connected with use of the aggregate in concrete and road construction.

In February 1990 the pilot plant was subject to an environmental audit which characterised inputs to and outputs from the process. The audit has been supplemented by a testing programme designed to build a database of results under varying conditions, particularly with respect to the quality of the aggregate. These results are outlined below, following a description of the process.

2. PROCESS DESCRIPTION

The major unit operations and process streams in a Neutralysis plant are shown schematically in Figure 1. There are three main sections - physical processing, kiln firing and gas cleaning.

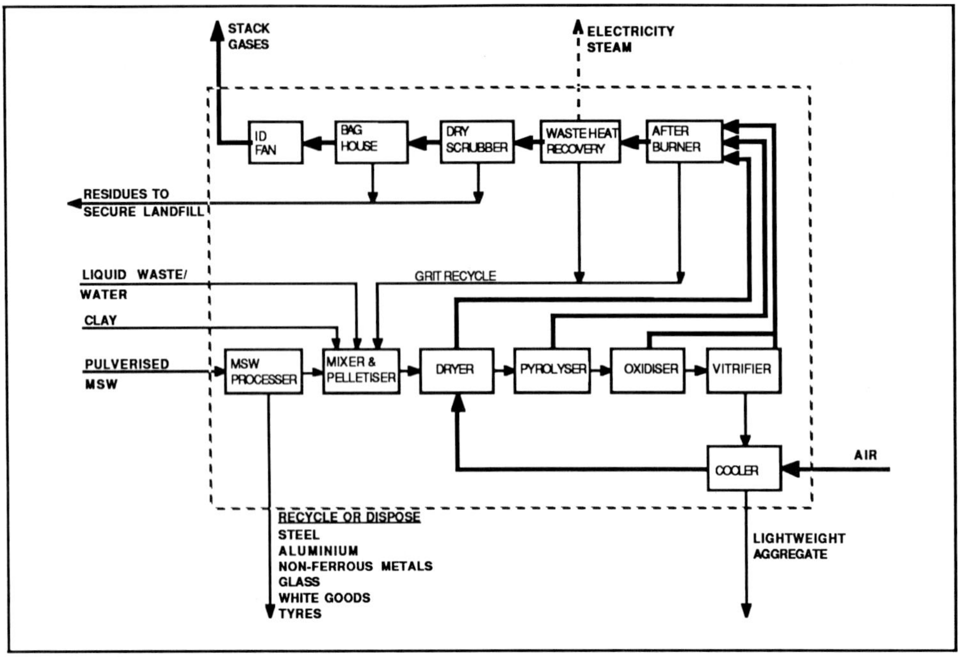

Figure 1. Schematic of a commercial Neutralysis plant showing the main unit operations. Solid streams are shown using a thin line, and gas streams using a broad line.

2.1 Physical Processing

The physical processing section uses technology which has been proven on a commercial scale in refuse derived fuel (RDF) and clay processing plants. Incoming MSW is pulverised and ferrous and non-ferrous metals and some glass are removed to produce a "fluff" RDF. This RDF is mixed with a similar weight of milled clay and liquid waste and the mixture is extruded to produce pellets of a size suitable for the required aggregate application. The rate of non-hazardous liquid waste (and/or water) addition is controlled so as to produce cohesive pellets. The pellets are passed through a rotary drum tumbler which smooths their shape and, in conjunction with a trommel screen, removes fines which are recycled back to the mixer. A dryer using hot air sourced from the product aggregate cooler is used to reduce the moisture content of the pellets.

2.2 Kiln Firing

The partially dried pellets are passed through a series of rotary kilns which in turn pyrolyse, combust and vitrify the pellets to produce the product aggregate. A similar kiln system has been used on a commercial scale at facilities which produce lightweight aggregate from standard materials.

In the first (pyrolysis) kiln, the pellets are brought to a temperature in excess of 500 °C and pyrolysed under starved air conditions. Porous pellets containing unburnt carbon are formed as the pyrolysis gases are driven off. In the second (oxidation) kiln, air is added in a staged manner to partially or fully burn off the carbon in the pores and raise the pellet temperature to approximately 900 °C. The porous pellets are then passed to a third (vitrification) kiln where their temperature is raised to approximately 1100 °C. Under these conditions the pellet surface is sealed and gas generated by oxidation of the residual carbon causes bloating, thus producing a lightweight aggregate with low moisture absorption. The oxidation and vitrification processes can also be carried out in a single kiln. Aggregate discharged from the vitrifier is cooled by air, thus providing heated air for the pellet dryer and combustion.

2.3 Gas Cleaning

In a commercial plant, all gases exiting the kilns are passed to a high temperature afterburner to destroy residual organics. Pyrolysis gas is the main fuel used in the afterburner, however supplementary fuel is provided if combustion of the pyrolysis gas alone cannot maintain the required minimum temperature of 1000 °C. Gases exiting the high temperature afterburner pass through a heat recovery step (to, for example, generate electricity for internal plant use and export) prior to being subject to, as a minimum, acid gas neutralisation by a dry scrubber and particulate removal by a baghouse prior to stack discharge. Some markets may require additional gas cleaning equipment, such as de-NO_X, to be installed.

The air pollution control equipment generates solid residues consisting of fly ash and spent and excess lime. These residues will be treated as necessary (e.g. by solidification) to minimise the leachability of heavy metals, and then disposed of to a dedicated landfill. In the longer term it is proposed to develop a method whereby these residues will be recycled back to the process or incorporated into construction materials. This requires development of a suitable pre-treatment process to reduce the quantity of heavy metals or to fix them in a non-leachable form.

2.4 Mass Balance

A solids mass balance for a "typical" Neutralysis plant rated at 500 tonne(RDF)/day is presented in Table 1. The actual mass balance will depend on factors such as the content of moisture, recyclables and ash in the MSW, and the desired RDF to clay ratio.

TABLE 1
Solids mass balance for a "typical" Neutralysis plant rated at 500 tonne of RDF per day.

Process stream	Flowrate (tonne/day)	Operating Basis
MSW to RDF plant	776	1 process line, 5 day/week
RDF to mixer	500	1 process line, 7 day/week
Clay to mixer	440	1 process line, 7 day/week
Liquid waste to pelletisers	140	2 process lines, 7 day/week
Raw pellets to kiln lines	1080	2 process lines, 7 day/week
Aggregate from vitrifiers	556	2 process lines, 7 day/week
Dry scrubber and baghouse residues	23	1 process line, 7 day/week

3. ENVIRONMENTAL CONSIDERATIONS

3.1 Philosophy

The marketplace for technologies such as the Neutralysis process is such that the environmental acceptability of the process, its emissions and products, and indeed the philosophy of the company, will be closely scrutinised by regulatory agencies, financiers and other bodies such as environmental pressure groups. Provision of relevant and adequate information to these groups is a key aspect in demonstrating the acceptability of the process and in siting and maintaining the operation of commercial Neutralysis plants. It is also important to ensure that licensees of the technology continue to operate in an acceptable manner in order to avoid the situation which has arisen with MSW incinerators, where adverse publicity associated with old or badly operated plants has handicapped the siting of modern facilities.

Neutralysis Industries has established an independent Environmental Review Committee (ERC) to initiate action and advise on environmental matters affecting the process and its commercialisation. The ERC reports directly to the Board of Directors and is made up of representatives from universities and independent consultants.

3.2 Intrinsic Environmental Advantages of the Neutralysis Process

The Neutralysis process has a number of intrinsic environmental advantages as compared with traditional methods for the combustion of MSW:

1. Waste material is recycled into useful products, i.e. recovered secondary materials such as steel and aluminium, lightweight aggregate and energy. This results in overall material and energy savings by a community. When used in concrete instead of conventional aggregate, lightweight aggregate may result in lower material and energy use during construction and provide superior acoustic and insulation properties.
2. There is no "bottom ash", the material which normally constitutes such ash is either destroyed or incorporated into the aggregate. The aggregate is exposed to extreme oxidising conditions in the vitrification kiln (approximately 1100 ^{o}C for some 40 minutes); conditions necessary to produce the desired physical properties and which (as is shown below) also destroy residual organics and volatilise or fix heavy metals.
3. Stack gas emissions are more controllable because:
   - The feed to the kilns is relatively homogeneous in terms of composition and particle size and the presence of clay provides the pellets with a thermal inertia which minimises intra-pellet temperature variations during thermal processing.
   - Thermal processing occurs under staged firing conditions with well mixed solids.
   - All off gases from the kilns pass to a high temperature afterburner where temperature, residence time, mixing and oxygen content are set so as to maximise destruction of organics, and in particular dioxins/furans and their precursors. The afterburner operating conditions are "decoupled" from those of the kiln in that they can be set independently without affecting the quality of the product aggregate.

3.3 Testing Programme

Since 1988 a number of intensive studies on the pilot plant and a programme of product testing work have been undertaken in order to build a database of the type of information required by regulatory authorities and product users. Established standard protocols have been used where possible and a manual of standard operating procedures for environmental testing work on the plant has been developed. The protocols used in the stack gas testing work have closely paralleled those used by the US EPA and Environment Canada in the National Incinerator Testing and Evaluation Programme (NITEP) (1). The work reported here concentrates on results achieved for the aggregate.

The following aspects must be taken into account when considering the environmental acceptability of the aggregate:

a. Leaching from aggregate stored in uncovered stockpiles.

b. Leaching from concrete under different environmental conditions, e.g. oxidising atmospheric conditions relevant to exposed structures and reducing conditions relevant to some foundations.

c. Leaching from demolition waste containing the aggregate.

d. Occupational health aspects related to handling of the aggregate and working the concrete in which it is incorporated (e.g. breathing in dust generated by drilling activities).

Any aggregate should be tested to demonstrate its acceptability in each of the above situations. Unfortunately there is a lack of internationally recognised standard regulatory tests for applications such as (a)-(d). The definition of such tests and acceptance criteria should be a priority for standards authorities. The existence of such tests will aid the commercialisation of all processes designed to produce building materials from wastes. In their absence a programme of the following leaching test procedures has been undertaken on Neutralysis lightweight aggregate:

1. The US EPA Toxicity Characteristic Leaching Procedure (TCLP, EPA Method 1311).
2. A distilled water extraction test (ASTM D3987).
3. A 10% (v/v) nitric acid extraction test using the same liquid/solid ratio (20:1) as the TCLP, but different contacting methods and crushed or uncrushed sample.
4. The US EPA Multiple Extraction Procedure (MEP, EPA Method 1320). This test involves initial use of the US EPA Extraction Procedure (EP, EPA Method 1310) followed by 9 subsequent extractions of the same solid using synthetic acid rain.

In addition, the aggregate has been analysed to determine the oxides present and the total concentrations of heavy metals and trace organics.

The TCLP test is, in general, relatively aggressive compared with the proposed applications because it involves leaching with an organic acid, whereas application (a) involves leaching of intact aggregate with, at worst, a moderately acid solution (acid rain) and applications (b) and (c) involve leaching of intact aggregate incorporated in a concrete matrix under alkaline conditions. Most heavy metals are less soluble under alkaline conditions, although under some conditions metals such as Zn can be more soluble (2). However, it should be noted that the acceptance criteria set for leachate concentrations in the TCLP test (i.e. the concentration levels above which the material becomes categorised as a hazardous waste) have been set at values which produces an acceptable risk for substances leached from a co-disposal landfill (containing the solid under test and MSW) through an aquifer to the biosphere. This is a different pathway to that in applications (a) and (b) and thus the acceptance criteria must be viewed cautiously, particularly if they are approached by test results.

Planned work aims to address application (d) by development of a method to investigate occupational health respirability aspects of use of the aggregate in concrete, particularly during subsequent operations such as cutting, sawing and drilling.

3.4 Total Analysis of Aggregate

Neutralysis aggregate is typically a matrix of SiO_2 and Al_2O_3 with lower concentrations of other metal oxides such as Fe_2O_3, CaO and Na_2O. The total concentration of uncombusted carbon (as measured by the loss on ignition at 550 °C) is less than 1%.

Aggregate produced in the audit run was soxhlet extracted for organics analysis. The concentrations of 9 key polynuclear aromatic hydrocarbons (PAHs) were all below the analytical detection limit (<0.01 μg/g), as were those of 8 key chlorinated benzenes (<0.01 μg/g), chlorinated phenols (<0.06 μg/g), polychlorinated biphenyls (PCBs) (<0.01 μg/g), 2,3,7,8 tetrachloro dibenzo-p-dioxin (<0.006 μg/g) and 2,3,7,8 tetrachloro dibenzofuran (<0.006 μg/g). The failure to detect these organics in the aggregate is not surprising given the extreme oxidising conditions in the vitrification kiln. Detection limits for the analysed organics were generally similar or lower than the level "A" Dutch criteria for contaminated soil investigations, i.e. the level which conforms to the anticipated average background concentration that can be found in various uncontaminated Dutch soils. Examples of "A" levels are 1 μg/g for PAHs, 0.05 μg/g for chlorobenzenes, 0.01 μg/g for chlorophenols and 0.05 μg/g for PCBs (3). Aggregate organic levels less than the Dutch "A" criteria should present no threat in terms of soil or aquifer contamination.

Table 2 presents total heavy metal concentrations measured in Neutralysis aggregate produced during the audit run and the range of values measured for other "normal" runs. Also shown are the concentrations of cadmium, chromium, lead and mercury in aggregate produced before and after the addition of heavy metals into the kiln feed during one pilot plant run. The metals were added as an acetate salt solution and were equivalent to approximately 120% more cadmium than in the MSW, 45% more chromium, 10% more mercury and 300% more lead. This exercise was designed to simulate an incident in which additional heavy metals were present in the MSW feed.

TABLE 2
Heavy metal concentrations in Neutralysis aggregate produced by different production runs

Element	Audit run (mg/kg)	Range in other runs* (mg/kg)	Spiking run (before addition) (mg/kg)	Spiking run (after addition) (mg/kg)
Arsenic	14	<1-6	nm	nm
Barium	490	nm	nm	nm
Cadmium	<1	<0.5-2.5	<0.9	<1.4
Chromium	340	28-200	144	180
Copper	2020	nm	nm	nm
Lead	940	12-210	223	460
Mercury	0.25	0.006-0.24	<0.01	<0.01
Nickel	130	nm	nm	nm
Selenium	4	<1	nm	nm
Silver	9	0.9-34	nm	nm
Zinc	3590	nm	nm	nm

nm not measured * Range of results obtained from 17 samples, various laboratories

Table 2 shows that there are significant variations in concentrations of less volatile heavy metals in the aggregate, and the spiking run results indicate that this at least in part reflects the variation in heavy metal content of the MSW feed. The aggregate concentrations of cadmium and mercury do not show such a variation because these metals are almost totally volatilised to the kiln off gases.

3.5 Leaching Tests on Aggregate

Table 3 summarises the metal concentrations found in leachate obtained by TCLP, water, nitric acid and MEP extraction of aggregate samples. Analyses of TCLP extracts have shown the concentrations of regulated organics to be below detection limits, a finding consistent with the results for total organic analysis of the aggregate presented above.

TABLE 3
Leachate concentrations obtained using TCLP, water, nitric acid and MEP extraction of aggregate.

Element	US EPA TCLP criteria (mg/l)	TCLP (audit run) (mg/l)	TCLP (other runs[1]) (mg/l)	Water (audit run) (mg/l)	Nitric acid[2] (mg/l)	MEP[3] (1st extract) (mg/l)	MEP[3] (2nd extract) (mg/l)
Arsenic	5.0	0.015	<0.001-0.08	0.012	<0.1	0.89	0.09
Barium	100.0	0.21	<0.2-1.4	0.02	nm	<0.5	<0.5
Cadmium	1.0	<0.01	<0.001-<0.02	<0.01	<0.003-0.01	<0.01	<0.01
Chromium	5.0	<0.01	<0.002-<0.06	<0.01	0.03-0.22	0.03	0.02
Copper	none	1.29	nm	<0.01	0.83-5.4	nm	nm
Lead	5.0	<0.01	<0.01-0.17	<0.01	<0.1-0.23	0.32	0.16
Mercury	0.2	<0.001	<0.0005-<0.002	<0.001	<0.0002	<0.0005	<0.0005
Selenium	1.0	<0.05	<0.001-0.02	<0.05	nm	<0.05	<0.05
Silver	5.0	0.015	<0.001-<0.2	<0.1	nm	<0.02	<0.02
Zinc	none	0.32	nm	0.03	0.39-0.58	nm	nm

1 Range of results obtained from 21 runs, various laboratories — nm not measured
2 Range of results obtained from 2 runs, uncrushed aggregate extracted for 1 hour
3 Result from single run

The data in Table 3 shows that in all cases the TCLP extracts were substantially below the US EPA criteria. This was also the case when water, nitric acid and the EP leaching media were used. Subsequent extractions in the MEP procedure continued to demonstrate a decline in leachate concentrations to below analytical detection limits. The aggregate produced in the heavy metal spiking run also produced leachate concentrations of cadmium, chromium, lead and mercury which were below detection limits. Thus variation in the concentration of these heavy metals in MSW and aggregate does not appear to result in a significant increase in heavy metal leachability.

As was mentioned above, the TCLP criteria is not strictly applicable to all aggregate applications. However, the margin by which the aggregate passes the TCLP test suggests that the leachability of the regulated heavy metals is environmentally acceptable. This conclusion does, however, require further confirmatory work using leachate tests designed to better simulate conditions found in the final aggregate applications. These further tests should also explore the environmental significance of metals such as copper and zinc, for which no TCLP criteria are specified. The nature of these tests, and the appropriate acceptable concentration criteria, are matters best considered within the context of development of new and internationally accepted standards.

4 CHARACTERISTICS OF CONCRETE CONTAINING AGGREGATE

Neutralysis lightweight aggregate is less dense than conventional aggregate. Its use produces concrete with a density approximately 75% of that made with normal dense aggregate. This results in lighter structures, thus reducing the scale of the required foundations and other supporting work. The aggregate has superior thermal and acoustic insulation properties as compared with conventional aggregate. It wears to present a rough anti-skid surface rather than a smooth surface, a characteristic making it suitable for incorporation into road surfaces. The operating conditions of the Neutralysis process can be adjusted to achieve aggregates with characteristics suitable for different applications.

A programme of physical testing work using Australian standard methods is underway to assess the engineering properties of a Portland cement concrete made from Neutralysis aggregate. The optimal operating conditions for a premium product are in the course of being defined, however the following results have been achieved at this stage:

- Compressive strengths up to a maximum of 53 MPa at 28 days.
- Hardened concrete densities in the range 1800-2080 kg/m^3, depending on the strength level of the concrete. This compares with the range for normal dense concrete of 2250-2400 kg/m^3.
- Indirect tensile splitting strengths which indicate a similar relationship to normal dense concrete.
- Drying shrinkages at 56 days of 600-850 μs and at 112 days of 750-1000 μs (a normal dense concrete control recorded 680 μs at 56 days and 780 μs at 112 days).
- Limited creep testing on early product indicated rapid creep to levels ranging between 2200-2700 μs at 3 months, followed by a subsequent levelling off to 2500-3000 μs at 12 months.
- The sensitivity of the concrete to the alkali/silica reaction has been tested on early product using accelerated methods. Cements with alkali contents ranging from 0.5% (normal) up to 2% all had an expansion at 12 months. This was less than the shrinkage at 112 days. Further tests are now in progress using a new and patented accelerated test method.
- Youngs Modulus (E) values on various products ranging from 2.0-2.3 GPa.

The physical testing work has shown that Neutralysis lightweight aggregate can be used to produce structural and lower grades of concrete, and in particular concrete masonry block, which satisfies current design and quality standards with considerable weight savings.

5 CONCLUSIONS

1. The Neutralysis process produces a marketable lightweight aggregate which is acceptable from the point of view of both environmental and civil engineering properties.
2. There is a need for internationally recognised tests to assess the environmental and occupational health acceptability of all construction materials (not just those derived from waste).

REFERENCES

1 Environment Canada (1988) National Incinerator Testing and Evaluation Programme: Environmental Characterisation of Mass Burning Incinerator Technology at Quebec City. Summary Report EPS 3/UP/5.

2 van der Sloot, HA (1990) Leaching Behaviour of Waste and Stabilised Waste Materials: Characterisation for Environmental Assessment Purposes. Waste Mgt & Res, 8, pp 215-228.

3 Netherlands (1983) Implementation of the Soil Clean-up (Interim) Act. The Netherlands Ministry of Housing, Planning and the Environment, April 1983.

Waste Materials in Construction.
J.J.J.R. Goumans, H.A. van der Sloot and Th.G. Aalbers (Editors)

THE EFFECTIVENESS OF GRANULATED BLASTFURNACE SLAG

M. HANAFUSA[1] and T. WATANABE[2]

[1] Chiba Riverment and Cement Corporation, 1 Kawasaki - cho Chiba City (Japan)

[2] Tech. & Eng. Dept, Cement Division, Ube Industries Ltd, 1978 - 2 Kogushi Ube City (Japan)

ABSTRACT

Effective utilization of Blastfurnace Slag is very important for the steel indusry. In Japan almost one hundred percent of Blastfurnace Slag is utilized. Above all, its use as cementitious materials is most valuable and environmentally advantageous.

This report presents the history of Blastfurnace Slag utilization in Japan, including an example of a manufacturing enterprise using Ground Granulated Blastfurnace Slag (GGBS) as a cementitious material.

Features of GGBS from an environmental viewpoint - especially concerning CO_2 emission - together with the necessity of international co-operation to promote slag utilization is also discussed in the Report.

1. INTRODUCTION

The production of iron and steel generates large amounts of by - products and waste materials of which blastfurnace slag accounts for the greater proportion of same.

In Japan, blastfurnace slag was mainly disposed of as waste or otherwise used only for reclamation fill in former days. However, almost one hundred percent of blastfurnace slag is presently, for example as material for cement, road making and fertilizer, etc. Above all, its use as cementitious materials has proven to be most valuable and environmentally advantageous.

2. HISTORY OF BLASTFURNACE SLAG UTILIZATION IN JAPAN

2.1 Treatment Process of Blastfurnace Slag and its Uses.

Figure 1 shows the generation and treatment process of blastfurnace slag. It will be noted that the slag generation ratio is about 300 kgs/tonne - iron.

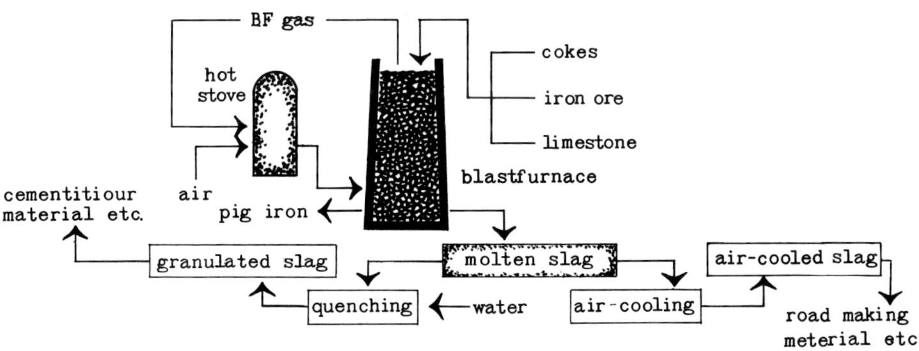

Fig. 1 Generation and treating process of blastfurnace slag

There are two cooling methods for molten slag :

(i) slow air - cooling, and
(ii) rapid water - cooling.

Air - cooled slag is rockform and used as a material for road making, concrete aggregare, fertilizer, etc. , after the sizing process has been completed.
On the other hand, water - cooled slag is sandform called Granulated Blastfurnace Slag (GBS) , and is mainly used to make cementitious material.

2 . 2 Supply and Demand of Blastfurnace Slag.

Figure 2 shows annual amounts of slag generation and utilization. (1) As seen in this figure, they are almost balanced.

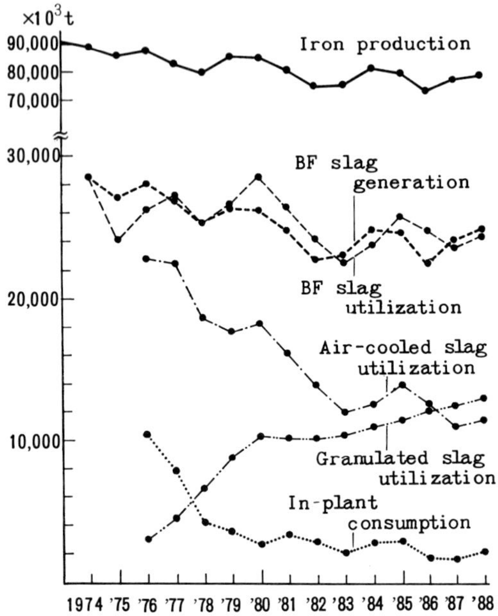

Fig. 2 Annual amounts of slag generation and utilization

The GBS has been rapidly increasing while the air - cooled slag has decreased. The reduction of in - plant consumption means that the reclamation of steel works has almost come to an end.

Figure 3 shows the ratio of slag granulation and sales for cement use. (2) Both ratios have been steadily increasing.

The factors influencing this tendency are as follows :

(i) In 1979 admixing of GBS in Portland cement of up to 5 % was approved by revision of the JIS (Japanese Industrial Standards) .

(ii) The demand of blastfurnace cement has increased due to the necessity for energy saving after the oil crisis.

(iii) In 1984 degradation of concrete structures due to alkali - aggregate reactoin became a serious problem and the mitigating effect of blast - furnace cement on this reaction was recognised by the government.

(iv) In 1986 the use of Ground Granulated Blastfurnace slag (GGBS) as an admixture in concrete was recognised in JSCE (Japan Society of Civil Engineering) standard.

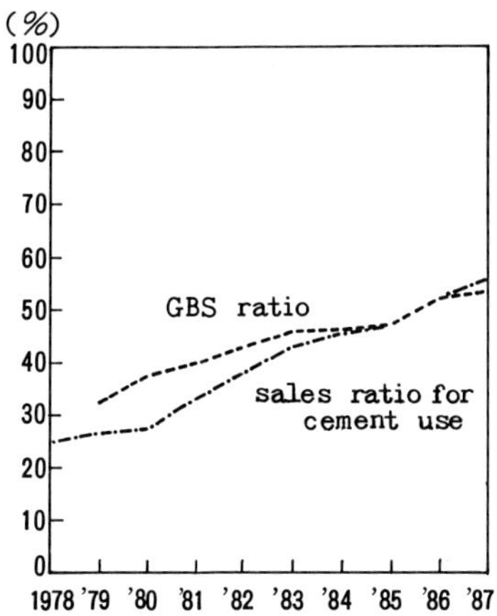

Fig. 3 Ratio of slag granulation and sales for cement use.

3. AN EXAMPLE OF GGBS MANUFACTURING ENTERPRISE

(The experience of Chiba Riverment & Cement Co.)

3. 1 The Motives to start New Business.

Many Japanese steelmakers have begun to diversify their business. Since the early 1980 ' s Kawasaki Steel also commenced studies on the more effective use of by - products and wastes from their steelworks. As a result, in 1984 a (new) joint venture company was established with Ube Industries, Ltd. (one of the major chemical and cement manufacturers in Japan) and known as Chiba Riverment & Cement Co. (CRC) .

The purpose of this company was to make GGBS in their steelworks and to sell the product through the cement makers' networks. The operation has proved to be successful for the following reasons:

* GGBS is one of the most value - added materials as utilization of blastfurnace slag;
* GGBS has sufficient quality as compared with Portland cement;
* New energy - saving technology has evolved around the grinding of slag with the development of the "vertical mill";
* Chiba Steel Works is located in the Tokyo metropolitan area where large quantities of cement are consumed;
* GGBS is very beneficial in terms of environmental impact.

3. 2 General Description of Equipment.

As seen in Figure 4, this plant consists of five major pieces of equipment:

(i) raw material storage:

hoppers, belt conveyors, etc., for transporting slag and gypsum to the mill.

(ii) grinding system:

vertical roller mill, hot air furnace, air exhaust fan, etc., for drying, grinding and classifying.

(iii) product collection system:
bag filters, etc., for collecting GGBS

(iv) finished products storage:

GGBS storage silos, feeders and weighing machines, etc., for shipment.

(v) mixing:

tanks, mixers, feeders, etc,. for mixing of GGBS with other materials prior to shipment.

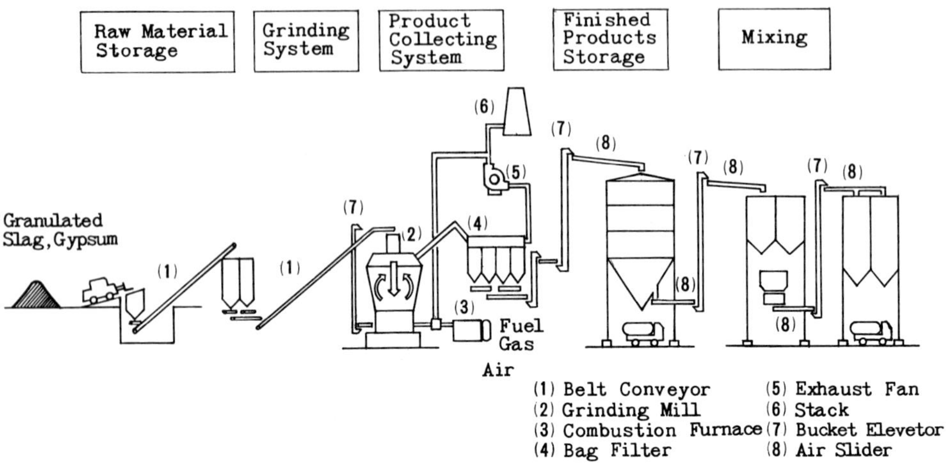

Fig. 4 Schematic Diagram of CRC Equipment

3.3 <u>The Feature of the Grinding System.</u>

Figure 5 shows the section of the Roller Mill used in CRC.

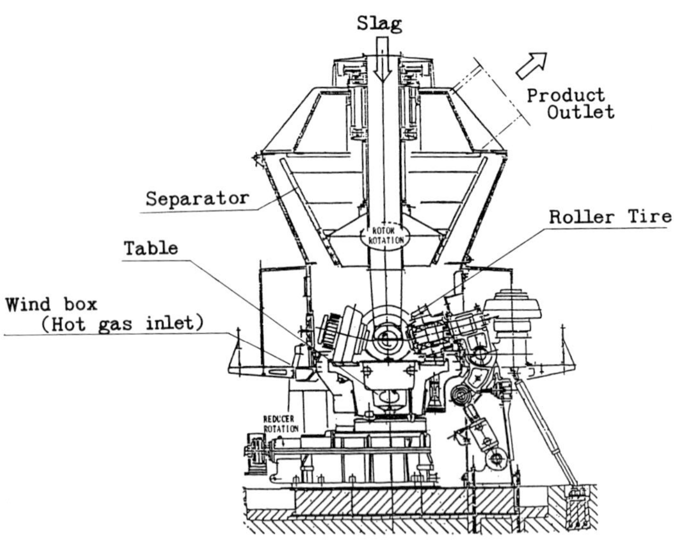

Fig. 5 Vertical Roller Mill

The most unique piece of equipment in CRC is its Vertical Roller Mill (as shown in Fig. 5). Roller Mills have been widely used for grinding raw materials or cement howevcr, prior to the introduction at CRC, the use of a Vertical Roller Mill exclusively to grinding slag had not been attempted. The reason for this was that although the Roller Mill had the advantage of saving energy as compared with grinding by a Ball Mill, there were some doubts about the quality of products and the ability of the roller and table to withstand abrasion.

However, Ube Industries in conjunction with Loesche (Germany) resolved both these problems and as a result CRC commissioned the first Vertical Roller Mill for grinding slag in the world.

3.4 Operational Results.

The capacity of this plant is 28,000 t/month, and production has been steadily increasing since the start of operation in 1985.

Table 1 shows the comparison of energy consumption between Riverment (the name of GGBS produced by CRC with Vertical Roller Mill), GGBS ground with conventional Ball Mill and Portland cement. The production of GGBS with Vertical Mill is therefore very effective in energy saving terms.

Table 1 Comparison of energy consumption

	GGBS with vertical mill	GGBS with ball mill	Portland cement
Electric power (KWH/t)	45	80	100
Fuel (KCal/t)	90,000	130,000	1,020,000

Table 2 and 3 show chemical and physical properties respectively.

Table 2 Chemical composition of Riverment

	(%)								Basicity	Moisture (%)
	ig•loss	SiO_2	Al_2O_3	Fe_2O_3	CaO	MgO	SO_3	S		
Riverment	0.5~0.9	31~35	13~16	0.2~0.4	40~43	5~8	1.7~2.3	0.6~0.8	1.9	0
JSCE Standard	≤3.0	–	–	–	–	≤10.0	≤3.0	≤2.0[1]	≥1.4	≤1.0

Table 3 Physical quality of Riverment

	Appearance	Specific gravity	Fineness			L-value
			Specific surface (cm^2/g)	90 μmR (%)	45 μmR (%)	
Riverment	white powder	2.9	4,200~4,600	0	≤1.0	85
JSCE standard	–	≥2.8	≥2,750	–	–	–

Table 4 shows the test examples of mortars with a mixture of Portland cement and GGBS. In general, the medium and long term strength is equal or higher than that of ordinary Portland cement.

Table 4 Test example of mortar with mixture of portland cement and Riverment

Mixing ratio (%)		Compressive strength (kgf/cm²)			
Portland cement	Riverment	3 day	7 day	28 day	91 day
100	0	150	264	429	491
90	10	145	250	430	515
70	30	132	227	435	540
50	50	99	181	431	589
30	70	67	151	356	548

3.5 Promotion of Official Recognition of Blastfurnace Slag as Cementitious Material.

There are two methods of slag utilization as cementitious material: one is to use GBS in the process of cement manufacturing in the cement plant, and the other method is to use GGBS as an admixture in the concrete mixing and cement products manufacturing processes. The former method has been authorised as BF (Blastfurnace) cement in JIS, whilst the latter has not yet been appproved in JIS, although under work by the authorities. It has already been recognised in JSCE standard as previously mentioned.

4. THE FEATURES OF GGBS/BF CEMENT

GGBS and Blastfurnace cement have many features as compared to ordinary Portland cement. Some of these features are noted hereunder.

(1) Superior Quality of Concrete using GGBS/BF Cement:

* Sufficient strength;
* Alkali - aggregate reaction mitigating effect;
* High durability - especially for marine structures;
* Low hydration heat.

(2) Low Cost:

* Low production cost - due to simple and energy saving process etc. ;
* Low transportation cost (GGBS) - due to direct transport to nearby consumption area.

5. CONTRIBUTION OF GGBS/BLASTFURNACE CEMENT TO ENVIRONMENT

By replacing Portland cement with GGBS /BF cement, the reduction of adverse environmental effects can be achieved.

5.1 Prevention of Greenhouse Effect.

In the cement industry emission of CO_2 is due to not only combustion of fossil fuels, but mostly to heat decomposition of limestone, viz. : $CaCO_3 \rightarrow CaO + CO_2$.

Figure 6 shows the CO_2 emission ratio in Japan in 1988. (3) The cement industry is second only to the steel industry in terms of CO_2 emissions.
On the other hand, the processing of GGBS generates only 10 percent of CO_2 emissions compared with the Portland cement process because blastfurnace slag has already been calcined in the ironmaking process. Energy consumption of GGBS production is also much less as previously mentioned.

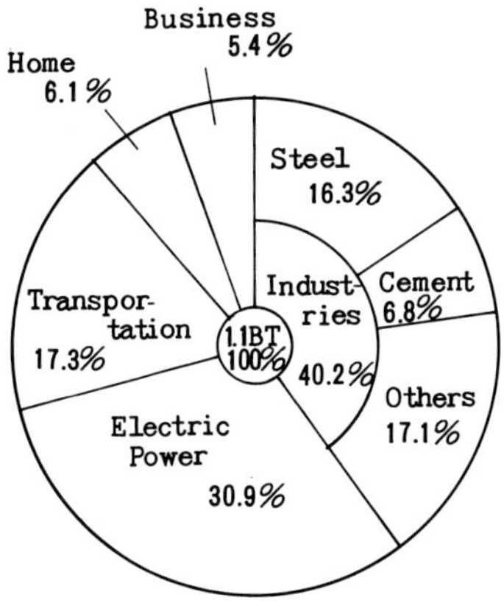

Fig. 6 CO_2 emission ratio in Japan in 1988

In Japan annual generation of Blasstfurnace slag is about 25 million tons, of which approximately 50% is utilized as cimentitious material. By our calculation, CO_2 emission from the cement industry would decrease by about 10 million tons or 13% if all slag could be utilized as cementitious material.

5.2 Preservation of Resource and Conservation of Natural Landscape.

* Limestone mines can be preserved.

5.3 Effective Utilization of Industrial Wastes.

* Blastfurnace slag is the largest waste product in the steel industry;
* Gypsum : most of the GGBS products contain about 4 % of gypsum, which is also waste from the desulphurising process of emission gas in iron ore sintering plant.

5.4 Minimization of Pollutants.

Quite different from Portland cement, the GGBS manufacturing system is very simple and blending, drying, grinding and separating are done in a mill simultaneously, which means effective prevention of fugitive dust and other pollutants. Furthermore, low energy consumption leads to low gas emission such as SO_2 , NOx, etc.

As mentioned above it is obvious that GGBS/BF cement are very beneficial, environ mentally.

The Japanese government recognised GGBS/BF cement as "Eco - mark" goods in September, 1990, which means that they are seen to be excellent from the environmental point of view. Furthermore, the government decided on "the action programme for the prevention of the global warming" at the Cabinet Council in October, 1990, in which two items were concerned with the utilization of blastfurnace slag as cementitious material. They were :

* To promote the use of blastfurnace cement in the construction industry :
* To promote popularisation of CO_2 reducing products through the Eco - mark system.

6. INTERNATIONAL OVERVIEW OF BLASTFURNACE SLAG UTILIZATION AS CEMENTITIOUS MATERIALS.

6.1 Present Condition of Blastfurnace Slag Utilization as Cementitious Matarials.

Annual production and utilization rates in twelve countries in 1984 are shown in Table 5. The utilization ratios largely vary between the countries and the average value is 34%. If all slags were utilized as cementitious material, CO_2 emission from these twelve countries would decrease by about 55 million tons per year.

Table 5 Annual production and utilization rates of Blastfurnace slag as cementitious materials in 1984 ($\times 10^6$ tons)

COUNTRY	Production (A)	Utilization (B)	% B/A
AUSTRALIA	4.7	0.12	3
CANADA	2.9	0.2	7
CHINA	22	16	73
FRANCE	10.4	1.9	18
GERMANY FR	15	2.8	19
INDIA	7.8	2.8	36
JAPAN	24	8.2	34
NETHERLANDS	1.1	1	91
NORWAY	0.1	none	0
SOUTHAFRICA	1.5	0.6	40
SWEDEN	0.1	0.03	30
UNITED KINGDOM	1.5	0.25	17
UNITED STATES	13	1	8
Total	104.1	34.9	34

6. 2 International Cooperation to Promote Slag Utilization.

In order to increase effective slag utilization, the authors have been cooperating in the introduction of GGBS plants not only in Japan but abroad, also (e. g. Australia, Korea and Taiwan) .

However, the following actions are considered to be very important for further progress in this field :

* Promotion of official recognition for GGBS in each country :
* Establishment of international information exchange mechanism on cementitious slag, including both makers and users.

7. Concluding Remarks.

In Japan, almost all generated blastfurnace slag is utilized, with the blastfurnace slag used as cementitious material being considered the most efficient from the standpoint of technical quality, economic cost and environmental protection.

Above all, contribution to a better global environment through the decrease of CO_2 emission is very significant, and the authors believe it is a good example of the coexistence of economy and environment, i. e. "Sustainable Development". Governments and the industries should therefore be encouraged to promote such utilization on the basis of national and international cooperation for "Our Common Future".

Referenece

1. Nippon Slag Association, Utilization of Blastfurnace Slag for Cementitious Material, 1990, P. 46. (In Japanese)
2. Nippon Slag Association, History of the Decade, 1988, PP.12. (In Japanese)
3. Asahi Newspaper, Co_2 Emission Rates in Japan, 29 October 1990. (In Japanese)
4. P. K. Mehta, Pozzolanic and Cementitious By - Products in Concrete, Proceedings of the 3rd International Conference on Fly Ash, Silica Fume, Slag and Natural Pozzolans in Concrete, Trondheim, Norway 1989, P. 35.

Waste Materials in Construction.
J.J.J.R. Goumans, H.A. van der Sloot and Th.G. Aalbers (Editors)

THE GRANULATED FOUNDRY SLAG AS A VALUABLE RAW MATERIAL IN THE CONCRETE AND LIME-SAND BRICK PRODUCTION

J. MAŁOLEPSZY, W. BRYLICKI and J. DEJA

Academy of Mining and Metallurgy 30-059 Kraków, al. Mickiewicza 30, paw.B-6
POLAND

SUMMARY

The properties of granulated foundry slags were studied in order to evaluate their usefulness as a cementitious material in the alkali activated slag concretes and autoclaved building materials. This work is a part of the extended studies dealing with the utilization of several industrial by-products and wastes, particularly metallurgical slags (1-7). The most important and successful investigations were carried out in the alkali activated slag concrete technology.

1. THE FORMATION AND THE PROPERTIES OF GRANULATED FOUNDRY SLAGS.

The granulated foundry slags are formed in the cast iron production. During the outflow of the metal from the cupola through the shaft runner the separation of slag occurs. The slag flowing continuously to the granulation chamber undergoes the granulation proces under the water jet.

The sample of the granulated foundry slag from the Śrem work was averaged dried and ground in a laboratoratory mill to the Blaine specific surface 3200 cm^2/g (density equal to 3.04 g/cm^3).The chemical composition is given in table 1.

TABLE 1
Chemical composition of the Śrem granulated foundry slag.

Component	CaO	MgO	SiO_2	Al_2O_3	Fe_2O_3	SO_3	l.o.i.
Percentage % by weight	34.46	1.56	38.86	7.57	10.48	0.65	6.22

As it results from the table 1 this slag should be classified as an acid and poorly active material. The natural radioactivity measurements prove that this slag shows very low radiation and can be used as a raw material in the building materials production following the Polish standard requirements.

The phase composition and vitreous phase content were estimated by means of the polarizing microscope. The main substance consists of the colourless silica glass. The crystalline components such as diopside $CaO \cdot MgO \cdot 2SiO_2$, fayalite $2Fe_2O_3 \cdot SiO_2$, rankinite $3CaO \cdot 2SiO_2$, mullite $3Al_2O_3 \cdot 2SiO_2$, magnetite Fe_3O_4, hematite Fe_2O_3 and metallic iron are non-uniformly distributed. The weight fractions of the vitreous and crystalline phase are 96.70 % and 3.30 % respectively. The XRD pattern shows, apart from the shoulder in the range 25 - 35° 2θ corresponding to the high vitreous phase content, only the peaks originating from the quartz (d = 4.23, 3.34, 2.462, 2.127 Å).

2. PROPERTIES OF CONCRETES.

The ground granulated foundry slag (B-1) and the mix consisting of the 70% granulated blast furnace slag and 30 % granulated foundry slag (B-2, B-3) were used as cementitious materials in concrete preparation. The following activators were used : NaOH added as 5 % of binder and the mix consisting of 3.3 % NaOH and 2.5 % Na_2CO_3. The fractioned gravel aggregate and sand were introduced to the concrete as the fillers. The composition and properties of concrete mix are given in table 2.

TABLE 2

Component	Proportions kg/m^3 in		
	B-1	B-2	B-3
Foundry slag	319	99	99
Blast furnace slag	-	224	224
Sand (0-2 mm)	719	719	719
Aggregate (2-20 mm)	1173	1152	1164
Water	141	143	143
NaOH	16	16	9.8
Na_2CO_3	-	-	8.3
W/S	0.44	0.44	0.44

The concrete cubes 15 x 15 x 15 cm were casted, vibrated on the vibrating table and subsequently stored at natural conditions (temperature 20°C,90% RH). The properties of these concretes are given in tables 3, 4.

TABLE 3

Concrete properties

Type of concrete	Compressive strength (MPa) at various ages				Absorbability
	3 d	7 d	28 d	90 d	%
B-1	3.0	7.6	17.8	20.4	6.8 %
B-2	7.0	12.4	21.4	26.2	5.2
B-3	18.2	23.2	28.2	31.2	3.8

TABLE 4

Shrinkage of concretes

Type of concrete	Shrinkage (mm/m) at different ages				
	3 d	7 d	28 d	90 d	180 d
B-1	0.321	0.427	0.543	0.618	0.621
B-2	0.281	0.327	0.374	0.421	0.428
B-3	0.181	0.224	0.281	0.284	0.301

3. THE LIME-SAND BRICK PROPERTIES

The slag-lime mixtures of different composition were prepared with aim to evaluate the usefulness of foundry slags in the lime-sand brick technology. The mixture proportions are given in table 5.

TABLE 5

Mix proportions and properties of lime-sand brick

No	Mix components % by weigth			Properties		
	quick lime	foundry slag	sand	compressive strength (MPa)	absorbability (%)	frost resistance
0	8	-	92	13.4	11.7	
1	8	1	91	14.5	11.3	fully
2	8	2	90	17.1	11.4	resistant
3	7	2	91	13.2	11.9	
4	7	1	92	12.8	12.4	
5	6	2	92	11.5	12.2	

The mixtures were homogenized with water (7.5 %) and subjected to the maturing proces at 80°C within 4 hours. Subsequently the cubes 4 x 4 x 4 cm were casted under the psessure 20 MPa and autoclaved under the pressure 16 MPa during 4 hours. The results of tests are shown in table 5.

4. DISCUSSION

The results of tests and investigations prove that the granulated foundry slag can be utilized as a component of alkali activated concretes. The better results can be achieved as the foundry slag is mixed with the granulated blast furnace one because of the acidic character of the former. The granulated blast furnace slag addition gives more beneficial phase composition and microstructure of hardened concrete enriching the hydrated material in the microcrystalline CSH phase. Apart from the higher compressive strength, the concrete shows relatively low shrinkage.

The foundry slag can be also used as a sand replacement in the lime-sand production giving the 25 % compressive strength increase. This effect is the consequence of the partial hydration of slag component with the formation of CSH and tobermorite which are responsible for the compressive strength of final product. It is also clear that the slag can play the role of lime replacement substituting 15 % of lime component in these materials.

5. REFERENCES

1. A. Derdacka, J. Małolepszy., Cement-Wapno-Gips 10 (1975). 291-295
2. J. Małolepszy., Proc. 8 th I.C.C.C. 1986 Rio de Janeiro Vol. IV,104-107.
3. J. Małolepszy., J. Deja, Silicates Industriels 12 (1988) 179 - 186.
4. J. Małolepszy , Proc. 10 th Baustoff und Silikattagung "Ibausil" 1988 Veimar vol II.
5. J. Deja, J. Małolepszy, Proc. 3 th Int. Conf. of Fly Ash Silica Fume, Slag and Natural Pozzolans Concrete. 1989 Trondheim Vol. II.
6. W. D. Głuchowski, Gruntosilikatni wirobi i konstrukcji. Budiwilnik. Kijów 1967.
7. J. Małolepszy,Scientific Bulletins of the Stanisław Staszica Academy of Mining and Metallurgy, Section Ceramics Cracow, Poland.

Waste Materials in Construction.
J.J.J.R. Goumans, H.A. van der Sloot and Th.G. Aalbers (Editors)

TECHNICAL EXPERIENCE IN THE USE OF INDUSTRIAL WASTE FOR BUILDING MATERIALS PRODUCTION AND ENVIRONMENTAL IMPACT

K. Popović[1], N. Kamenić[1], B. Tkalčić-Ciboci[1], V. Soukup[2]

[1]Civil Engineering Institute, University of Zagreb, J. Rakuše 1.
41000 Zagreb, Croatia (Yugoslavia)

[2]Institute for Public Health of the City of Zagreb, Mirogojska 16,
41000 Zagreb, Croatia (Yugoslavia)

SUMMARY

Due to the shortage of portland cement and to the intention of decreasing its production costs, use of industrial waste materials is very common in Yugoslavia, and over 90 per cent of cements contain hydraulic active mineral admixtures such as blast furnace slags, fly ashes and pozzolanas. Long term practical experience has confirmed well known facts that besides the above mentioned reasons for the addition of these materials to the cement, their use can improve its properties i.e. increase chemical durability, reduce heat of hydration and even augment strength development (e.g. artificial pozzolana such as condensed silica fume - a waste from ferro alloys production).

So far the experts have considered mostly the technical consequence of such industrial wastes application as secondary materials and have paid very little attention to the environmental aspects of that re-use, i.e. health hazzards caused by water and soil pollution as a consequence of leaching, or by radioactivity of some industrial wastes etc.

This paper gives a short presentation of some specific characteristics and long term experience in practical use of waste materials for cement production, together with the results of the first attempt to determine heavy metals content and their leaching from slag, ash and mortars. This will serve to compare the health hazard caused by fly ash and slag addition to portland cement with those of conventional building materials.

1. GENERAL

Thirty years ago there was a lack of cement kiln capacities in Yugoslavia to satisfy the needs of intensive construction activities. The simplest way for increasing the production was to make blended cements admixing hidraulic active materials such as natural pozzolanas and blast furnace slag. In the meantime the cement plants have developed their possibilities and even surpassed the needs and nowadays only two thirds of cement production capacities are used. But the custom of using mineral admixtures to cement is still present in order to save energy for clinker burning and to reduce cement production costs. For that reason only about 10 per cent of 8 million tons of cement produced in Yugoslavia in 1990 was without mineral admixtures and the rest contains between 15 and 25 per cent of inexpensive components.

The assortment of mineral admixtures has been changing with time. In the begining there was a fifty-fifty ratio between natural pozzolanas and blast furnace slag. The mentioned tendency energy saving and production costs reduction in some cases brought about the use of inadequate natural pozzolanas which exhibit poor hydraulic activity and/or increasing water demand. Since standard specifications and customer requirements for the selection of mineral admixtures has become stronger, the proportion of inadequate pozzolanas was significantly reduced. Blast furnace slag addition to portland cement is not accompanied with such harmful appearences but the available quantities of granulated slag conforming to requirements for such use are not sufficient. Because of that, possibilities of introducing other hydraulic active admixtures to portland cement have been investigated.

Owing to their chemical composition and particularly the structure which is a consequence of their formation process, a number of industrial wastes exhibit hidraulic properties and can be used as admixtures to portland cement. The most frequent and important of such materials besides blast furnace slag, which has the longest tradition for these purposes, is fly ash from coal burning in power plants, but some other types of slags, ashes and artificial pozzolanas are also used. Experience from industrial use as well as some results from laboratory investigations obtained with different industrial wastes are described in the following pages:

2. SLAGS

As already mentioned blast furnace slag is the most commonly used industrial waste in cement production. If its chemical character is basic and if appropriately cooled(quick enough)to obtain non crystalline structure, it will exhibit satisfactory hydraulicity when the necessary activators are present During hydration process portland cement paste contains enough lime and sulphates for hydraulic activation of slag.

The presence of slag somewhat decreases cement strength development, but in accordance with experiences from long term practice in Yugoslavia, cements containing up to 20 and sometimes even up to 30 per cent of B.F.S. meet the requirements for strength of 45 MPa after 28 days when clinker of adequate quality is used and fineness of grinding is sufficient. Cements marked PC 15 z 45 and PC 30 z 45 (portland cements containing up to 15 or 30 per cent of B.F.S.) developing 45 MPa compressive strengths after 28 days and 15 MPa compressive strengths after 3 days of hardening are the most common in Yugoslavia. Blended cements containing appx. 50 per cent of B.F.S. and developing 35 MPa compressive strength after 28 days are also produced though in limited quantities.

The other beneficial effect of B.F.S. in preventing damages caused by sulphate attack has been neglected in practice so far although blended cement with addition of more than 70 per cent of B.F.S. is considered as sulphate resistant according to yugoslav standard JUS B.C1.014. As a consequence of such cement assortment and composition, all available quantities of B.F.S. in Yugoslavia are consumed by cement factories, and a small quantity is even imported. For the same reason some cements contain slag and pozzolanic material in the mix.

Only blast furnace slag is defined as mineral admixture to cement by regulations and standards, whereas other slags are not allowed for these purposes, although some of them also have satisfactory hydraulic activity. The reason for that is primarily the possible risk of unsoundness caused by delayed hydration of free Cao and MgO. Slag from steel production which is available in great quantities is a typical example for that. But there is a **slag from Si-Mn** alloy production which, when appropriately quenched, exhibits good hydraulic activity and cannot be dangerous in the sense of unsoundness with regard to the chemical composition (SiO_2 40.5%, CaO 26.4%, Al_2O3 18.7% Fe_2O_3 0.3%, MgO 2.2%, CaO_{free} 0%) and temperature of forming. Laboratory examinations have shown that the addition of Si-Mn slag to portland cement does not decrease its strengths more than other conventional mineral admixtures do and that any harmful influence to other cement properties was not noticed. Grindability of that slag is fair. Industrialy produced cement in 70 t/h mill confirmed those results. (Table 1).

3. FLY ASH

In spite of large quantity of fly ash in Yugoslavia, the use of this admixture to portland cement is limited for two main reasons:

- variations in composition - especialy in sulphate content
- some fly ashes increase water demand of concrete and mortar thus decreasing strength and durability of these cement composites

There is however a power plant in **Kakanj** near Sarajevo which produces very specific **fly ash.** Its chemical characteristics (SiO_2 42, Al_2O_3 19, Fe_2O_3 9, CaO 23, SO_3 2 per cent etc) and pozzolanic activity (approx. 10 MPa) are in the usual range but the ash significantly decreases water requirement of portland cement (and concrete as well). Cement factory which is situated next to the power plant produces two types of cement: one with addition of approx. 20 per cent fly ash for ordinary purposes with 28 days compressive strengths of 45 MPaa, and low heat cement containing about 50 per cent fly ash and developing 28 days compressive strengths of 25 MPa.

TABLE 1

Properties of cements with Si-Mn slag admixture produced in laboratory and industrial mill

Sample	Composition(%) Clin.	Slag	Gips.	Blaine (m^2/kg^{-1})	Water stand(%) consist.	Setting time (min) init.	fin.	Compressive strengths(MPa) 3d	7d	28d
Labor.	95	0	5	408	24.0	70	160	31.9	40.6	47.4
Labor.	85	10	5	395	24.5	60	140	24.8	36.0	45.5
Labor.	75	20	5	387	24.6	60	140	24.0	32.3	43.6
Labor.	65	30	5	395	25.0	70	140	21.0	28.1	40.8
Industr.	95	0	5	340	27.1	110	180	24.3	38.9	50.2
Industr.	75 + 10 pozzolana	10	5	380	27.3	100	160	21.9	29.5	43.9
Industr.	75	20	5	340	27.0	100	170	19.8	27.9	41.6

Due to the mentioned influence of the fly ash, water requirement of ordinary portland cement is between 23 and 24 per cent, and that of low heaat cement is around 21 per cent.

The explanation for such behaviour of the fly ash is the particle shape which is spherical due to the burning process temperature of 1600°C.At this temperature fly ash melts, particles become spherical and keep this shape after quenching by air. Such particles act as a ball bearing in the fresh mixes of concrete or mortar.

That enables the preparation of concretes with significantly lower water to cement, ratios which improve strengths and durability. For instance, by using low heat cement which in standard cement mortar (w/c=0.5) develops 28 days compressive strengths of 25 MPa, concretes having over 40 MPa 28 days strengths and even 50 MPa 90 days strengths are produced without using water reducing agents. At the same time the hydration heat of this cement is limited to 250 kJg^{-1} after 7 days and to 295 kJg^{-1} after 28 days (1)

So in addition to decreased cement production costs and energy saving for clinker burning, significant improvement of cement properties is achieved. Portland cement from the Kakanj factory is used for ordinary purposes and low heat cement has been used for several construction works for water power plant dams on the rivers Vrbas, Neretva, Drina etc. Concretes of very good strengths vere produced and the heat of hydration did not exceed permitted limits.

It is also worth to mentioning the fly ash from power plant Kosovo, which is characterized by high sulphate content (up to 16 per cent of SO_3).

Though some professionals have refused to use this fly ash as admixture to cement just because of the mentioned SO_3 content, the experiments have shown that this sulphate can be used for retarding hydration of mineral C_3A i.e. for the regulating cement setting time. Although a small quantity of raw gypsum (1 per cent) has to be added into fresh cement paste when ash addition in only 10 per cent. Also total sulphate content in portland cement remains in the range defined by standards if fly ash proportion in the cement is up to 20 per cent.

After solving the problem of SO_3 content this fly ash can be succesfully used as mineraal admixture because it has satisfying pozzolanic activity (10 MPa) and does not show the tendency to increasing cement water requirement. (Table 2) shows the influence of Kosovo fly ash and of SO_3 content on cement setting time and strength development.

TABLE 2

Properties of laboratory produced cements with admixture of Kosovo fly ash

Composition (%)			Setting time (hours, min)		Compressive strengths (MPa)		
Clink.	Ash	Gypsum	init.	final	3d	7d	28d
99	0	1	flash set		19.1	21.7	32.0
96	0	4	$3,^{00}$	$6,^{10}$	21.0	32.0	47.0
90	10	0	$0,^{50}$	$2,^{30}$	19.9	27.2	38.8
90	20	0	$2,^{00}$	$4,^{00}$	24.5	33.3	42,5
70	30	0	$3,^{40}$	$6,^{20}$	18.0	27.2	38.7
89	10	1	$1,^{10}$	$4,^{20}$	22.7	33.1	47.0
86	10	4	$3,^{50}$	$5,^{20}$	24.0	35.0	46.6
79	20	1	$3,^{10}$	$5,^{00}$	24.0	32.2	43.4
76	20	4	$3,^{20}$	$4,^{50}$	16,2	24.2	38.0

Additional energy savings can be realized with both of these fly ashes on the basis of their particle size distribution. Since the original specific surface of those fly ashes is similar to that of portland cement (280-300 m^2/kg) a significant improvement in the production process can be achieved by introducing the fly ash into the grinding process between the ball mill and the air separator. In this way only the coarser particles of fly ash return to the ball mill and the greater part of ash goes to the cement silo withaut charging the mill. This enables energy saving for grinding (11-25 per cent), and also more efficient comminution of clinker. As a result (satisfatory cement properties are achieved as showed by experimental study. (2).

Besides ordinary fly ashes from burning coal in thermo power-plants mixes of fly ash and gypsum occur during incinerating coal and limestone mix in fluidized bed furnaces combined with desulphurisation. Gypsum is formed by combining SO_3 from fuel with CaO from limestone. Possibility of using such mixes has also been explored and results show that the mentioned mix can be usefully applied as hydraulic active admixture and set retarder for portland cement at the same time. Only very small quantities of gypsum for initial setting time regulation have to be added to portland cement and the rest of SO_3 is supplied from this "sulphurized" fly ash. Results obtained with that fly ash - gypsum mix are similar to those from experiments with the above mentioned fly ash from Kosovo.

4. OTHER ARTIFICIAL POZZOLANAS

No doubt that technically the most interesting industrial waste, used as a secondary raw material in building materials production is **condensed silica fume**, a by-product from ferro-alloys production. This artificial pozzolana exhibit extremely high hydraulic activity owing to its chemical composition (over 90 per cent SiO_2), amorphous structure and very small particles. Approximately 30,000 tons of condensed silica fume are collected yearly in yugoslav ferro-silicon plants. Only a part of it is used as an admixture to concrete in order to improve the durability against chemical agression, but broader application was not possible earlier since its inconvenient physical characteristics (bulk density approx. 250 kg m^{-3}, and BET spec. surface approx. 20 m^2 g^{-1}) cause problems in handling and transport. In addition to this, the beneficial effect is reduced by negative influence of very fine c.s.f. particles on the water requirement of cement and concrete, so the use of (super)plasticizers is necesseary.

In our opinion and experience the most suitable way for broad application of this waste material is agglomeration of c.s.f. in pellets with addition of water and production of cement by intergrinding these pellets together with portland cement clinker. Very special properties of cement and concrete are achieved in presence of c.s.f. as demostrated in tables 3-7.

The long term industrial use of c.s.f. as cement admixture has shown significant improvement in cement strength in spite of lower clinker content in the cement containing condensed silica fume. (Table 7).The beneficial effects are the consequence of physical and chemical influence of c.s.f. Chemically very active SiO_2 binds the most vulnerable part of portland cement paste $Ca(OH)_2$ into additional CSH gel, filling pores in the paste. The porosity is also diminished by the presence of very fine particles. Both changes significantly improve strenghs and chemical durability (4) of cement composites

TABLE 3

Physical and mechanical properties of cements containing silica fume

Silica fume Content	Silica fume Form and way of addition	Blaine ($m^2\ g^{-1}$)	Water stand consist. (%)	Consist. flow-table (cm)	Compressive strength (MPa) 3 d	Compressive strength (MPa) 28 d	Shrinkage 28 days ($mm\ m^{-1}$)
0	(comparative)	333	27.3	11.5	27.1	46.9	0.502
7,5 %	original dust	3000 (dust)	29.7	11.3	28.7	62.0	0.661
7,5 %	industrial inter grinding of pellets	389	26.5	12.0	31.4	51.5	0.549

TABLE 4

Composition and properties of concrete with silica fume

Content (%)	Form and way of addition of c.s.f.	Cement content ($kg\ m^{-3}$)	w/c ratio	Slump (cm)	Compressive strength (MPa) 7d	28d	90d	Efficiency* at 28 days cement	clink.
0		300	0.58	6.5	23.6	29.6	32.8	0.99	0.99
7,5	industrial inter grinding of pellets	300	0.61	7.0	26.0	33.8	37.3	1.12	1.22

$$* \text{ Efficiency} = \frac{\text{Strength}}{\text{Cement content}} \quad \frac{(dN\ mm^{-2})}{(kg\ m^{-3})}$$

against acid and sulphate aggression (Table 6) the effectiveness of c.s.f. in preventing excessive expansion of concrete due to the alkali aggegate reaction was confirmed by ASTM C 441-69 test. The freeze-thaw durability is also improved by addition od c.s.f.to cement or concrete (5).

The ash obtained by burning **rice husks**, straw and similar bio-wastes is an artificial pozzolana very similar to condensed silica fume regarding chemical composition (high SiO_2 content), noncrystalline structure and very high specific surface. Added to portland cement in small quantities it increases significantly its strengths and chemical durability in spite of higher water to cement ratio necessary when this ash is present (Table 8).

Finally in incinerators for municipal and special wastes noticeable quantities of fly ashes and slags also occur. In spite of significant variations in

TABLE 5

Influence of c.s.f. on the properties of low heat cement

Cement composition (%)				Strength (MPa)				Heat of hydration ($J\ g^{-1}$)	
Clin.	Gyps.	Admix.	c.s.f.	7 days flex.	7 days compr	28 days flex.	28 days compr.	7 days	28 days
				3.0	14.0	4.0	25.0 *	250	295*
47	3	50 Ash	0	3.6	16.3	4.5	23.0	242	283
44.6	2.9	47.5 "	5	4.4	16.2	5.5	28.6	221	270
40	5	55 Slag	0	5.2	20.5	8.3	36.9	247	311
33	5	55 "	7	5.0	18.8	8.7	36.4	218	277
37	5	51 "	7	5.1	21.4	9.1	44.4	242	291

* Minimal strength and maximal hydration heat required by yugoslav standard.

TABLE 6

Sulphate expansion of cement mortar according to ASTM C 452

Type of cement (composition)	Expansion at 14 days (%)
Ordinary portland (C_3A = 9.7%)	0.075
Ordinary portland containing 7% of c.s.f. (interground pellets-industrial mill)	0.054
Ordinary portland containing 12% of c.s.f. (interground pellets-industrial mill)	0.035
ASTM Type II - moderate sulphate resistance (C_3A = 6.2%)	0.048

Note: ASTM C 150 requirement for sulphate resistant cement is maximal expansion of 0.045% at 14 days.

chemical composition it is generally possible to admix those wastes to portland cement and concrete. Laboratory experiments to explore the technical possibilities of that use are under way.

TABLE 7

Statistical characteristics of portland cement with and without c.s.f. produced in cement factory Koromačno in period October 1983 through October 1987.

Cement propertie		Number of sampl.	Mean value	Max. value	Min. value	Standard deviation	Coef. of variation
Specific surface (m^2kg^{-1})	no c.s.f.	199	365	399	340	146	4.0%
	5% c.s.f.	758	380	428	334	249	6.6%
Water for standard cons. (%)	no c.s.f.	199	28.8	29.4	28.2	o.4	1.4%
	5% c.s.f.	758	28.2	29.2	26.8	o.4	2.1%
Compressive strengths (MPa)							
after 7 days	no c.s.f.	199	22.0	23.7	20.0	0.8	3.6%
	5% c.s.f.	658	21.6	26.2	19.3	1.7	7.9%
after 28 days	no c.s.f.	199	42.6	44.3	41.2	0.8	1.9
	5% c.s.f.	758	46.4	50.9	42.3	1.9	4.0%

TABLE 8

Influence of rice husks ash on strengths and chemical durability of portland cement mortars

Propertie	portland cement	PC with 10% ash	PC with 20% ash
w/c ratio*	0.50	0.55	0.55
Compressive strengths (MPa)			
after 7 days in water	29.1	43.7	31.1
after 28 days in water	52.1	56.1	46.7
after 90 days in water	53.4	59.2	49.3
after 28 days in water and additional 62 days in NH_4NO_3 (10% solution)	16.1	46.8	41.0

*mortars prepared to same consistency determined by flow table

5. ENVIRONMENTAL ASPECTS OF USE OF WASTE MATERIALS IN BUILDING INDUSTRY

Besides the economic and technical importance of using waste materials in building industry, this activity is important from the environmental protection point of wiew. The first effect of such use of wastes is decreasing of stock piles of those materials. Secondly in this way natural resources which serve as raw materials for cement clinker burning are preserved.The reduced consuming of fuels for clinker burning also preserves natural resources and smaller quantities of carbon dioxide and sulphur dioxide pollute atmosphere.

Although the proportion of used waste is small in comparison to the total produced quantities of these materials, the mentioned effects on environment protection are not neglibile. Numerically expressed 1.5 million tons of cement clinker substituted by those admixtures (approx. 20 per cent of 8 million tons of cement produced yearly in Yugoslavia) means 2.6 million tons less quarrying, saving about 150,000 tons of liquid fuel or about 250,000 tons of typical coal (20,000 kJ kg^{-1}). And finally this will result in 470,000 tons of CO_2 and approx.7,500 tons of SO_2 less in the atmosphere (for liquid fuel containing about 2.5 per cent of sulphur).

On the other hand, the use of industrial wastes especially fly ashes and slags from coal burning in power plants and incinerators for waste increases health hazards while some of their components, such as heavy metals, poisonous and radioactive substances etc. show the tendency to concentrate in these wastes. (6).

In order to get a basic idea about harmful influence on the environment caused by the use of such waste material preliminary investigations have been undertaken to determine radioactivity and heavy metals contents of slag and fly ash from municipal waste incinerating plant. These determinations were followed by leaching tests of the burning residues and of cement composites containing slag and fly ash in technically applicable and reasonable concentrations. The experiments and results obtained are briefly described below.

Blended portland cements were laboratory produced by intergrinding PC clinker, 4 per cent of gypsum for set regulation and 10 per cent of slag or fly ash from the waste incinerating plant. Cement sample produced without waste addition served for comparison. Standard cement mortars (cement to sand ratio 1:3, water to cement ratio 0.50) were prepared using the three mentioned cements.

Total contents of heavy metals in the original slag and fly ash samples and in the ground mortar samples were determined by atomic absorption spectrometry. Digestion with aqua regia for determination of acid soluble portion of metals was applied according to DIN 38414 - S7 Table 9). Results of leachability by water from original fly ash and slag samples according to DIN 38414 - S4 are

TABLE 9 Chemical characteristics of slag, fly ash and cement mortars (mg/kg)

COMPONENT	Leaching by strong acid (DIN 38414-S7)					Leaching by water 72 h (DIN 38414-S4)	
	Fly Ash	Slag	Mortar with Ash	Mortar with Slag	Comparativ mor.	Fly Ash	Slag
Cr	124	395	96	113	127	3.70	0.2*
Cu	540	4015	30	82	22	0.2*	0.2*
Zn	2034	1867	68	56	21	0.1*	0.1*
Ni	113	512	39	97	54	0.5*	0.5*
Pb	932	1177	60	60	30	0.5*	0.5*
Cd	16	3	1.4	3	4	0.40	0.10
Hg	0.27	0.35	0.26	0.05	0.05	0.01*	0.01*
NO_3^-	-	-	-	-	-	28.50	93.00
Cl^-	-	-	-	-	-	708.00	652.00
F^-	-	-	-	-	-	0.96	4.79
SO_4^{--}	-	-	-	-	-	19742.00	1737.00

* results with mark * are under detection limit of the instrument

TABLE 10 Results of water leaching of cement mortars (mg/kg) 14 days, (DIN 38414-S4)

COMPONENT	Ground mortar (particles < 90 μm)			Coarse mortar (2 - 4 cm)		
	10% Ash in cem.	10% Slag in cem.	Pure cem.	10% Ash in cem.	10% Slag in cem.	Pure cem.
Cr	0.2*	0.2*	0.2*	0.2*	0.2*	0.2*
Cu	0.2*	0.2*	0.2*	0.2*	0.2*	0.2*
Zn	0.1*	0.2	0.2	0.1*	0.1*	0.1*
Ni	0.5*	0.5*	0.5*	0.5*	0.5*	0.5*
Pb	0.5*	0.5*	0.5*	0.5*	0.5*	0.5*
Cd	0.3	0.05*	0.05*	0.05*	0.05*	0.05*
Hg	0.01*	0.01*	0.01*	0.02	0.01*	0.01*
NO_3^-	46.20	47.60	48.0	0.00	1.10	1.10
Cl^-	67.00	57.80	53.20	4.60	6.20	4.40
F^-	6.60	8.00	7.40	0.34	0.57	0.43
SO_4^{--}	80.00	67.60	71.20	12.80	19.70	13.70

* results with mark * are under detection limit of the instrument

also presented in Table 9.
The compositions of water eluates after 14 days leaching 28 days hydrated cement mortars (DIN 38414 - S4) were determined using coarse specimens (2-4 cm) and the ground mortar samples (particles mostly under 90 μm).
The results are shown in Table 10.

Considering the results in Table 9 it can be said that, due to the very low content of fly ash and slag (2.5 to 3.0. mass per cent) the concentrations of heavy metals in cement mortars is significantly lower than in observed admixtures and show no difference in comparison with mortars made by "pure" cement.

Leaching with water shows very small concentrations of heavy metals in the eluate even when performed on original fly ash or slag samples.Concentrations (anions, Zn, Cd) in eluates from mortars are even lower especially in the case of coarser i.e. less permeable specimens.

REFERENCES

1 V. Korać and V. Ukrainčik, in: Fly Ash from Kakanj Power Plant in Cement Production, Proceed. of the 7th Intern. Congress on the Chemistry of Cement, Paris, France, 30 June-4 July 1980, Septima,Paris 1980, Vol. IV, pp. 242-249.

2 K. Popović, and B. Tkalčić-Ciboci, in: Separate Grinding of PC Clinker Versus Intergrinding with Fly Ash, Proceed. of 3rd Intern. Conf. on Fly Ash, Silica Fume, Slag and Natural Pozzol. in Concrete, Trondheim, Norway, 18-23 June, 1989, CANMET, Ottawa, 1989, Suppl. Papers, pp 252-258.

3 K. Popović, A. Đureković, and V. Ukrainčik, in: Blended and Special Cements Incorporating Condensed Silica Fume, Proceed. of the 8th Intern. Congr. on the Chem. of Cement, Rio de Janeiro, Brasil, 22-37 September 1986, Alba Grafica e Editora, Rio de Jaineiro, Vol. IV, pp. 137-142.

4 K. Popović, V. Ukrainčik, and A. Đureković, Durab. of Build. Mat.2(1984) 171 - 186.

5 A. Đureković, V. Čalogović, and K. Popović,Cem. and Concr. Res.19 (1989) 267 - 277.

6 R.B. Dean, Incineration of Municipal Waste, Academic Press, London, 1988,

Waste Materials in Construction.
J.J.J.R. Goumans, H.A. van der Sloot and Th.G. Aalbers (Editors)

FEASIBILITY OF THE MANUFACTURING OF BUILDING MATERIALS FROM MAGNESIUM SLAG

M. Courtial, R. Cabrillac and R. Duval

Laboratoire d'Energétique et d'Economie d'Energie
Université PARIS X NANTERRE.

IUT, Département Génie Civil, 1, Allée des Chênes Pourpres
95014 CERGY PONTOISE CEDEX, FRANCE

SUMMARY

Magnesium manufactured by "MAGNETHERM" process leads to obtention of two by-products: a "powdered" slag which is similar to a cement and a "granulated" slag which has the granulometry of a sand and possesses latent hydraulic properties. This research consists in investigation about the variation of the properties of these slags and valuing the characteristics of mortars made from them. The final aim of this work is valorization of magnesium slags by manufacturing building blocks.

INTRODUCTION

By-products, object of our study, are magnesium slags produced by the PECHINEY ELECTROMETALLURGIE company in a plant located in MARIGNAC in the Southwest of FRANCE. Magnesium is obtained by the "MAGNETHERM" process and the waste is composed of two products: "powdered" slag (1/3 of the production) and "granulated" slag (2/3 of the production) obtained from a same liquid but differently cooled. In 1990 [1], the factory of MARIGNAC, 6th producer of magnesium in the world, produced 14,500 t of magnesium leading to 85,000 t of by-products, 28,000 t of "powdered" slag and 57,000 t of "granulated" slag. But only 20,000 t of slags were used for occasional applications and the stocking of these wastes is a real problem for the direction of MARIGNAC factory which hopes to find a durable process to valorize its whole slag production.

Although the utilization of granulated ferrous slags as cementitious materials [2] has been developped in North America and Europe, the use of non ferrous slags is not well established in concrete manufacturing. Some copper, nickel and lead slags have been evaluated for performance as cementitious components in more backfill and in concrete, but utilization of magnesium slags is not quoted in literature.

However magnesium slags have cementitious properties at ordinary temperature [3][4]which make them interesting raw materials to manufacture building blocks [5]. Our work consists in studying these possibilities. Therefore we studied the nature of magnesium slags, the regularity of their properties and the strength development of mortars made from these by-products. Our final goal is to define a composition and a manufacturing process which enables enhancement of the value of magnesium slags.

PRODUCTION AND PROCESSING OF MAGNESIUM SLAGS

Magnesium slags are by-products of a metallurgical process called "Magnetherm" [6] which has significant savings in energy.

Three raw materials are used in the operation: dolomite, bauxite and ferrous-silicon.

The dolomite is decarbonated in a rotative furnace at the temperature 1200°C and gives a calcium and magnesium oxyde. Bauxite is the flux stone for the processing and ferrous-silicon is the reductor. The blended raw materials are heated between 1,600°C and 1,700°C in a partial pressure (60 mm mercury). Magnesium is, reduced and the following reaction takes place:

$$2(\mathrm{MgO, CaO}) + \mathrm{Si} \rightarrow \mathrm{SiO_2, 2CaO} + 2\mathrm{Mg}$$

Magnesium emits metallic fumes which condense in a crucible (figure 1).

* *Pechiney Electrométallurgie BP n° 1 31440 MARIGNAC FRANCE*

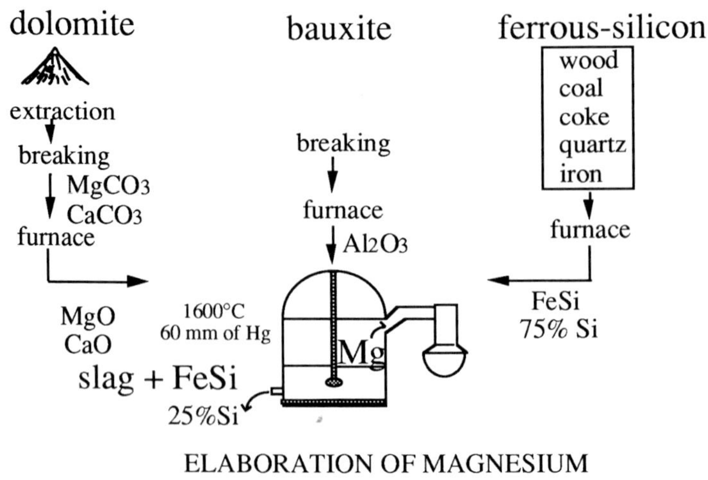

figure 1

Then the magnesium is transported to the fondery where it is refined at high temperature.

The molten materials are poured in three slag ladles by means of a hole in the bottom of the furnace. Ferrous-silicon is first drawn off. The molten slag is quenched in water and evacuated to a deep pit. Lastly the slag is drawn out of the pit and stored with a high residual moisture content. This method of quenching produces 2/3 slag called "granulated" slag.

The residual slag which is sticked on the surface of the crucible is slowly air-cooled. During the slow cooling a polymorphic transformation of the main compounds, the dicalcium silicate, takes place with an expansion volume. The air-cooled slag disintegrates to give a very fine powder. This powdered slag constitutes 1/3 of the production of magnesium slags and has real cementitious value. See figure 2.

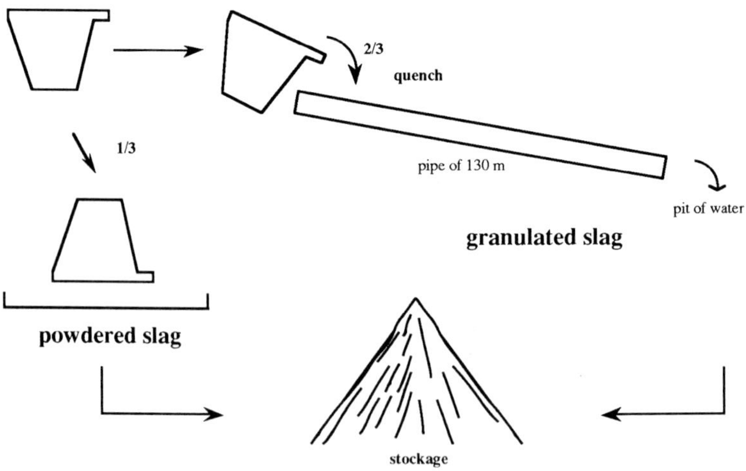

figure 2

NATURE OF MAGNESIUM SLAGS

The factors affecting magnesium slags hydraulicity are mainly the glass content, the chemical and mineralogical composition and the physical properties.

From a physical point of view, granulated slag consists of white grains very crumbly with a hard core. The particle size distribution varies between 0.5 and 5 mm. After spin-drying granulated slag sets after fifteen days. The water content of granulated is 11-13% after eight days subsequent to the draining.

Powdered slag consists of very fine grains and develops a specific surface of 2,000 cm2/g below the finess of a portland cement.

The density of the two compounds is similar and about three.

The bulk chemical composition is determined by chemical analysis (X Ray fluorescence and atomic absorption) and the results are shown in table 1.

table1

compound	Lime CaO	Silica SiO_2	Alumina Al_2O_3	Magnésia MgO
average	57.7%	25.85%	11.6%	4.8%
standard deviation	0.87	0.72	0.66	0.76

There is few variation in chemical composition between slag grains. The magnesium slags are represented in the $CaO-SiO_2-Al_2O_3$ system (figure 3).

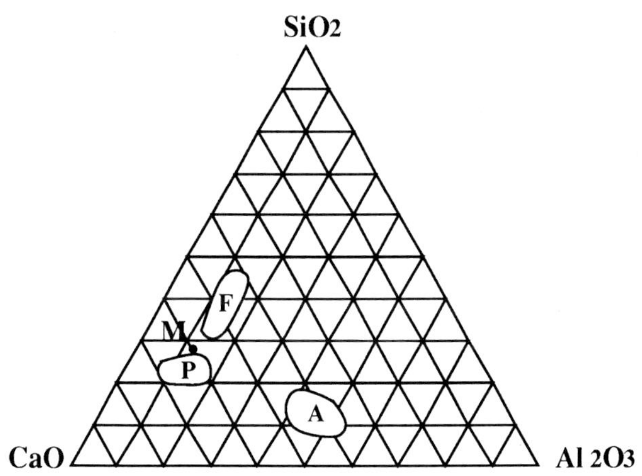

figure 3

The magnesium slag is nearer to ordinary Portland cement P [7] than ferrous slag (F) [8]; it is on the border of the zone of Portland cement (P). Magnesium slag (M) ,in particular, is more basic than ferrous slag (CaO/ SiO_2= 2,1). Compared with aluminous cement (A), it is richer in SiO_2 and CaO.

The mineralogical composition rather than the chemical composition determines the hydraulic properties [8] of magnesium slags. In the system $CaO-SiO_2-Al_2O_3$, the main silicate which appears is C_2S*. See figure 4.

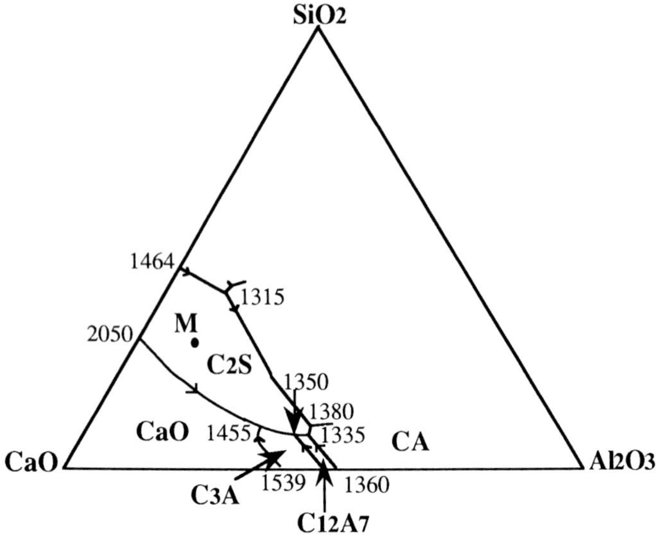

figure 4

Granulated slag contains the dicalcium silicate C_2S β which presents hydraulic properties. On the contrary powdered slag contains C2S γ which is a polymorphic transformation of the metastable β form of C_2S into its stable γ modification during the slow air-cooling. This cristalline mineral is generally considered as inert toward water. The other crystalline compounds found in powdered slag are MgO and the calcium aluminate $C_{12}A_7$.

The glass content of the granulated slag was reported to be 40%. There is also a few content C_3A which hydrates quickly during the quenching to form a stable cubic hydrate C_3AH_6.

STRENGH DEVELOPMENT OF HYDRATED SLAGS

Since we aim at enhancing the value of magnesium slags in building blocks form, we must study compressive strengths of mortars made with mixture of powdered slag (P) and granulated slag (G).

The manufacturing process consists in mixing solids with water during three minutes. Then the mixture is placed in standard moulds 4x4x16 cm^3 without vibration. Demoulding at the end of a day, the samples are stored in a moist room at 20-22°C.The compressive strengths are tested in the direction of pouring and in the perpendicular direction and we report the average of the values (Cs).

The production of magnesium is two third granulated slag and one third of powdered slag. It is important to study the valorization of this production, but we study first mixtures of different ratios of powdered-granulated slags. We have also studed the effect af adding hydraulic lime (Standard specification XHA in France) and gypsum on the hydrated properties.

* *C=CaO* *S=SiO_2* *A=Al_2O_3* *H=H_2O*

1. Studies of mixtures powdered-granulated slags

The following figures show the relationship between compressive strength at 14 days, 28 days and 7 months and mixing water. The composition of mixtures is 100%, 80% and 33% powdered slag. The mixing water percentage (W/P+G) is calculated according to the whole mass of slag.

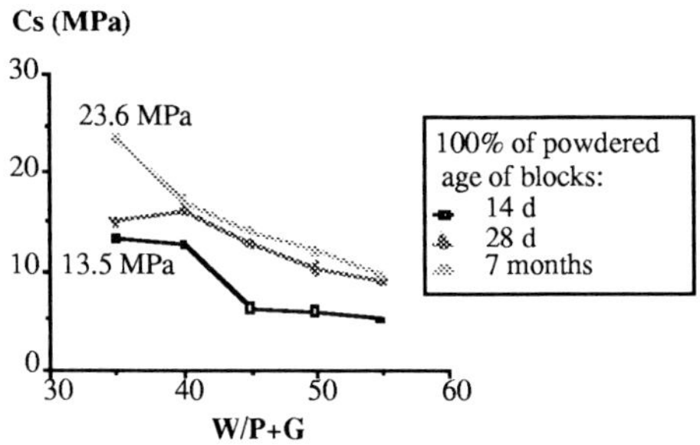

figure 5

The compressive strength of pure powdered slag decreases with increasing mixing water. The ratio of strength between 14 days and 7 months is about two and the best results are obtained with 35% of water. The increasing strength is significant between 14 days and 28 days (figure 5).

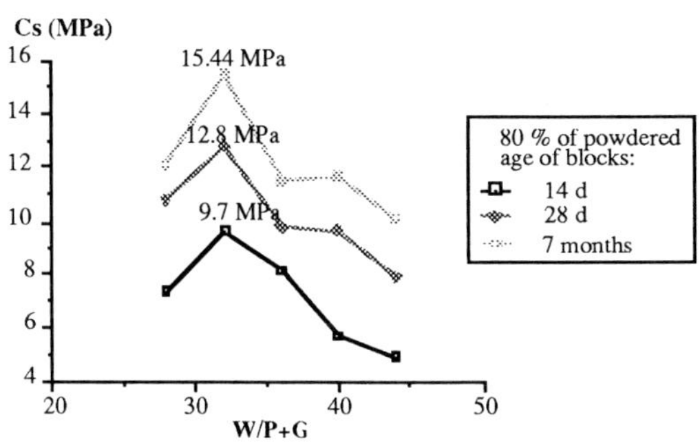

figure 6

Test specimens with 80% of powdered slag have lower compressive strength than pure powdered. The hydraulic activity of the granulated is slower than that powdered and the best water content is about 32%. (Figure 6). Yet increasing strength between 28 days and 7 months is more important because of the latent hydraulic properties of granulated slag.

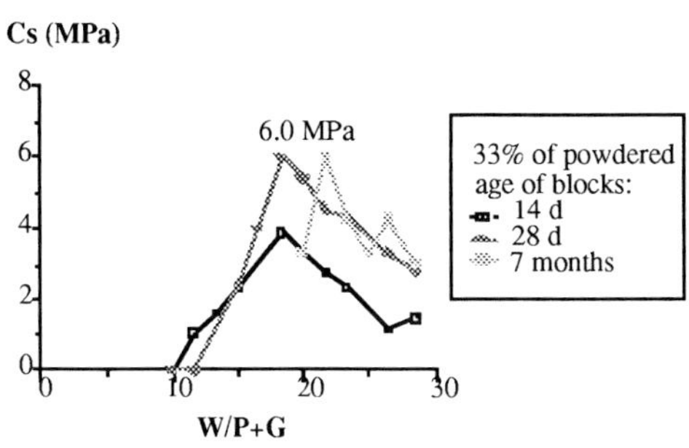

figure 7

Significant decreasing strength for mixtures with 33% powdered slag is shown in figure 7. The compressive strength of these specimens is three times lower than that of pure powdered. The best results are obtained with about 18% of water. The strength development of mortars is more pronounced with the high percentages of water.

2. Influence of addition of lime and gypsum

In order to utilize the whole magnesium slag production, we added hydraulic lime XHA or gypsum for increasing the compressive strength of mortars containing 33% powdered and 67% granulated slag.

The effect of various lime additions on compressive strength is presented in figure 8. Lime percentage is calculated according to the whole mass of slag.

The strength increases with increasing lime content. The highest strength gain at 14 days and 28 days is obtained for samples containing 20% lime. At this level the compressive strength is multiplied by three compared with the mixture without lime. Nevertheless the strength is lower than that obtained with pure powdered.

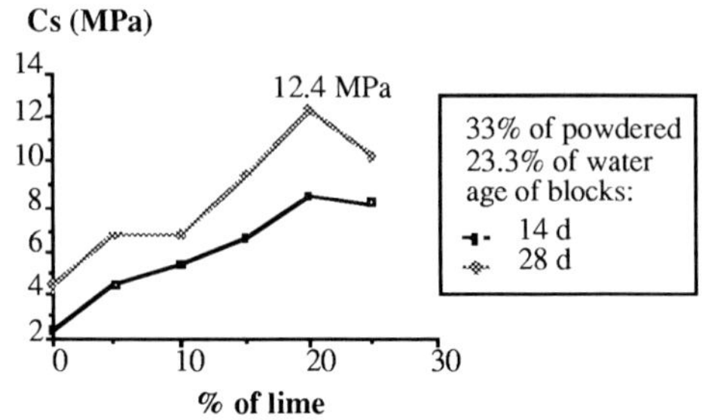

figure 8

From an economical point of view, in order to manufacture classical building blocks, the addition of hydraulic lime is limited to a maximum content 10%.

A sulfatic activator such as gypsum has also been used. The effect of a 5% gypsum content on the compressive strength is shown in figure 9, in fonction of mixing water content (W/P+G+Gy), which is evaluated according to the whole quantity of solid.

We note an increase of strength when the water quantity decreases but significant strength development due to gypsum activity is observed: at 7 days, mixtures with gypsum produce similar performances compared with those at 28 days without gypsum. Moreover strengths at 28 days are twice than these at 7 days.

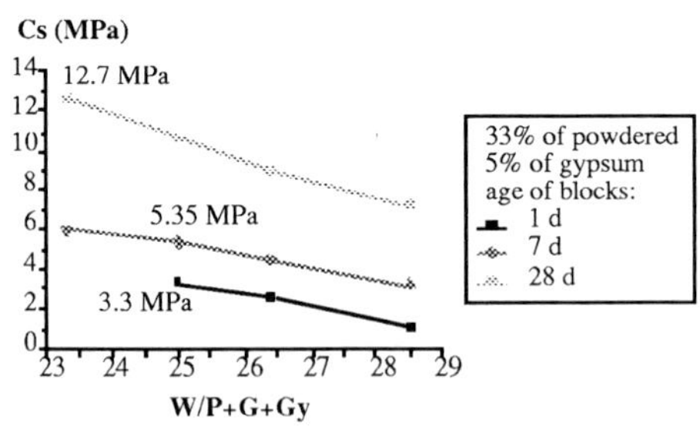

figure 9

3. Influence of vibration

Improvement in compaction method of hydrated cementitious mixtures should be an important goal, because early strength development is required. The effect of vibration (one minute) on the compressive strength is reported in the figure 10, for instance, with 33% of powdered slag with different mixing water quantities calculated according to the whole mass of slag.

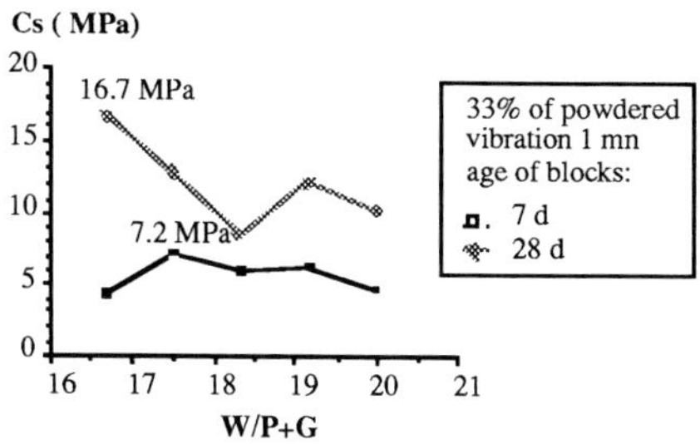

figure 10

Strengths are about three times higher than these obtained without vibration.The vibration allows to decrease the quantity of mixing water and to increase the density of samples about 20%.

PROBLEMS OF REPRODUCIBILITY

Magnesium slags are composed of a vitreous phase of microporous texture with amounts of cristalline phases. We observed differences in cristallization of the aluminate product $C_{12}A_7$ during the air-cooled quench. We noted that the compressive strength increases with the degree of cristallization. To reduce these difficulties of manufacturing, we used magnesium slag after blending.

The curing of slags is another parameter which has an influence on the compressive strength.

When powdered slag is subject to aeration, it takes up water and gradually becomes lumpy. The effect of aeration on the compressive strength is shown in figure 11. The strength of powdered slag decreases about 30% when it is subject to air-setting.

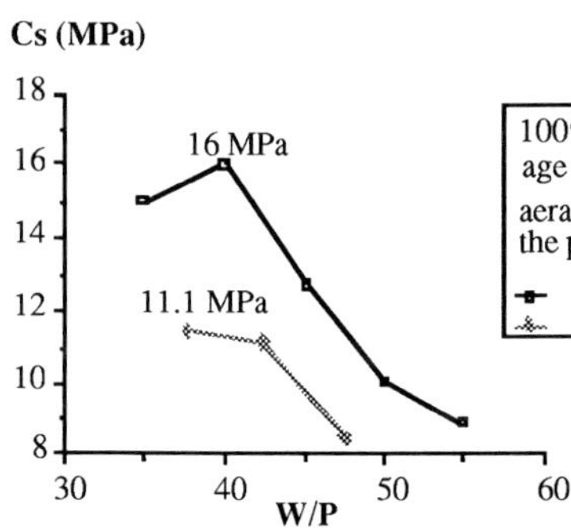

figure 11

For granulated slag, air setting is also a problem. The relationship between hydraulic lime content and strength development for two series of experiments is illustrated in the figure 12. The first serie is subject to air-setting (1% moisture content) and not the second which is utilized directly after quenching (13%). Mixing water quantity has been calculated considering these moisture contents.

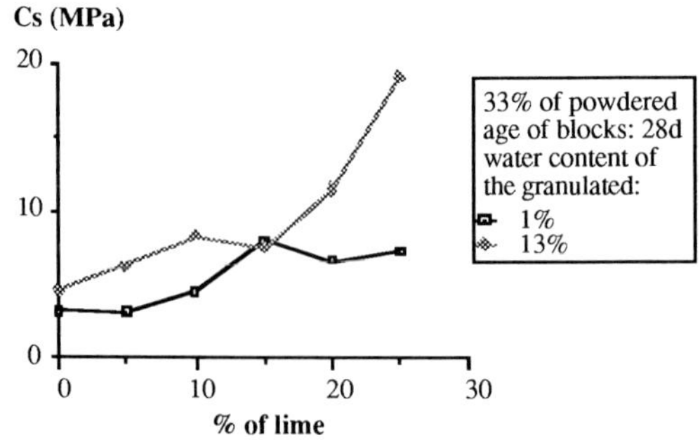

figure 12

Another difficulty results from a very quick setting time (a few minutes) of mixtures of powdered-granulated slags after hydration, which leads to problems of reproducibility of the moulding process. The manufactoring process has been changing for two years. The granulated slag is quenching with water containing powdered slag, while before it was with pure water.That seems the main cause for the setting of granulated slag, because the setting time observed was not so quickly [3] [4] [5]. It is to prevent this set and to control hydration that we added 5% gypsum in the mixtures. An increasing initial setting time of mortars ranged from twenty to thirty minutes results from the incorporation of gypsum. Moreover mortars incorporating gypsum require less vibration during placing operations in the moulds and improve workability.

CONCLUSION

To use these magnesium slags as building blocks, the experiments must correspond to industrial specifications: rapid demoulding, satisfactory initial strength after demoulding, compressive strength required, limited dry shrinkage. The mixtures 33% powdered-67% granulated slags with 5% gypsum have promising potential, since the early compressive strengths of magnesium slag mortars are enough sufficient (3 MPa at 1 day and 8 MPa at 7 days).

Further investigations are necessary to develop data drying shrinkage, long term mechanical properties and durability of magnesium slag mortars, before beginning an industrial production of building materials. By way of precaution it is necessary to realize a very complete study; several years ago, a few tests in order to realize directly industrial blocks failed because of the insufficient knowledge of hydraulic properties of magnesium slags and behaviour of mortars made from them.

From an ecomomical point of view, valorization of magnesium slags by manufacturing building blocks is very important not only for "PECHINEY ELECTROMETALLURGIE" company but also for building materials producers. In that way the use of the whole production of French magnesium slags could lead to manufacture 2,000,000 blocks each year representing a volume of 70,000 m^3 and the building of 3,000 individual flats.

REFERENCES

[1] J. Rollet and M. Ménétrier, Pechiney Electrometallurgie, Usine de Marignac

[2] Proceedings.*Colloque International sur l'utilisation des sous-produits en Génie-Civil.* Paris 1978.

[3] A. Carles-Gibergues. *Les ajouts dans les microbétons. Influence sur l'auréole de transition et les propriétés mécaniques.* Thèse d'état. UPS Toulouse. 1980.

[4] R. Cabrillac. *Etude prospective en vue de l'élaboration d'un matériau isolant porteur de grande diffusion.* Thèse de troisième cycle. UPS Toulouse. 1984.

[5] R. Cabrillac, W. Luhowiac, R. Duval. *Study of mechanical properties of mortars made of magnesium slags and future prospect of uses.* Third CIB Rilem Simposium Materials for low income housing. Mexico 1989.

[6] Fascicule documentaire sur l'usine de Marignac, Pechiney Electrométallurgie, 1989.

[7] F.M. Lea. *The Chemistry of Cement and Concrete.* Third Ed.Chemical Publishing Company, New-York. 1971.

[8] J. Alexandre, J.L.Sebileau. *Le laitier de haut-fourneau.* Ed CTPL. 1988.

[9] H.F.W. Taylor. *The Chemistry of Cements.* Academic Press. London and New-York. 1984.

Waste Materials in Construction.
J.J.J.R. Goumans, H.A. van der Sloot and Th.G. Aalbers (Editors)

SPRAY DRY ABSORPTION RESIDUE IN CONCRETE PRODUCTS

H.A.W. CORNELISSEN

CBP Department, N.V. KEMA, P.O. Box 9035,
6800 ET Arnhem (The Netherlands)

SUMMARY

The application of Spray Dry Absorption (SDA) residue in plain concrete has been investigated. For this purpose a typical SDA residue was selected containing 70% PFA and 30% SDA product. This product was mainly composed of sulphates and sulphites. SDA products were involved having sulphate-sulphite ratios between 4% and 50%. Tests were performed on pastes, mortar and concretes in which 20% of the cement was substituted by SDA residue. Basic concrete technological properties as well as durability and environmental properties were determined. The results were compared with the appropriate standards and requirements. It could be concluded that the compositions investigated can be applied successfully to plain concrete products. However, verification under actual conditions will be necessary.

1. INTRODUCTION

At coal fired power stations sulphur dioxide can be removed from the flue gas by using flue gas desulphurization (FGD) units. By means of the limestone process gypsum will be formed as an end product. Because the offtake of gypsum may not be sufficient, the option called spray dry absorption process (SDA) is under investigation in The Netherlands. This process produces SDA products which mainly consist of a mixture of sulphates and sulphites. If no precollection of fly-ash occurs an "SDA residue" will be formed containing fly-ash and SDA product.

SDA units have been operated in the USA but also in Austria, Sweden and Denmark.

At KEMA an extended research programme was carried out with respect to various applications of SDA residue. The application in sand lime bricks proved to be very successful (1). This paper presents the results of a second promising application, viz. SDA residue in concrete products with no steel reinforcement because of the chloride content of the SDA product. This paper provides an overview of the results, while detailed information can be found in (2).

2. MATERIALS AND COMPOSITIONS

Various SDA products were investigated. The SDA product obtained from Fläkt proved to contain a sulphate-sulphite ratio of 4%. By adding gypsum, artificial SDA products were composed having ratios of 25% and 50%. The SDA product provided by Stadtwerke Düsseldorf showed a sulphate-sulphite ratio of 27%. Further on these SDA products will be indicated by their sulphate-sulphite ratio, such as SDA-04, SDA-25, SDA-50 and SDA-27. Some details of the compositions of these products are given in Table 1.

In order to simulate typical SDA residues, mixes were prepared with 70% PFA and 30% SDA product. In all tests a similar type of PFA was used (48.2% SiO2, 27.1% Al2O3, 4.8% CaO, 8.7% Fe2O3, 3% MgO, 0.7% Na2O and 2.5% K2O).

The effect of SDA residue was investigated in pastes, mortars and concretes. For this purpose the results were compared with reference mixes having either no cement substitution or substitution by PFA only. In general parallel tests were performed with ordinary Portland cement and normal hardening Portland Blast furnace cement, abbreviated as pc-A and hoc-A respectively.

In the case of the pastes 20% (by weight) of the cement was substituted. This was also true for the mortar specimens which were produced according the EN 196 standard (67% sand, 22% binder, 11% water).

The concrete specimens were made in accordance with the Dutch recommendations (issue no. 12) drafted and published by CUR (3). The binder content of the concretes was 320 kg per m^3, the maximum grain size 31.5 mm and the slump was for all mixes 70 mm plus or minus 10 mm. For the comparative tests 20% of the cement was substituted by SDA residue.

TABLE 1

Compositions of the SDA products investigated

component (%)	Fläkt	Stadtwerke Düsseldorf
$CaCl_2$	10.6	6.3
$CaCO_3$	6.5	39.6
$Ca(OH)_2$	<0.2	2.8
$CaSO_3.0.5H_2O$	71.8	39.0
$CaSO_4.2H_2O$	3.0	12.0

3. CONCRETE TECHNOLOGICAL TESTS

The effect of cement substitution by SDA residue was investigated for pastes, mortars and concretes. With respect to pastes setting time and soundness (EN 112) were determined. The results are presented in Table 2.

As can be seen in this table, addition of SDA residue results in a retardation of setting time compared with the reference pastes and the pastes with cement replacement by PFA. In all cases, however, the requirements of CUR recommendation 12 were fulfilled (3).

The requirements were also fulfilled with respect to soundness.

TABLE 2

Results of tests on pastes

Composition	initial setting time (hours)	final setting time (hours)
<u>with pc-A</u>		
reference	2-3	3-4
PFA	2-3	3-4
SDA-04	4-5	6-7
SDA-27	4-5	6-7
SDA-25	4-5	5-6
SDA-50	4-5	6-7
<u>with hoc-A</u>		
reference	2-4	4-5
PFA	3-4	5-6
SDA-04	5-6	6-7
SDA-27	5-6	8-9
SDA-25	4-5	7-8
SDA-50	6-7	7-8

On fresh mortars, all having similar binder and water contents, the spread was measured by means of the Hägermann consistency apparatus. After wet curing at 20 °C, compressive strengths were determined at ages of 28 days and 182 days. The corresponding results are given in Table 3.

It can be seen that the effect of cement substitution on the spread is only marginal. For the interpretation of the strength results, the reference mix strengths were taken as 100%. It follows from the figures that the reduction of strength after a curing period of 182 days is very small in general. At an age of 28 days especially the combinations with pc-A turned out to be less favourable.

TABLE 3

Results of tests on mortar

composition	spread (mm)	f'm (28 d) (N/mm^2)	(%)	f'm (182 d) (N/mm^2)	(%)
with pc-A					
reference	163	43.0	100	64.7	100
PFA	169	35.9	83	59.7	92
SDA-04	158	38.3	89	64.5	100
SDA-27	166	36.7	85	59.9	93
SDA-25	156	29.8	69	61.1	94
SDA-50	168	24.8	58	63.7	98
with hoc-A					
reference	176	45.9	100	68.2	100
PFA	175	38.5	84	64.1	94
SDA-04	167	41.4	90	60.9	89
SDA-27	174	31.7	69	50.4	74
SDA-25	167	37.0	81	57.0	84
SDA-50	180	43.7	95	59.8	88

Comparative tests were conducted on fresh and hardened concrete as well. An overview of the results can be found in Table 4, in which the water content for 70 mm slump, and compressive strength values, are given after wet curing periods (at 20 °C) of 28 days and 182 days. Only minor differences in water content were observed. All compressive strength properties, except for pc-A and SDA-50, were in accordance with CUR recommendation 12, being at least 80% of the 28-day reference strength (3). Excellent strength development was found especially for the pc-A concretes.

TABLE 4

Results of tests on concrete

composition	water content (l/m^3)	f'c (28 d) (N/mm^2)	(%)	f'c (182 d) (N/mm^2)	(%)
<u>with pc-A</u>					
reference	167	42.9	100	51.4	100
PFA	162	36.7	86	54.5	106
SDA-04	163	45.6	107	64.6	126
SDA-27	162	43.4	101	65.3	127
SDA-25	166	39.1	91	62.9	122
SDA-50	165	34.0	79	61.4	119
<u>with hoc-A</u>					
reference	159	51.5	100	66.7	100
PFA	157	43.7	85	62.4	94
SDA-04	157	50.8	99	60.5	91
SDA-27	159	43.6	85	54.9	82
SDA-25	162	44.1	86	59.4	89
SDA-50	159	45.3	88	60.7	91

4. DURABILITY TESTS

Pore volume and pore size distribution were analyzed by mercury porosity tests on concrete specimens after wet curing periods of 8-12 weeks and 2 years. Table 5 gives the main results, which are expressed in terms of average pore size and pore volume (< 5 mu).

Especially hoc-A substitution by PFA or SDA residues showed, after 8-12 weeks of curing, a significant reduction of pore diameter. After a curing period of 2 years this was found only for the hoc-A and SDA-50 combination. The differences in pore volumes proved to be small.

TABLE 5

Pore size and pore volume of concrete specimens

composition	age = 12-8 weeks		age = 2 years	
	pore diameter (10^{-9} m)	pore volume (%)	pore diameter (10^{-9} m)	pore volume (%)
with pc-A				
reference	64.8	10.2	62.1	11.3
PFA	66.1	9.1	40.3	9.8
SDA-25	57.5	11.4	42.8	6.8
SDA-50	55.1	8.4	51.8	7.9
with hoc-A				
reference	39.1	6.7	26.2	6.1
PFA	12.9	7.8	38.5	8.8
SDA-25	14.5	6.7	30.9	6.4
SDA-50	13.0	5.9	13.9	7.5

Because of the sulphate supply of SDA residue, expansive ettringite formation might be possible. However, up to an age of 2 years no destructive reactions were observed for the tested concrete and mortar specimens. This can be derived from the strength results as well.

Also the possible oxidation of sulphites into sulphates was studied. Tests revealed that at the governing pH values above 10, this oxidation can be ignored (4).

All combinations indicated in Table 5 were subjected to standard freeze-thawing tests (NEN 2872). These tests were performed on concrete cubes with an age of 3 months and similar cubes having an age of 2 years. All concretes proved to be resistant.

5. ENVIRONMENTAL PROPERTIES

With a view to the application of building materials, environmental properties have to be taken into account. For this purpose the Dutch authorities are developing requirements. In the draft version (November 1990) of the regulation order concerning building materials (BSB), a distinction is made between two categories: one for (residual) raw materials (category N), and another for products made from these raw materials (category V). A first step in the assessment of products is the determination of their composition. If this composition is not within certain limits, additional diffusion tests are necessary in the next step.

The chemical composition was measured for pc-A and hoc-A concretes with PFA and SDA-25. The results are given in Table 6, which also contains the corresponding class V1 and V2 requirements. It can be seen that the V2 values are not exceeded, while with respect to the V1 requirements the Ni content was too high for pc-A concretes, while the Sb content was too high for pc-A reference concrete and PFA concrete. The SO4 content was too high for all hoc-A concretes and pc-A with SDA-25.

It is noted that the requirements of the BSB could not fully be satisfied. However, this is also true for accepted compositions. Compared with the accepted pc and hoc concretes the addition of SDA residue in these relatively small amounts is expected to be of minor influence on the overall properties. It must be realized that for normal concrete compositions the SDA residue will be about 6% of the cement weight or less than 0.75% of the concrete mass.

TABLE 6

Chemical analysis of concretes (in mg/kg)

	pc-A concretes			hoc-A concretes			BSB class	
component	reference	PFA	SDA-25	reference	PFA	SDA-25	V1	V2
As	3.1	5.6	4.1	<3	3.6	<3	30	45
Ba	146	222	178	150	235	213	600	3000
Cd	<0.15	<0.15	<0.15	<0.15	<0.15	<0.15	0.8	4
Co	4.4	5.8	5.4	1.9	4.8	4.1	20	100
Cr	25	25	24	15	19	18	100	500
Cu	23	25	27	14	25	19	30	150
Hg	<0.3	<0.3	<0.3	<0.3	<0.3	<0.3	0.4	2
Mo	<3	<3	<3	<3	<3	<3	10	50
Ni	27	24	24	9	15	13	20	100
Pb	9.2	18	11	7.1	11	9.6	100	500
Sb	4.9	4.3	3.7	2.4	2.5	2.6	4	20
Se	<3	<3	<3	<3	<3	<3	4	20
Sn	<2	<2	<2	<2	<2	<2	20	100
V	58	54	48	19	32	28	100	500
Zn	53	55	52	26	50	35	100	500
SO_4	6000	4800	9600	9000	8100	13500	7000	20000

6. CONCLUSIONS

Except for slightly retarded setting times, concretes in which cement was partly substituted by SDA residues showed strength and durability performances comparable to or better than reference or PFA concretes. Also no damaging expansive ettringite reactions were observed to an exposure time of two years.

In some aspects SDA concretes did not comply with the Dutch regulation order concerning building materials (relating to environmental properties). However, this was also the case for generally accepted pc-A and hoc-A concretes.

The utilization of SDA residues in plain concrete products shows promising perspectives. Complementary research, however, is recommended for given SDA residues under specific actual conditions including long-term durability aspects.

REFERENCES

1 P.J.C. Bloem and B.J.G. Sciarone, Application of pulverized fly-ash and spray-dry absorption products in sand-lime-brick production. KEMA Scientific and Technical Reports 7 (1), 1989 pp. 35-45.

2 H.A.W. Cornelissen, Application of SDA residue in concrete. Part 3: durability aspects and applications in the concrete products industry. KEMA report 60911-CBP 90-914, 1990, pp. 98 (in Dutch).

3 PFA as a filler for mortar and concrete. Recommendation 12, Center for Civil Engineering Research, Codes and Specifications, Gouda, The Netherlands, 1987 (in Dutch).

4 P.J.C. Bloem and B.J.G. Sciarone, Application of spray-dry products in building materials. Proceedings of the second international conference on the FGD and chemical gypsum, Toronto, May 1991.

Waste Materials in Construction.
J.J.J.R. Goumans, H.A. van der Sloot and Th.G. Aalbers (Editors)

FGD Gypsum and self-levelling floor screeds

Leo Moonen

Novem BV

P.O. Box 17, 6130 AA SITTARD, The Netherlands

In The Netherlands a technique has been developed under direction by Novem to process Flue Gas Desulphurisation gypsum into an anhydrite of a high and constant quality suited for the application in self-levelling floor screeds.
On short term this application stands for approximately 50% of the Dutch production of FGD gypsum and on the longer term a complete processing of FGD gypsum can be assured.
For Europe this application offers on the longer term the possibility of processing 7 million tons of FGD gypsum annually. A quantity which practically equals the production of FGD gypsum to be expected by the turn of the century.

In my lecture I will provide you with the information which has led up to this point of view.

1. NATIONAL GOVERNMENT POLICY AND ACTIVITY

1.1 Environmental Management in use of coal

Because of the energy crisis in the 1970s the Dutch authorities developed an energy policy aimed at energy conservation and the diversification of energy supply.
An important part of this diversification was the re-introduction of coal use, particularly to support the public electricity supply, and which was required to take place in an environmentally acceptable manner.

1.2 Role of Novem

To address the logistic, technological and environmental problems of the use of coal the National Research Program Coal (NOK) was established. Under funding by the Netherlands government, the Dutch Agency for Energy and Environment (Novem BV) was charged under NOK with responsibility for establishing means of ensuring recovery of desulphurisation residues.

2. QUALITY AND QUANTITY OF FGD GYPSUM SUPPLY

Requirements by the government for control of sulphurdioxide emissions from large boilers have led to commercial generation of FGD gypsum in the Netherlands beginning in 1985. The FGD facilities are typically wet lime or limestone scrubbers using in-situ forced oxidation and operating without use of pre-scrubber.

2.1 FGD Gypsum Quality

Within the framework of the NOK activity some Dutch gypsum-using firms have done research into the technical limits and the use of FGD gypsum in their production lines. As a result of this research, the use of FGD gypsum for supply to existing wallboard manufacturing plants was judged possible. Although colour and chlorine demands further attention.
Besides this research in the Netherlands and in Germany has aslo shown that FGD gypsum is equivalent to natural gypsum so far as health and environmental considerations.

2.2 FGD Gypsum Quantity

In The Netherlands on the basis of prior experience with imported low-sulphur coal it can be assumed that a typical unit of 600 MWe will yield approximately 45,000 tons of FGD gypsum per year including a nominal 10% surface moisture.
The planned coal consumption for electric utility supply in the Netherlands in 1995 will amount to 9 million tons. This implies that, with the use of present boiler types and desulfurization processes, approximately 400,000 tons of FGD gypsum will be produced.

3. OVERVIEW AND PRIORITIZATION OF MARKET DEVELOPMENT

3.1 Market survey on the possibilities of application of FGD gypsum

Because of the lack of available gypsum supplies one cannot speak of a strongly developed gypsum industry in The Netherlands. The expected quantities of FGD gypsum were a reason to Novem to explore in 1986 the favourable applications of FGD gypsum. This exploration pointed to a number of interesting possibilities, together amounting to a potential consumption of 6 to 700,000 tons of FGD gypsum per year.
Seeing the development of the application of FGD gypsum, already started by several firms, and taking into consideration Novem's tasks and instruments, Novem's attention went out particularly to the possible use of FGD gypsum in self-levelling floor screeds.

3.2 Market for self-levelling floor screeds

In buildings a floorconstruction in The Netherlands is provided with a floor screed with a thickness of 3 to 5 centimetres. This floor screed is made of a mortar consisting of sand, cement and water and is applied by hand.
The application of this requires a lot of physical effort under working conditions which are often very poor. Because of this many employees become unemployable and they have to leave their employment at an early stage.
That's why the authorities insist on the development and use of methods with which these problems might be avoided. Self-levelling floor screeds are such a method.
The production of these floor screeds takes place on the basis of self-levelling mortars. The mechanically produced mortar is fluid and is poured out onto the construction floor by means of pumps.

Because of the self-levelling capacity of this mortar a smooth floor screed is created. The mortar mainly consists of sand, water and a binder. As a binder can be used: natural anhydrite, synthetic anhydrite and anhydrite on the basis of FGD gypsum.
However, up to now the limiting factor is the availability of sufficient anhydrite with a satisfying quality. The production of anhydrite on the basis of FGD gypsum offers the solution to this.

4. ANHYDRITE PRODUCTION TECHNOLOGIES

4.1 Overview of options

The calcination of FGD gypsum into anhydrite is not conducted on a large scale because of the lack of a production technique that is technologically and economically suitable. In principle it is possible to convert FGD-gypsum into anhydrite by:

- Thermic processes: that means the dihydrate is converted into beta-hemihydrate and after that into anhydrite by means of contact with a hot air current or by means of a hot surface.
- Hydrothermic processes: that means the dihydrate is first converted in an autoclave into an alpha hemihydrate, which later is calcined into anhydrite by means of a thermic process.

From an inventory made by direction of Novem it turned out that the German firms Krupp Polysius and Babcock could each deliver a suitable production technique.

4.2 Laboratory research

By direction of Novem a thermic process of Krupp Polysius and a hydrothermic process of Babcock have been investigated on a laboratory scale.
Based on the laboratory tests it was judged that the thermic process by Krupp Polysius is the preferred technique from the standpoint of product and process technology and cost effectiveness.

4.3 Pilot research

Immediately after the laboratory research a pilot research was started by direction of the firms Gyvlon, Ankersmit and Vliegasunie in cooperation with Novem.
Seeing the economic perspectives the attention was in the first instance focussed on the Krupp technique.
From this research it soon turned out that the qualities of the FGD gypsum to be expected for The Netherlands were such that an excellent and constant quality of anhydrite can be obtained by the Krupp technique.

4.4 Production line for FGD gypsum anhydrite

An essential component of the production line for FGD gypsum anhydrite consists of the Polcal-installation, delivered by Krupp Polysius.

It is a simple construction with few moving parts, in principle consisting of some cyclones, a combustion chamber, a dust removing plant and a ventilating fan.
The operation of the installation is as follows:

- the wet FGD gypsum cake feed is dried and converted into hemihydrate and anhydrite by being injected into a hot airstream issued by the combustion chamber.
- The dried gypsum, the hemihydrate and the anhydrite is separated from the hot airstream in the cyclones.
- The anhydrite is cooled by being injected into the inlet fresh airstream which thereby becomes preheated.

The solids residence times in the diverse process steps are only a few seconds.
After that the anhydrite is milled and mixed with additives to become a binder suited for self-levelling floor screeds.

4.5 **Demonstration project**
Also within the policy of the European Community attention is paid to energy and environment. In connection with this financial support is offered to the development and implementation of relevant technologies.
By timely recognition of these developments it was possible with Novem's support to hand in as early as April 1989 a 'proposal for a demonstration project' to the Commission of the European Communities. Officially the request took place by the firms Ankersmit, Gyvlon, Krupp Polysius and Vliegasunie.

4.6 **Construction and exploitation of the demo**
By the end of 1989 the request was granted by the E.C. The firms Ankersmit, Gyvlon, Krupp Polysius and Vliegasunie have started the preparations for the construction. The rounding off of the building activities is scheduled for the second half of 1992 and the monitoring programme inherent to a demonstration will be started than.
If the demonstration goes as desired The Netherlands will have the disposal of a production line with an annual production of 80 to 90,000 tons of anhydrite, in which a use of 110 to 130,000 tons of FGD gypsum is involved.
The E.C. herewith has the disposal of a proved technique with which an essential contribution can be made to the application of FGD gypsum.

4.7 **Economics**
The demonstration of this anhydrite production line sustained by FGD gypsum will achieve its principal goal only if it leads to broad commercial utilization of the technology in western Europe. For this to occur the commercial production of anhydrite must be economically attractive in the principal markets to be served by the product.
The economic feasibility has been examined for a production line with a capacity of 10 tons anhydrite per hour. The investment is approximately 12 million Dutch florins.

For the Dutch situation the production line turned out to be economically justified at an annual production of 60 to 65,000 tons of anhydrite, a FGD gypsum price of approximately 0.-- Dfl. per ton and an anhydrite price of at least 110.-- Dfl. per ton (free buyer).

5. MARKET INTRODUCTION OF SELF-LEVELLING FLOOR SCREEDS

5.1 Dutch replacement potential for self-levelling floor screeds

By direction of Novem an examination was performed into the Dutch market perspective for self-levelling floor screeds.
Self-levelling floor screeds can particularly be applied in residential and non-residential buildings.
The demand for floor screeds has substantially increased from a level of 16 million square metres in 1985 to almost 22 million square metres in 1988.
The existing market for self-levelling floor screeds in 1989 was estimated at 350 to 750,000 square metres. The substitution potential of self-levelling floor screeds depends on the sort and size of the building projects. Taking this into account in The Netherlands still remains a potential annual market for self-levelling floor screeds of at least 9 million square metres or approximately 0,6 square metre per head of the Dutch population.
The use of anhydrite per square metre amounts to approximately 25 kilogrammes, so that The Netherlands represent a potential demand of at least 225,000 tons annually, which is equivalent to 315,000 tons of FGD gypsum or the quantity which is release by about seven 600 MWe coal-burning electricity plants.
For the use of self-levelling anhydrite mortars similar levels may be expected for the other E.C.-member states, so that the potential E.C.-market annually amounts to at least 7 million tons of FGD gypsum.
The demand for anhydrite in behalf of self-levelling floor screeds is at the moment complied with by synthetic anhydrite from the fluorine-hydrogen production and by natural anhydrite.
The price of suitable anhydrite, inclusive the additives required, runs from 80 to 120 Dfl. per ton, delivered in The Netherlands.

5.2 Quality control

One of the possibilities to promote the application of self-levelling floor screeds is certification.
In this case it is important to have a good look at the entire cycle, this means from the raw materials to the finished product.
An inventory of the quality-level present was rounded off in 1990.
In the meantime several application demonstrations have started in connection with the drafting of the level to be realized and of the establishment of a quality system.
An important demonstration is the application of self-levelling floor screeds in the new accommodation of the Dutch Ministry of Housing, Physical Planning and the Environment in The Hague. It is an office-building of a reinforced concrete construction with a floor surface of about 75,000 square metres.

6. CONCLUSION

An important potential extension of the processing of FGD gypsum is the use of it in self-levelling floor screeds.
For this the production of a suitable anhydrite is required. By direction of Novem the technique required for this has been inventorized and examined for its suitability on a laboratory and pilot scale. Said technique has been granted as a demonstration project by the E.C. The preparations of the building activities have been started. The rounding off of the building activities is scheduled for the second half of 1992. During a period of time of 1 to 2 years a monitoring programme will be executed in order to be able to verify the aims and purposes.
If these aims and purposes are obtained then we have a technique at our disposal to process FGD gypsum into an anhydrite with a quality and at a price suited for self-levelling floor screeds.
For The Netherlands this means on short term an annual processing of 110 to 130,000 tons of FGD gypsum with at longer term a possible extension up to 315,000 tons of FGD gypsum.
For the E.C. with approximately 320 million inhabitants said application involves a potential processing of about 7 million tons of FGD gypsum, which is approximately the annual production of FGD gypsum possibly to be expected in the year 2000.
To be able to realize this, an enormous effort will still be required to the introduction of the processing technique to be demonstrated, and the use of self-levelling floor screeds as a labour-friendly alternative for the present floor screeds.

Waste Materials in Construction.
J.J.J.R. Goumans, H.A. van der Sloot and Th.G. Aalbers (Editors)

PRODUCTION AND APPLICATION OF A USEFUL SLAG FROM INORGANIC WASTE PRODUCTS WITH A SMELTING PROCESS

F.J.M. Lamers[1], H.M.L. Schuur[1], A.J. Saraber[1] and J. Braam[1]

[1]PBI, The Projectbureau for Industry and Environment, P.O. Box 1187, 6201 BD Maastricht (The Netherlands)

SUMMARY

By means of a smelting process, a range of inorganic waste materials can be converted into products that are fit to be utilized within strict environmental regulations. Smelting is a cleaning method, because measures are taken to separate heavy metals; in this aspect it differs from melting or vitrification, which can be called an immobilization method. Depending on the processing during cooling of the smelt, glassy or crystalline slags can be produced, which among others can be applied as cement raw material. After a general description of this PBI smelting method, results are given of a test on municipal waste incineration fly ash and of the specific cement properties of a glassy slag, produced from coal fly ash.

1. INTRODUCTION

Heavy metal containing inorganic wastes present a growing problem for our environment. Partly this is due to their increasing volume and partly because of stricter environmental requirements which will be enforced by legislation in the near future, such as the Dutch Building Materials Decree. Furthermore the export of inorganic waste materials for disposal abroad becomes almost impossible. As a result good prospects can arise for reprocessing schemes which offer the possibility to convert inorganic waste materials into environmentally acceptable products. One of such schemes is a smelting process, developed by PBI. In this article an overview is presented on this smelting process and on the slags that can be derived from it. In the second part of the article research results are given on: i) Technical and environmental characteristics of synthetic slags, produced from Municipal Waste Incineration fly ash. ii) Technical properties of alkali-activated slag cement, produced with synthetic slags, and coal fly ash filler.

2. DESCRIPTION

2.1 Description of the PBI smelting process

If during melting of waste materials, extra measures are taken to separate and discharge heavy metals, the technology is called smelting (1). "Melting" or "vitrification" can be considered an immobilization process, whereas "smelting" is in fact more a cleaning method.

The PBI smelting process is shown in fig. 1: After possible drying and preheating a waste material is fed into a smelting oven; in this smelting oven two heavy metal separation and discharge processes are stimulated, to reach the refining effect pursued in the smelting technology:

i) **Volatilization;** chalcophyllic -low fusion temperature- heavy metals such as Cadmium, Mercury, Lead and Zinc and also Arsenic, will volatilize from the high temperature melt and will precipitate on the flue dust in the flue gases during cooling. There they can be separated as metal dust with conventional techniques. Volatilization occurs allways with high temperatures; in smelting it is extra stimulated.

ii) **Gravimetric separation;** when reducing conditions are imposed onto the smelting batch, siderophyllic -high fusion temperature- heavy metals such as Cu, Cr, Ni, will loose their oxygen and pass into the native metal form. The metal droplets thus formed, have a

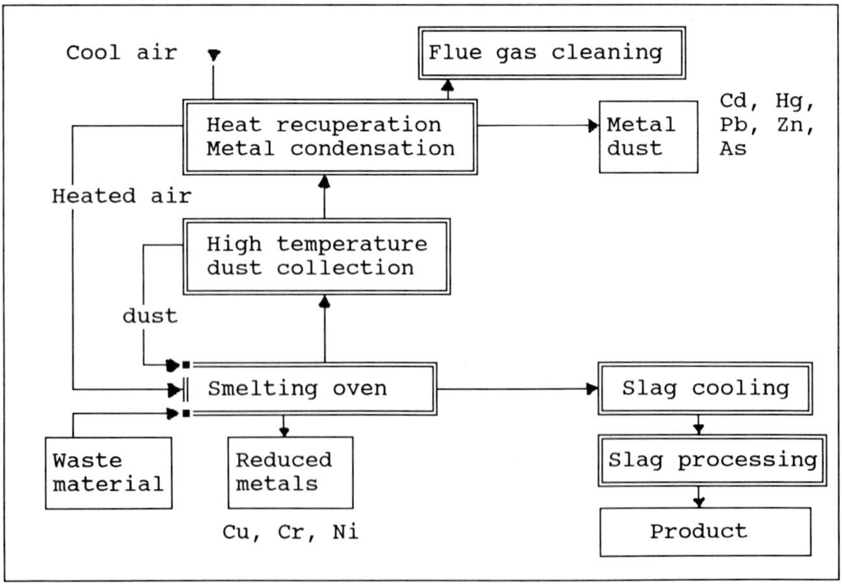

Fig. 1 PBI smelting and heavy metal separation scheme

considerably higher specific density than the rest of the melt. They will beconcentrated at the bottom of the oven and can be separated. The gravimetric separation mechanism is characteristic of the PBI smelting process.

The slags that are produced through smelting are cleaned of heavy metal contamination, as will be shown further in this article. The leaching behaviour is favourable.

2.2 Limiting conditions for smelting of waste materials

For the smelting of waste materials the following limiting conditions exist: i) The smelting temperature must be at an acceptable level, this means a maximum temperature of about 1500°C. ii) The viscosity of the smelt must be sufficiently low for the separation of heavy metals, both by volatilization and by gravimetry.

Both prerequisites can be reached if the contents of SiO_2, Al_2O_3 and CaO fall within the marked compòsitional area field (Fig. 2), where melting takes place below 1500°C. As shown in fig. 2, a lot of waste materials meet with these demands (2)

Furthermore by combination of different types of waste materials (e.g. calcium rich and calcium poor), optimizing of the smelting composition is possible.

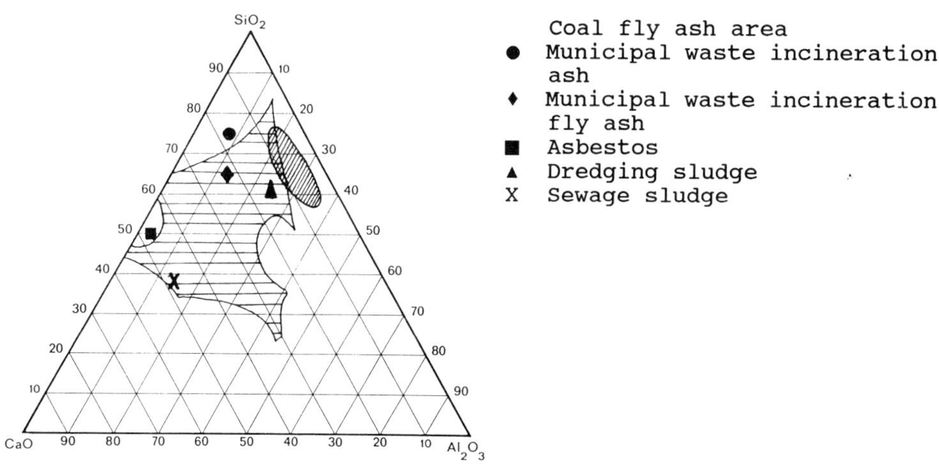

Fig. 2. Rankin triangle diagram with compositions of several waste materials. The area with melting below 1500°C is marked.

2.3 Types of synthetic slags and their quality

Depending on the applied cooling method, crystalline slags (slow cooling) or glassy slags (quenching) can be produced after smelting. After suitable treatment both types of slag comply with strict environmental standards. In fig. 3. the product groups are shown where theoretically, synthetic slags derived from several waste materials could be utilized.

2.3.1. Application of glassy slag

Technically, one of the properties of glassy slags is their pozzolanic or even semi hydraulic reactivity after fine milling, which makes them suitable as a cement half product. The reactivity of the slags is largely dependant on their chemical composition. An other possible application of glassy slag is in the production of glass fibres.

2.3.2. Application of crystalline slags

The crystalline slag is, compared to glassy slags, a more stable product, fit for building purposes, which seems less influenced by chemical composition. With the aid of shaping processes, theoretically several product types could be produced.

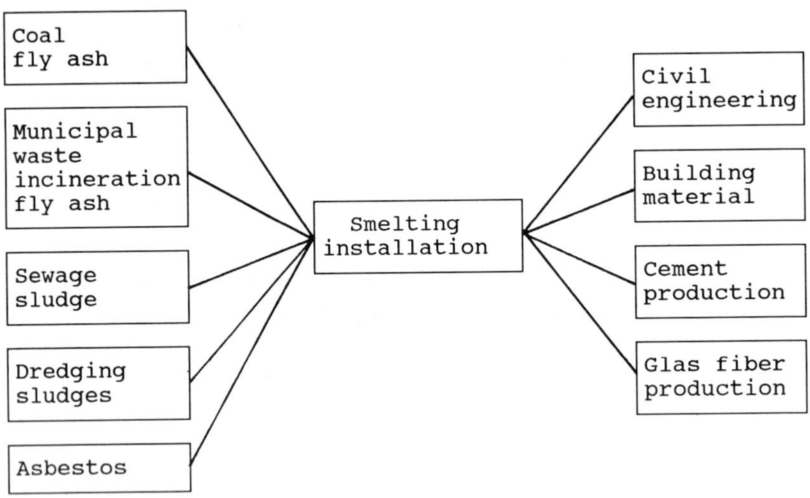

Fig. 3. Production of secundary materials through smelting

2.4 Concluding remarks

A smelting process offers possibilities to convert inorganic, heavily contaminated waste materials into cleaned slags. These cleaned synthetic slags can both be produced in the glassy and in the crystalline form. The synthetic slags can potentially be applied in the cement industry and in several sectors of the building industry. The disposal prices in the Netherlands are at a level that make smelting economically interesting.

3. SMELTING OF MUNICIPAL WASTE INCINERATION FLY ASH: ENVIRONMENTAL AND TECHNICAL PROPERTIES OF RESULTING SLAGS

3.1 Introduction

Supported by the Dutch Institute of Health care and Environment, tests have been performed on the smelting and metal separation behaviour of heavily contaminated municipal waste incineration fly ash. The main goal was the determination of the environmental behaviour of the new synthetic slag after smelting. Both glassy slags and crystalline slags were succesfully produced and were tested.

3.2 Description of the test smelt procedure

The following smelting procedure was used:
After analysing the main components, to ensure that the fly ash meets the given compositional limits for smelting, the samples have been molten in a 30 l oil fired smelting oven with carbon inner side. The residence time was long enough to ensure the complete separation of heavy metal phases. To catch the flue dust, on top of the oven a bowl cooler has been installed in which the heavy metals volatilize and precipitate. After smelting, the glassy slag was poured down in a waterbath, to stimulate granulation. To collect the separated metal alloys, the last part of the melt (bottomside in which the alloys are precipitated) was poured out separately. With a magnet, alloy particles still present in the main slag, are taken out to prevent influences during analysis.

3.3 Compositional changes of municipal waste incineration fly ash after smelting according to the PBI process

Compositional analyses have been performed on the synthetic slags; the results were compared to the contents in the original municipal waste incineration fly ash. See table 1.
The results indicate that during the PBI smelting process, the following components have been separated from the molten municipal waste incineration fly ash: As, Cd, Cl, Co, Cu, F, Hg, Mo, Ni, Pb, Sb, Se, Sn, V, Zn and iron. After smelting, the components all comply with the compositional requirements of the Building Materials Decree. For comparison, the contents of some components in an average basalt are shown (3). The contents of this components in synthetic slag are mostly comparable or lower.

3.4 Maximum leaching behaviour

Maximum leaching tests according to Dutch prenorm NVN 2508 were performed on the slags. (Fine milling, 2 x L / S 100, initial pH 4 and pH 7). Contents in the leaching liquid after the test procedure,

TABLE 1 Chemical composition of municipal waste incineration fly ash before and after processing in the smelting oven, compared to that of the natural material basalt. Chemical analyses after solution in HCl / HNO_3; (results in ppm).

Component	original fly ash	% Reduction	Slag	Basalt	S 1 ** limit
As	20	> 97.5	< 0.5		375
Ba	430	87.5	53		7,500
Cd	240(*)	> 99.8	< 0.2	< 1	10
Cl	57,000(*)	> 99.8	< 15		5,000
Co	33	60	1.1		250
Cr	240	40	140	255	1,250
Cu	1,040(*)	92	83	61	375
F	3,600	99.4	21		4,500
Hg	1.0	> 80	< 0.2		5
Mo	30	> 93	< 2	< 50	125
Ni	70	> 93	< 5	195	250
Pb	5,700(*)	> 99.8	< 10	< 10	1,250
Sb	660(*)	> 99.8	< 1		50
Se	9.5	> 80	< 2		50
Sn	300(*)	> 98	< 5		250
V	54	92	4.3		1,250
Zn	14,000(*)	99.1	130	115	1,250

(*) The content surpasses the compositional limit S1 (**) of the Building Materials Decree

were only detectable for the components Ba, Cr, V and Zn. Taking into account that the glassy slags preferably will be used as cement secondary raw material, or as an aggregate, it can be concluded that the materials produced from them will comply with all leaching limits from the Building Materials Decree. Tank leaching tests are in progress.

3.5 Technical properties of the synthetic slags

Several technical properties of the slags were determined:

- Pozzolanicity of glassy slags, in progress during writing.
- Compressive strength and water absorbtion (as a measure of freeze thaw behaviour) of crystalline slags. The strength of 368 MPa and a negligible water absorbtion are comparable to the best quality basalt.
- Strength of concrete with crystalline slag aggregate, in progress.

3.6 Conclusions

After processing of municipal waste incineration fly ash with the PBI smelting and metal separation method, an environmentally clean and technically good quality product arises. The production of several types of slag (glassy and crystalline) is technically possible. The properties of this slags offer interesting utilization possiblities.

4. PROPERTIES OF ALKALI ACTIVATED SLAG - FLY ASH CEMENT, WITH SYNTHETIC SLAG PRODUCED FROM COAL Fly ash

4.1 Introduction

Between 1988 and 1990 our company PBI, together with the Limburg Electricity generating Company PLEM, has carried out the so called KoReL project. Continuing on the ideas of Trief (4), a design for an 80,000 ton scale alkali activated slag cement (ASC) factory was developed, based on synthetic slag produced from coal fly ash. The project was originally intended to give a solution to possible future marketing problems for Dutch coal fly ash. Although the assumption of coal fly ash marketing problems in the Netherlands has still proven premature, in this presentation, the cement properties of ASC based on coal fly ash derived synthetic slag can be used to illustrate the cementing properties of optimized synthetic slag from all kinds of waste materials.

4.2 Alkali - activated slag fly ash cement ASC

Trief proposed the production of a high quality synthetic slags from an optimized mixture of coal fly ash and limestone (and if necessary an alumina addition) (Fig. 4) by melting the mixture at 1500°C and quenching the melt. The reactivity of the resulting slag is superior to granulated blastfurnace slag. The slag composition chosen by Trief corresponded with the optimum composition determined earlier by Keil and Locher(5). During a testing programm we empirically confirmed their findings (fig. 5). After fine milling, fly ash filler and an alkaline activator are added to the semi hydraulic synthetic slag. A binder results, that contains more than 70% coal fly ash. When water is added, a structure with cement

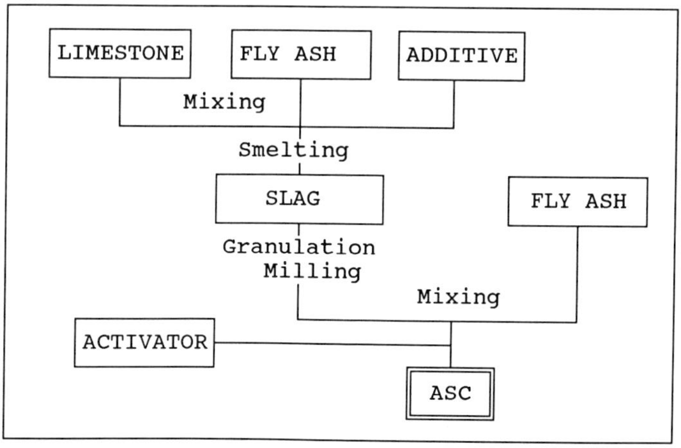

Fig. 4
Mixing scheme for ASC preparation

hydrates grows, like in portland clinker activated cements. ASC could therefore possibly, partly replace portland clinker activated cements.

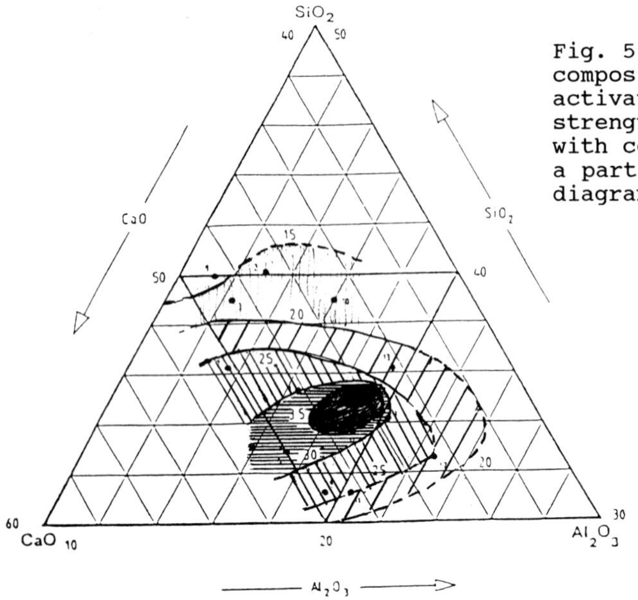

Fig. 5 Fields of optimum composition for 4% NaOH activation: 28 d compressive strength development of mortars with coarse slag, depicted in a part of the the Rankin diagram

4.3 <u>Slag production and ASC production</u>

In the project, semi hydraulic slag with an optimum composition was produced from fly ash, limestone and an alumina source in a 750 kg/hr electrically and oil heated glass furnace. Contents of heavy metals and maximum leaching behaviour were determined and proved to be on a very low level.

During the course of the KoReL project a wide range of different ASC mixtures have been tested. Variations in the mixes which were taken into account are for example: a. The type of activation; b. milling fineness of the slag; c. type and amount of coal fly ash used as a filler. The basic mix consisted of 50% fly ash, originating from Low NOx second generation combustion. Table 2 shows the 2 main ASC types, chosen for further research.

	TYPE 1	TYPE 2
Slag content	50 %	50 %
Fly ash content	50 %	50 %
Activator type:		
* Na2SO4 / Ca(OH)2	x	-
* Waterglass	-	x
Slag milling fineness	7,000 cm^2/g	5,000

TABLE 2
Composition of the 2 main ASC types, chosen for further research

4.4 Properties of alkali-activated slagcement

During the research project, properties have been determined on mortar prisms, cement pastes and (for ASC 1) on concrete. Because of the restricted volume of this article, only a general outline is given. Additional results will be presented in a future paper (6). Properties of ASC type 1 and Type 2, compared to Portland clinker activated cements are shown in table 3.

TABLE 3 Properties of ASC 1 and ASC 2 compared to portland clinker activated cements.

Property	ASC 1	ASC 2	PCA
Water demand	0.46 - 0.48	0.38 - 0.40	0.48 - 0.50
Setting times	2 - 4 hrs	30 min - 1 hr	2 - 3 hrs
Heat of hydration	120 kJ / kg	120 kJ / kg	340 kJ / kg
Compr. Strength 24 h	15 - 25 MPa	25 - 40 MPa	10 - 15 MPa
28 d	35 - 50 MPa	60 - 90 MPa	40 - 55 MPa
Tensile strength			
28 d	8 MPa	11 MPa	9 MPa
Color	Gray	Dark Gray	Light grey
Efflorescence	Little	Little	No
CO_2 resistance	Moderate	Moderate	Good
H_2SO_4 resistance	Good	Excellent	Bad
Sulfate resistance	Good	Excellent	Fair
Freeze - thaw resist	Good	Good	Good
Leaching behaviour	Good	Good	Good
Radio activity	Good	Good	Good
Influence of curing	large	large	Little

The interesting features of ASC lie in:
a. A high early strength combined with a very low heat of hydration. b. A good resistance to acid and other aggressive environments.
The strength development of ASC 1 and 2 and of Dutch portland- and blastfurnace cement is shown in fig. 6.
The general picture is that mortars and concrete with ASC 1 have properties that are comparable to normal portland clinker activated cement mortars and concretes. Mortars with ASC 2 have superior properties to normal portland clinker activated cements, but have a fast setting time and rheological effects which require extra attention.

4.5 Conclusions

The synthetic slag produced from coal fly ash and limestone, shows a high reactivity in ASC cements. This leads to the conclusion that with a optimized production proces, synthetic slag is an excellent alternative for granulated blastfurnace slag, as raw material for cement production. ASC can be produced with interesting properties such as a combination of a high early strength and a low heat of hydration. Concrete with ASC complies with strict requirements regarding composition, leaching behaviour and radio activity.

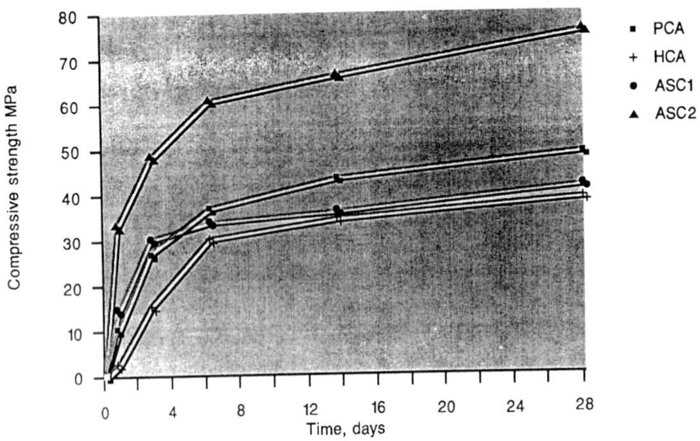

Fig. 6 Strength development of ASC mortars compared to portland cement and blastfurnace cement activated mortars

5. EVALUATION

In paragraphs 3 and 4, 2 different slag production projects were outlined. The outcomes from both testing projects confirm that smelting is an interesting option for the reprocessing of waste materials. Firstly, from heavily contaminated waste materials, an environmentally clean product can be produced and secondly, slags can be produced with interesting technical properties for several sectors in the building industry. A special preparation is the production of optimized semi hydraulic synthetic glassy slag for application in the cement industry.

Of course ultimately economical factors will decide wether a waste material should be treated by smelting technology.

REFERENCES

1 C. Broadbent, Pers. comm., 1991
2 PBI, Verwerking van afvalstoffen middels een smeltprocédé, 1989, R 89.018
3 TAUW Infra Consult, Laboratoriumonderzoek naar de samenstelling en uitloogbaarheid van diverse wegenbouwmaterialen, 1985, Rapportnummer 51014.04/RO - 02
4 L. Trief, in: Proceedings 4th Int. Ash Utilization Symposium, St Louis, U.S., March 24 - 25, 1976, pp 599 - 608
5 F. Keil and F.W. Locher, Zement Kalk Gips, Vol 11 (1958), pp 245-253
6 F. Lamers, in preparation, Materialen, 1991

Waste Materials in Construction.
J.J.J.R. Goumans, H.A. van der Sloot and Th.G. Aalbers (Editors)

The IR Process

Leonard S. Sarko and Harold Greenberg
Inorganic Recycling Corporation
7469C Worthington-Galena Rd.
Worthington, Ohio 43085, USA

SUMMARY

Inorganic Recycling Corporation has developed a process which chemically fixates inorganic hazardous waste in order to manufacture inert, nonleachable ceramic products. All materials exiting this System are usable. None are placed in a landfill, into water treatment systems or will pollute the air. The waste components are not encapsulated or sintered but chemically bound as part of a ceramic silicate matrix. The waste constituents are both functional and necessary within the IR Products.

1. SYSTEM APPLICATION

1.1 Amenability

Any material which is predominately composed of inorganic compounds such as metal sulfates, metal carbonates, metal phosphates, etc., would be amenable to this system. Metallic compound mixtures, drag-ins or variations of either concentration or constituency will not adversely affect this technology. The physical state of the inorganics (liquid, slurry, sludge or solid) is not chemically significant. This system will not process scrap metal materials or organic wastes.

1.2 Waste Stream Examples

The examples provided here are not all inclusive but will provide an understanding as to the range of inorganic wastes for which this technology may be employed. The IR Process has been applied to the following waste streams:

a. Filter Cake from Waste Treatment of Plating and Coating Operations;
b. Residual Ash produced by Municipal Incinerator Operations;
c. Electric Arc Furnace Dust from Steel Manufacturing Operations;
d. Waste Products from Glass or Frit Manufactures;
e. Soils contaminated with Inorganic Materials;
f. Contaminated Metal Sludges dredged from Lagoons or Pits;
g. Leachates collected from Landfill Sites; and
h. Waste Treatment Products from the Automotive Industry.

2. LABORATORY PHASE

2.1 Waste Stream Categorization

An initial screening is accomplished between the waste Generator and IR Sales Personnel. General information is gathered such as the process which creates the waste residuals, what waste treatment, if any, has been applied (i.e. precipitation), volume produced, location, etc. This screening eliminates waste streams which knowingly will not be amenable to the IR Process and supplies the base line data for initial laboratory work.

2.2 Laboratory Analysis

A 20 liter sample is sent the the IR Laboratory, provided the residual material is not rejected during initial screening. The representative sample is subjected to a series of analytical tests creating a chemical profile. The profile includes but is not limited to:

a. General Chemical Information

pH, density, moisture content, particle size, etc.

b. Major Constituencies

elemental concentration greater than 5000 mg/1

c. Mid-Range Constituencies

elemental concentrations between 100 mg/1 and 5000 mg/1

d. Leach Level Constituencies

elemental concentration between 1 mg/1 and 100 mg/1

e. Ceramic Parameters

constituencies divided into ceramic characteristic functions

The IR Process is a chemical reaction series designed to result in a ceramic product. As such, the data collected during the analytical stage will define the parameters needed to produce the proper reaction sequence. This reaction sequence is determined during the Formulation Stage of the Lab Protocol.

2.3 Formulation

On completion of the chemical profile, the lab develops acceptable formulae for the specific waste stream. Multiple formulae are possible for any one stream. The Laboratory analyzes, collates and calculates a chemical sequence taking into consideration the product requirements, available chemicals, engineering requirements, the waste chemical profile and other factors which may come into play. It will then test any particular formula, on a bench scale, to verify that the specific formula developed will accomplish what is required for the product desired. On completion of the formulation sequence a lab report is issued detailing the results of the analytical and formulation tests.

3. ENGINEERING

3.1 Small Quantity Generators

The laboratory report is provided to the Engineering Department. This department is responsible to determine the optimum processing route for any specific waste stream. Small volume generators will, most likely, be directed to a common processing facility. Material handling issues, transportation costs, and processing costs will be reviewed in order to develop a price schedule.

3.2 Large Quantity Generators

Large volume customers may require a processing unit built at the inorganic material generation site. The Engineering Department will develop the plans and specifications for the large volume facility tailoring the design to meet the generator and product requirements. A final cost sheet is created detailing both the construction and processing costs.

4. THE IR SYSTEM

4.1 Capacities

Although the IR System can be designed to process most any quantity of generator feedstock, the average IR System will process 40 to 50 tons per day. Additional capacities can be handled for larger generators by adding key processing components. Fixed base systems can also be downsized to handle as little as 20 tons per day. For those who produce less than 20 tons per day, it is more economical to ship these materials to a central processing facility.

4.2 Construction Requirements

A 50 ton per day unit requires the following:

1. Building - 10,000 to 30,000 square feet
2. Tractor-trailer space for loading and unloading
3. Bulk Storage Tanks - process chemical storage
4. Utilities -

   Water - 20 to 30 gallons per minute

   Electrical - approximately 150 KW

   Gas - Natural or Propane, quantity varies on input materials

4.3 Quality Assurance/ Quality Control

The IR Laboratory creates a detailed test sequence and logs all pertinent data during production operations. This is to verify that both the system chemistry and equipment is operating properly. Logs are primarily used to monitor recycling equipment parameters; tests are conducted on both incoming chemicals, including generator feedstock materials, and outgoing products, to assure consistent production results.

4.4 Mixer/Reactor Section

The recycling equipment is divided into two major categories. The first of these categories is the Mixer/Reactor Section. All incoming generator materials are handled on a batch basis. Typical mix batches range from one to five tons. A sample of the generator batch material is tested prior to entering the reactor. This verifies that the material matches the chemical parameters established during the laboratory formulation series. Once approved the material is entered into the reactor. A series of chemicals is then added to the generator feedstock. This series reacts the feedstock for the purpose of placing the constituents into a chemical mode that will bind within the silicate matrix. On conclusion of this reaction sequence, silicates are blended into the vessel contents. Potential silicates used are silica sand, clay, cullet and a number of other sources. The silicate type, concentration and blend will determine the specific product desired. On completion of blending, the material is pumped to a surge system. The surge system compensates for any time delays between the two main process sections.

4.5 Vitrification

The vitrification phase is operated on a continuous basis. The main component of this phase is the kiln. The function of the kiln is to provide the energy needed to complete the endothermic reaction resulting in the ceramic products. The surge system places the reacted/blended material into the kiln at a controlled rate. Kiln temperature is determined during the formulation sequence in the laboratory and is a function of the feedstock constituents. The reacted/blended material becomes vitrified and molten. No combustion or destruction of generator components occurs during this operation. This molten material flows out of the kiln and is captured by an Exit System. This system may be operated on a dry or wet basis. The dry system functions with the molten material falling into molds, where it is cooled and ejected. In the wet system, the molten material falls into a water tank, fracturing the material, after which it is exited by an auger. The product, derived from either the wet or dry system, is tested for leachability and to assure product specifications and quality. It is then packaged and shipped to the purchaser.

4.6 Scrubber

Part of the vitrification phase includes a comprehensive air scrubber system. Particulates may be picked up by the kiln air system. Micron size particulates and above are captured via knock-out boxes and other technologies such as venturies. These are filtered through standard filter presses with the resultant cake recycled to the front of the mixer/reactor system.

IR Mobile Processing Unit at General Motors Plant

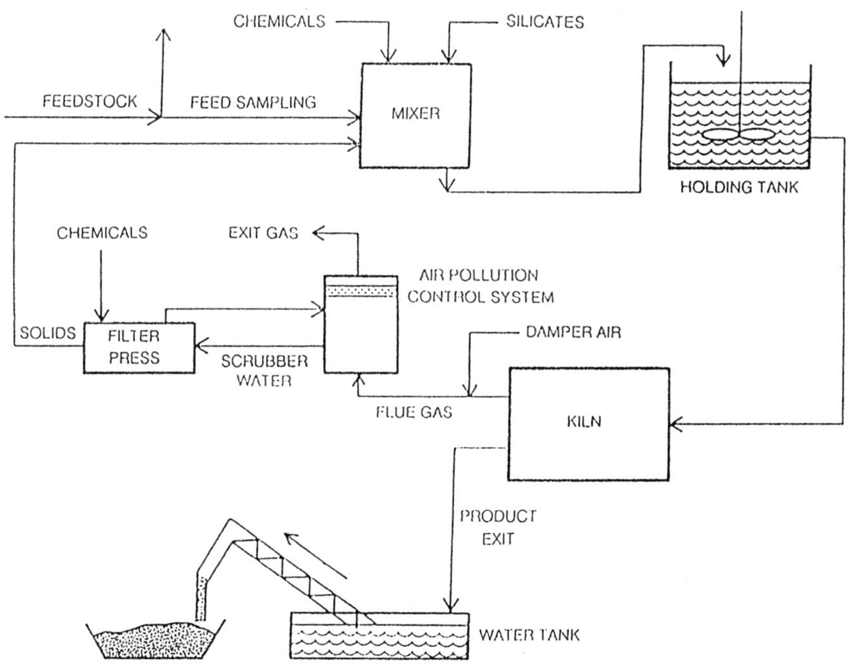

FIGURE 1
FLOW DIAGRAM OF THE IRI RECYCLING PROCESS

Submicron particles are captured using an irrigated filter and mist eliminator. This stream is also recycled back to the front of the system.

Anions may fume during kiln operations and may be picked up by the air system as well. Packed tower technologies are used to react and capture these anions. Secondary products are produced from these streams. As an example, plating and coating waste streams often contain a percentage of sulfates. During kiln operations a percentage of the sulfates will gasify to sulfur dioxide. The sulfur dioxide is reacted in the packed tower with sodium hydroxide producing sodium metabisulfite, which is packaged and sold.

5. PRODUCTS

5.1 Ceramics

The key products produced by the IR Process are considered ceramic. Ceramic may be defined as any material composed of inorganic elements, which when exposed to high temperature, creates a solid matrix. At a molecular level, these materials may be crystalline or noncrystalline, in which silica or alumina acts as the matrix building block.

Not all materials that contain silica or alumina are ceramic. Cement is composed of silica elements, however it is based on calcium chemistry, therefore it is not a ceramic. Ceramic materials are nonreactive, corrosive resistant and very strongly structured at molecular levels. Cement materials have none of these characteristics.

The IR Products are presently used as feedstock equivalents for a number of ceramic products. The material directly produced by IR is presently sized to one-half centimeter or less, bulk bagged in one ton increments and shipped to specialty aggregate manufacturers. These firms size and blend these materials providing the ingredients to manufacturers of the final end-products made. In essence, IR is replacing the mining operations in the mineral marketplace. IR, however, is able to provide a much more consistent product as a result of the chemical control of our operations. The examples provided in the subsequent sections are materials which have been produced using the IR products. In comparison with other minerals, IR can produce products analogous to pure aluminas or corundum on the upper end to gypsum or calsite on the lower end. The IR Products have been crushed to 600 mesh (powder) without leaching its constituents on exposure to corrosive environments.

5.2 Architectural

Product produced within this category include:

1. Wall tile
2. Floor tile
3. Sinks and tubs
4. Laboratory countertops

5. Patio stones
6. Mosaics
7. Vanities
8. Fireproof wallboards
9. Brick, Block
10. Roofing Media

There are many small markets which have applied the IR Products, i.e. pen set bases, picture frames, statues, etc. which are too numerous to detail. Basically, any ceramic material, used as an architectural product, could use the IR Product. The transition metals, which began as the hazardous material of concern, enhances crystal formation and as the pigmentation source. Unlike organic dyes, these colorants will not fade with time or be affected by corrosive environments.

5.3 Abrasives

Products produced within this category include:

1. Sandpaper
2. Shot blast
3. Grinding wheels
4. Grinding media
5. Buffing compounds
6. Polishing compounds

Product chemistry will define specific abrasivity of a material. Changes in that chemistry will alter the abrasivity producing a different product. Most formulations will produce medium type abrasives. An example of a medium abrasive is a media which cleans a tool without damaging the tool itself. The potential abrasive range which can be produced is quite wide. The metallic transition metals significantly defines the abrasivity characteristic through its function in crystal development.

5.4 Refractories

Products produced within this category include:

1. High temperature bricks
2. High temperature blocks
3. Insulation Material

Refractories are specifically designed to operate in high temperature environments. IR Products have very high melting and deformation points, therefore can withstand the temperatures required for insulator type applications. The transition metals within these Products provide a conductivity/emissivity characteristic which is required for kiln brick and block functions. The inertness of the IR Products also help function in shield application around heat treat operations when corrosive chemicals are used.

6. ALTERNATIVE TECHNOLOGIES AND THEIR FLAWS

6.1 Landfill

The most common application is to place inorganic waste materials into a landfill. Normal ground conditions are acidic. Organic wastes, placed in the landfill, will intensify this acidity through their decomposition. The acidic environments mobilize the metal inorganic waste constituents. These metals travel through the landfill to the aquifer or leachate collection system located underneath. The resultant metal concentration in those leachates will surpass acceptable governmental standards.

6.2 Encapsulation

Technologies exist which encapsulate inorganic wastes in a plastic or cement derivative material. The object of this technology is to place the hazardous constituents into a protective cage, thereby reducing the exposure of the constituents to the surrounding environment. Plastic methodologies are excellent against chemical degradation but are very susceptible to physical degradation. Cement methodologies are excellent against physical degradation but the acid conditions, always prevalent in landfills, will decompose the cement. When the cage is damaged, the hazardous components become exposed once again. Encapsulation technologies can only delay the metal migration problem, not eliminate it.

6.3 Sintering

Technologies exist which place inorganic hazardous wastes directly into high temperature environments such as incinerators. Incinerators were designed to combust organics. Organics can be totally destroyed by placing them into this environment for a proper amount of time. Inorganic elements are not combustible. At most, the incinerator can cause a phase change to a liquid or gas state. Upon cooling, the elements will drop out as an ash residual. If this ash is placed in a landfill, the metal constituents will leach.

7. EXAMPLES

7.1 Generator Feedstock

IR has provided data from four different waste feedstocks:

1. Automotive Chromium Waste Filter Cake
2. Electric Arc Furnace Baghouse Dust
3. Mixed Transition Metal Filter Cake
4. Fly Ash from a Municipal Incinerator

The numbers listed will be used as the reference definition within the subsequent charts.

7.2 Constituent Analysis of Waste Materials

The constituents are measured on an elemental basis using the nondestructive analytical technique of X-Ray Fluorescence-Energy Dispersive Technology. The following machine settings were used for all tests:

1. Accelerating voltage 20.0 KeV
2. Beam- sample incidence angle 90.0 degrees
3. X-ray emergence angle 30.0 degrees

4. X-ray- window incidence angle 0.0 degrees
5. Window thickness 15.0 microns
6. Totals normalized to 100 percent

	Weight Percent			
Feedstock	1	2	3	4
Element				
Na	23.98	0	31.56	8.5
Mg	3.41	1.18	11.10	6.6
Si	0.17	2.07	0	9.4
P	1.40	0	3.35	1.80
Cl	0	2.41	4.04	15.56
K	0.35	2.70	1.98	2.5
Ca	8.36	17.18	6.95	18.86
Cr	32.69	0.57	3.35	1.52
Mn	0	3.41	1.76	0
Fe	0.84	29.84	7.24	8.45
Cu	0	0.84	3.04	3.56
Ni	0	0	2.53	4.25
Zn	0	34.89	7.24	2.68
Pb	0	3.33	0	7.89
Al	0	0.46	0	7.40
S	28.80	1.12	15.86	1.03

7.3 Constituent Analysis Product

The constituents are measured on an elemental basis using the nondestructive analytical technique of X-Ray Fluorescence-Energy Dispersive Technology. The following machine settings were used for all tests:

1. Accelerating voltage 20.0 KeV
2. Beam- sample incidence angle 90.0 degrees
3. X-ray emergence angle 30.0 degrees
4. X-ray- window incidence angle 0.0 degrees
5. Window thickness 15.0 microns
6. Totals normalized to 100 percent

	Weight Percent			
Feedstock	1	2	3	4
Element				
Na	3.53	2.34	26.05	8.23
Mg	1.03	0.84	7.50	2.50
Si	52.45	41.54	50.10	51.10
Cl	0	0	0.87	2.28
K	0.68	2.18	0.22	1.06
Ca	16.20	11.44	1.34	14.66
Cr	12.24	0.67	0.87	.90
Mn	0	1.37	0	0
Fe	5.37	10.58	0	2.20
Cu	0	0.56	0	1.25
Ni	0	0	1.10	2.10
Zn	0	19.19	3.17	1.22
Pb	0	0.79	0	3.17
Al	8.50	8.25	8.02	8.75
S	0	0.25	0.76	0.58

7.4 Leachate Results after Vitrification

The test used to measure leachability of the IR Products was the USEPA TCLP Test. This procedure takes a known weight and size of the product and exposes it to a specific concentration of acetic acid over a specific time frame. The solute is filtered and analyzed using atomic absorption methodologies for specific metal species.

	Leachate (mg/l)				
Element	1	2	3	4	USEPA Limit
Cadmium	< 0.01	< 0.01	0.066	< 0.01	1.0
Chromium	0.1	< 0.06	< 0.06	< 0.06	5.0
Mercury	< 0.001	< 0.001	< 0.001	< 0.001	0.2
Barium	0.23	< 0.10	< 0.10	0.17	100.0
Lead	< 0.03	< 0.03	0.5	0.1	5.0
Selenium	< 0.01	0.06	0.05	< 0.01	1.0
Silver	< 0.01	< 0.01	< 0.01	< 0.01	5.0
Arsenic	0.5	< 0.2	< 0.2	< 0.2	5.0

8. CONCLUSION

The IR Process was developed over a ten year period of time. This system is fully operational on a commercial basis. Our services are being used by Fortune 100 corporations as well as United States governmental agencies. The USEPA has thoroughly analyzed the IR Process and has recognized and acknowledged its ability to recycle inorganic hazardous waste.

Waste Materials in Construction.
J.J.J.R. Goumans, H.A. van der Sloot and Th.G. Aalbers (Editors)

QUALITY IMPROVEMENT OF RIVER SEDIMENTS AND WASTE WATER TREATMENT SLUDGES BY SOLIDIFICATION AND IMMOBILIZATION

J.H. DIJKINK [1], K.J. BRABER and R.F. DUZIJN [2]

[1]Research & Engineering Consultants B.V., Beneluxlaan 9, 3527 HS Utrecht (The Netherlands)

[2]TAUW Infra Consult B.V., P.O. Box 479, 7400 AL Deventer (The Netherlands)

SUMMARY

Investigations were conducted to assess the possibilities of quality improvement of river sediments using the DOMOFIX process, a process of solidification and immobilization. The primary goal of this approach is to enable reuse of the immobilized products as building material. The chemical composition and the leaching behaviour of treated and untreated matter were determined. In addition, some civil engineering parameters such as compressive and tensile strength were measured. Though no differences were measured between the composition of treated and untreated sediment, it appeared from the leaching experiments, that Zn and As were (partly) immobilized. Other elements such as Cu seem to be mobilized, possibly due to the formation of ammonium complexes, the NH_3 being liberated by the high pH values which arise during the DOMOFIX treatment. The DOMOFIX process is not yet able to immobilize organic compounds such as PAHs and oil. The power to immobilize hazardous compounds would seem to depend on the type of sediment involved.

1. INTRODUCTION

The Netherlands is a country with a great number of rivers, canals and harbours, situated at the slow flowing river deltas near the North Sea. Over the last few decades industrialization has shown a rapid growth, resulting in an enormous expansion of activities, especially in the harbour areas. Since special (legislative) measures were not (yet) taken, huge amounts of untreated industrial waste water were discharged into rivers, canals and harbours. As a result, polluted matter accumulated in high concentrations. Since the rivers and harbours have to be deepened, highly contaminated river sediments have to be disposed off. These residues are now being deposited in special depots and special precautions have to be taken due to the leaching of heavy metals. If this leaching could be minimized or stopped it may be possible to reuse the treated material. Immobilization and solidification are likely to provide an answer to this question. The DOMOFIX process is one of these systems. It is a Swiss invention and was introduced into The Netherlands by Research & Engineering Consultants B.V. Together with the participants, Ankersmit Verenigde Maalbedrijven (Ankersmit United Milling Company) and N.V. Industriebank LIOF

(N.V. Industrial Bank LIOF), a test programme was set up and carried out in Switzerland in the summer of 1989.
TAUW Infra Consult B.V. was responsible for sampling the river sediment (in June, 1989) and the environmental test programme, which started in October, 1990 and finished in May, 1990. Eight different materials were tested for immobilization, which were divided into three groups:-
(1) 4 samples taken from river/harbour sediments;
(2) 2 samples taken from residue matter from soil treatment plants;
(3) 2 samples taken from sewage plants (industrial and domestic).
Only the first group is dealt with in this paper as they form the main subject of the programme.

DOMOFIX PROCESS

The DOMOFIX process is a process used to treat river, harbour and industrial sediment.

One of the principles of the process is the mixture of special lime and other cement based products. Particularly Dolomite is able to "demolish" the structure of the river sediment by its electrolytical behaviour and to set intercellular and capillary water free. The immobilization of environmental hazardous waste matter, especially heavy metals, is achieved by minerals from a special group of silicone based minerals (minerals of the Al-type) and special components such as zeolites. Pre-investigation at laboratory scale is necessary before particular substances can be determined.

With the prescribed quantity of the various components the material is thoroughly mixed (see photograph 1) and homogenized for a certain period of time. At the second stage in the process the pre-treated matter may be pelletized. This depends largely upon the type of river sediment matter being used. The prepared matter then goes into a special type of compactor. This "Hi-Compact Anlage" (see photograph 2) is specially designed to compact sediment from sewage plants and works with a pressure of 50 bar. Within 1.5 - 2 minutes a large amount of water is extracted and a soil particle content of approximately 60 - 70 % by mass remains. After final hardening this figure has risen to 80 - 90 % by mass.

RESEARCH PROGRAMME

The aim of this preliminary research and test programme was to work out the effectiveness of the total process. The emphasis was put on the environmental aspects, i.e. the degree of immobilization. The main part of the programme was formed by the environmental tests.

Photograph 1 (above) :
Mixing Sediment
and Additives

Photograph 2 (left) :
The Hi Compact
Anlage

Different types of river sediment were used. This means a difference in composition, type of pollutants (organic or inorganic components) and the origin. The research programme had to show the effectiveness of the extraction (volume reduction), the mechanical behaviour of the solidified matter and the effectiveness of the immobilization.

MATERIAL CHARACTERISTICS

The research focused on the civil, mechanical and environmental aspects of the DOMOFIX process.

To check the mechanical properties of the solidified river sediments, tests were carried out which measured the tensile and compressive strengths. The civil engineering aspects of this research were limited to these parameters. Other process parameters such as the volume of centrifugal and delivery water were also determined.

Considering the primary goal of the investigations, namely the reuse of treated products as building material, particular attention was paid to the environmental aspects of the DOMOFIX process. The environmental orientated part of the research deals with the determination of the immobilization effect and the efficiency of the DOMOFIX process. This was executed by carrying out a total analysis and a characterization of the leaching behaviour of both treated and untreated sediment. Relevant components were heavy metals and organic compounds.

2. MATERIALS AND METHODS

2.1 Materials

Experiments were carried out using four different types of river and harbour sediments originating from the Geulhaven in Rotterdam, Malburgen in Arnhem, Bruinisse and Stein. These samples (approximately 200 l) were taken by specialists from TAUW Infra Consult B.V., who also supervised the pre-handling of the river sediments and advised on the execution of the "Hi-Compact" tests which were carried out in Switzerland. Samples were taken, for laboratory tests, from the untreated river sediments and the resulting solid compressed tablets.

2.2 Methods

2.2.1 Mechanical Research

Parameters and characteristics to be determined were the dry weight, mass and volume of the tablets, the compressive strength, the tensile strength and the figure of increase for dry residue components. Therefore, specimens were stored under special circumstances in a cool dark place for 28 and 90 days. Specimens were also stored under water for 7 days and were tested afterwards.

2.2.2 Environmental Aspects Tests

Both treated and untreated river sediments were submitted to a total analysis (the analysis of heavy metals was preceded by a total destruction of the matter using HF) and several leaching tests. The leaching tests were carried out in accordance with the Dutch NVN 2508 and O-NVN 5432 and included column tests, diffusion tests and single shake tests.
The leachates were submitted for analysis on the relevant components depending upon the material under investigation. The choice of relevant components was based upon previous research carried out by the Dutch National Institute for Waste Water Treatment DBW/RIZA. Table 1 gives an overview of tests carried out.

TABLE 1

Overview of tests carried out for treated and untreated river sediment.

Test	Treated River Sediments	Untreated River Sediments	Analysis
Total analysis	X	X	Heavy metals Organic compounds
Column test (NVN 2508)	X	X	Heavy metals
Diffusion test (O-NVN 5432)	X		Heavy metals
Single Shake test (NVN 2508)	X	X	Organic compounds

The column and diffusion test were carried out, on the one hand, considering their predictive value for emissions under practical circumstances, on the other hand, considering the Provisional Building Material Decree (ref. 1). This Provisional Decree prescribes the diffusion test for form-given and the column test for non form-given materials (form-given material has a compressive strength exceeding 2 MPa and a volume per element > 50 cm^3). The single shake test was carried out in order to determine the leaching of organic compounds, which cannot be carried out accurately in the column test in accordance with NVN 2508. Contrary to the standard procedure (NVN 2508), the leachates from the single shake tests have been collected by centrifugation instead of filtration. Hereafter, the bottles were rinsed with acetone/petroleumether so as to avoid any loss of organic compounds.

3. RESULTS

3.1 Mechanical Aspects

The following table shows the results of the mechanical characteristics.

TABLE 2

Mechanical characteristics of four different types of river sediment.

	Geulhaven	Bruinisse	Arnhem	Stein
% by mass dry residue river sediment (original samples)	55.1	56.4	55.6	41.4
% by mass dry residue after compression inc. additives	77.1-79.2	78.9-90.1	79.9-81.0	79.9-86.2
tensile strength:				
28 days (N/mm^2)	1.6- 2.2	2.7- 3.0	1.4- 2.3	0.9- 1.5
90 days (N/mm^2)	- 2.6	2.4- 3.7	1.5- 2.7	1.3- 2.4
compressive strength:				
28 days (N/mm^2)	6.1- 7.4	7.9-13.7	8.4-11.1	3.1 -5.5
90 days (N/mm^2)	5.7-20.2	8.4-13.6	5.5-11.9	5.7- 7.0
+7 days under water (N/mm^2)	2.4- 7.0	6.8-13.3	6.9- 7.0	3.6- 6.3

From this table it is obvious that a very good strength can be achieved, although the optimum mixture was not yet determined.

3.2 Environmental Aspects - Total Analysis, Column Test, Single Shake Test

The total analysis and the results of the leaching tests are presented in 4 histograms (Figures 1-4), one for each river sediment treated using the DOMOFIX process. Only the most striking results are presented this way, full details are given in ref. 2. The units used in the histograms are as follows:-

- total analysis : quantity in mg/kg d.m.
- column test : quantity in mg/kg d.m.
- single shake test : quantity in mg/kg d.m.

From figures 1-4 it appears that the results of this preliminary research fluctuate somewhat. The success of immobilization depends upon the type of river sediment and the components considered. Some notable results:-

- The treated and untreated matter show no significant difference regarding the total analysis. This indicates that dilution of components, by additives, does not play any role in the DOMOFIX process.

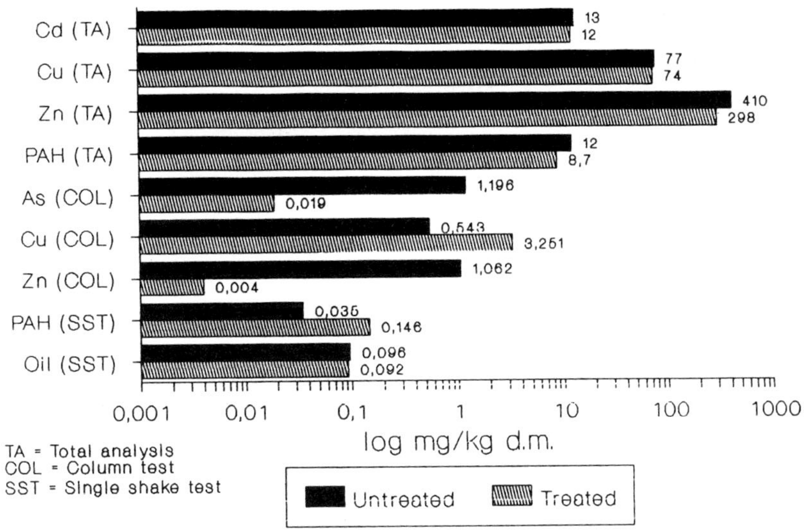

Figure 1. River sediment Bruinisse

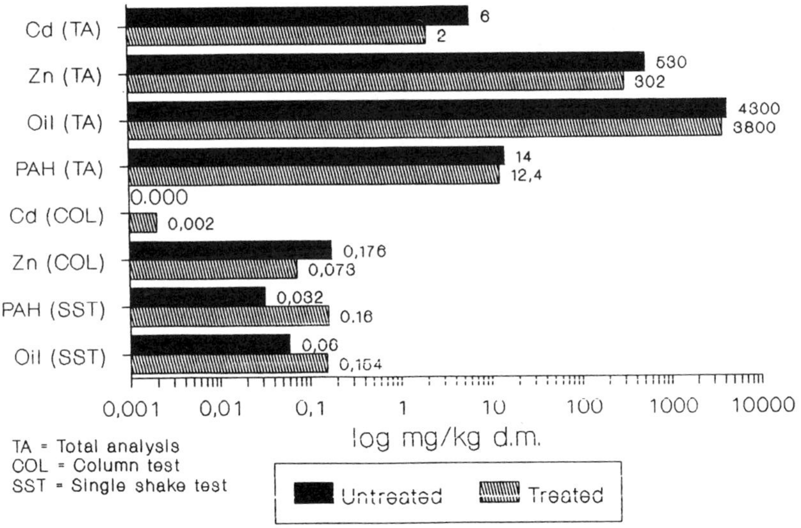

Figure 2. River sediment Geulhaven

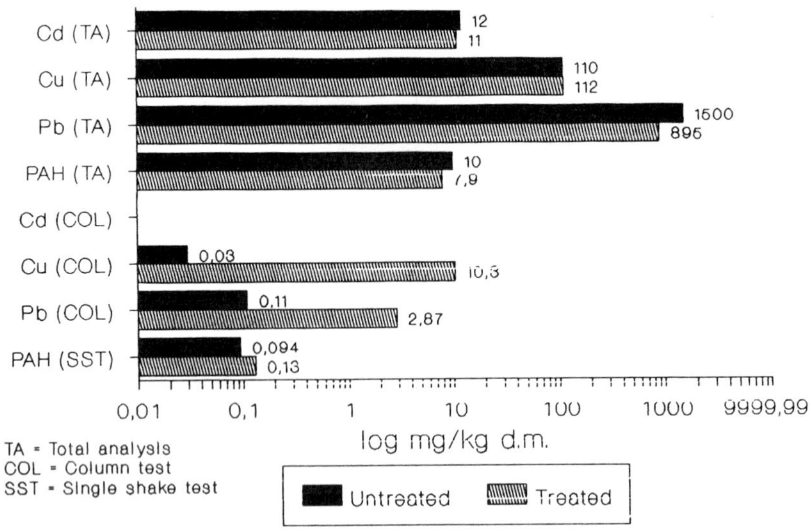

Figure 3. River sediment Arnhem

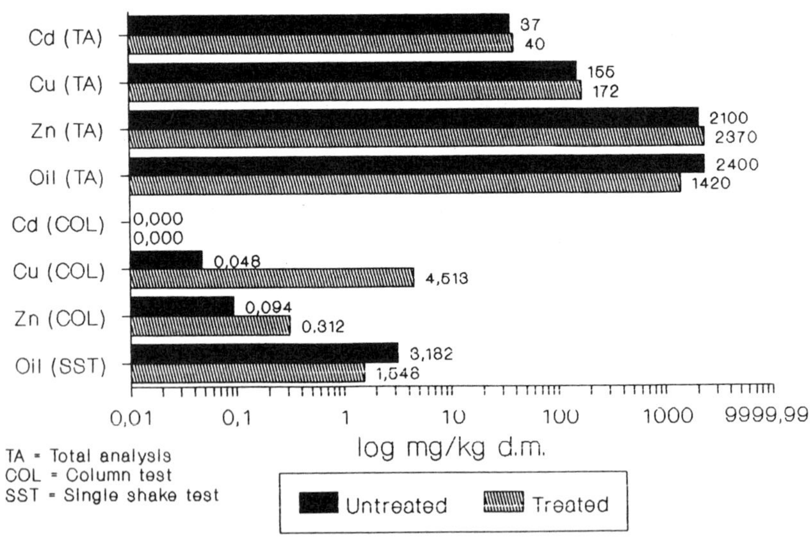

Figure 4. River sediment Stein

The average dilution factor, defined by the formula D = 1 + weighed additives (d.m.)/ weighed river sediment (d.m.) was calculated 1.4. This, together with the above mentioned, indicates, in general, that the additives contain heavy metals in about the same magnitude as the original sludges.

- From the results of the column test it appears that some components, such as Zn and As, are immobilized using the DOMOFIX process, whereas others, such as Cu, appear to be mobilized. The last phenomenon may be due to the presence of ammonium in the sediments and/or in the additives, in combination with a high pH, which may cause $Cu(NH_3)_4^{2+}$ complexes to leach out easily (this was obvious from the odour from the tablets during and after production).
- The results of the shake test indicate that the DOMOFIX process is not yet capable of immobilizing organic compounds.
- The success of immobilization depends upon the river sediment treated; for Bruinisse and Geulhaven the results look promising but for Arnhem the treatment gives relatively poor results.

EFFICIENCY OF IMMOBILIZATION

From the results of the column test the efficiency of the immobilization (Ri) was calculated. Ri is defined using the formula:-

$$Ri = 100\% \times (Q_u - Q_t)/Q_u$$

Q_t and Q_u are the total quantities (in mg/kg d.m.) leached during the column test for treated and untreated sediment respectively.

The efficiency of immobilization for As and Zn were calculated to be 98 and 54-100% respectively, which is promising. On the other hand however, for Cu, Ni, Cr and Pb for some of the sediments negative efficiencies (<0%) were calculated, which points to mobilization rather than immobilization. This may be explained by the high pH levels (10-13) caused by alkaline additives and the presence of NH_3, leading to easily leached complexes.

DIFFUSION TEST

From the diffusion tests, carried out on the treated river sediment, the fluxes could be calculated for relevant (inorganic) components. The flux (unit:mg/m^2) is a guide as to the extent of the diffusion and is used as a test parameter for form-given building materials in the Provisional Building Material Decree (ref. 1). The results are given in table 3.

TABLE 3

Fluxes (mg/m^2) in diffusion tests

Component	Bruinisse	Geulhaven	Arnhem	Stein
Cd	0.0	0.0	0.0	0.0
Cr			2.4	
Cu	24.4		42.5	53.3
Ni	4.7			
Pb			3.3	7.0
Zn	0.0	0.0	0.0	0.0
Hg	0.0	0.0	0.0	
As	2.1		3.4	

The fluxes have been compared with the norms for form-giving building materials (the so called V-norms).

Based upon the results given in table 3, the treated river sediments should be categorized as V2 materials, which means that the use in construction work will be approved under certain conditions (not considering the total analysis, see Section 4).

4. CONCLUSIONS AND RECOMMENDATIONS

4.1 Conclusions

In this contribution some results are discussed, from an orientating investigation to the possibilities of stabilization and immobilization of river sediments using the DOMOFIX process. The following conclusions may be drawn:-

1. No significant differences were measured between the composition (total analysis of relevant components) of treated and untreated river sediments. Thus, a decrease of the concentration in the leachates, if any, after immobilization would not be due to dilution.
2. When the leaching behaviour of both treated and untreated matter is studied, some components (such as Zn and As) appear to be immobilized, whereas others (such as Cu) seem to be mobilized. This may be accounted for by the formation of $Cu(NH_3)_4^{2+}$ complexes and desorption of adsorbed (to clay mineral) metals by the action of hydroxide ions that arise during the DOMOFIX treatment (high pH levels).
3. According to the results obtained from the single shake test, it appears that the DOMOFIX process is not yet able to immobilize organic compounds such as PAHs and oil.
4. The ability of the DOMOFIX process to immobilize hazardous components would seem to depend upon the type of river sediments involved.

5. Based upon the total analysis, the column test and the diffusion test the processed material may be considered for use in construction work. This will only be approved of if certain criteria are met. The results of this preliminary investigation have therefore been compared with the norms stated in the Dutch Provisional Building Material Decree (ref. 1). The Decree makes a distinction between non form-given (indicated "N") and form-given (indicated "V") materials. The comparison is shown in Table 4.

TABLE 4

DOMOFIX processed material compared with the Provisional Building Material Decree

DOMOFIX processed river sediments	Assessment of non form-given material (N)		Assessment of form-given material (V)	
	Total Analysis	Leaching behaviour (1) column test	Total Analysis	Leaching behaviour (1) diffusion test
Bruinisse	X (oil)	N3 (Cu,Ni)	V2 (Cd,Cu,Ni,PAHs)	V2
Geulhaven	X (oil,EOX)	N2	V3 (PAHs)	V2
Arnhem	X (Cd,EOX)	N3 (Cu,Pb)	X (Cd)	V2
Stein	X (Cd,oil)	N3 (Cu)	X (Cd)	V2

<u>Legend</u> : (1) only metals considered

X material will not be approved of as a suitable building material

N2/N3 non form-given material may be approved of under (strict) conditions

V2 form-given material may be approved of under (strict) conditions

( ) the important components mentioned are given in brackets.

In view of the total analysis (table 4), only the treated sediments from Bruinisse and Geulhaven may be approved of as form-given materials in construction works. In view of the leaching behaviour, reuse appears to be permissible for all of the treated sediments (form-given or otherwise) under severe restrictions (see ref. 1) of isolation etc. Due to both the total analysis and leaching behaviour being considered in the Provisional Decree, the total analysis will, as yet, be decisive in the assessment of whether the treated sediments will be approved of in constructions works or not.

4.2 Recommendations

In order to improve the DOMOFIX process for river sediments the following recommendations should be taken into account for future research:-

1. Further insight into the chemical composition and physical characteristics of the additives used in the DOMOFIX process and the sediments is of the utmost importance.
2. Additives containing ammonium salts should be avoided and ammonium resident in the sludges should not be set free or should be removed.
3. Additives leading to high pH levels (11-12) should be avoided.
4. Adapting the additives (lower heavy metal content and different properties) and the stabilization process in such a way that hazardous components will arise in the delivery water (which has to be treated) and thus be removed from the sludges. Owing to this the chemical composition of the treated material will be improved and accordingly the leaching behaviour.
5. Pre-treating the sediments by extraction or thermal treatment in order to remove the organic compounds. This, however, will soon be too expensive, therefore, additives treating both inorganic and organic compounds are preferred.

REFERENCES

1 Concept - Voorontwerp Bouwstoffenbesluit Bodembescherming, 25 April 1989. (Provisional Building Material Decree).

2 TAUW Infra Consult B.V. Immobilisatie - effekt van het DOMOFIX - procédé op uitlooggedrag baggerspecie. Rapportnummer 3116735 (1990).

Waste Materials in Construction.
J.J.J.R. Goumans, H.A. van der Sloot and Th.G. Aalbers (Editors)

Coal Fly Ash Slurries for Back-filling

Sumio Horiuchi, Takuro Odawara and Hirato Takiwaki
Institute of Technology, Shimizu Corporation
(3-4-17, Etchujima, Koto-ku, Tokyo, Japan)

1. INTRODUCTION

In Japan, more than 4 million tons of coal ash are annually discharged from power plants, and 60% of the total coal ash is disposed of in spite of enthusiastic studies on methods for it's utilization. It's use as a construction material is the main focus of such investigations. The authors are studying the use of coal ash slurries as reclaiming material[1)], and have utilized 80,000 tons of coal fly ash for preparing 100,000m³ of self-hardening slurry to construct man-made islands.

Because of the high cost of land in Japan, there are many plans for big civil engineering projects to construct large underground structures. Tunneling and shielding are used for excavation, and back-filling is an important technology for minimizing environmental impact, such as settlement of ground or leakage of ground water.

As a conventional back-filling grout of shields and tunnels, cement/water slurry (the primary slurry) and sodium silicate solution are injected together by the 1.5 shot method, however occasionally do not meet the following requirements for reliable back-filling;

(1) Since the primary slurry is generally pumped a long distance from mixing point to injection point, slurry volume in the pipe line is so great that it can not be discharged for clean up. It is, therefore, necessary to prevent hardening of the primary slurry in the pipe line for several hours. On the other hand, the primary slurry should start to harden after mixing with a hardening initiator, and develop long-term strength more than 2MPa.

(2) Requirements for strength and hardening speed of the back-fill are dependent on the surrounding soil. Since it is sometimes necessary to adjust the strength several times at one site, strength adjustment should be easy to perform.

(3) Since the structures are to be used for a long time, high durability of the grout against environmental factors is necessary.

Because coal fly ash has a high pozzolanic activity and good fluidity, it has been used as a component of a variety of grouts. It seems well-suited for being a major component of the primary slurry for back-filling grout, however few studies have been reported. This paper discusses the applicability of coal fly ash as the main component of back-filling grout, based on results obtained in a series of laboratory tests.

2. EXPERIMENTS

2.1 Materials

Two types of fly ashes from Australian coals were used in this study. As shown in Table 1, both fly ashes contain a small amount of calcium, and type UB is an acid fly ash.

Table 1 Properties of fly ashes

Fly Ash	Chemical components (%)				Ig. loss (%)	pH	Gs	Mean Dia. (μm)
	SiO_2	Al_2O_3	Fe_2O_3	CaO				
UB	76.4	18.9	1.91	0.14	1.75	4.2	2.19	10.4
WA	65.9	19.3	3.85	2.71	4.06	11.7	2.21	12.2

To decrease segregation, 300 mesh bentonite was added to the primary slurry. Ordinary portland cement was used to develop strength. Two retarders, a hydroxycarboxylic acid compound (HA) and a naphthalene sulfonic acid compound (NS), were selected and added to the primary slurry at 4% of cement weight. Sodium silicate III (ASTM D3400-85) solution was added to the primary slurry for initiating strength development.

2.2 Test procedure

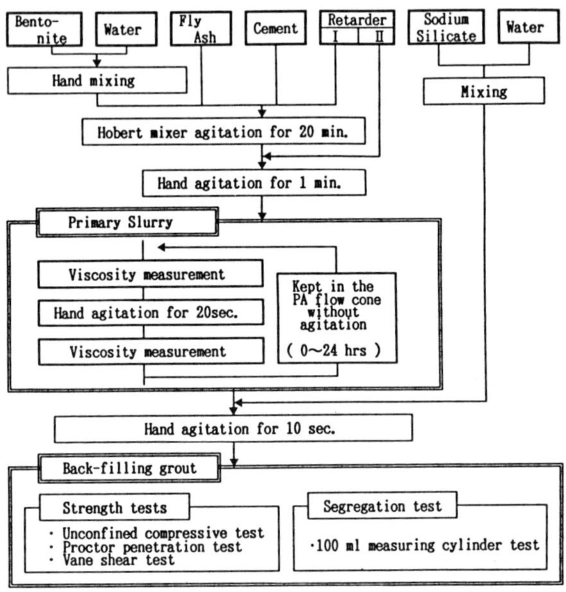

Figure 1 Test procedures

Figure 1 shows the experimental procedure. Bentonite/water suspension was prepared by one minute hand agitation using an egg beater. Cement, fly ash and retarder were added to the bentonite suspension and mixed by a Hobert type mixer at a rotary speed of 140rpm and a planetary speed of 60rpm. Additional retarder was added to the slurry in most of the cases for the best quality primary slurry.

Viscosity was measured using the PA flow cone (ASTM C939-87). Segregation rate (Sr: ratio of decreased volume at 24hrs to the initial slurry volume) was measured by using a 100ml measuring cylinder.

Condition of the primary slurry stocked in the pipe line was predicted by the following two kinds of flow measurements;

(1) flow of the slurry stocked for a given period in a plugged PA cone, and

(2) flow of the stocked slurry re-agitated for 20 seconds by the egg beater.

The sodium silicate solution was mixed to the primary slurry with 10 seconds agitation using the egg beater. Strength test samples were then prepared in plastic moulds and cured at 20°C in tap water. Strength development of the grout was determined by unconfined compressive test, proctor mortar penetrometer test or vane shear test.

The specifications of the primary slurry and the grout are as follows;

1) Viscosity: For the better pumpability, PA flow should be 10-20 sec.
2) Segregation: For the reliable filling, the segregation rate of the grout should be less than 3%.
3) Strength: Long-term strength of the grout should be more than 2MPa. For use as a flash setting filling, the grout should start to harden just after mixing with the initiator and develop more than 0.1MPa unconfined compressive strength within one hour.

Terms used in this paper are defined as follows;

*Fly ash content (Cf (Kg/m^3)): Flyash weight in $1m^3$ primary slurry
*Cement content (Cc (Kg/m^3)): Cement weight in $1m^3$ primary slurry
*Bentonite content (Cb (Kg/m^3)): Bentonite weight in $1m^3$ primary slurry
*Sodium silicate addition rate (As (%)): Weight ratio of sodium silicate III to cement
*Water in the sodium silicate solution (Aw (%)): Weight ratio of water to sodium silicate III

3. RESULTS

3.1 Retarder addition

To prevent the hardening of the primary slurry, it is essential to add hardening retarders to the slurry. Table 2 shows the

Table 2 Comparison of retarder addition

FA	Retarders		PA Flow (sec)					Sr	qu (MPa)		
	I	II	0hr	1hr	3hr	6hr	24hr	(%)	7 day	28 day	70 day
UB	HA4%	----	11.0	12.1(10.6)	12.9(10.6)	14.0(10.6)	----(14.9)	2.06	0.038	0.445	0.885
	HA2%	HA2%	10.8	11.7(10.5)	12.7(10.6)	14.4(10.5)	----(15.5)	2.06	0.043	0.443	0.913
	NS4%	----	9.87	10.3(10.0)	12.5(10.5)	----(20.9)	----(----)	2.06	0.534	2.328	5.842
	NS2%	NS2%	8.85	9.64(8.81)	9.86(8.89)	----(17.3)	----(----)	2.06	0.608	2.436	5.570
	HA2%	NS2%	9.54	10.0(9.44)	----(10.2)	----(10.4)	----(12.1)	0.30	0.247	0.676	2.311
	NS2%	HA2%	9.64	10.1(9.62)	10.4(9.67)	11.7(9.66)	23.7(9.85)	7.53	0.269	0.579	2.165
WA	HA4%	----	9.89	9.86(9.50)	10.6(9.84)	11.9(9.75)	40.5(9.83)	6.46	0.026	0.392	0.747
	HA2%	HA2%	10.2	11.4(10.1)	11.6(10.2)	12.2(10.1)	18.1(10.2)	6.74	0.030	0.367	0.660
	NS4%	----	9.67	9.86(9.50)	10.4(9.44)	15.9(10.0)	----(----)	3.70	0.362	1.710	4.566
	NS2%	NS2%	9.56	119 (8.91)	44.3(8.88)	12.4(9.12)	----(----)	12.3	0.413	1.732	3.424
	HA2%	NS2%	9.06	10.0(9.37)	----(9.78)	13.7(9.80)	133 (9.94)	6.92	0.149	0.550	1.902
	NS2%	HA2%	9.64	10.2(9.53)	10.5(9.52)	11.4(9.56)	209 (9.59)	10.9	0.188	0.415	1.499

* PA flow data in ( ) : PA flow after re-agitation
---- : Impossible to measure

effect of retarder on properties of the primary slurries and grouts.

Addition of NS is favorable to the strength development of the grouts, however diminishes the good pumpability; the re-agitated slurries are not flowable after 24hrs. On the contrary, HA addition is very effective for maintaining the good pumpability, however it has a negative effect on strength development. The combination of NS and HA is effective to get a good pumpability and minimizes the negative effect on strength development. In the point of segregation, HA addition followed by NS addition is better than vice versa, and therefore this adding method is used to prepare the primary slurry throughout of this study.

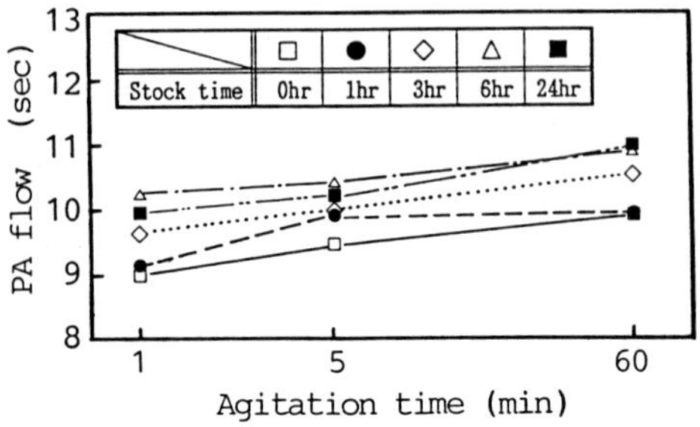

Figure 2 Effect of agitation period of bentonite/water suspension on the primary slurry PA flow (FA:WA, C_f=408Kg/m^3, C_c=233, C_b=31Kg/m^3)

3.2 Bentonite addition

The methods to prevent slurry segregation may be classified into the two categories; decreasing the free water and increasing the viscosity. Both purposes can be accomplished by bentonite addition to slurry, however the followings should be taken into account;

i) Optimum bentonite content:
Bentonite addition increases the slurry viscosity, and therefore it is important to determine the optimum bentonite content, at which both the segregation and the viscosity are within the specifications.

ii) Viscosity change:
In the actual preparation, fly ash and cement are added to a bentonite/water suspension, whose viscosity increases with time. It is therefore important to evaluate the effect of the viscosity increase on the properties of the primary slurry.

Figure 2 shows effect of the agitation time in preparation of the bentonite/water suspension

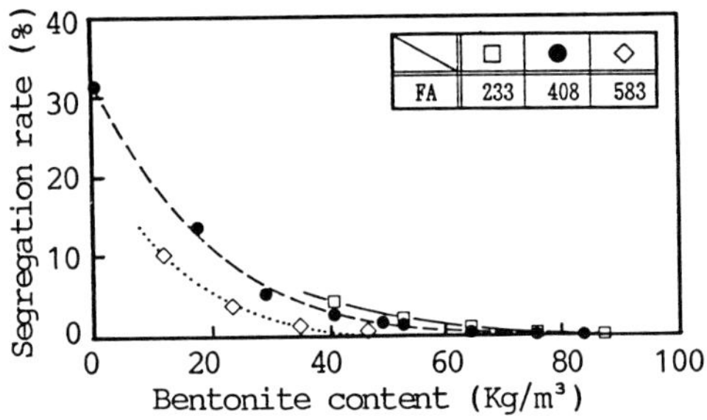

Figure 3 Effect of bentonite addition on segregation rate (FA:UB, C_c=233Kg/m^3)

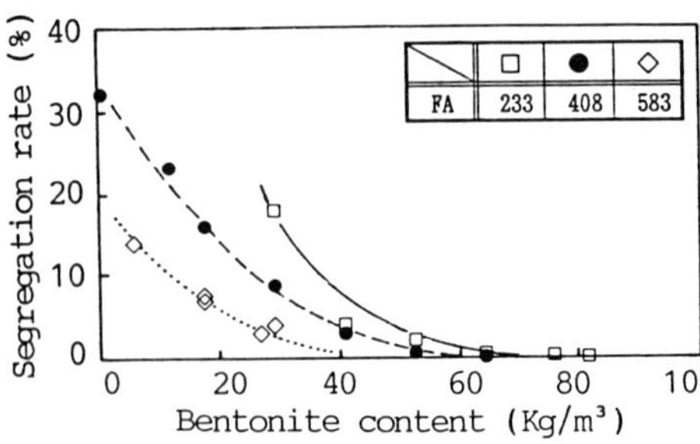

Figure 4 Effect of bentonite addition on segregation rate (FA:WA, C_c=233Kg/m^3)

on the PA flow of the primary slurry. The primary slurry viscosity increases with the agitation time, however the change is very small within 60min agitation. In the test procedure, the agitation time was, therefore, determined as one minute hand mixing by egg beater.

Figures 3 and 4 show the relation between the bentonite content and the segregation rate, and an increase of the bentonite content is reflected in a decrease of the segregation rate. In the figures, increase of fly ash content results in lowering the segregation, and 15Kg/m³ of bentonite content can be decreased by 100Kg/m³ increase of fly ash content. This means that fly ash addition reduces the backfiller cost when the price of fly ash is lower than 15% of the bentonite price.

The relation between bentonite content and PA flow is shown in Figures 5 and 6. Increasing of the bentonite content increases the PA flow, however this effect is not so significant as the effect on segregation. The relation between bentonite/water ratio and the PA flow is shown in Figure 7, where the PA flow cannot be expressed by a simple function, because some part of the water is trapped on the surface of the cement and fly ash. Since the water required for making fly ash slurry differs with fly ash type[1], the bentonite content required depends on the fly ash type.

According to the test results, the optimum bentonite

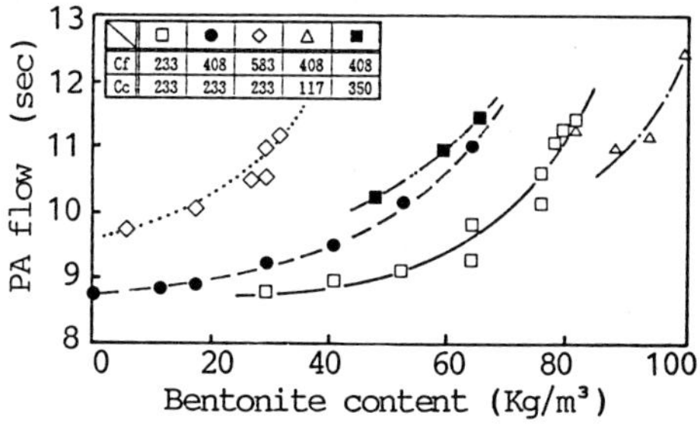

Figure 5 Effect of bentonite addition on PA flow (Fly Ash: UB)

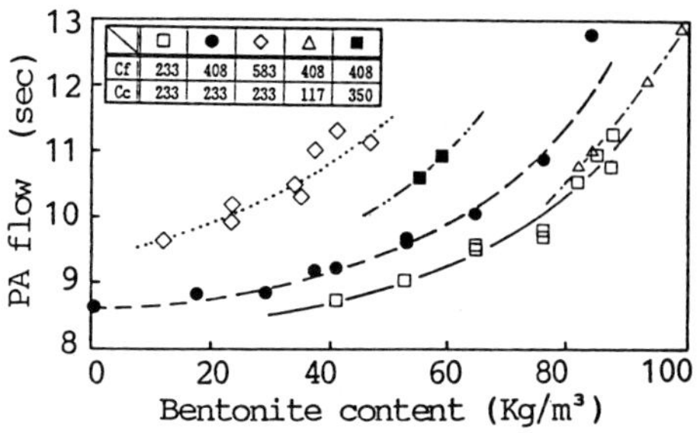

Figure 6 Effect of bentonite addition on PA flow (Fly Ash: WA)

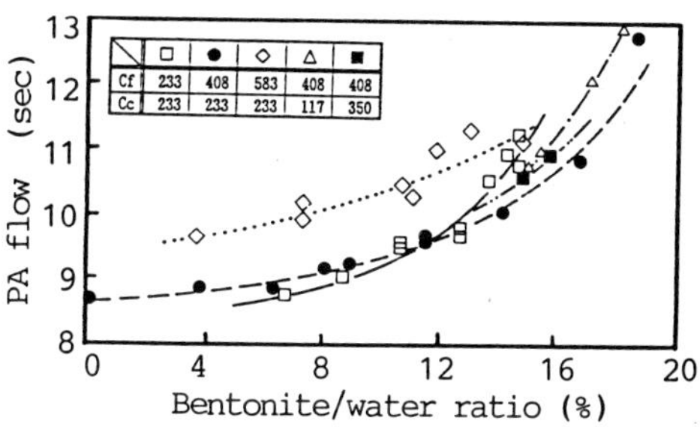

Figure 7 Relation between PA flow and bentonite/water ratio (FA:UB)

Table 3 Optimum primary slurries

Fly Ash	Primary slurry composition (Kg/m³)					
	Fly ash	Cement	Bentonite	Water	Retarders I (HA)	Retarders II (NS)
UB	233	233	76	783	4.7	4.7
	409	233	53	713	4.7	4.7
	584	233	23	645	4.7	4.7
WA	233	233	64	788	4.7	4.7
	409	233	41	719	4.7	4.7
	584	233	18	649	4.7	4.7

contents can be determined as shown in Table 3, and more than 500Kg/m³ fly ash can be mixed into the primary slurry.

3.3 Strength development of the grouts

In Figures 8 and 9, strength development of the grouts with curing time (t) are shown. Without sodium silicate addition, the grouts do not develop strength for more than four weeks. However, a small sodium silicate addition can initiate the strength development of the grouts, and increase of sodium silicate quantity results in the higher early strength. When the sodium silicate addition rate (As) is beyond 50%, the grouts start to harden within several seconds after mixing. The relation between the time required for developing 0.1MPa unconfined compressive strength (qu) and As is shown in Figure 10. It is shown to be possible to control strength development by adjusting only the sodium silicate quantity.

Figures 11 and 12 show the effect of fly ash content on strength development. In the

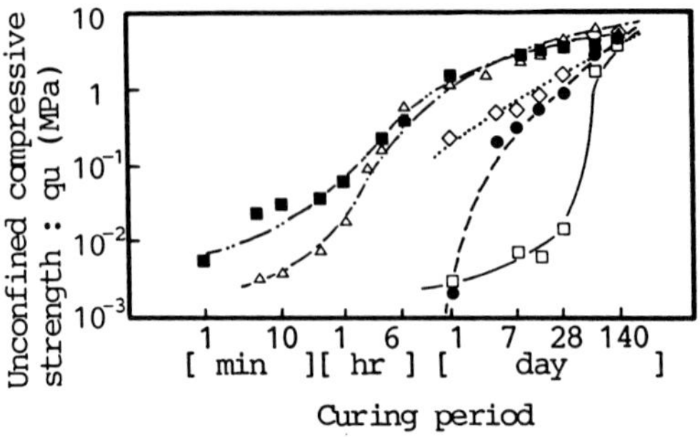

Figure 8 Strength development of back-filling grouts (FA:UB, Cf=408Kg/m³, Cc=233Kg/m³)

	□	●	◇	△	■
As	0%	20%	60%	100%	200%
Aw	0%	14%	43%	71%	143%

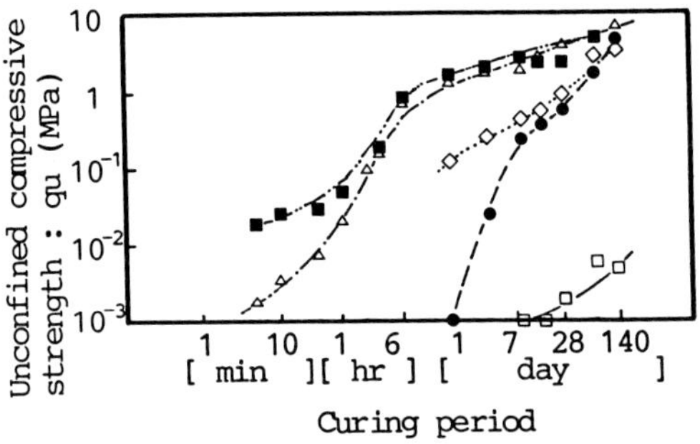

Figure 9 Strength development of back-filling grouts (Fly Ash:WA) (FA:WA, Cf=408Kg/m³, Cc=233Kg/m³)

	□	●	◇	△	■
As	0%	20%	60%	100%	200%
Aw	0%	14%	43%	71%	143%

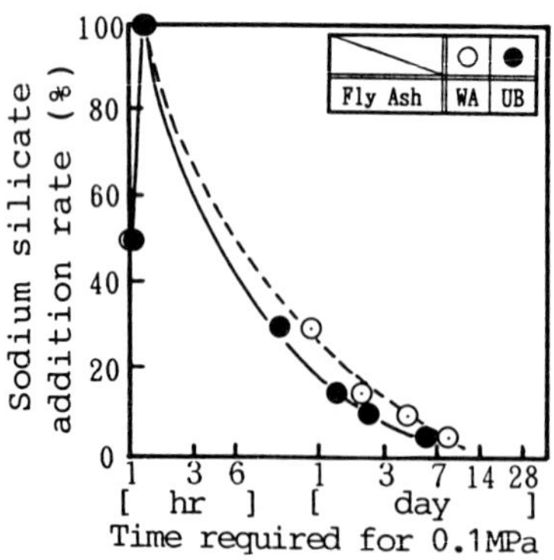

Figure 10 Effect of As on time required for 0.1MPa qu (Cf=408Kg/m³, Cc=233Kg/m³)

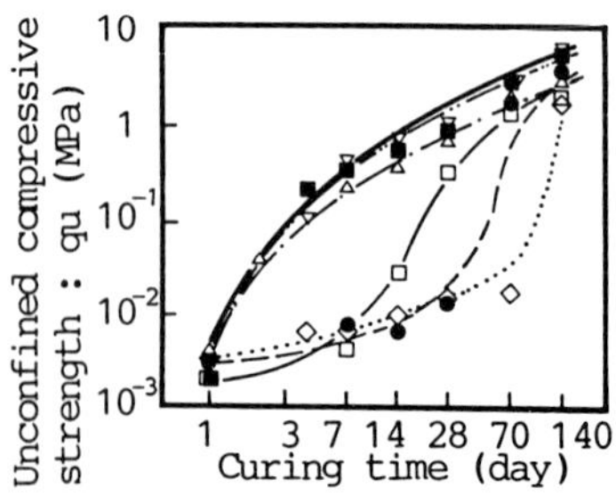

Figure 11 Effect of Cf on strength development (FA:UB, Cc=233Kg/m³)

	□	●	◇	△	■	▽
Cf	233	408	583	233	408	583
As	0%	0%	0%	20%	20%	20%

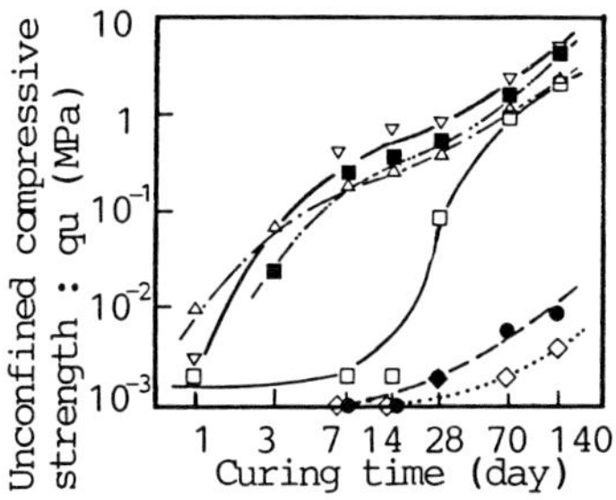

Figure 12 Effect of Cf on strength development (FA:WA, Cc=233Kg/m³)

	□	●	◇	△	■	▽
Cf	233	408	583	233	408	583
As	0%	0%	0%	20%	20%	20%

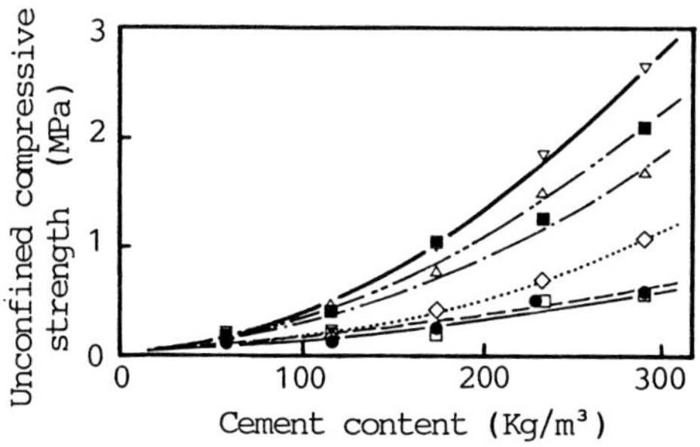

Figure 13 Effect of cement content on strength development (FA:UB, As=20%, Aw=650%)

	□	●	◇	△	■	▽
Cf	233	408	583	233	408	583
t	28 day	28 day	28 day	70 day	70 day	70 day

cases of sodium silicate free grouts, the higher the fly ash content, the later the strength development starts. In these primary slurries, the bentonite content is decreased with increasing of the fly ash content. Since the specific surface area of the bentonite is 1000 times greater than fly ash or cement, the amount of retarder absorbed onto bentonite surface must be much greater than that on fly ash or cement. Efficiency of retardation is therefore considerably reduced at higher bentonite content. When the sodium silicate is added, strength of the grouts are increased with increasing of the fly ash content. This strength increase results from the higher density of the grout.

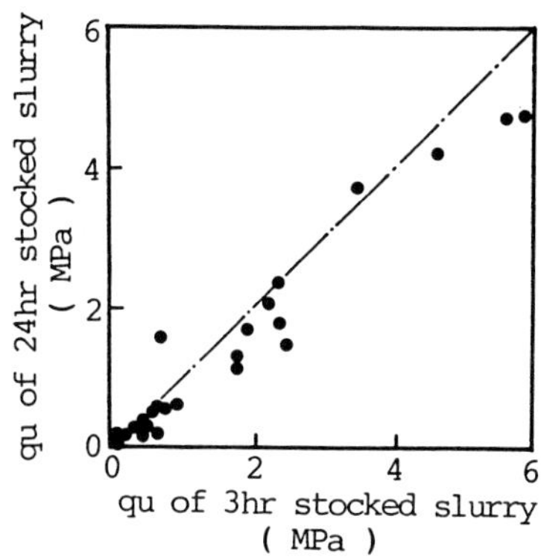

Figure 14 Effect of stock time of the primary slurry on qu

Figure 13 shows the effect of cement content on strength development. The higher the cement content, the greater the grout strength, and increase of the fly ash content increases the strength. As shown in Figure 11-13, fly ash addition is an efficient method of increasing strength.

The effect of stock time of the primary slurry on strength development is shown in Figure 14, and there is no strength reduction caused by elongated slurry stock.

4. CONCLUSIONS

Results obtained show the high potential of coal fly ash as a major component of back-filling grout as follows;

1) More than 500 Kg/m^3 of fly ash can be mixed to the primary slurry, however segregation does occur. The addition of bentonite is effective in preventing segregation but increases viscosity, therefore, we must determine the optimum quantity of bentonite. The amount of bentonite required can be decreased by increasing the fly ash content.
2) To prevent hardening of the slurry, retarders ought to be added amounting to more than 4% of cement weight. In this slurry, viscosity does increase while stocking, however it can be easily recovered by a calm agitation.
3) The primary slurry itself does not develop strength even after four weeks, however begins to harden just after mixing with sodium silicate. Early strength development can be controlled by adjusting the amount of sodium silicate added. When the amount of sodium silicate is beyond 50% of cement weight in the primary slurry, the hardening process can be quickened and the grout can be used in circumstances which require flash-setting.
4) Increase of fly ash content increase the strength development.

Referring to the results, the primary slurry containing 410 Kg/m^3 coal fly ash, 230 Kg/m^3 cement, 53 Kg/m^3 bentonite, 9 Kg/m^3 retarders and 720 Kg/m^3 of water was used at an actual construction site and its high potential as a back-filler was confirmed. Since strength development of the slurry can be easily controlled by the amount of sodium silicate added, the grout can be used not only for back-filling but also for a variety of construction works.

REFERENCE

1) Horiuchi,S et al.: 7th Int. Ash Utilization Symp., 1985,pp.907-917

Waste Materials in Construction.
J.J.J.R. Goumans, H.A. van der Sloot and Th.G. Aalbers (Editors)

THE FEASIBILITY OF RECYCLING SPENT HAZARDOUS SANDBLASTING GRIT INTO ASPHALT CONCRETE

Jeffrey Means[1], Jeffery Heath[2], Edwin Barth[3], Kenneth Monlux[4] and Jeffrey Solare[4]

[1]Battelle Memorial Institute, Columbus, Ohio

[2]Naval Civil Engineering Laboratory, Port Hueneme, California

[3]U.S. Environmental Protection Agency, Center for Environmental Research Information, Cincinnati, Ohio

[4]R&G Environmental Services, San Jose, California

SUMMARY

The recycling of spent sandblasting grit, commonly referred to as spent abrasive blast material (ABM), into asphalt concrete is being investigated by the U.S. Navy as an alternative to disposing the spent ABM in a landfill. This paper discusses issues related to the technical feasibility and regulatory acceptability of this concept. These issues include the chemical characterization of spent ABM, asphalt mix design criteria, the results of bench-scale tests, regulatory compliance issues, and a discussion of the advantages and disadvantages of recycling spent ABM into asphalt concrete. The merits of recycling versus some other option should be evaluated on a case-by-case basis.

1. BACKGROUND

The U.S. Navy generates spent ABM as a result of its ship-cleaning operations. The spent ABM generally contains low concentrations of metals from the paints, antifouling compounds, and other coatings that are applied to ship hulls. In the past, much of this spent ABM has been disposed in landfills -- nonhazardous waste landfills for spent ABM having very low metal concentrations and hazardous landfills for spent ABM containing relatively high metal contents. However, landfill disposal is being curtailed because of rising disposal costs, land ban restrictions imposed by the Resource Conservation and Recovery Act (RCRA), and the growing emphasis on waste minimization. Spent ABM appears to be a good candidate for recycling in asphalt concrete or other composites because its textural characteristics are compatible with the composites. Also, it was shown in a previous study that certain spent ABM does not respond well to stabilization/
solidification technology to insolubilize the metallic constituents[(1)].

Battelle, along with the Naval Civil Engineering Laboratory and R&G Environmental Services (an environmental subsidiary of an asphalt production company), is investigating the feasibility of recycling spent

ABM into asphalt concrete for subsequent use in road and parking lot paving applications. Spent ABM from three different Navy bases in California is currently being studied. These bases include Hunters Point Annex (HPA) in Hunters Point, California; Mare Island Naval Shipyard (MINS) in Vallejo, California, and Long Beach Naval Shipyard (LBNS) in Long Beach, California. The spent ABM at HPA and LBNS is from previous ship-cleaning operations, while the grit from MINS is from an active sandblasting operation. Recycling by-product materials is encouraged by U.S. environmental regulatory agencies, as long as the practice does not represent "use constituting disposal" or sham recycling. This paper describes the status of these three ongoing investigations.

2. CHEMICAL CHARACTERIZATION OF SPENT ABM

Several different types of sandblasting grit exist. A complete listing is beyond the scope of this study, however, the different types include the following:

- A ground-up nickel smelting slag called "green diamond"
- A ground-up copper smelting slag referred to as "clean blast" or "clean slag"
- Beach sand
- Steel shot.

The spent ABM from MINS is "clean blast" and from HPA is beach sand. The spent ABM from LBNS is from an unknown source but texturally resembles beach sand with a small admixture of steel shot.

The spent ABM from all three sites has been chemically characterized for total and soluble (leachable) metal content, and the results are summarized in Table 1.

Note that the MINS <u>unspent</u> ABM contains approximately 500-2,000 mg/kg copper (before sandblasting) because this material is produced from copper smelter slag.

The total metals analyses is based on an extraction in hot concentrated mineral acid. The California Waste Extraction Test (WET) is a 48-hour batch leach test using a 0.4 M sodium citrate solution. The Toxicity Characteristic Leaching Procedure (TCLP) is an 18-hour batch leaching procedure using a dilute sodium acetate solution. Table 1 shows that the primary metallic contaminants in the ABM are copper and lead.

TABLE 1

Ranges of Metal Contents in Spent ABM

	Total Content (mg/kg)	WET-Soluble Content (mg/L)	TCLP-Soluble Content (mg/L)
<u>HPA</u>			
Cu	1,000 - 2,600**	100 - 200**	*
Pb	90 - 300 **	10 - 35**	0.6 - 14
Zn	70 - 2,500	85	*
Ni	10 - 270	3.5	*
Cr	30 - 180	8 - 14	<0.5
<u>MINS</u>			
Cu	1,500 - 7,500**	up to 300**	*
Pb	15 - 120	up to 9**	<3
Zn	up to 1,000	NA	*
Ni	up to 100	NA	*
Cr	up to 25	NA	
<u>LBNS</u>			
Cu	2,000 - 4,800**	50 - 165**	*
Pb	30 - 630	2 - 43**	up to 9**
Zn	640 - 760	NA	*
Ni	45 - 76	NA	*
Cr	78 - 84	NA	up to 0.6

* Not a TCLP metal.
** Exceeds pertinent regulatory threshold and therefore defined as a hazardous waste.
NA = Not analyzed.

3. REGULATORY HAZARDOUS CRITERIA

ABM material that exceeds the Total Threshold Limit Concentration (TTLC) and/or Soluble Total Limit Concentration (STLC) threshold values for any metal is considered hazardous in the state of California. As a general rule, ABM that exceeds the TCLP threshold for any metal is referred to as exhibiting a "toxicity characteristic" and is considered hazardous by the U.S. Environmental Protection Agency (USEPA). If ABM exceeds either a TTLC or STLC but not a TCLP threshold, it is referred to as "California-only" hazardous -- that is, it is considered hazardous in the state of California but not by the USEPA. All of the ABM examined thus far in this study is either nonhazardous or "California-only" hazardous based on copper and/or lead TTLC and/or STLC exceedances (Table 1); none is "EPA-hazardous." Therefore, the material is regulated as hazardous waste only in the state of California and not by the USEPA.

Moreover, the hazardous ABM in this study is regulated only under California regulatory policy and rules, which as discussed below are in the process of being modified to permit the recycling of certain types of spent ABM currently defined as hazardous in the state of California.

Current California Department of Health Services (DOHS) policy on recycling spent ABM into asphalt concrete or other composites allows nonhazardous and certain types of hazardous (TTLC exceedance only) by-products such as ABM to be recycled. In September 1990, DOHS issued a draft modification to existing policy which, if implemented, will allow the recycling of a broader range of hazardous by-products, including those with STLC exceedances. Until the policy is formally promulgated, DOHS is reviewing recycling projects on a case-by-case basis, and issuing variances for recycling if certain conditions can be met. While a detailed discussion of these conditions is beyond the scope of this paper, the principal conditions are as follows:

A. The policy pertains to California-hazardous spent ABM only (EPA-hazardous may not be recycled). The asphalt concrete or other composite that the spent ABM is recycled into must conform to the California TTLC and STLC criteria.

B. The asphalt concrete must have suitable physical characteristics and must meet standard road paving structural criteria. Also, the spent ABM must contribute to the structural integrity of the concrete rather than detract from it.

C. A risk analysis must be performed to indicate that no significant risk is posed by either the recycling operation or the introduction of the spent ABM into the public domain.

This paper deals principally with requirements A and B above. Requirement C is being satisfied by conducting a risk analysis using standard DOHS and EPA risk analysis methodology.

4. BENCH-SCALE AND PILOT-SCALE TESTING

To comply with requirements A and B above, Battelle is coordinating bench-scale testing where the spent ABM is being incorporated into asphalt concrete and selected physical and chemical properties of the product are being measured. Compliance with the TTLC and STLC thresholds in the asphalt concrete as indicated in requirement A above can be achieved simply by using the appropriate proportion of spent ABM in the asphalt concrete product. For example, if the spent ABM exceeds the TTLC and/or STLC criteria by a factor of 5, then the TTLC or STLC criteria can be achieved by simply ensuring that the spent ABM comprises no greater than 20% (by weight) of the asphalt concrete product.

The physical or structural integrity criteria specified in requirement B above mandate that a series of bench-scale tests be performed, wherein asphalt concrete test specimens are produced in the laboratory under carefully controlled and documented procedures. First, loose asphalt is produced using specified proportions of spent ABM, aggregate of varying particle sizes, and asphalt (oil) content. Then the loose asphalt is heated and compressed into cylinders under simulated field compaction conditions. The cylinder is then tested for various physical parameters to determine the structural integrity of the asphalt concrete product.

The physical parameters of principal concern are as follows:

- Stabliometry (resistance to deformation due to load)
- Cohesiometry (cohesion)
- % voids (degree of compaction)
- Swell (water resistance).

After an asphalt concrete product that satisfies requirements A and B is produced in the bench-scale tests, we will conduct a pilot test at the hot plant to verify that the materials have a comparable physical integrity and can be produced at full-scale production.

5. TEST RESULTS

Based on the data collected to date, the asphalt concrete test samples containing spent ABM consistently meet all the structural integrity criteria except for stabilometry values, which are sometimes lower than the criterion of 35 for this project. (A stabliometry value of 30 corresponds to acceptable usage on light-traffic roads, 35 for medium-traffic roads, and 37 for heavy-traffic roads). Stabilometry values of 27-28 were produced when the spent ABM comprised approximately 50% of the asphalt concrete; however, this is a very high proportion of spent ABM, and these tests were run for research purposes only. At a more normal spent ABM loading proportion of 7-10% (by weight, approximately 10% less by volume), the stabilometry values increased to 31 to 46. The lower values in this range are acceptable for light-traffic roadways but do not meet the desired criteria of 35 for medium-traffic roadways. However, these tests were based on the use of the spent ABM in standard production-grade asphalt concrete at the facility performing the tests. It is possible to optimize the mix of ingredients to each specific spent ABM material using a procedure called centrifuge kerosene equivalent (CFE). These tests are currently being performed for each of the spent ABM materials in this study. The expected outcome is that the asphalt concrete resulting from the design mix specification will consistently meet or exceed the stabilometry criteria of 35.

6. DISCUSSION

6.1 Asphalt Integrity

Our bench-scale test data collected thus far suggest that a high quality asphalt concrete product can be produced at a spent ABM proportion of about 7-10% by weight. This is consistent with the results of a similar study for a different group of spent ABM, which demonstrated that a suitable quality asphalt concrete was produced at a spent ABM proportion of 10-20% by weight[2]. This latter study also showed that as the spent ABM content increased from 10 to 20%, the bitumen or oil requirement decreased by 0.5%, thereby reducing product cost. However, our data indicate that at 50% spent ABM, product quality drops to below the level needed for application on even light-traffic roadways.

6.2 Advantages of Recycling Spent ABM into Asphalt Concrete

1. The cost of recycling spent ABM into asphalt concrete is much lower than the cost of disposal; for example, at MINS the estimated cost of continued disposal in a hazardous waste landfill is $1,452K/year versus a maximum cost of $220K/year for the recycling option. This is based on an estimated annual production of 2,200 tons of spent ABM.
2. The recycling and reuse option is much higher in the hierarchy of hazardous waste management than disposal with or without treatment options. Furthermore, waste minimization credit may be given to the generator of the spent ABM because the spent ABM is not manifested as hazardous waste when it is transported to the hot plant for recycling.
3. The recycling option does not consume valuable landfill space, which can be reserved for higher-level hazardous wastes. Most spent ABM contains relatively low metal concentrations. The three grits in this study were hazardous in the state of California but were nonhazardous according to the EPA TCLP leaching test.
4. Some spent ABM contains an elevated aluminum content, which could lead to swelling and cracking if recycled into other construction materials such as cement or concrete.

6.3 Disadvantages of Recycling Spent ABM into Asphalt Concrete

1. If the spent ABM is hazardous, the material needs to be handled as a hazardous material (although not as a hazardous waste) and must comply with cognizant transportation and storage regulations. Also, cognizant regulatory requirements must be satisfied or a permit or variance may be required. In this study, these requirements include a bench-scale and pilot-scale test program and a risk analysis to show that the planned activity will not adversely affect human health and the environment.
2. Different types of spent ABM have varying particle sizes and differing capacities to absorb oil. Therefore, some bench-scale or laboratory testing and analysis are recommended to design the optimal mix of ingredients that will yield the highest stability product.

3. Certain constituents will interfere with the production of high quality asphalt. For example, high organic content (such as from paint chips and other organic coatings) is detrimental to stability, and sulfate may cause swelling upon contact with water.
4. If bench-scale testing is performed to design a mix, then it is important that the feeder sand/aggregate used in the bench-scale tests be the same as that used in the full-scale operation at the hot plant. Otherwise, the bench-scale test will not provide a true representation of the full-scale process. Feeder sand and aggregate are frequently purchased on the open market and physical characteristics such as particle size, shape, and density can vary significantly from batch to batch.

REFERENCES

1 J.L. Means, G.L. Headington, S.K. Ong, K.W. Nehring, and J.C. Heath. The Chemical Stabilization of Metal-Contaminated Sandblasting Grit at Naval Station, Treasure Island, Hunters Point Annex: A Summary of Waste Characteristics, Bench-scale Treatability Procedures, and Bench-scale Treatability Data. Final report to Naval Civil Engineering Laboratory by Battelle Memorial Institute (3 volumes), 1991.

2 California Office of Transportation. Copper Oxide Blasting Slag as an Aggregate in Asphalt Concrete, final report memorandum to R.N. Doty on research project dated 2 August 1989, File Number 631140-30040, 1989.

Waste Materials in Construction.
J.J.J.R. Goumans, H.A. van der Sloot and Th.G. Aalbers (Editors)

EFFECTIVE UTILIZATION OF COAL ASHES IN ROAD CONSTRUCTION

K. Torii and M. Kawamura

Department of Civil Engineering, Faculty of Technology, Kanazawa University, 2-40-20 Kodatsuno, Kanazawa (Japan)

SUMMARY

The effective utilization of coal ashes, which are industrial waste products from coal-burning power stations, as a material in road construction has been assessed. The results have been obtained from a laboratory study of compacted fly ash-furnace bottom ash mixtures stabilized with and without chemical additives. From the results, it was found that the compacted fly ash-furnace bottom ash mixtures stabilized with small amounts of hydrated lime or cement fulfilled the requirements of the strength and the resistance against the immersion in water as base course and structural fill materials in road.

1. INTRODUCTION

In Japan, a lot of coal burning power stations constructed in the last decade after oil crisis have actively been working. The total amount of coal ashes annually produced from these power stations is estimated to be some 4 millions ton. About 85 percent of coal ashes which are now produced in Japan is fly ash and 15 percent of them is furnace bottom ash. Some of fly ashes have been used as a raw material for cement production or as a mineral admixture for concrete-making. However, at present, the percentage of effective utilization of coal ashes is about 30 percent. About 70 percent of coal ashes has been abandoned into the sea nearby power stations. On the other hand, the supply of traditional road-making materials such as crushed stone and river sand is considerably limited, and considerations have therefore been given to the use of alternative materials including industrial waste products. This approach has the advantage of conserving natural resources, whilst, at the same time, complying with the environmental need of the disposal and treatment of industrial waste products.

Fly ashes produced in Japan mostly originate from domestic, overseas bituminous and sub-bituminous coals. Some workers point out that some of fly ashes can exhibit a self-hardening property when compacted at a proper moisture content (1),(2). This property of fly ashes is very beneficial for utilizing fly ashes as a material for base course in road and a fill in embankment. Furnace bottom ash particles, which have not been utilized in Japan at all, large, angular and porous. Therefore, it is considered that the fly ash-furnace bottom ash mixtures show a better compacting effect and hardening property compared with compacted fly ashes because particle size distributions and structure of mixtures are improved (3). Fly ash-furnace bottom ash mixture is being used extensively in the road construction. However, these ashes have different properties from conventional materials.

The special interest of our study is put on the strength development and the resistance against the immersion in water in the unstabilized and stabilized materials. The suitability of compacted fly ash-furnace bottom ash mixtures was investigated from the view points of compaction effect, pH, compressive strength, CBR value, and resistance against the immersion in water. Furthermore, the hardening process and microstructural feature of compacted fly ash-furnace bottom ash mixtures were elucidated by X-ray diffraction analysis and SEM observations.

2. MATERIALS AND EXPERIMENTAL PROCEDURE

2.1 Materials

The bituminous fly ash (FA) and furnace bottom ash (FBA) used in this study are discharged from Takasago power station in Japan. They are produced in the burning process of pulverized domestic coals. Their chemical compositions and physical properties are presented in Table 1. The specific gravity of the fly ash and the furnace bottom ash is 2.18 and 2.15 respectively, which are about 15 percent lower than soil normally used in the road construction. The X-ray diffraction diagrams given in Fig.1 show that the major components of the fly ash are glass, quartz, mullite, free lime and anhydrite. It is also found from Fig.1 that the major components of the furnace bottom ash are quartz, mullite and feldspar, and that its glassy phase is relatively small. SEM micrographs given in Fig.2 show that spherical fly ash particles are mostly in the range between 10 μm and 100 μm, and that the shape of porous furnace bottom ash particles from 50 mm down to dust in size is irregular. In the fly ash-furnace bottom ash mixtures, the ratios of fly ash to furnace bottom ash are selected at 2, 1 and 1/2. Their typical particle size distribution curves are shown in Fig.3. Generally, most of fly ash particles are within the silt size range, whilst furnace bottom ash within the sand size range. Therefore, favorable particle size distributions as a material of base course in road can be achieved by mixing both materials. The chemical additives added are hydrated lime, portland cement and gypsum (flue gas desulfurization by-product). The additive content ranges from 2.5 % to 15 % by the weight of dry sample.

2.2 Experimental Procedure

2.2.1 Unconfined Compression Test

Cylindrical specimens, 5 cm in diameter and 10 cm high, were statically compacted by an oil jack to obtain the optimum moisture content and the maximum dry density for each mixture from the Japanese standard compaction test (JIS 1210, 1.1-a). The specimens, sealed tightly in polyethylene sack, were cured

Table 1 Chemical compositions and physical properties of coal ashes.

	Fly ash (FA)	Furnace bottom ash (FBA)
Ig. loss (%)	5.7	9.9
SiO_2 (%)	54.2	53.1
Al_2O_3 (%)	27.9	22.9
Fe_2O_3 (%)	6.1	6.8
CaO (%)	3.7	3.9
MgO (%)	2.5	1.6
Na_2O (%)	0.2	2.1
K_2O (%)	0.4	1.1
Residual carbon (%)	4.6	6.9
Specific gravity	2.18	2.15
Bulk specific gravity	----	1.88
Bulk density (t/m^3)	1.06	0.84
Residual of 88 μm sieve (%)	8.1	86.8
Methyl-blue absorption (mg/g)	0.40	0.63

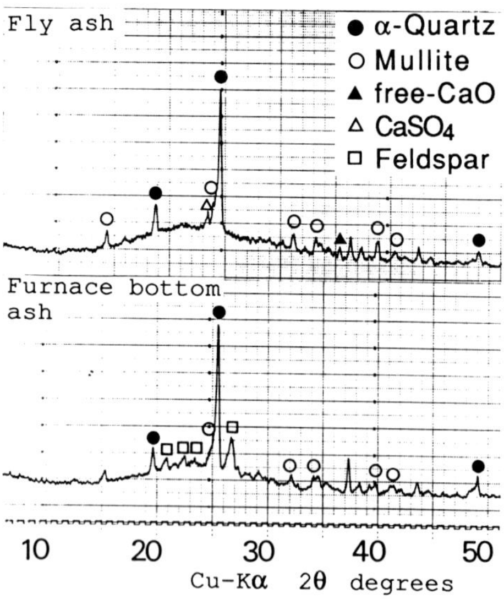

Fig.1 X-ray diffraction diagrams of fly ash and furnace bottom ash.

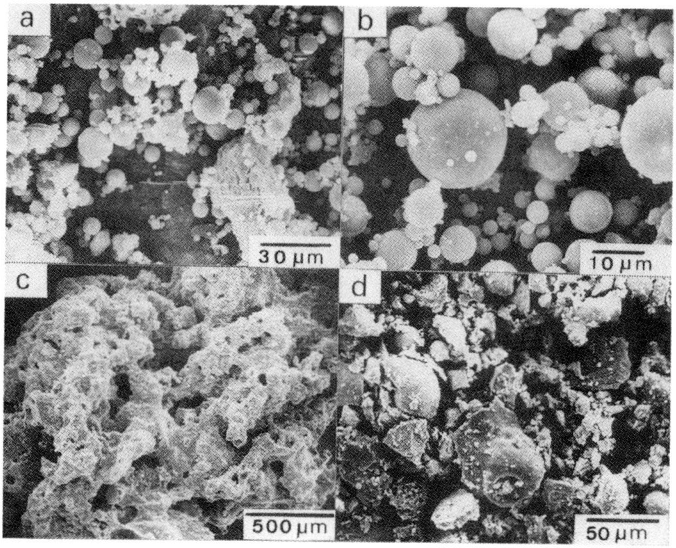

Fig.2
SEM micrographs of fly ash and furnace bottom ash
(a, b : fly ash particles,
c, d : furnace bottom ash particles).

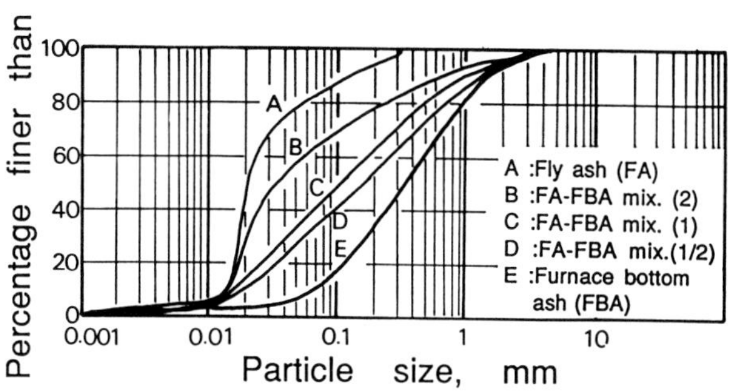

Fig.3
Particle size distributions of fly ash-furnace bottom ash mixtures.

at 20 °C for the prescribed curing time. Unconfined compression test using an autograph was carried out under the constant strain rate of 1 % per minute.

2.2.2 CBR Test

Specimens for the CBR test were prepared at the optimum moisture content according to the Japanese standard CBR test (JIS 1211). The change in CBR value and expansion percentage during the immersion in water were measured.

2.2.3 Immersion Test

Immersion test was carried out based on the procedure given in BS 1924:1975. After the specimens were given standard curing for 7 days, the specimens were removed from the polyethylene sack and immersed in water for 7 days. The immersion ratio (IR) was calculated by the following equation : IR = (14 day strength of immersed specimens / 14 day strength of normally cured specimens) × 100 %.

2.2.4 X-ray Diffraction Analysis and SEM Observation

The ground samples vacuum-dried at a room temperature were used for X-ray diffraction analysis and pH measurement. The fracture surface of fragments of the specimens was observed by using scanning electron microscope. Samples for SEM observations were also vacuum-dried at a room temperature, and then coated with carbon and palladium.

3. RESULTS AND DISCUSSION

3.1 Density and Compaction Properties

The liquid limit of fly ash used was 26 % and its plastic limit could not be measured. As most of fly ash particles are spherical and their particle size is fine and uniform, fly ash has no plasticity. These physical properties show that the compaction effect of fly ash is fairly good.

Compaction curves of fly ash-furnace bottom ash mixtures are shown in Fig.4. Fly ash can be easily handled and compacted at a proper moisture content. Weak and friable particles in the furnace bottom ash broke down during compaction. The maximum dry density of compacted furnace bottom ash was as low as 1.1 g/cm^3 due to its low specific gravity. Compaction curves of both the fly ash and the fly ash-furnace bottom ash mixtures are similar to those obtained in cohesive soils. The maximum dry density of the fly ash and fly ash-furnace bottom ash mixtures ranged from 1.30 g/cm^3 to 1.35 g/cm^3, which was about 20 percent greater than that of the furnace bottom ash itself. The increase in dry density may be attributed to both the increase in compaction effect by improving particle size distributions and the decrease in large voids by filling them with small particles. The unit weight of compacted fly ash-furnace bottom ash mixtures was 20 % to 30 % less than that of soils normally used in road construction. The use of these materials is particularly effective in reducing the settlement of embankments on soft grounds or decreasing the earth pressure on retaining walls.

3.2 Strength Characteristics

The time-dependent changes in compressive strength of compacted furnace bottom ashes with hydrated lime and cement are shown in Fig.5. No increase in compressive strength with curing time was found in compacted furnace bottom ashes without chemical additives. The compressive strength of compacted furnace bottom ashes stabilized with hydrated lime increased between 28 days and 90 days due to the progression of pozzolanic reaction between fine

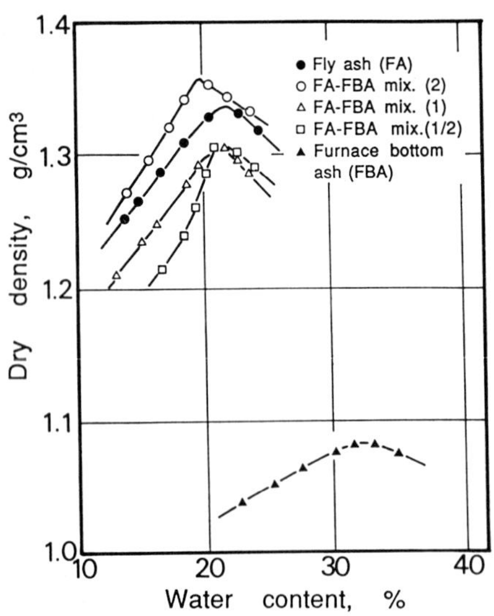

Fig.4 Compaction curves of fly ash and fly ash-furnace bottom ash mixtures.

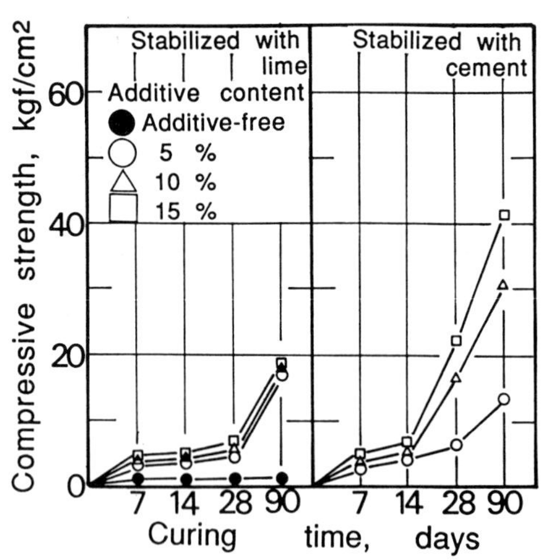

Fig.5 Compressive strength of compacted furnace bottom ash stabilized with hydrated lime and cement.

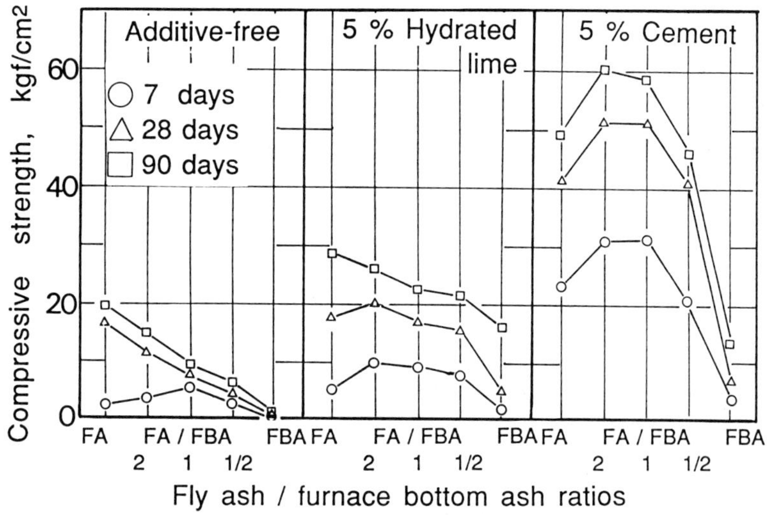

Fig.6 Compressive strength of compacted fly ash-furnace bottom ash mixtures stabilized with hydrated lime and cement.

particles in furnace bottom ash and hydrated lime added. The compressive strengths of compacted furnace bottom ash stabilized with cement were relatively high compared with those of compacted furnace bottom ash with hydrated lime at all ages, and its strength gain with curing time was in proportion to the cement content after 14 days. However, stabilization of furnace bottom ashes with hydrated lime and cement is not effective.

The compressive strengths of compacted fly ash-furnace bottom ash mixtures with and without hydrated lime and cement are given in Fig.6. All of compacted fly ash-furnace bottom ash mixture showed a great strength gain with the curing time, and the maximum compressive strength in the additive-free mixtures at 90 days was found in the compacted fly ash. This strength gain in the compacted fly ashes may be attributable to the self-hardening property of the fly ash itself. Namely, when the fly ash is compacted at an appropriate moisture content, the pozzolanic reaction of fly ashes can occur by the use of free lime supplied from the fly ash itself. SEM micrographs of fracture surfaces of the specimens showed that fine reaction products such as C-S-H gel and ettringite were produced on the surfaces of fly ash particles at 28 days, and that these fine products connected fly ash particles (Fig.7). The content of water solubles and pH of the fly ash and the furnace bottom ash are presented in Table 2. Some workers showed that the compressive strength of the mixtures due to hardening property increased with increasing content of free lime and water solubles in specimens prepared under the controlled conditions of density, moisture content and curing (4),(5). As shown in Fig.8, by adding the hydrated lime and cement, pH of compacted fly ash and fly ash-furnace bottom ash mixture became high, which was maintained at high level during long curing periods. The strengths of 7 days old compacted fly ash-furnace bottom ash mixtures with 5 % cement ranged from 20 to 30 kgf/cm^2, those of 28 days old compacted fly ash-furnace bottom ash mixtures with 5 % hydrated lime from 15 to 20 kgf/cm^2. These strengths fulfill the specification of Japan Road Association which requires 30 kgf/cm^2 for cement-stabilized soils at 7 days and 10 kgf/cm^2 for lime-stabilized soils at 10 days.

Table 2 Water solubles and pH of coal ashes.

	Fly ash (FA)	Furnace bottom ash (FBA)
pH	12.1	8.8
CaO(mg/g)	0.30	0.05
Na_2O(mg/g)	0.07	0.01
K_2O(mg/g)	0.003	0.002
SO_3(mg/g)	0.41	0.09

Table 3 CBR values of compacted fly ash-furnace bottom ash mixtures with and without additives (%).

	Additive -free	5 % cement	5 % lime
Fly ash(FA)	8	489	274
FA/FBA = 2	49	555	289
FA/FBA = 1	29	401	188
FA/FBA = 1/2	130	419	262
Furnace bottom ash(FBA)	66	173	87

Fig. 7 SEM micrographs of the fracture surface of compacted fly ash and fly ash-furnace bottom ash mixtures without additives at 28 days (a, b : compacted fly ash, c, d : compacted fly ash-furnace bottom ash mixtures (FA/FBA=1)).

3.3 Expansion in Immersion and CBR value

The rainfall or underground water may flow through fly ash mixtures and disintegrate them. The results of expansion test using CBR mould are shown in Fig.9. The expansion of compacted fly ash occurred immediately after the immersion in water, and its expansion percentage reached as high as 1.5 % at the immersion time of 4 days. Thus, the CBR value of the compacted fly ash decreased from 25 % to 8 % during the immersion in water. The coefficient of permeability of compacted furnace bottom ash ranges from 10^{-2} to 10^{-3} cm/s, which is 1000 times larger than that of compacted fly ash (3). The expansion of compacted fly ash-furnace bottom ash mixture drastically decreased with increasing content of furnace bottom ash in the mixtures. The replacement of a half of fly ash by furnace bottom ash showed little expansion during the immersion in water. In the compacted fly ash-furnace bottom ash mixtures without additives, the mixtures with the ratio of fly ash to furnace bottom ash of 1/2 showed the maximum CBR value of 130 %. From the results of CBR values given in Table 3, it was also found that the addition of small amounts of hydrated lime and cement was very effective not only in preventing the expansion in immersion in water, but also in increasing the CBR value.

3.4 Resistance against Immersion in Water

The addition of small amounts of hydrated lime and cement to fly ash can effectively reduce their water and frost susceptibility. In particular, the addition of both hydrated lime (or cement) and gypsum is more effective for improving the strength and durability of compacted fly ash by promoting the formation of ettringite at early ages (6).

The results of immersion test of compacted fly ash and fly ash-furnace bottom ash with chemical additives are presented in Table 4 and 5, where both the strength after immersion and the immersion ratio (IR) for various mixtures are given for each material. As pointed out by some workers (7), the IR value of about 80 % can ensure the adequate performance and the durability as a material of base course in road. Some of compacted fly ash specimens with small amounts of hydrated lime (or hydrated lime plus gypsum) were collapsed by slaking during the immersion in water. However, no disintegration was found in compacted fly ash-furnace bottom ash mixtures with hydrated lime (or

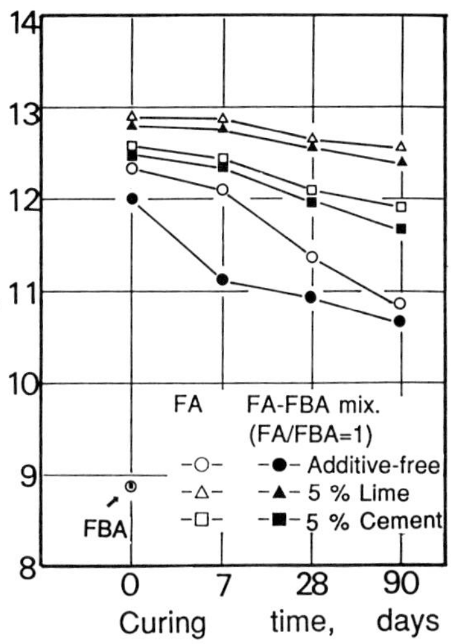

Fig.8 Change in pH of compacted fly ash and fly ash-furnace bottom ash mixtures with curing time.

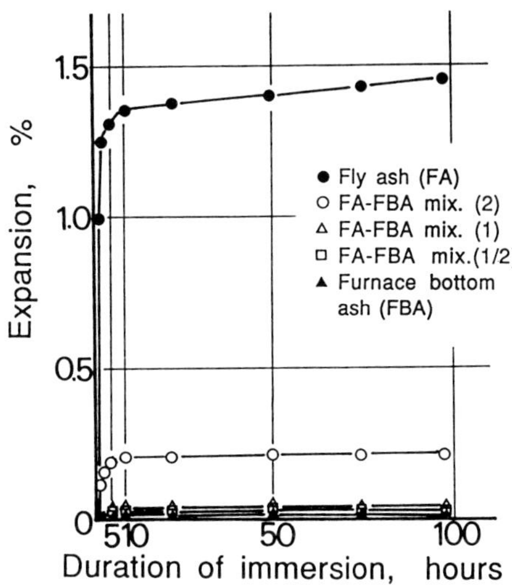

Fig.9 Expansion percentage of compacted fly ash and fly ash-furnace bottom ash mixtures in immersion in water.

Table 4 Results of the immersion test of compacted fly ash-furnace bottom ash mixtures with lime and lime plus gypsum (kgf/cm^2).

Additives	Fly ash	Fly ash-furnace bottom ash mix. (FA/FBA=1)
Lime 2.5%	Collapsed	12.7 (86%)
5 %	8.1 (81%)	15.5 (96%)
10 %	16.8 (71%)	17.3 (88%)
Lime-gypsum		
5% (L/G=3)	Collapsed	22.4 (88%)
5% (L/G=1)	Collapsed	12.7 (88%)
10% (L/G=3)	12.5 (89%)	25.3 (89%)
10% (L/G=1)	8.6 (95%)	15.4 (81%)

( ): Immersion ratio

Table 5 Results of the immersion test of compacted fly ash-furnace bottom ash mixtures with cement and cement plus gypsum (kgf/cm^2).

Additives	Fly ash	Fly ash-furnace bottom ash mix. (FA/FBA=1)
Cement 2.5%	25.5 (83%)	17.4 (99%)
5 %	37.8 (89%)	47.1 (97%)
10 %	61.5 (99%)	68.7 (88%)
Cement-gypsum		
5% (C/G=3)	29.4 (88%)	41.0 (103%)
5% (C/G=1)	19.6 (80%)	28.3 (111%)
10% (C/G=3)	58.6 (98%)	68.5 (99%)
10% (C/G=1)	39.0 (98%)	46.9 (101%)

( ): Immersion ratio

hydrated lime plus gypsum). It was confirmed that compacted fly ash-furnace bottom ash mixtures showed a high resistance against the immersion in water compared with compacted fly ash when they had the same strength. On the other hand, compacted fly ash or fly ash-furnace bottom ash mixture with cement (or cement plus gypsum) showed a considerable strength gain during the immersion in water, when their IR values were all more than 80 %. It was found from the results of immersion test that a good performance could be achieved when the strength of compacted fly ash or fly ash-furnace bottom ash mixtures was more than 5 kgf/cm^2 at 7 days.

4. CONCLUSIONS

In Japan, from the environmental point of view, it gradually becomes difficult to ensure the large disposal area where coal ashes are stored. Furthermore, coal ashes loosely dumped in disposal area sometimes result in land slide, surface erosion and dusting, which possibly pollutes the environment nearby the disposal area. Therefore, it is an important and urgent problem to establish the recycling system of industrial waste products such as coal ashes and iron slags.

From the experimental results, it was confirmed that coal ashes possessed favorable engineering properties as a material of base corse in road and a structural fill material. When fly ash-furnace bottom ash mixtures with hydrated lime and cement were properly compacted, they showed a good compaction effect and great strength gain with curing time. Compacted fly ash-furnace bottom ash mixtures stabilized with 5 % cement (or cement plus gypsum) successfully fulfilled the requirements of the strength and the resistance against the immersion in water as a material in road.

REFERENCES

1. R.C. Joshi, et al., J. of ASCE, 101, No.TE4 (1975) 791-806.
2. A.M. DiGioia and W.L. Nuzzo, J. of ASCE, 98, No.PO1 (1972) 77-98.
3. K. Torii and M. Kawamura, Soils & Foundations, 373, (1989) 67-72 (in Japanese).
4. H.B. Sutherland, et al., J. of the Institution of Highway Engineers, 5, 6 (1968) 19-27.
5. D.J. Throne and J.D. Watt, British Coal Utilization Association, No. DCH/3 (1964).
6. M. Kawamura and K. Torii, Proc. of 8th Int. Conf on the Chemistry of Cement, 3 (1986) 32-36.
7. R.J. Kettle and R.I.T. Williams, Roads and Road Construction, 559 (1969) 200-206.

Waste Materials in Construction.
J.J.J.R. Goumans, H.A. van der Sloot and Th.G. Aalbers (Editors)

THE USE OF INCINERATOR SLAG IN ASPHALT FOR ROAD CONSTRUCTIONS

D.J. Nonneman[1], F.A. Hansen[2] and M.H.M. Coppens[3]
[1] Productgroep Wegen en ADV, Ingenieursbureau Amsterdam, Wibautstraat 3, 1091 GH Amsterdam (The Netherlands)
[2] Kwaliteit en Adviezen, Koninklijke Wegenbouw Stevin b.v., P.O Box 8330, 3503 RH Utrecht (The Netherlands)
[3] Kwaliteitsdienst, Hollandsche Wegenbouw Zanen, Hogebrinkerweg 19 , 3871 KM Hoevelaken (The Netherlands)

SUMMARY

Up to now slag from household refuse incinerator plants are only used as a fill material in road constructions. In that case precautions are necessary as to isolating a slag layer from ground water and covering the slag with either sand-bentonite or bituminous material. Just recently in a German publication (1) the use of an asphalt mix with slags was mentioned.

In 1988 the City of Amsterdam, Road-construction Laboratory, together with HWZ and KWS contractors carried out an extensive study into the possibilities of using incinerator slag of the waste incinerator AVI-Amsterdam Noord in asphalt base course mix. The incinerator combusts domestic wastes of the City of Amsterdam.

In this publication the results of that research and the results of the site investigations are made public on a larger scale than can be reached by internal publication only.

1. INTRODUCTION

In the early eighties man became aware something should be done to decrease the amount of waste materials loading the environment. Out of that it was self-evident the city of Amsterdam started to find a useful solution to the problem of the residue of the municipal waste incinerator plant. In 1984 the city of Amsterdam started a research to find possibilities for the use of fly-ashes in asphalt for base course layers. After laboratory research by the Roadconstruction Laboratory some site investigations proved asphalt could be made and processed when 50 % of the filler has been replaced by untreated fly-ashes of the waste incinerator.

Also the use of 20 % crushed asphalt in the mix gave no problems what soever. Those investigations resulted into fly-ashes of the waste incinerator AVI-Amsterdam Noord to be used as replacement for 50% of the filler material (smaller than 63 μm) in the asphalt mix used in Amsterdam.

Out of this history of trying to find a suitable solution for the use of secondary starting material it is not that strange to see in 1988 investigations get started on how to use incinerator slag in asphalt for base course layers. The use in the relatively thin upper layer was looked at as an impractical solution to reduce the remaining great amounts of incinerator slag produced. Preliminary laboratory investigations were carried out to predict the asphalt properties and the amount of incinerator slag that could be used. Based on those investigations a test site at the Burgemeester Stramanweg was made where a cycle-road was constructed in which asphalt for base coarse layers was used containing 40 % incinerator slag and 40 % crushed asphalt. The guidelines worked out in the CROW-publication nr. 15 "Resten zijn geen afval (meer)" (2) were taken into consideration during this study.

2. MATERIAL PROPERTIES

2.1. Incinerator slag

The slag used was normal production slag i.e. crushed, sieved and deironed, in a size of 0 to 16 mm which meet in general the requirements in the guideline (2) for use of incinerator slag in road constructions. No extra treatment other then a global iron extraction was performed. The slag was more than six weeks old and did have a mean moisture rate of 19,5%. Due to its production process and the origin of its ground materials the slag contain a diversity of materials among which metals, glass and broken bricks. Table 1 gives the results of the percentages of those materials in three fractions.

TABLE 1 (3)
Percentages by weight of metals, glas and rubble in incinerator slag.

Fraction Sieves NEN 2560	Metals	Glass	Rubble	Remaining slag
> C 11,2	8	13	30	49
C 11,2 - C 8	2	33	29	36
C 8 - C 5,6	4	43	17	35
Total > C 5,6	4	31	25	40

Other laboratory investigations gave the following gradation, table 2, of the incinerator slag.

TABLE 2 (3)
Gradation of incinerator slag in percentage by weight.

Sieves NEN 2560	cum. % by weight
C 16	1,8
C 11,2	7,4
C 8	15,4
C 5,6	24,2
C 4	34,2
2 mm	47,9
1 mm	58,2
500 μm	67,3
250 μm	78,2
180 μm	84,3
125 μm	87,9
63 μm	91,4
< 63 μm	8,6

TABLE 3 (3)
Density of incinerator slag by gradation.

Fraction	Density kg/m³
C 8 - C 5,6	2686
250 μm - 180 μm	2783
> 2 mm (without C 8 - C 5,6)	2564
< 2 mm (without 250 μm - 180 μm)	2640

More batches were tested to find the density of the incinerator slag. We defined the density for some gradations. The values are gathered in table 3. The density of the total material in accordance with the Aashto test was defined to 2625 kg/m³. In the test mentioned the material is stored under water during at least 15 hours. Due to the porosity of the slag a more exact way to define the weight by volume is bulk density ovendry. For the material used it was defined to 2180 kg/m³. In bulk the density measured was 2330 kg/m³, this results in a absorption of 6,9 %. Due to the way the slag is stored, in open air without any shelter, the moisture rate to be found count up to 20 %

2.2 MARSHALL STABILITY

The three laboratories of the research partners carried out a study on three groups of mixtures. The first consists of 100 % incinerator slag. The second group consist of 25 % gravel 4/32 mm and 75 % incinerator slag. A third consisted of 20 % gravel 4/32 mm, 40 % incinerator slag and 40 % crushed asphalt.

Those mixtures were tested according to a strength test for asphalt mixtures named after Marshall (4).

The results of this Marshall test showed for the first group a normal stability combined with high consumption of bitumen and a relatively high Marshall flow. Therefore the Marshall quotient is relatively low. This combination of values was considered as non relevant for a research to find a usefull application for incinerator slag in base coarse layers.

The results of the Marhall test of the second group showed relatively high stabilities combined with a relatively low flow and a relatively high use of bitumen.

The normal mix for base coarse layer in Amsterdam consisted at that time of 40 % crushed asphalt. The use of 40 % crushed asphalt found its origine in political pressure to come to a decrease of environmental loading by crushed asphalt. The combination of the results mentioned above and the fact that solving one problem should not lead to problems on other levels the third mixture constisted of 80 % recycled materials of which 50 % consisted of crushed asphalt and 50 % of incinerator slag. Lucky enough the results where promissing. We found a little higher use of bitumen in comparison with other Amsterdam mixtures. In comparison with those mixtures a normal Marshall flow and Marshall quotient was found. Therefore it was considered a mixture some site investigation had to be carried out upon.

In table 4, 5,and 6 the results of those three mixtures are gathered.

TABLE 4 (3)
Results of Marshall test for mixtures containing only slag and bitumen.

Marshall results Group I				
Slag	93,0	92,0	91,0	% by weight
Gravel 4/32	0,0	0,0	0,0	% by weight
Crushed asphalt	0,0	0,0	0,0	% by weight
Bitumen 45/60				
in 100% aggregate	7,0	8,0	9,0	% by weight
on 100% aggregate	7,5	8,7	9,9	% by weight
Density tablet	1928	1915	1968	kg/m^3
Density mix	2358	2289	2270	kg/m^3
HR (voids)	18,2	16,3	13,3	% by volume
HR" (voids aggreg.)	41,9	47,7	56,3	% by volume
Pm (stability)	8087	6612	7321	N
Fm (flow)	3,0	3,2	3,4	mm
Qm (stab./flow)	2692	2097	2251	N/mm

TABLE 5 (3)
Results of Marshall test for mixtures containing 75 % slag 25 % gravel and no crushed asphalt.

Marshall results Group II				
Slag	75,0	75,0	75,0	% by weight
Gravel 4/32	25,0	25,0	25,0	% by weight
Crushed asphalt	0,0	0,0	0,0	% by weight
Bitumen 80/100				
on 100% aggregate	8,0	9,0	10,0	% by weight
Density tablet	2145	2133	2144	kg/m^3
Density mix	2361	1334	2307	kg/m^3
HR (voids)	9,2	8,6	7,0	% by volume
HR" (voids aggreg.)	63	67	73	% by volume
Pm (stability)	10060	8940	7650	N
Fm (flow)	2,3	2,6	3,0	mm
Qm (stab./flow)	4360	3440	2550	N/mm

TABLE 6 (3)
Results of Marshall test for mixtures containing 40 % slag, 40 % crushed asphalt and 20 % gravel 4/32.

Marshall results Group III				
Slag	39,0	39,0	39,0	% by weight
Gravel 4/32	22,0	22,0	22,0	% by weight
Crushed asphalt	39,0	39,0	39,0	% by weight
Bitumen 80/100 on 100% aggregate	5,0	6,0	7,0	% by weight
Density tablet	2169	2200	2224	kg/m^3
Density mix	2401	2371	2343	kg/m^3
HR (voids)	9,7	7,2	5,1	% by volume
HR" (voids aggreg.)	51	63	74	% by volume
Pm (stability)	10870	10620	9360	N
Fm (flow)	1,9	2,3	2,9	mm
Qm (stab./flow)	5660	4680	3260	N/mm

The percentage bitumen contains 1,8 % bitumen out of crushed asphalt

3. TEST CHARGES

Before starting the test site at the Burgemeester Stramanweg in Amsterdam on October 12 and October 13, 1988 some test charges with variable bitumenpercentage were made at the asphalt production plant in Amsterdam. These test batches were produced to study the effect of the addition of incinerator slag at a 300 tons per hour asphalt production rate. The slag together with the gravel 4/32 was sent through the drying drum and further on through the by-pass. The crushed asphalt was put into the parallel drum. Within the normal mixing time, the normal mix-temperature and mix-moisture content were obtained.

The first charges made contained 7 and 9 % bitumen in total. Both where greasy so the amount of bitumen was decreased to 5 % in total. A test charge looked relatively good visually, but it was decided that the test site production should be made adding 5,5 and 6 % total bitumen. On October 14, 1988 two charges of 90 tons were produced and processed at the test site at the Burgemeester Stramanweg in Amsterdam where a cycle-road in asphalt was constructed. The bitumen added was an 80/100.

For both mixes both producing and processing where above expectation. In the next chapter the results of plant and site investigations on material properties are compared.

Probably due to the high moisture rates - 19 till 20 % - and the high percentage of material smaller than 63 μm of the slag, the cloth-filter got clogged and the production rate decreased rapidly.

Relatively great amouts of dust were separated in the baghouse and about 0,5 % dust was added again to the mix. From asphalt mix analysis it was concluded that in fact some 0,5 till 1,0 % more dust could have been added to the mix without any problem. But it is expected that this would not give a solution for both the dust problem and the filter problem.

4. RESULTS TEST SITE

During the production of the mix at the asphalt plant of both types 8 samples were taken. After the processing of the mix on site out of each type 4 cores were drilled. All those samples were tested. In table 7 the results are gathered.

TABLE 7 (3)
Results of samples taken during production and out of the construction.

	Mix at plant		Mix at site		
bitumen	5,5	6,0	5,5	6,0	% by weight
samples	8	8	4	4	-
Pm	9060	10010	-	-	N
Fm	2,3	2,3	-	-	mm
Qm	4000	4320	-	-	N/mm
Density	2200	2250	2225	2190	kg/m^3
HR (voids)	7,4	4,3	10,8	9,6	% by volume
HR" (voids aggr.)	60	75	97,9	98,5	% by volume

The aggregate was completely coated by bitumen and no separation of the mix was found. The workability and compactability of the mix were good.

After the laying of the base coarse the cycle-road was completed with a upper layer of 50 mm. red asphalt. At this moment the cycle-road is more than two years old and no problems have been found.

In addition to this test site the KWS on november 4, 1988 constructed a private road in Ouderkerk aan de Amstel. For this construction two different productionmethodes have been used. Also the amount of slag was reduced to 25 % by methode I and 30 % by methode II of the total aggregate. Reason for this reduction was the expectation that the use of less slag would have a positive influence on the production rate of the asphaltplant.

In methode I the slag and the crushed asphalt past the parallel drum. The mix-temperature reached 155 degrees centigrate. The mix still contained an amount of moisture which showed by steaming and a penetrating smell during processing. The processing temperature was 130 °C. In the second methode gravel and slag went by the main drum. The crushed asphalt went through the parallel drum to reach the appropriate temperature. This way of production gave a remarkably higher mix-temperature of 174 °C. The mix can be processed notably better and also the compaction found was higher then reached by methode I. The processing temperature was 160 °C.

In both processes the reduction of the production rate was 60 % so reduction of the percentage slag did not have the expected result.

5. CONCLUSIONS

Out of the laboratory research and the results of the two test sites the conclusion may be drawn that the production and processing of asphalt in wich incinerator slag has been used is technically possible restricted though by a considerable loss of productivity. Drying and heating of the slag is best performed by using the main drum. Due to the production process of the slag the slag consists of a great amount of dust.

When the new production plant of the municipal waste incinerator is ready producing a slag with a lesser percentage of dust (< 63 μm) is relatively easy. At that moment it is possible to separate the slag and the kettle-ashes within the production process. The use of that new slag or at this moment the use of slag with a lesser moisture rate or in an amount of about 10 % could give opportunities to set bounds to the productionloss. Further investigations in those directions are desirable.

REFERENCES

1 Eugen Haas, Feldversuch zur Untersuchung des Elutionsverhaltens von Müllverbrennungsschlacken, Baustoff Recycling + Deponietechnik, Heft 4, 1990, page 15 - 18

2 Resten zijn geen afval (meer), afvalverbrandingsslakken, publication nr. 15, Stichting Centrum voor Regelgeving en Onderzoek in de Grond-, Water- en Wegenbouw en de Verkeerstechniek, Ede, oktober 1988.

3 ir D.J. Nonneman, Eindrapportage hergebruik AVI-slak in asfalt en betonwaren voor de wegenbouw, Ingenieursbureau Amsterdam, rapportnr. 91-SR-001, Amsterdam, 1991. (in preparation)

4 RAW Standard Conditions of Contract for Works of Civil Engineering construction 1990, Stichting Centrum voor Regelgeving en Onderzoek in de Grond-, Water- en Wegenbouw en de Verkeerstechniek, Ede, mei 1990.

Waste Materials in Construction.
J.J.J.R. Goumans, H.A. van der Sloot and Th.G. Aalbers (Editors)

POTENTIAL REUSE OF WASTE MATERIALS IN HYDRAULIC ENGINEERING IN THE NETHERLANDS

E.F.M. Nieuwenhuis[1], L. de Quelerij[1], J.K. Vrijling[2], G.J.H. Vergeer[3]

[1] Fugro BV, P.O. Box 63, 2260 AB Leidschendam (the Netherlands)
[2] Rijkswaterstaat, Civil Engineering Division, P.O. Box 20000, 3500 GA Utrecht (the Netherlands)
[3] Centre for Civil Engineerig Research and Codes, P.O. Box 420, 2800 AK Gouda (the Netherlands)

SUMMARY

Waste materials from industrial processes can be re-used as secondary construction materials and thus replace conventional materials. The potential re-use of these materials in civil engineering depends on the physical and environmental properties of the materials in relation to the functional requirements of the sub-structures in which the materials are to be used. In addition the price and the availability of these materials, compared to those and the required quantaties of the conventional materials, are of importance. In order to determine priorities in the applications for optimal re-use of the waste materials in hydraulic engineering in the Netherlands a multi criteria analysis has been carried out, based on the physical, environmental and economical aspects of the secondary materials.

1. INTRODUCTION

In 1987 the Dutch Centre for Civil Engineering Research and Codes (CUR) in the Netherlands since 1987 commissioned an investigation into the potential applications of secondary materials in hydraulic engineering. The investigation was devided into five phases, namely:

1. literature study on material properties and applications;
2. laboratory tests in order to determine unknown material properties;
3. development of potential new combinations of applications;
4. execution of experimental and demonstration projects;
5. assessment of recommendations, guidelines, standards and requirements.

The secondary materials considered in the investigation, are granulated concrete and brick rubble, coal fly ash and coal slag, asphalt rubble, minestone, steel and phosphor slag, incinerator slag, dredging sludge and gypsum.

The investigation is based on an economical approach, considering both quantity and quality aspects. The quantity aspects concern the supplied quantities of secondary materials and the demanded quantities of construction materials for hydraulic constructions. The quality aspects deal with the structural and environmental properties of the secondary materials and the functional requirements of the sub-structures in which the materials potentially can be applied.

Phase 1 of the investigation resulted in 1989 in a final report containing the properties of the secondary materials, the present applications of these materials in general and the potential applications of these materials in hydraulic engineering (1). Phase 2 resulted in several supplementary reports on the structural and environmental properties of the secondary materials. The investigation of new combinations of applications (phase 3) has been postponed, since the current applications of the untreated materials seem enough promising. At present preparations are being made for several experimental projects. Further more an economic model for the successful recycling of waste materials has been developed as well as a probabilistic model of the environmental impact of alternative materials in hydraulic engineering. In addition the procedures for decision-making, which have to be met when using alternative materials in hydraulic engineering (sub)structures, have been determined.

To determine which applications in hydraulic engineering are most promising and in order to be able to select the most appropriate experimental projects priorities have to be assessed for the potential applications. The priorities have been determined by using the method of multi criteria analysis (MCA). This paper describes the performed MCA and presents the results of the priorities for the application of the waste materials in relation to the specific sub-structures.

2. SUPPLY OF SECONDARY MATERIALS AND PRESENT APPLICATIONS

Table 1 shows the re-used and the surplus amount of a number of waste materials in million tonnes per year in the Netherlands (1988). The sum of the given quantities presents the total supply of the secondary materials.

TABLE 1

Surplus amount of secondary materials in million tonnes per year (1988).

Waste material	re-used	surplus
Concrete and brick rubble	3.50 only granulated	3.50
Granulated asphalt rubble	0.50 warm regeneration	0.50
Incinerator slag	0.29 only dry	0.36
Steel slag	0.14	0.17
Phosphor slag	0.29	0.36
Gypsum	0.00	2.50
Dredging sludge	0.50 class 1 sludge (Euroclay)	20.00
Coals fly ash	0.59 only bounded	0.02

The majority of projects with the use of secondary materials in civil engineering apply to building and road engineering structures. Some examples are granulated concrete and brick rubble and coals slag, which are mainly used as

foundation materials for roads. The main market for fly ash is the cement industry, although a large quantity was recently used as replacement for sand in a land fill.
Asphalt re-used through warm regeneration is mainly aplied in road construction. In hydraulic engineering minestone is used as bunds and filterlayers while steel and phosphor slag is applied as top- and filterlayers of shore protection structures.

3. DEMAND FOR CONSTRUCTION MATERIALS IN HYDRAULIC ENGINEERING

Table 2 shows the demand for primary construction materials in million tonnes per year for hydraulic engineering constructions such as dikes, dams, shore protections and slope revetments (see figure 1). Hereby several hydraulic substructure have been considerd, such as the armour layer, protective layer, filterlayer, core and bunds. To each of these sub-structures special functions can be attributed such as floodprotection, soil retention, filtering and resisting shear forces. These functions finally were transformed to specific requirements for material properties such as strength, stiffness, particle size, density and permeability.

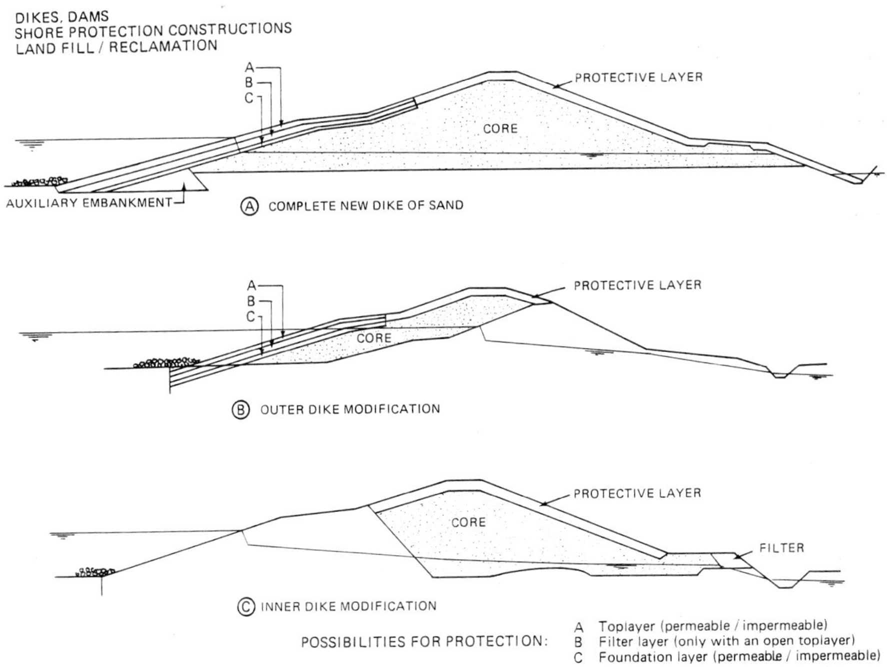

Fig. 1. Examples of considered hydraulic engineering (sub)structures.

TABLE 2

Demand for primary construction materials in hydraulic engineering (million tonnes per year) (2).

Primairy Material	Application	Yearly Maintenance	New Constr.	Possible application of secondary material
Sand	Core/land fill	11.5	54	fly ash, gran., gypsum
	Bituminous Bounded	0.07	0.09	fly ash
	In cement-concrete	0.04	0.05	fly ash
Gravel	Filter/toplayer	0.04	0.03	granulates, slag
	Bituminous Bounded	0.04	0.01	granulates, asphalt
	In cement-concrete	0.05	0.07	granulates
Crushed stone	Toplayer	0.8	0.14	granulates, asphalt, slag, minstone
Clay	Protective layer	---	2.2	dredging sludge

4. POTENTIAL APPLICATION OF SECONDARY MATERIALS IN HYDRAULIC ENGINEERING

Based on the material properties on the one hand and the structural functional requirements on the other, secondary materials have potential for applications in hydraulic engineering constructions. Examples of secondary materials, which can be used in the prenamed constructions as replacement of conventional materials, are shown in table 2.

In some cases a material property of a secondary material can even be better than the same property of the conventional material. For instance in constructions on soft soil, coal fly ash, having comparable strength and stiffness properties to sand, may be preferred over sand because of its lower density. However the environmental properties of seconday materials create the main problem for the potential application.

The way priorities for the applications of secondary materials in the substructures are determined is demonstrated by using the toplayer as an example. Among others, the following secondary materials can be utilized in the toplayer of a dike or dam in case of low current speeds: granulated concrete or brick rubble, granulated asphalt rubble, steel and phosphor slag and minestone.

5. ASSESSING PRIORITIES OF APPLICATIONS OF SECONDARY MATERIALS

Two types of priorities can be distinguished. It is possible that for a certain sub-structure several secondary materials can be used. On the other hand when a secondary material is available the most eminent application has to be found in a construction.

The MCA method is a tool which can be used to compare alternative solutions by determining the degree in which the solutions meet the previously stated judgement criteria. The determination of this degree is based on the relative importance of the selected main- and subcriteria. The method basically consists of the following six steps, which will be explained by the example of a toplayer:

1. Defining the main judgement criteria: these criteria have been related to respectively constructional, environmental and economical aspects.
2. Defining the subcriteria for the prestated main criteria: the subcriteria for the toplayer are shown in table 3.
3. Determining the degree of importance (weight) of the main criteria on a base, in such a way that the sum of these weights equals 100%: this can be done by considering two main criteria and evaluating which of the two is found the most important and in what degree (equal, more important, far more important). Doing this for all main criteria in relation to each other results in a degree of importance per main criteria, as is shown in Table 3 for a toplayer.
4. Determining the degree of importance of the subcriteria on a base up to 100% per main criterion: the procedure is analogue to the procedure described at step 3.

TABLE 3

Main criteria, subcriteria and degree of importance of these criteria.

Main criteria	Degree of Importance	Subcriteria	Degree of Importance
Structural aspects	44	Particle size	25
		Density	25
		Stiffness and strength	10
		Freeze/thauw resistance	15
		Abrasion	15
		Workability	10
			100
Environmental aspects	28	Chemical composition	20
		Contaminant migration	80
			100
Economical aspects	28	Quantities	40
		Social importance	40
		Price	20
			100

5. Determining the degree in which the selected materials meet the selected subcriteria (see Table 4 for the toplayer). This is done on a scale of for instance 1 to 5, where 1 indicates that the material meets the subcriterion badly, while 5 indicates that the material meets the subcriteria very well. A

score of 0 can not occur because such a material should already be eliminated for the considered application in an earlier stage.

6. A computer program can be used in order to determine the priorities in material applications for a certain hydraulic sub-stucture. The results of such an analysis are given in Table 4. It is indicated that the score of the first five prefered materials for the toplayer is close to each other. The low score of minestone is due to a low environmental score (conservative assumption) and a low economical score (no surplus in the Netherlands).

TABLE 4

The degree in which the selected materials meet the selected subcriteria in case of application in a toplayer

	SUBCRITERION												
	Particle size	Density	Stiffness and strength	Freeze/thauw resistance	Abrasion	Workability	Chemical composition	Contaminant migration	Quantities	Social importance	Price	SCORE	PRIO-RITIES
Gran. concrete rubble	5	4	5	3	4	5	4	4	5	3	4	83	(2)
Gran. brick rubble	5	3	5	4	2	5	4	4	5	3	4	79	(3)
Asphalt rubble	3	4	3	5	5	5	3	4	4	2	2	73	(5)
Minestone	5	4	4	2	3	4	3	3	1	1	4	59	(6)
Steel slag	4	5	5	5	5	5	1	5	3	4	3	84	(1)
Phosphor slag	5	4	4	5	4	5	1	3	4	5	3	78	(4)

The following remarks are made refering to the prescribed usage of the MCA method:

- The degrees of importance (steps 3 and 4) and the scores (step 5) are not strictly objective and should be accepted by multi disciplinary agreement, in order to make sure all involved parties have delivered an acceptable contribution to the choice of a final solution.
- The degrees of importance of the subcriteria can be different for other constructional applications. Beside that contradictional influences can occur when determining the degrees of importance. For instance: the 'social importance' indicates the social necessity for the need to re-use waste materials as secondary construction materials, while the evaluated degrees of importance can depend on the environmental aspects 'composition' and 'percolation'. All

quantative aspects as for instance price were translated to a level of quality like cheeper or more expensive.

- It is mentioned that the score at step 5 will depend on the application, for instance the density is very important when the material is used in a toplayer, while a material with a low density scores high for use as dike core-material, because then the settlements will remain limited.
- For the determination of the degree in which the selected materials meet the environmental aspects criteria are used which still are under development. In case of lack of information on these aspects conservative assumptions may result in low priorities.

In addition to the toplayer the MCA has been used to determine priorities of secondary materials in the other applications such as a filter layer, foundation layer, bunds and for a dry and a wet dike core.

On the one hand the results of the MCA can indicate which material has priority in a certain application, while on the other hand the most eminent application can be found for a material by dertermining the replacement value in relation to the conventional material for that application. For instance the use of concrete or brick rubble in a dike core as sand replacement is economically less interesting than the application in a filter layer as gravel replacement.

6. CONCLUSIONS

The study indicates that the most eminent applications of secondary materials in a top layer, filter layer, foundation layer and bunds are granulated concrete and brick rubble, steel slag (not for a filter layer) and phosphor slag (see Table 5), considering both the priorities of the materials for each application and the most eminent application per material.

For a wet dike core or land fill Euroclay (manufactured from dredging sludge) seems to be the most eminent application, while for the dry application in a dike core of land fill also fly ash and incinerator slag are applicable.

Supplemental data on mainly the environmental aspects can easily be used as a new input of the MCA.

For the next stage of the investigation (phase 4) primarily the most eminent applications, resulting from the performed MCA, will be selected. At this stage some experimental projects on prototype scale are being prepared. Examples are the application of fly ash in a land fill as replacement of sand and the use of granulated concrete rubble as a toplayer of a canal shore protection.

TABLE 5

Results of Multi Criteria Analysis: priorities in application for secondary materials in hydraulic engineering sub-structures

Group number	Circum-stances	Sub-structure	Prefered applicated secondary material
I	Wet	Toplayer Filterlayer Foundationlayer Auxiliary dam	Granulated concrete rubble Granulated brick rubble Steel slag (not in filter) Phosphor slag
II	Dry	Core Land fill Ground reclamation	Fly ash Euroclay Incinerator slag
III	Wet	Core Land fill Ground reclamation	Euroclay
IV	Wet	Toplayer monolite (Warm regeneration)	Recycled asphalt rubble

The aim of the investigation is to result in guidelines, standards and recommendations (phase 5) and stimulate the use of secondary material in hydraulic engineering structures. This development may even lead to innovative structures, which are mainly built up of secondary materials, as is shown in figure 2.

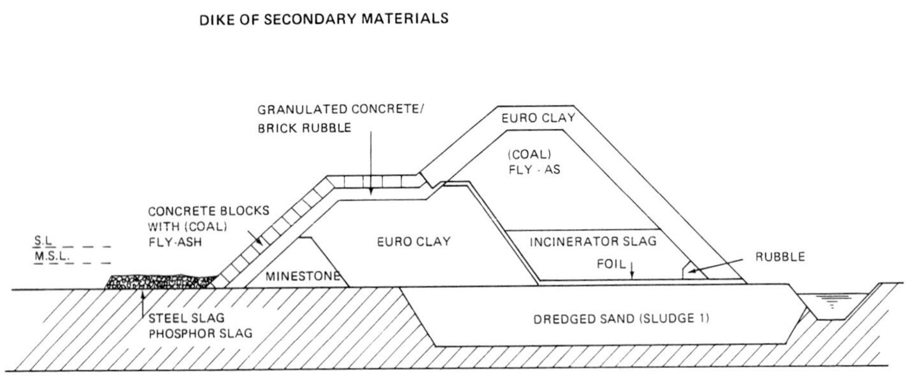

Fig. 2. Future impression of an 'alternative dike'.

REFERENCES

1 Application of secondary materials in hydraulic engineering, Literature study, CUR-report 89-1, Gouda, the Netherlands, 1989 (in Dutch).

2 Present and potential applications of secondary materials in hydraulic engineering, Fugro B.V., Leidschendam, the Netherlands, 1989 (in Dutch).

Waste Materials in Construction.
J.J.J.R. Goumans, H.A. van der Sloot and Th.G. Aalbers (Editors)

THE USE OF INDUSTRIAL RESIDUES IN THE DUTCH CEMENT INDUSTRY

W. van Loo, Central Laboratory, ENCI Nederland B.V., P.O. Box 1, 6200 AA Maastricht (The Netherlands)

0. ABSTRACT

The Dutch cement company ENCI N.V., with three manufacturing plants (Maastricht, IJmuiden and Rozenburg) processes industrial residues to such an extent and in such a way that it is useful and advantageous for the environment as well as for society and industry. Since several decades this has already led to significant savings of natural minerals, thereby contributing to the fact that in this field the Netherlands clearly distinguish themselves from the surrounding countries. It is in this context that the increasing use of non-fossil fuels should be mentioned as well.

1. CEMENT MANUFACTURE - GENERAL

Historically cement is manufactured from natural minerals, e.g.:

- limestone
- sand
- clay and
- iron ore.

These raw materials are ground to a fine dry meal and then sintered at high temperatures (up to 1,450°C) in huge rotating kilns which have a capacity of 3,000 to 4,000 tons per day.

Cement kilns are usually fired with fossil fuels such as natural gas, fuel oil or pulverized coal.

During this process the so-called portland clinker is manufactured. This clinker consists of solid particles which, due to the sintering process, are very compact.

Table 1 shows the chemical composition of portland clinker and indicates which sources are commonly used for the main components.

Table 1

Component	Content (%)	Source
Calcium oxide (CaO)	67	limestone
Silicon oxide (SiO_2)	22	sand
Aluminum oxide (Al_2O_3)	5	clay
Iron oxide (Fe_2O_3)	3	iron ore

Cement is produced by intergrinding this coarse clinker together with approx. 5% setting regulator (for which gypsum and/or anhydrite is used) to a very fine powder (300-500 m^2/kg cement). This powder is commonly known as the classical portland cement.

Portland cement is a hydraulic binding agent. This means that it will harden as a consequence of a reaction with water, resulting in a cement stone which is insoluble in water.
This cement stone is the binding agent for sand and gravel in concrete. To obtain one cubic meter of strong, durable concrete, 2,000 kg of sand and gravel are bound by approximately 300 kg of cement and 150 kg of water.

2. THE PROCESSING OF RESIDUES IN THE CLINKER MANUFACTURE

About 60 percent of the portland clinker used by the Dutch cement industry is produced in Maastricht. The remaining 40 percent are imported.
In the clinker production as much residues as possible are applied. This holds for both the raw materials and the fuels.

a. Raw materials

In Maastricht the following raw materials are being used as a starting point for the clinker manufacture:

* 79% natural minerals (limestone and sand)

* 21% residues: - fly ash from burned pulverized coal (60,000 tonnes)
- coal shale ash (40,000 tonnes)
- ferriferous additives (9,000 tonnes)
- blast furnace slag (70,000 tonnes).

Fly ash from burned pulverized coal is the same material that is also applied in the cement grinding process (see chapter 3).
Coal shale ash originates from the coal shale (or mine stone) burning in the kilns; the ash content is approximately 65%. The mine stone is a residue from the coal mine industry that used to be found in South Limburg and in the surrounding regions.
The processed ferriferous additive is iron oxide. This iron oxide is a by product from chemical industrial processes such as the production of titanium oxide or phosphoric acid.

Recently ENCI started to use also blast furnace slag in the clinker production as a substitute for natural limestone.
In the future an amount of 200,000 tonnes per year is forecasted.
The application of the above mentioned raw materials results in an essential reduction of the consumption of natural resources.

b. Fuels

The fuels fired in the kiln of ENCI at her plant in Maastricht are only partly of fossil nature: lignite and a minor amount of natural gas which give together approximately 30% of the thermal energy needed for the sintering process.
The remaining 70% are obtained by burning the above mentioned shale (18%), residual coke (48%) and a glycol blend (approx. 4%).

Most of the residual coke comes from the Flexicoker of the ESSO oil refinery at Rotterdam. This byproduct contains relatively high concentrations of the micro elements nickel and vanadium that originate from the crude oil.

For reasons of environmental protection this product cannot be burnt without special precautionary measures.
But in a cement kiln these micro elements are chemically bound to the clinker with an extremely high efficiency (over 99,5%).
This argumentation holds for the glycol blend as well.

In this blend some tenth of percents of antimony are found.
Leaching of these trace elements from hardened concrete does not occur because of the immobilization capacity of the hydrated cement stone.

3. THE PROCESSING OF RESIDUES IN THE CEMENT MANUFACTURE

In addition to the classical portland cement with a typical portland clinker content of 95%, various types of blended cements have been developed, of which the most important are:

- blast furnace slag cement (typical clinker content 30%)
- portland fly ash cement (typical clinker content 70%).

In these processes the industrial residues blast furnace slag or fly ash from pulverized coal respectively are ground together with the portland clinker and the setting regulator.

Blast furnace slag is a residue from the production of pig iron in blast furnaces. Provided the slag melt is cooled quickly enough (by granulating with water) it has latent hydraulic properties which means that if it is activated in a suitable way (for example together with portland clinker) a cement stone is produced with properties that are similar to those of cement stone produced out of portland cement.
It is expected that in 1991 ENCI, ROBUR and CEMIJ will process approx. 1.5 million tonnes of slag into cement.

Fly ash from pulverized coal is a residue from the burning process of pulverized coal in power stations. This material mainly consists of spherical particles. Due to its pozzolanic properties it can contribute to the strengthening of concrete. Furthermore the speciale shape of the fly ash particles positively influence the workability or consistency of fresh concrete.

The increase in the use of coal for the generation of energy in the Netherlands has resulted in the development of various applications of the fly ash.

In 1991 ENCI will process about 82,000 tonnes of fly ash for her cement production in addition to the 60,000 tonnes which will be used in the clinker production.

It is expected that the fly ash consumption for cement production purposes will increase even more as ordinary portland cement will be gradually replaced by portland fly ash cement.

Finally large amounts of gypsum ($CaSO_4.2H_2O$) and anhydrite ($CaSO_4$), mainly residues from the chemical industry (e.g. from the production of hydrogen fluoride), are used as setting regulators (175,000 tons/year in total).

4. THE TYPES OF CEMENT IN THE NETHERLANDS AND IN SOME OTHER EUROPEAN COUNTRIES

Table 2 shows a survey of the cement types (share of portland cement and blended cement types respectively) used in The Netherlands and in some other West-European countries (1989). From these data for each country the average content of portland clinker has been derived.

Table 2

Cement type	Netherlands (ENCI share 1990)	5 West-European countries (domestic sales) 1989				
		B	FRG	F	UK '88	I
portland cement (%)	25	35	73	32	91	55
composite portland cement (%)	10	24	8	57	-	38
blast furnace slag cement (%)	63	41	14	4	2	5
special cement qualities (%) (variable clinker content)	2	-	5	11	7	2
estimated clinker content (%)	48	64	82	80	91	83

B = Belgium UK = United Kingdom F = France
FRG = West-Germany I = Italy

This table clearly shows how large the efforts of the Dutch cement industry are to minimize the content of portland clinker.

In the future the clinker content of the ENCI cements will continue to decrease from 48 to 46%.

The genuine portland cement qualities are special binding agents with a high to very high initial strength which are applied in the concrete products industry.

5. SUMMARY

At present 2.1 million tonnes of industrial residues are used in the production process of 3.5 million tonnes of cement in The Netherlands. It is the intention to increase the part of industrial residues in cement production during the course of the nineties.

Residues are also amply used for the generation of the thermal energy needed for the sintering of the portland clinker. Only 32% of the thermal energy needed for the ENCI plant in Maastricht is obtained from fossil fuels (lignite and a small amount of natural gas); all other energy is obtained from so-called secondary fuels.

This large-scale consumption of industrial residues in the cement manufacture is unique and without equal in the industrialized world.

Thanks to her policy in this field the Dutch cement industry has essentially contributed to:
- a minimization of the use of natural minerals;
- the realization of a useful way of processing large amounts of industrial residues without any harm for the environment.

She will continue this policy in the future.

Nevertheless portland clinker remains an essential component of all our cements. Because only by using portland clinker it is possible to produce concrete of sufficient strength and - what is as important - concrete of a good and long term durability and safety.

Waste Materials in Construction.
J.J.J.R. Goumans, H.A. van der Sloot and Th.G. Aalbers (Editors)

MUNICIPAL SOLID WASTE RESIDUES IN THE NETHERLANDS.

P. Leenders
VEABRIN (Association of Operators of MSW Plants in the Netherlands).
Coolsingel 6, 3011 AD Rotterdam, The Netherlands

SUMMARY

Some 3 million tons of household and industrial wastes are incinerated in the Netherlands each year. In 2000 this amount will be about 7.6 million tons.
The residues of incineration in 1989 were 720,000 tons of bottom ashes and 90,000 tons of fly-ash. In the last five years more than 2.5 million tons of bottom-ashes have been used (mainly in embankments) and more than 150,000 tons of fly-ash in asphaltic concrete.
Quality control of the bottom-ash is carried out since 1988. Three incineration plants now have a certificate of quality assurance.
Advanced treatment possibilities for the residues as well as the consequences of separate collection of waste-components are investigated.

1. INTRODUCTION.

In the Netherlands about 3 million tons of municipal and industrial solid wastes (MSW) have been incinerated in 8 installations (1990).
This amount will increase to about 7.6 million tons in the year 2000, taken into account the possibilities of waste prevention and separate collection/recycling.
Additional MSW incineration capacity of more than 1 million tons is in preparation this moment.

Some 720,000 tons of MSW bottom-ash and 90,000 tons of MSW fly-ash have been produced in 1989 as well as 880,000 Mwh of electricity.
The amount of residues from flue-gas cleaning was limited in 1989, but will increase rapidly in the next few years to more than 15,000 tons/year due to tighter regulations.
The bottom-ash and fly-ash are collected separately in all plants.

Scrap is removed from the bottom-ash as well as parts larger than 40 mm. The scrap, some 70,000 tons/year, is recycled. In a single case the non-ferro parts are separated. The fly-ash is landfilled for about 60%. The remaining 40% is used in filler material for asphaltic concrete.

Bottom-ash has been used as road base material for roads and industrial sites, as material for embankments and as aggregate in concrete, concrete products and asphaltic concrete.
More than 2.5 million tons of bottom-ashes and some 150,000 tons of fly-ash have been used in the Netherlands in the last few years.
Guidelines for the use of bottom-ashes in earth and roadworks, dealing with material and environmental aspects are available.
Quality control of the bottom-ashes on a frequent base is carried out since 1988 in nearly all the MSW plants.

2. MANAGEMENT ASPECTS OF MSW RESIDUES.

In the last few years the following management aspects of MSW residues have been developed:
- quality aspects
- logistics
- research and development
- regulations

These activities have been strongly influenced by the founding of the VEABRIN (Association of Operators of MSW Plants in the Netherlands) in 1985.

2.1. Quality aspects.

An extensive program of quality control of the MSW bottom-ashes was started in 1987 with four plants (Amsterdam, The Hague, AVR Rotterdam, and Dordrecht/GEVUDO). The amounts of bottom-ash produced at that time were respectively 107,000 - 86,000 - 221,000 and 42,000 tons/year.
In 1988 several other plants joined this programm. Physical and environmental properties have been determined of samples taken in production periods of two weeks during a whole year.
The main conclusions of the program are:
- a sample frequency of 1-2 samples a day during a production period of two weeks give sufficient accuracy for the environmental quality of the

bottom-ashes;
- the environmental quality of the bottom-ash meets the regulations;
- the physical quality of the bottom-ash meets the regulations in nearly all the cases;
- there is a comparable behaviour of the environmental quality in time between the MSW plants;
- variations in the environmental quality are caused probably by variations in the amount and compostion of industrial wastes that are co-incinerated;

On the bases of the quality control plan a quality assurance certificate for bottom-ashes to be used in earth and roadworks has been granted by KIWA to AVR Rotterdam (1989), Dordrecht/GEVUDO and Amsterdam (1990).

The results of the quality control have played an important role in the decisions to use the bottom-ashes in usefull applications.

Preparations for a similar program for the flue-gas cleaning residues have been started in 1991. The quality control of MSW fly-ash on an regular basis will follow later.

2.2. Logistics.

The amount of bottom-ashes produced varies from some 10,000 tons to 230,000 tons/year per plant.
The main application for bottom-ash in the last few years was in embankments. These projects varied from 30,000 tons to more than 660,000 tons. Due to technical and environmental restrictions these amounts have to be delivered in a short time: the embankment for a motorway near Rotterdam, with about 500,000 tons of bottom-ash, had to be realized within 15 months. This means that enough storage facility at the MSW plants is necessary as well as a good timing of production of ashes, the (essential) storage time of at least six weeks and the construction planning of the application.
It is desirable that each MSW plant does have storage space for a minimum of one year production of the ashes. This space can also be used for the quality control program, in which the possibility to separate batches with different qualities is necessary.
In the embankment of the motorway mentioned before, bottom-ashes of several MSW plants have been used. Logistic coordination and quality information are essential to fullfill the demands for such large applications.

The same is applicable for those residues that have to be landfilled. In the Netherlands the available and suitable landfill space is scarce.
For industrial waste materials special measures are prescribed (liners, mono-landfill, water-treatment, handling/packaging etc.).
Logistic coordination and quality information mean a more efficient and effective use of the space available.

2.3. Research and development.

In the last few years about 6 million guilders have been spent by the plant owners for research and development in the field of the MSW residues. These activities varied from material research, small scale experiments, full scale field tests, feasability studies to general research. A few examples will be mentioned.

To improve the quality of bottom-ash several treatment methods have been and are investigated: increased separation of ferro and non-ferro components, separation of residual combustable parts and material with a low crushing-resistance factor, stabilization and washing methods, etc.
More fundamentally is the research into the speciation of heavy metals in the bottom-ashes and fly-ash.
These investigations should give an answer on the questions how to remove or immobilize specific elements in an efficient and effective way. A research program has been prepared but not yet started.
Melting processes for bottom-ash and fly-ash have been studied for their technical, financial and product aspects.
Field tests were executed with road foundations underneath concrete block paving, aggregate replacement in concrete blocks and asphaltmixtures, in an embankment project and in a pilot application to use bottom-ash as a fill material below the waterlevel in a harbour project.
More in general the consequences of the separate collection of garden-, fruit- and vegetable waste at the source on the incineration process and the residues have been studied in full scale incineration test experiments. Also tighter procedures of acceptance of wastes at the MSW plants and their influences on the quality of the residues are investigated.
A specific study was carried out to estimate the consequences of the use of bottom-ash in embankment projects in the long term: what kind of assurances (financial and organization) are necessary to maintain the level of

environmental protection as installed at the start.
Under the International Energy Agency an Expert Working Group (EWG) is busy since 1990.
The aim of this international group of individuals from academia, the private and public sectors, is to provide a scientific evaluation of current information on MSW ash issues, their utilization or disposal. The program of the EWG will be concluded by the end of 1992 with a symposium and an extensive report.
National and international coordination of research and development activities in the field of MSW residues is essential to achieve effective and efficient treatment methods and ways of utilization or disposal of these residues.

2.4. Regulations.

As mentioned before in the Netherlands guidelines are available for the use of MSW bottom-ash in earth and roadwork construction. Environmental regulations are revised this moment.
In practice it has been noticed that the existence of guidelines and regulations is very important to make clear decisions possible on utilization, way of disposal and investment in research and development programs as well as treatment facilities. Uncertainty or guidelines that are to be changed on a short notice are fatal for these activities.
On the other side regulations based on insufficient scientific data and (too) worst case approaches will lead to a decreasing enthousiasm for quality treatment and utilization as well as an increasing demand for disposal facilities.
In this field international research and discussions between scientists, producers and legislators is insufficient. The EWG mentioned in 2.3. is a good example of cooperation on the scientific level. Producers of MSW residues are not yet organized in different countries.
The international legislation on the use of these materials in different applications lacks of coordination.
From the management point of view, research and development activities on the side of the producers and in other sectors should be coordinated and at least known to each other. In preparation of revised regulations this knowledge should be used as much as possible.
A special item is the responsibility for the utilization or disposal and possible consequences in the long run of MSW residues.

With the number of measures that has to be taken at utilization/disposal there always will be the question of maintenance of these measures and residual risks for the environment. At this moment there are thoughts about assurances, funds, special maintenance organizations etc.
Also the question of legislation on this point is investigated.
Participation of the MSW owners is essential in this subject to prevent one-side measures that could be contra-productive to the wish of responsible utilization/disposal and quality improvement activities.

3. FUTURE DEVELOPMENTS.

The management of MSW residues in the future will be mainly influenced by environmental regulations. In the development of these regulations a strong movement is seen towards 'no risk' (or 'marginal risk') situations. However the scientific founding of the risks for the environment of the use of different materials or products is scarce. Besides that comparison of these possible risks with those of other events or human activities is hardly ever considered. The use of simplified modelling and the lack of validation and verification of test results with actual field conditions lead to 'worst-worst-worst...-case' approaches.
When users of building materials and products have a free choice - and that is nearly always the case - their choice will be obvious:
those materials/products with the least environmental restrictions.
That means that primary materials (natural materials) will be in favour and secundary materials will have to be more and more disposed of (as long as those industrial processes with which those secundary materials arise still exist).
However disposal will require protective measures, mostly for a very long period of time, based on the same 'worst-worst-worst.....case' approaches mentioned before.
So apart from the severe difficulties of finding disposal space (the NIMBY syndrom) the lack of scientific founding of the environmental risks of utilization of secundary materials lead to an economic and public spillage of our means.
The same can be said of research and development, as well as investment in technologies to improve the quality of treatment processes and the residues of those processes.
In the field of waste technology and waste treatment, processes without residues are unrealistic. Clean technology in one environmental

'compartment' will inevitably lead to residues in another 'compartment' (air - water, air - soil, water - soil, etc.).
This means that environmental regulations should not be restricted to one compartment but have to be integrated.
Risk assesment and risk comparison between activities and events are essential tools to determine the level of regulation.
It is obvious that international cooperation and coordination on expert and legislative level is necessary to achieve the wanted goals.

When faced with justified and scientific founded regulations the industry, the MSW plants included, is prepared to be involved in research and development, quality improvement and investments in treatment facilities and quality control/assurance. In that case the money available will be spent in the benefit of the environment in the most effective and efficient way.

REFERENCE:

Leenders, P. Management of Solid Waste Incinerator Residues in the Netherlands.
Müllverbrennung und Umwelt 3, EF-Verlag Berlin, 1989.

Waste Materials in Construction.
J.J.J.R. Goumans, H.A. van der Sloot and Th.G. Aalbers (Editors)

AN ECONOMIC MODEL FOR THE SUCCESSFUL RECYCLING OF WASTE MATERIALS

J.K. VRIJLING

Ministry of Transport and Public Works, Civil Engineering Division, Department of Structural Research, P.O. Box 20.000, 3502 LA Utrecht (The Netherlands)

ABSTRACT

For a successful substitution of classical building materials by alternative or waste materials an insight in the economic structure of the respective markets is necessary.
Alternative materials are generally waste or by-products of economically viable production processes. The supply of these materials is in a fixed relation to the primary product and independent of the demand.
If there is no positive demand for the waste materials, the producer will be forced to store the products on scarce land. Thus the waste material gets a negative price which affects the cost of the primary product.

If the waste material can be marketed as an alternative building material, the difference between the negative price of the waste material and the market price of the equivalent classical material is available to cover extra processing costs.

This means that the recycling of waste materials can be stimulated by increasing the cost of storage and by research into possible applications of (processed) waste materials in civil structures.

In the paper two case studies of the economic reuse of waste materials will be presented.
Possible gouvernment measures to stimulate alternative materials are analysed.

1 ANALYSIS OF THE PROBLEM

For a successful policy that aims at the partial substitution of classical building materials by alternative or waste materials an insight in the economic structure of the respective markets of primary and secondary materials is of utmost importance.

The first observation is that alternative materials are generally waste or by-products of economically viable production processes. The supply of these materials is related in a fixed ratio to the quantity of the primary product that is produced. This last quantity is gouverned by the demand in the primary market. Assuming that the quantity produced on the average equals the quantity sold , the amount of secondary or waste material that

results, is a function of the equilibrium in the primary market. The important conclusion is that the supply of the secondary or waste material is independent of the demand in the secondary market, where it has to be disposed off.
If there is no positive demand for the waste material, the producer will be forced to store the waste on scarce land. Thus the waste material gets a negative value, due to the cost of storage, which in principle affects the cost of the primary product.
An other solution for the waste problem is the transportation of the material to the sea, to dump it there. If the cost of transportation is less than the cost of storage and the dumping is acceptable within national regulations, this solution will be chosen. But the waste material still has a negative value.
The best path is however to discover a market where the waste material has some positive value. This might be as an input in a new primary process or as a substitute for a classical natural material in an existing primary process. If the value in the new market exceeds the costs of transportation and the possible costs of improvement of the waste material, the positive nett value will lead to a cost reduction for the original prime product, that generated the waste material.

In some cases,depending on the cost structure of the primary product, the positive or negative value of the waste material will influence the cost price of the primary product and therewith its supply curve and the quantity sold. This in turn influences the amount of waste material produced.
So, although the supply of waste material is in-elastic over a wide range of the market value, an sufficiently negative value will, via reduced sales of the primary product lead to a reduction in the supply of the waste material.
Along the same lines, an extremely high value of the waste material will reduce the cost of the primary product and stimulate its sale, thus causing an increased supply of waste material.
The resulting supply curve for the waste material and the demand curve for the three situations is depicted in Fig. 1.

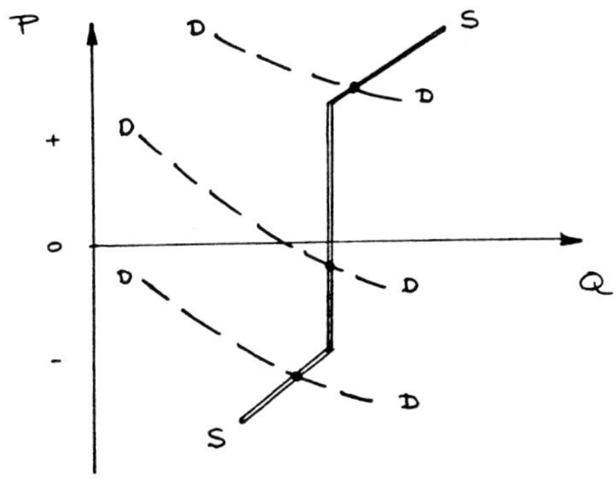

Fig. 1. The supply curve S-S of a waste material

For extremely low (i.e. negative) prices the quantity of waste material will be reduced. Finally the producer will be forced out of the primary market by the negative value of the waste. In this case the reward for the producer will be large if he find a market where the waste material can be sold as an alternative material. It should be realised that in many cases the extreme negative price, for instance caused by stringent regulations for waste storage, will not become a reality. The negative price stays a future threat, that the producer tries to evade by developing the now more attractive market for alternative use.

In case the demand for the primary product increases or decreases, the entire supply curve of the waste material will shift to the right or the left respectively, with scant regard for the consequent development in the value of the waste material.

2 FROM WASTE TO ALTERNATIVE MATERIAL

The change of waste materials into useful materials by economic forces is very old. Adam Smith describes in his famous book "The Wealth of Nations", how "among nations of hunters and shepherds, whose food consists chiefly in the flesh of these animals, every man, by providing himself with food, provides himself with the materials of more clothing than he can wear. If there was no commerce, the greater part of them would be thrown away as things of no value. However commerce raises their price above what it costs to send them to wealthier neighbours."
Today fur and genuine leather are luxury items.
Many examples can be given of materials that started their life as

the waste of a primary process, but developed in the course of time into valuable alternative materials. In some cases nobody even thinks of the materials as original waste (Table 1).

PRIMARY PROCESS	PRIMARY PRODUCT	SECONDARY MATERIAL
rearing cattle	milk,meat	dung,hides
refining oil	gasoline	asphalt
coal burning	electricity	ash,gypsum
dredging	safe channel	spoils
demolition	office site	rubble
smelting	steel,copper	slag

Table 1. Some examples of waste products changed into alternative or secondary materials

Especially leather and asphalt are at present high valued commodities.

The previous paragraph showed that the supply of waste material is completely inelastic.

If the producer of the waste material wants to market it as an alternative material for another primary process, the material has generally speaking to be processed or screened in order to improve its quality. After this processing the material has to be sold in the market of the classical material, where it will be sold as an alternative material.

To stimulate the market the producer has to prove that the material is technically equivalent with the classical material and environmentally safe. The research, performed by third parties, needed to proof this theoretically, will add to the cost. The first application will not come easily in the conservative civil engineering market. Additional incentives or risk sharing with prospective clients will further add to the cost.

However, the difference between the negative price of the waste material resulting from storage or dumping and the positive value in the alternative materials market will completely accrue to the producer. This is caused by the inelasticity of the supply curve of the waste material. The difference must exceed the cost, that has to be incurred to bring the material to the market.

If it is assumed for the moment, that the material market, where the alternative material is offered, can be characterised by the perfect competition of many sellers than the entrance of the new material will not influence the market equilibrium and the market price. Consequently the alternative material, being technically equivalent, will command the same price as the classical material (Fig. 2).

Any perceived unfavourable difference between the classical and the alternative material will however depress to some degree the value of the latter.

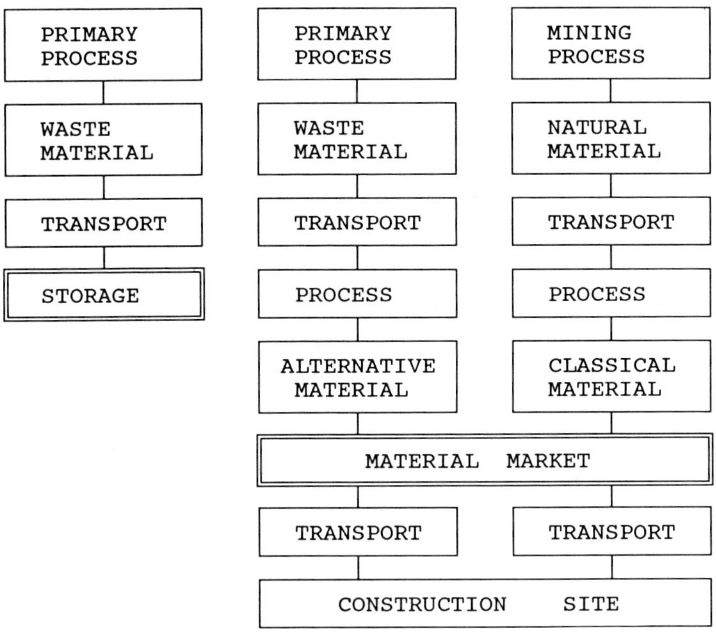

Fig. 2. The economical position of the alternative material

So the alternative material is wedged in between the negative price of storage and the value of the classical material in the market. As long as the difference exceeds the cost of research, promotion, processing and transport to bring the alternative material to the market, the flow will continue.

The example given in Table 2. shows the principle; although a loss of 1.0 is incurred by bringing the material to the market it is attractive compared with the cost of storage.

Due to the structure of the market (perfect competition) sudden changes in the supply of the alternative material, caused by changes in the demand for the primary product, will not influence the market price as the demand is infinitely elastic.

	storage	altern. mater.	classical mater.
market value	0.0	10.0	10.0
transport	- 1.0	-1.0	-1.0
storage	-10.0	0.0	0.0
process	0.0	-5.0	0.0
transport	0.0	-5.0	0.0
	-----	----	----
nett	-11.0	-1.0	+9.0

Table 2. The nett result of storage and the positioning as an alternative material in a bulk market.

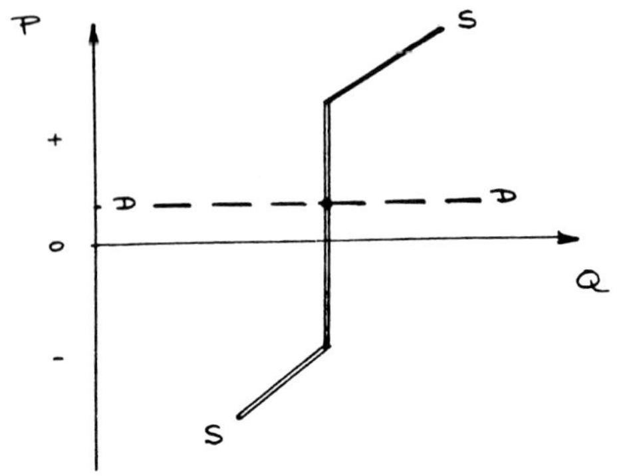

Fig. 3. The supply curve S-S and the demand curve D-D for an alternative material in a perfectly competitive material market

From the standpoint of the seller it might be attractive to differentiate the alternative material from the classical material on the basis of real or perceived favourable properties. Such properties could be e.g. a low specific weight or a specific chemical content. The differentiated product is offered in its own market niche, where it will command a higher price than the bulk of the classical material. The demand curve has now a finite elasticity (Fig. 4.).

	storage	altern. mater.	classical mater.
market value	0.0	15.0	10.0
transport	- 1.0	-1.0	-1.0
storage	-10.0	0.0	0.0
process	0.0	-5.0	0.0
transport	0.0	-5.0	0.0
	-----	----	----
nett	-11.0	+4.0	+9.0

Table 3. The nett result of storage and the positioning as an alternative material in a market niche

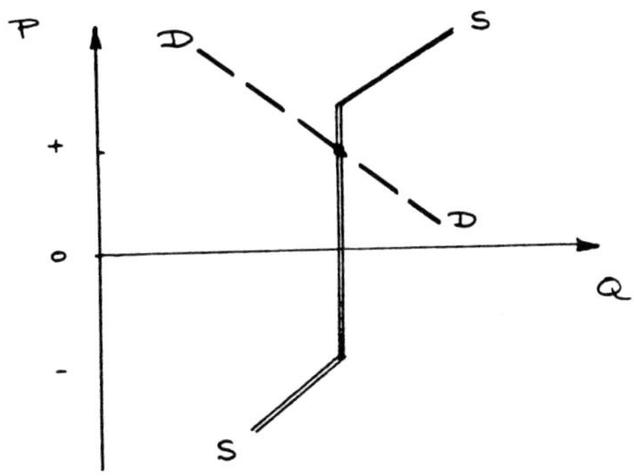

Fig. 4. The supply curve S-S and the demand curve D-D of a waste material in a niche market

In case of a finite elasticity of demand, the price is however sensitive to variations in the amount supplied to the niche market. An increase in the supply will shift the entire supply curve to the right, thus depressing the market price. This poses an difficult problem as the amount of alternative material depends proportionally on the amount of primary product produced. To maintain the position of the alternative material in its niche market, the surplus has to be stored or sold in the bulk market. The last possibility may clearly threaten the niche market, if the differentiation is weak and has no technical base.

In conclusion it can be stated that the positioning of the alternative material in one niche market is dangerous due to the non-manageable supply. It is necessary to spread the supply over more differentiated markets. This will however only diminish the problem, not solve it. Preferably a part of the supply should be sold in a perfectly competitive bulk market, that can accommodate surplus supply without price detoriation.

It should be noted that also the opposite case of a contracting supply to the niche market can be very damaging to the long term prospects of the alternative material. Although in the short term a price rise, favourable for the supplier, results, the buyers will come to doubt the stability of the supply and start looking for replacements.

3 THE IMPORTANCE OF TRANSPORT COSTS

As shown in Fig. 2. transport costs are an important part of the cost that has to be incurred to bring any material, waste or natural, to the site were it will be used. If two competing producers of the same material are located at the same distance of the site and have to use the same mode of transport the cost increase will be equal for both. And both will add the transport cost to the market price (Fig. 5.).As soon as one of the two producers is located nearer to the site, he will have a cost advantage.

This is however only valid, if there are more competing producers or if the transport market is characterised by perfect competition. Then the selling price will be equal to the market price plus the transport costs and the material will be bought from the nearest producer. When one producer is able to reach the site with a cheaper mode of transport (e.g. a ship), he may quote the lowest price although he is further removed from the site (Fig. 6). The market of the other producer will shrink in area.

In the case of only two producers the nearest will probably not pass this advantage to the buyer, if he owns the means of transport. The selling price (including transport) is likely to be constant over the whole area served by the two producers (Fig. 7 & 8.).

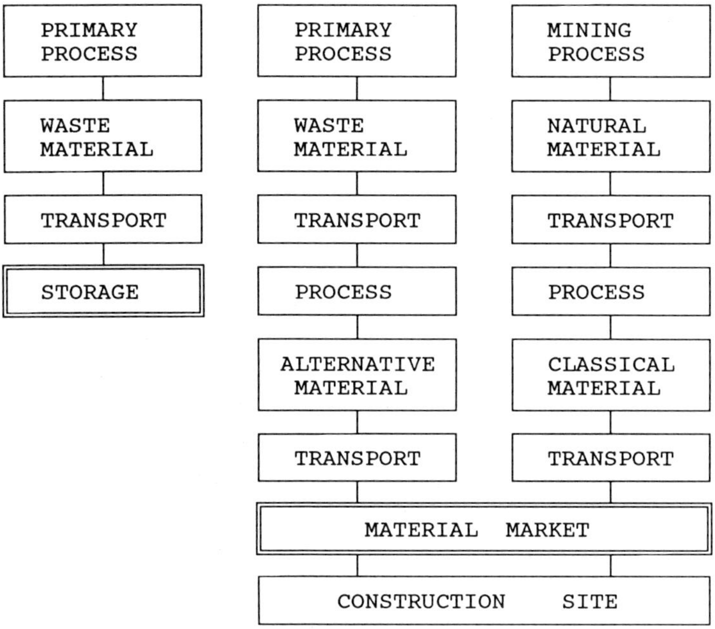

Fig. 8. The economical position of the alternative material if the means of transport are owned by the producer

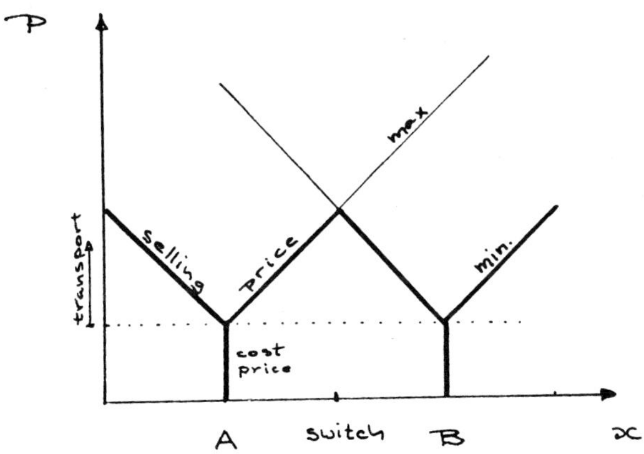

Fig. 5 The selling price as a function of the location of the site in case of perfect competition

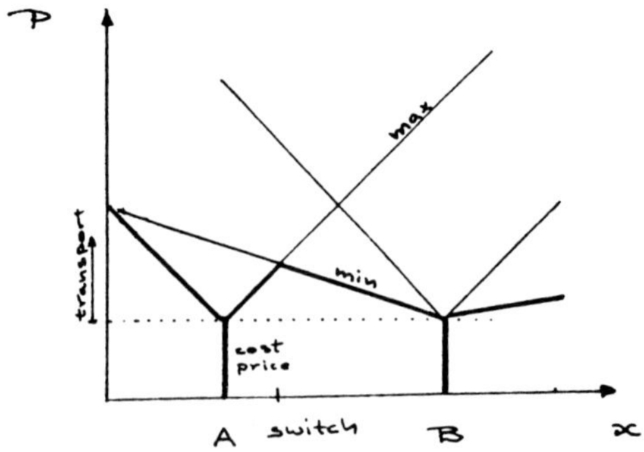

Fig. 6 The selling price as function of the location of the site, if one producer has a cheaper mode of transport.

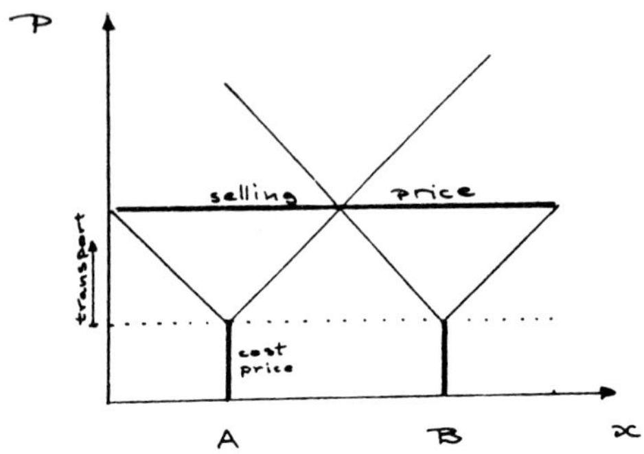

Fig. 7 The selling price in the case of two producers owning the means of transport

They are, in a certain sense, geographical monopolists, able to charge any price between the minimum and the maximum (cost price of the competitor) in their own area. In the perfectly competitive case a higher cost price will lead to a shrinkage of the market area (Fig. 9).

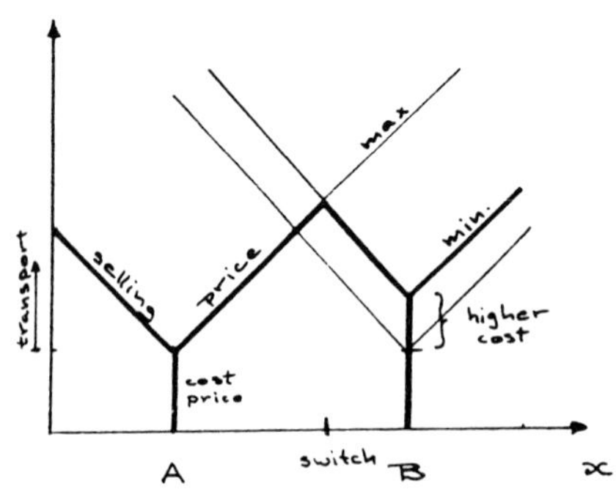

Fig. 9 The shrinkage of the market area due to a higher cost price

4. POSSIBLE GOUVERNMENT MEASURES TO STIMULATE RECYCLING

The gouvernment has in general two options to influence the economic life. It can pass laws and regulations that prescribe certain actions for its subjects or it can try to use market forces by introducing taxes, levies or subsidies. Experience seems to indicate that the second approach is more successful.

A law that forbids the emission of a certain waste will be successful if it can be enforced. The counter-forces might however be formidable because it is likely that the primary process, that generates the waste, has to be stopped. In some sense this has a similar effect as an infinite negative price.

A law that puts a ceiling to the amount emitted at some point in the future, is probably more effective as it gives room and time to adapt the primary process or to find other solutions for the waste problem. In fact the future threat enhances the negative value for the producer.

The introduction of a levy per unit of waste emitted falls in the second category. Such a levy attaches immediately a negative price to the waste, that will stimulate the reduction of the emission or the alternative use.

The same two approaches could be applied to the storage of the waste. Storage limits the emission to a predefined part of the environment making it more controllable. Generally storage, being more expensive than emission, leads to a negative value for the waste.

To make the storage impossible by law is the most radical measure, that puts a heavy strain on the industry that produces the waste. The imposition of a tax or a levy on storage increases the negative value of the waste.

Looking from an economic point of view, gouvernmental measures that cause a finite increase of the negative value are preferred. These measures, if based on an insight in the cost structure of the producer, can be better targeted on the goal, limitation of the emission or storage of waste.

The total prohibition by law of emission or storage may have unpredictable results as it is equal to an infinitely negative price in economic terms.

From Table 2 it can be concluded that the development of the waste material into an alternative material gains in attractiveness and is stimulated, if the cost of storage is increased by a tax.

The development of an alternative use for a waste material needs a considerable investment in environmental and technical research. Next, if a new application or a new market is discovered, investment in plant to process the material and in promotion is required. These investments add to the cost of the alternative material and make this route relatively less attractive.
Gouvernment subsidies could have a stimulating effect. Especially the technical and environmental research are fit for subsidies, because the gouvernmental influence can enhance the objectivity and the reliability of the studies. By attaching conditions, like cooperation with third parties, to the funding this could be reached in conjunction with a lower cost of the alternative material.

The gouvernment could also stimulate the use of alternative materials by influencing the price or the availability of the competing classical material (Fig. 2).
The ending of mining operations by law is a theoretical possibility with limited practical value due to its unpredictable and probably economically harmful consequences.
The imposition of a tax or levy on every unit mined material has more predictable effects as it will increase the market price of the classical material.
The levy could be chosen equal to the cost of restoring the environment after the mining operation.
The room for manoeuvering is however limited by the open international market for classical materials. If the price of the national producers is increased by a levy, they may loose a part or all of their market to foreign producers (Fig.9.). If the gouvernment wants to avoid this, the size of the levy should be carefully chosen.

5 CASE STUDY: FLY ASH

The generation of electricity by coal fired power stations produces mainly three solid waste materials: fly ash, slag, gypsum. For every ton of coal approximately 100 kg fly ash and 15 kg slag is produced.
Due to these ratio's a negative price of US$ 5.00 per ton fly ash will add only US$ 0.50 to the cost of a ton coal with a market price of US$ 50.00.
The influence on the price of electricity is even more limited as

the fuel cost forms only a part of the cost of a KWh. Thus only an extreme negative price of fly ash will so substantially increase the cost of electricity, that the supply will be limited (Fig. 1).

Initially the fly ash was stored. The atmosphere changed however when on the 1-st of April 1983 the storage site at Rumst in Belgium was closed.

When it became known that on the 1-st of July 1985 the storage site " de Staartjeswaard " near Beuningen would be closed too, these measures raised the spectre of an infinite negative price for fly ash. The development of an alternative material market for fly ash became a necessity.

In 1982 the Vliegasunie was founded by three cooperating electricity producers to develop markets for, among others, fly ash in the construction industry. The stated goal is to sell the fly ash in a responsible way against the highest possible price. A priority in the approach of markets is defined:

- commercially attractive markets
- commercially less attractive markets
- dumping in silo's

Many ideas were floated and tested:

- the replacement of expensive classical materials (e.g. cement)
- the improvement of the final product (e.g. improved watertightness of concrete)
- the replacement of high value materials by a product based on fly ash (e.g. artificial gravel, building blocks)

The approach of commercially attractive markets is facilitated by high value applications and strict requirements for the quality of the fly ash; constant composition, good grain size distribution and a uniform colour. BETONAS is offered as a high value filler for concrete with latent puzzolanic properties. Strict quality control is applied.

An additional requirement for a high value product is a guaranteed and stable supply as explained in $ 2.0. The Vliegasunie explains in several annual reports the need for storage and free transport through the country to keep the supply to the high value markets stable.

Also the importance of the acceptance of fly ash as part of the concrete mix by prominent customers and the inclusion of the BETONAS-product in the concrete codes, after extensive research, gets attention in the annual reports as a positive development, that will promote sales. The market price of BETONAS is approximately US$ 20.00 per ton.

Aerated concrete forms since 1987 an other important and growing market for fly ash.

The less attractive high volume market is served in the last place according to the Vliegasunie. However the largest part of the fly ash production by far is sold to the cement industry to be used in the cement production.
It is interesting to note that since 1987 the Vliegasunie contracted its own means of transport for fly ash. This guarantees safe transport of the negatively valued fly ash.

Still not all fly ash can be sold with sufficient certainty, so additional high volume applications have to be developed.
The application as "Baustoff" in the road embankment, covered by a normal earth finish, is accepted in Germany. In 1986 an embankment application in the road project A2 Oberhausen-Hannover was realised. The long transport distance caused high costs, but a lot of valuable experience was gained.
All civil engineering requirements were fulfilled and the environment was not contaminated. A similar demonstration project is developed in Holland at this moment.

In 1983 a decision was taken to build a plant to sinter fly ash into Lytag, a gravel substitute with a low specific weight. The product reached the market in november 1984. Although the market value is satisfactory at approx. U$ 12.00 per ton and exceeds the price of natural gravel, the Lytag product needs a negative
e fly ash value.

A constant search for high value niche markets is undertaken by the Vliegasunie in cooperation with entrepreneurs. Studies into the use of fly ash in building blocks and paving blocks are examples.

The uncertainty of the development of the market depends on

- the ability to export fly ash and the amount of import into the country
- change in quality due to new coal burning techniques aimed at lowering the NOx emissions
- environmental regulations that restrict civil engineering use
- competition of other waste materials as old construction materials (building waste) , slag and imports of fly ash

- slow acceptance by the conservative engineering profession

In 1985 the "Bouwstoffenbesluit" which equates the leaching behaviour of a.o. products containing fly ash with products based on the traditional materials, is proposed by the gouvernment. As such a comparison is clearly disadvantageous for the alternative materials, the reluctance of entrepreneurs to investigate new possibilities for these materials and to invest is clearly growing since that moment.

Recently a second version of the "Bouwstoffenbesluit" became known and appeared to be even more stringent because it takes the chemical content as a basis. Now the use of fly ash as an alternative material is hampered even more.

year	'83	'84	'85	'86	'87	'88	'89
QUANTITY (kt)							
cement	201.0	278.6	273.1	319.9	369.3	409.3	608.1
asfaltfiller	51.1	68.8	59.5	46.0	43.2	78.5	72.7
roads	46.2	2.5	2.2	18.3	29.0	38.2	150.4
lytag	0.0	0.0	35.6	84.0	110.4	126.9	121.5
filler in concrete	3.0	3.9	9.2	27.4	30.4	31.8	28.9
other	2.2	2.0	4.1	3.4	7.5	11.7	1.6
storage	127.5	137.1	68.5	32.5	23.9	16.0	-216.7
total fly ash	431.0	493.0	452.2	531.5	613.7	712.4	766.5
total slag				64	60	81.1	
INCOME (kfl)							
fly ash	169	309	463	1071	996	1837	2829
slag			198	318	215	283	251
other source	6453	4796	5543	6743	8996	12467	18576
total income	6622	5105	6204	8132	10207	14587	21656
VALUE (fl/t)							
neg. value	-15.36	-10.35	-13.28	-14.70	-16.28	-20.08	-27.93
av. market value	0.56	0.87	1.21	2.15	1.69	2.64	2.88
out of pocket	-14.97	-9.73	-12.26	-12.69	-14.66	-17.50	-24.23

Table 4. Overview of the development of the Vliegasunie

The sale of fly ash has progressed very successful since 1982. The percentage sold as an alternative material has grown from 39% to nearly 100% in recent years. Not withstanding the increasing supply the average market value has increased from Dfl 0.56 to Dfl. 2.88 in 1989. Nearly half of the turn over is realised in the high value concrete filler market. which is limited to 30,000 ton per year (Table 4.).

The negative value of fly ash is climbing from Dfl. -15,00 to Dfl. -27.93 in 1989 in spite of the efforts at the sales side. This is also caused by the fact that all costs of the waste materials are passed via the accounts of Vliegasunie since 1987.

5 CASE STUDY; ASPHALTIC CONCRETE

Research has proven that the broken asphaltic concrete from road pavements can be easily recycled without any loss of technical properties. From environmental point of view recycling is neutral in the sense that the composition stays practically unchanged.

In this case study the costs are compared of recycled old asphalt and of new asphalt applied in the base layer and the top layer of a road that has to be maintained.

The supply of old asphalt is in-elastic as it follows from the area of road surface that is broken up.

If new asphaltic concrete is applied, the broken old material has to be removed and stored. The storage cost is estimated at Dfl. 15.00 per ton. Therefore the cost of materials for a new asphaltic concrete base layer of Dfl. 33.52 has to be increased with the storage cost of the old material. Thus the total price of the new layer is Dfl. 48.52 per ton.

For a partial recycling of the old material in the new product, it has to be crushed first. The fines, that remain after the crushing, consist partly of soil and organic matter that has been taken up during the breaking up of the old road. These fines are classified as chemical waste and have to be stored as such against a tariff of Dfl. 200.00 per ton. The clean crushed material is stored separately in a dry place.

The old asphalt is recycled in a proportion of 1 part old at 3 parts new material. As the recycling process needs a higher temperature (275' C against 180' C normally) the cost of energy and repair is slightly higher. Above that an extra dosage installation is needed. The extra cost is estimated at Dfl. 1.00 per ton. Taking all together the material cost of recycled asphalt amounts to an attractive Dfl. 39.07 per ton.

This is the result of the relatively high negative value of stored old material.

An imperfection in the consideration above is that with the

recycling ratio of 1:3 4 tons of new asphaltic concrete result from every ton of old material. However 1 ton old material is generally replaced by 1 ton new material in the maintenance process. So the old material can only partly be reused and the rest has to be stored against Dfl. 15.00 per ton. After this correction the recycling is economically not attractive in this example (Table 5.)

COST (Dfl./ton)	base layer	top layer
new asphalt	48.52	58.57
old asphalt	49.62	46.05

Table 5. A comparison of the cost of new and recycled asphaltic concrete for base and top layers

For top layers a similar model is used. The main difference being that the top layer has to be removed by a fraising machine. Now the contamination of the broken material with organic matter is prevented.

The recycling of fraised asphaltic concrete top layers seems very attractive (Table 5.).

The results of this case study do not reach any further than the validity of the assumptions, but they illustrate, that the model can be an aid to decision making. So one can show that high storage costs and improved old-new ratio's stimulate the recycling. Also the avoidance of the contamination with organic matter, when taking up a broken base layer gives an important contribution to the economics of recycling.

In practice the market limits the recycling potential. Some clients prefer their roads to be built from new material without any technical reason. In other cases the recycling is hampered by the client, because he asks a positive price for the old asphalt.

6 CONCLUSIONS

The economic model for the recycling of waste materials developed in the paper shows its effectiveness in the analysis of the two case studies.

It gives insight in the measures the gouvernments can take to stimulate the recycling.

Attaching a negative value to the waste material, limited by the economic bearing capacity of the primary process, will stimulate recycling or reuse. Laws categorically forbidding emission or storage should be avoided as they give an infinite negative value to the waste. This may have unpredictable consequences.

To limit the cost of the development of waste into alternative materials for prospective producers, the gouvernment may stimulate and subsidize research. It can also guarantee the objectivity of the research by involving third parties.
As a client the gouvernment can order the first applications of the new materials accepting the risk.

The use of waste material as a replacement of classical natural material can be stimulated by increasing the price of the classical material by a levy or by limiting the supply. The room for price increases is limited by foreign producers that can use the price increase to pay their transport costs.

The gouvernment should stimulate, that building codes and contracts only mention required performance of materials, not required origin.

For the development of high value markets quality control of the product and acceptance by prominent clients is essential.

To bring the supply and the demand together without damaging the long term prospects of high value markets for alternative materials, storage is necessary and should be allowed.
In principle the possibilities of the market to solve the waste problem is limited if the country is segmented. Free flows stimulate reuse, as the market has the maximal flexibility to bring supply and demand together.

The paper shows, that a complete picture of the economics of the waste producing process, the economics of storage and dumping, the economics of the classical materials market; the difference in environmental harm of storing c.q. dumping waste or using it as an alternative material on the one hand and the environmental damage caused by mining of classical materials on the other, has to be sketched and understood, before successful policies can be formulated.

Waste Materials in Construction.
J.J.J.R. Goumans, H.A. van der Sloot and Th.G. Aalbers (Editors)

THE WASTES FROM POWER PLANTS AS SUBSTITUTE OF NATURAL RAW MATERIALS

Z. GIERGICZNY

Institute of Mineral Building Materials, 45-641 Opole 10, Poland

SUMMARY

The suitability of fly ashes as a substitute of natural raw materials may be enlarged by using grain size separation, selective collection in the dust collector, physical and chemical activation. The positive results was obtained at blended cements and building materials production.

The fly ashes and botton ashes disposed as byproducts in power generating stations are utilized not only from the ecological reasons but also because of their beneficious properties. The utilization of fly ashes in many technologies results from their granulometric characteristic such as particle size ranging from a few Å to 200 µm and pozzolanic properties. The suitability of fly ashes may be enlarged by the following operations: grain size separation, selective collection in the dust collector, physical and chemical activation.

The grain size separation gives the fly ash concentrates of different properties. Table 1 shows the variability of chemical composition as a function of grain size.

TABLE 1

Chemical composition of fly ash grain fractions

Type of fly ash	Grain fraction µm	L.O.I	SiO_2	Al_2O_3	Fe_2O_3	CaO	MgO	SO_3	CaO free
High-	0-20	3,7	19.2	11.5	6.6	43.0	0.8	12.8	7.2
calcium	20-40	4.2	35.2	16.2	7.4	31.2	1.4	4.9	5.6
fly	40-60	2.6	45.4	17.7	7.6	22.6	1.5	2.5	3.4
ash	>60	3.3	59.2	19.7	5.3	9.6	0.4	1.0	0.8

The finest fractions (dust) may be used as a calcium bearing fertilizer, as main components of cementing materials with a small additive of OPC or without OPC and as a neutralizing agent for liquid wastes. The sand fraction (coarse grains) may be successfully used as a sand replacement in concrete production and road making. Practically the fly ash concentrates are produced by selective collection from the particular sections of electrofilter.

The properties of the high-calcium fly ashes may be improved by grinding (physical activation). The free CaO content in this type of fly ash is high and as a consequence, the volume change of the hardened material would take place (CaO hydration to $Ca(OH)_2$). The grinding of fly ash brings about the intensive water penetration through the defected CaO and fly ash particles surfaces. The fly ash thus activated can be used in the low cement materials production (30% OPC clinkier) of the standart compressive strength in the range 33-50 MPa. The principal phases formed during the hydration proces are calcium silicate hydrate phases and ettryngite. This ground high-calcium fly ash can be successfully used in civil engineering (stockyards, embankments).

Selection of the low-calcium ($SiO_2+Al_2O_3+Fe_2O_3>70\%$) fly ash grain fractions gives the fly ash concentrates of good pozzolanic properties and fractions containing iron magnetic compounds and unburnt coal. The finest fraction is of the greatest importance in the cementing materials production owing to the active silica and alumina components reacting quickly with $Ca(OH)_2$ water solution to form calcium silicate and aluminate hydrated phases. The 30% substitution of portland cement in concrete by the fly ash of the specific surface exceeding 400 m^2/kg does not involve any strength characteristics change.

It should be concluded that the fly ash utilization development can be achieved additionally by separation and selection of particular tractions.

Waste Materials in Construction.
J.J.J.R. Goumans, H.A. van der Sloot and Th.G. Aalbers (Editors)

ADVANCED UTILIZATION OF FLY-ASH AS ARTIFICIAL AGGREGATES

T. Yamamoto[1], H. Mihashi[2] and K. Hirai[2]

[1]Electric Technology Res. & Dev. Center, Tohoku Electric Power Company Inc., 980 Sendai (Japan)

[2]Dept. of Architecture, Faculty of Engineering, Tohoku University, 980 Sendai (Japan)

SUMMARY

The outline of a newly developed artificial fine aggregates (coal-ash sand) is shown. This technique have made it possible to use a relatively large amount of fly ash. Some results to improve the material properties of coal-ash sand mortar are also presented: for example, strength, shrinkage and durability of freezing and thawing.

1. MATERIAL PROPERTIES

By mixing a large amount of fly-ash from coal fire electric power plants with a small quantity of Portland cement and water, these artificial aggregates (coal-ash sand) are produced by means of Irich mixer [1]. Various types of coal-ash sand can be made by changing the mix proportion, turning speed of the mixer and also adding some fibers. The characteristic properties of the coal-ash sand are shown in Table 1. The most desirable point is that this material is much lighter than usual sand. The most undesirable property as a material for concrete may be much higher absorption than usual one, which causes a large deformation due to shrinkage.

Strength properties of coal-ash sand mortar are shown in Fig. 1. These tests were carried out according to JIS R 5201 in principle. The size of

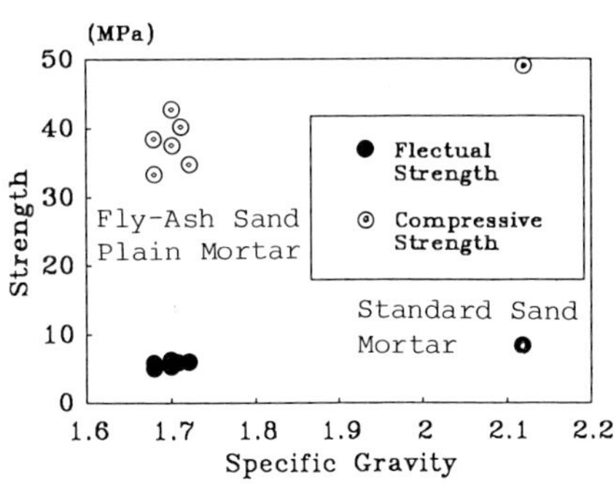

Fig. 1. Strength properties.

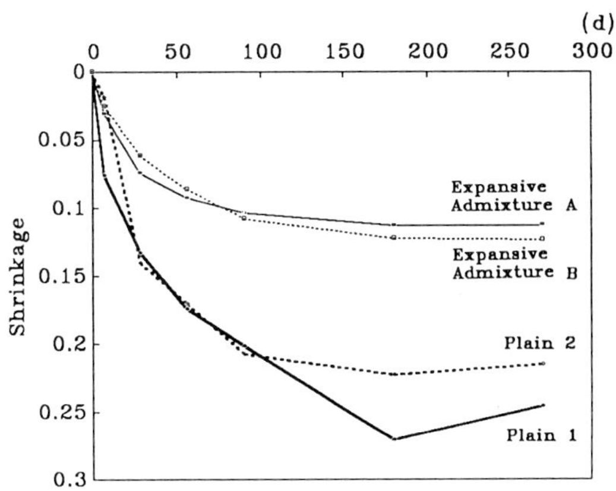

Fig. 2. Shrinkage properties.

TABLE 1
Relation between fundamental properties and mix proportion of coal-ash sand.

Cement to Fly Ash (weight ratio)	Absorption(%)	Specific Gravity under Oven-dry	Crushing Ratio(%)
1. : 9.	29.0	1.33	34.8
1.5 : 8.5	28.9	1.33	26.1
2. : 8.	25.9	1.42	17.8
2.5 : 7.5	26.9	1.41	17.7
3.3 : 6.7	25.9	1.47	9.1
Light W. Agg. A	15.4	1.44	23.2
B	15.9	1.52	16.6
(on the Market)			

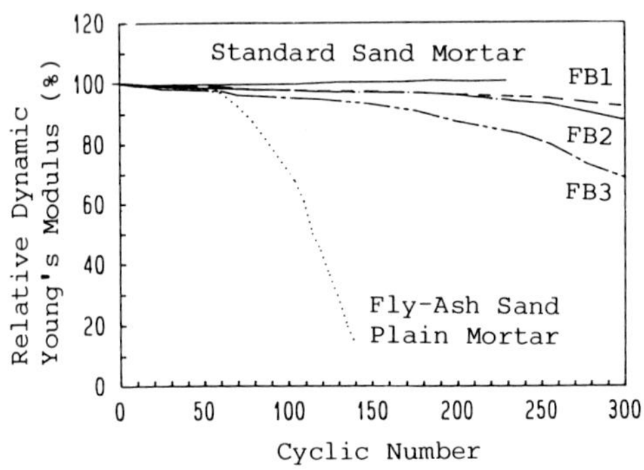

Fig. 3. Resistance to frost damage --- FB1: Vinylon; FB2: PAN-Carbon; FB3: Plant.

specimens was 40x40x160 for three point bend tests and two pieces separated after bend tests were used for compression tests. Water cement ratio was 65% and the volume per cent of sand was kept constant. The test results shown in Fig.1 are mean values of three prisms. For plain mortar of fly-ash sand mortar, six different types of sand were tested. Although the strength of fly-ash sand mortar was weaker than that of Toyoura standard sand mortar, the ratio of the specific gravity was about 80%.

Shrinkage properties are shown in Fig. 2. The shrinkage strain of fly-ash sand mortar was about twice larger than that of toyoura standard sand mortar, it was reduced up to the half by adding 4% of expansive admixtures on the market.

2. DURABILITY OF FREEZING AND THAWING

Although the durability of freezing and thawing of fly-ash sand plain mortar is not sufficient, it can be remarkably improved by adding some admixtures and fiber reinforcement [2]. Some examples of the test results are shown in Fig. 3. These tests were carried out according to ASTM C666B. In these tests, three different types of fiber were used as the reinforcement, whose content volume was 3%.

REFERENCES

1 T. Yamamoto, Summaries of Technical Papers of Annual Meeting, Architectural Institute of Japan, 1986, pp. 621-622.

2 T. Narita, H. Mihashi, K. Hirai and T. Yamamoto, Concrete Research and Technology, Japan Concrete Institute, vol. 2, No. 1, 1991, pp. 67-75.

Waste Materials in Construction.
J.J.J.R. Goumans, H.A. van der Sloot and Th.G. Aalbers (Editors)

HYDRAULIC CONSOLIDATION OF INDUSTRIAL BY-PRODUCTS AND RECYCLING MATERIALS EXAMINATION AND EVALUATION

Michael Schmidt, Oberklamweg 6, D-6906 Leimen, Germany
and
Paul Vogel, Heidelberger Zement AG, Research and Development Department, D-6906 Leimen, Germany

Recycling materials and industrial by-products can be used in construction if the structures built with them adequately withstand all loads occurring during use and if they have no unacceptable effects on the environment. The structural and, if applicable, the ecological properties may be improved by consolidation with hydraulic binders (see table 1). A widespread field of application is found in hydraulically bound road bases in road construction.

Table 1

material	Use of hydraulic binding agents for: structural engineering – load-bearing perfomance	structural engineering – durability	environment – environmental compatibility	environment – saving in raw materials	environment – waste reduction
surplus sands	+ +			+ +	
coal flue ashes	+ +	+	+ 1)	+ +	+ + 1)
incinerator residues	+	+ +	+ + 2)	+ +	+ + 2)
mining rubble	+	+ +		+ +	+ +
used concrete	+		+	+ +	+
used asphalt	+ +		+ + 3)	+ +	+ +

1) dependent on incineration processes
2) with combustion flue ashes
3) with tar constituents

Evidence for the structural suitabilty of these materials is generally given by existing road building specifications and testing procedures, which up to now, however, are only based on experience with natural minerals. The testing procedures and other evaluating criteria required for recycling materials and industrial by-products are either completely different or stiffer.
An example is the common and well-established frost test in Germany for hydraulically bound road bases made of natural mineral: Because some recycling

materials and industrial by-products are highly absorbent to water, the freeze-thaw-cycles of the hydraulically bound test specimens must be applied more often and with increased moisture curing [1], in order to determine with certainty the freeze-thaw-resistance.

The evaluation of industrial by-products and recycling materials for use in road building can be carried out according to the scheme in figure 1. In stage 1, the structural properties of the starting material, e.g., the strength, the freeze-thaw-resistance, and the structural constituents, are established and evaluated, after which an initial decision concerning its suitability is made. At best, it is suitable in an unbound state, e.g., for frost blanket layers. At worst, the material shows such poor technological properties that the only alternative is disposal.

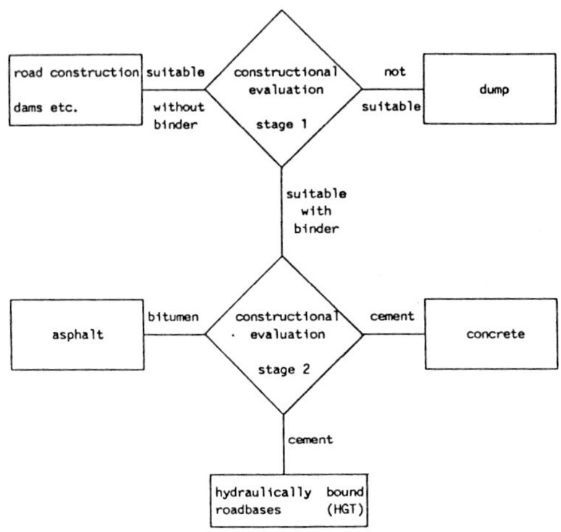

fig. 1: Evaluation of recycling- and secondary materials for use in structural engeneering with or without binders

In many cases the properties of recycling materials and secondary materials can be improved with hydraulic binders to allow for their subsequent use. The final decision in this regard, however, can only be made by an evaluation of the material in its bound state, i.e., in a hardened hydraulically bound mixture, designated here as stage 2 of the structural evaluation. For a recycling material or a secondary material to be suitable for building purposes, both strength and a sufficient resistance to weathering influences such as frost and moisture must be demonstrated. Moreover, the material should not contain any expansive constituents that might effect volume stability. The base mixture should have sufficient workability to obtain the high solidity necessary for satisfying ecological requirements and maintaining sufficient durability.

Evidence of environmental compatibility has only recently been required. The users and the responsible authorithies are considerably uncertain about specific testing and evaluation criteria due to the lack of practical experience:

Consolidation with selected hydraulic binders considerably reduces the leaching of pollutants by physicochemical bonding and/or by sealing the grain structure from water. The previous testing procedures applied to cracked materials do not sufficiently take these effects into consideration and are therefore unsuitable. On the other hand, the flow-through procedure described in [2], for example, can be applied to demonstrate the binding of pollutants. The so-called "trough processes", in which entire specimens are stored in water, are easier to carry out. Depending on the objectives of the test, the water can be stirred or passed through carbon dioxide. Provided such a testing procedure for hydraulically bound road bases proves its efficiency, it will become an integral part of FGSV standards. A final judgement on a road base bound with secondary material must ultimately be formed based on experience, and with consideration of the local conditions surrounding the finished structure, e.g., by installing a field lysimeter in the case of unknown substances.

References

[1] Schmidt, M.: Verwertung von Müllverbrennungsrückständen zur Herstellung zementgebundener Baustoffe. beton 38 (1988), H. 6, S. 238 - 254

[2] Sprung, S.: Rechenberg, W.: Einbindung von Schwermetallen in Abfallstoffen durch Verfestigung mit Zement. beton 38 (1988), H. 5, S. 193 -198

Waste Materials in Construction.
J.J.J.R. Goumans, H.A. van der Sloot and Th.G. Aalbers (Editors)

FIELD AND LABORATORY DENSITIES OF MUNICIPAL SOLID WASTE INCINERATOR ASH/WASTEWATER SLUDGE MIXTURES IN A CODISPOSAL ABOVE-GROUND LANDFILL

J. BENOIT[1] and T.T. EIGHMY[1]

[1]Environmental Research Group, Department of Civil Engineering, 236 Kingsbury Hall, University of New Hampshire, Durham, New Hampshire 03824 (USA)

SUMMARY

Geotechnical properties of a 5:1 (volume:volume) municipal solid waste bottom ash:dewatered wastewater sludge mixture were evaluated for the purposes of design consideration and management of an above-ground, side-slopped ash/sludge landfill. Laboratory proctor density testing and field density evaluation of the compacted ash/sludge mixture using a variety of construction equipment indicated that the mixture can be compacted to dry densities up to 1.25 t/m^3 at moisture content ranging from 30 to 60 percent. The water content appears to be the only variable controlling compaction of this well-graded material. Vibratory and kneading type compaction, lift thickness, number of passes, and age of the mixture did not significantly influence the compactibility of this generally wet-of-optimum mixture. However, compaction does increase disposal capacity 5 to 20% by simple use of a bulldozer to achieve maximum densities.

1. INTRODUCTION

Codisposal of municipal solid waste incinerator bottom ash and wastewater sludge has a number of potential advantages: two potentially problematic materials are codisposed in one facility, the geochemical environment is conducive to metal immobilization by sulfide precipitation (1) and, the ash can serve as a bulking agent for the sludge filling the voids in the ash matrix, thus maximizing disposal capacity (1). To properly utilize this scenario, information is needed on field compaction to maximize disposal capacity and ensure stability of the completed landfill.

2. MATERIALS AND METHODS

The bottom ash is produced from a two-stage, modular combustor (108 TPD, 36 TPD per module). Some cyclone fly ash is periodically added to the quenched bottom ash but the average percentage is very low (1-2%). The combined primary and secondary sludge is obtained from a 8,602 m^3/day activated sludge plant that is dewatered with a belt press filter to a solids content of 12%. Detailed methods are presented elsewhere (2,3).

3. RESULTS AND DISCUSSION

Grain size distributions and specific gravities [ASTM D4218 D1140, D854 and C127, (4)] of aged and fresh bottom ash, 5:1 laboratory mixtures, and 5:1 field mixtures (mixed with backhoes and bulldozers) as shown on figure 1 indicate the materials to be generally well-graded. 5:1 mixtures were evaluated because of design considerations. Coefficients of uniformity, coefficients of curvature, and specific gravity of the blends are: aged (old) ash - 38.8, 0.8, and 2:34; fresh (new) ash - 30.0, 1.3, 2.34; laboratory 5:1 - 9.6, 1.5, 1.92; and field 5:1 - 2.2, 0.2, and 2.46.

Compaction curves using Standard and Modified Proctor methods [ASTM D698, (4)] on laboratory and field 5:1 mixtures revealed compaction - dry density relations as shown in Figure 2. The results indicate that the dry density increases with a decrease in placement water content. These results strongly suggest that the 5:1 mixture remained at a wet of optimum stage during compaction. Water content appears to control achievable densities.

Full-scale field compaction was conducted using the following equipment and associated average contact pressures: static roller (120 kPa), vibratory roller (500 kPa), vibratory padfoot (1500 kPa), sheepsfoot roller (550 kPa), landfill compactor (300 kPa), and D85 bulldozer (85 kPa). Lift thickness (0.3, 0.45, 0.6 m) and number of passes (4, 6, 8, 12, 16) were evaluated for most of the compactors. Figure 3 shows the results. Consistent with the findings from the laboratory study, water content controlled achievable dry densities. The use of a bulldozer with lift thicknesses of 0.3 m seems more suitable in compacting material in the field because of its low contact pressure.

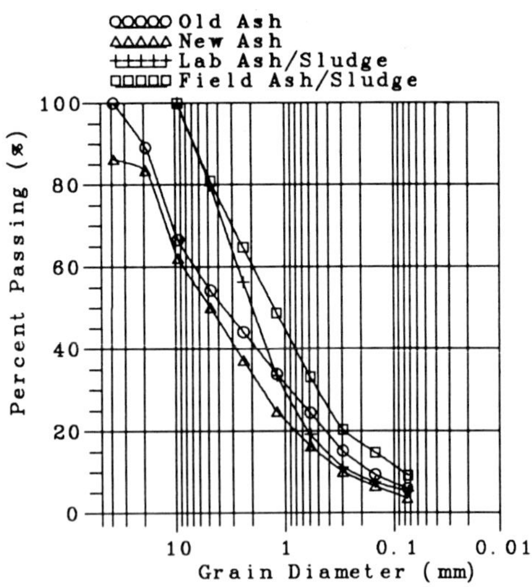

Fig. 1 Grain size distributions.

Such efforts can improve disposal capacity by 5 to 20%. Other compactors with higher contact pressures often displaced the mixture rather than compacting it.

Results from stability analyses using strength properties obtained from laboratory direct shear and compression testing suggest that codisposal of two materials that are synergistically utilized can provide a stable and compact structure.

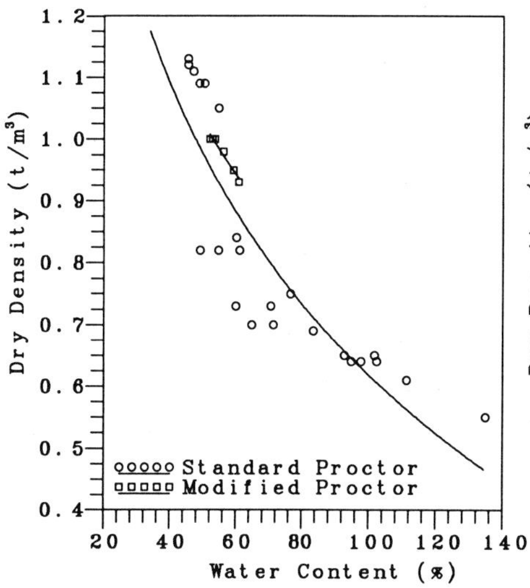

Fig. 2. Proctor compaction and measured dry densities for ash/sludge mixtures.

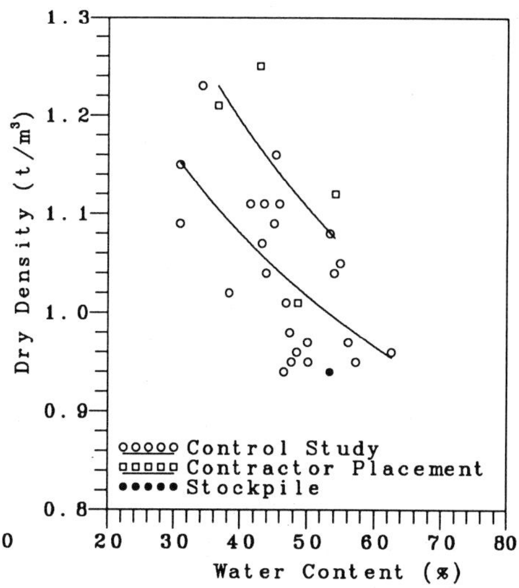

Fig. 3. Observed field densities using various compaction equipment.

REFERENCES

1 T.T. Eighmy, N.E. Kinner and T.P. Ballestero, Final Report - Codisposal of Lamprey Regional Solid Waste Cooperative Incinerator Bottom Ash and Somersworth Wastewater Sludges, ERG Publ., UNH, Durham, 1988.

2 J. Benoit, T.T. Eighmy, C. Dwinal and K. Sperry, Geotechnical Evaluation of the Dewaterability of Ash/Sludge Mixtures, ERG Publ., UNH, Durham, 1989.

3 J. Benoit and T.T. Eighmy, Final Report - Methods of Placement and Stability Analyses for Ash/Sludge Mixtures, ERG Publ., UNH, Durham, 1989.

4 ASTM, American Society for Testing and Materials, Philadelphia, Pennsylvania, 1988.

Waste Materials in Construction.
J.J.J.R. Goumans, H.A. van der Sloot and Th.G. Aalbers (Editors)

THE COMBINED USE OF INCINERATED HOUSEHOLD RUBBISH ASH AND SILICOALUMINIOUS ASH IN CONCRETE

A. VAQUIER AND S. JULIEN

Building Materials and Durability's Laboratory, INSA-UPS, Complexe Scientifique de Rangueil, 31077 Toulouse Cedex (France)

ABSTRACT

A number of problems occur when using incinerated ash, for instance loss of workability during moulding, considerable delay in hardening, a large decrease in mechanical strength and the freeing of aggressive ions. These parameters can be brought within acceptable limits mixing this type of ash with class F fly ash.

1. PHYSICO-CHEMICAL PROPERTIES OF THE ASH UNDER CONSIDERATION

1.1 Content by weight of main constituents

	SiO_2	Al_2O_3	CaO
incinerated ash	67	14	19
silicoaluminous ash	68	30	2

Dron's diagram proves that mixing an equal amount of these two pozzolanic ash types requires the addition of lime at the rate of between 65 and 100 %.

1.2 Mineralogical composition

Incinerated ash is highly cristallized, mainly quartz and alkaline chloride, but with some lime and anhydrite. F- type fly ash, on the other hand, is essentially vitrous.

2. BEHAVIOUR OF ASH WHEN ADDED TO MORTAR

2.1 - Rheological features

In the mortar samples used, (1/3) 30 % of ordinary portland cement 45 has been replaced by ash. Figs 1 and 2 show that silicoaluminous ash considerably improves mortar workability and brings setting time to within acceptable limits.

2.2 Fixation of dissolved ions

Replacing 15 % of incinerated ash by silicoaluminous ash reduces the amount of chlorine by a factor of 5 and halves the amount of dissolved sulphate : -these elements become fixed as

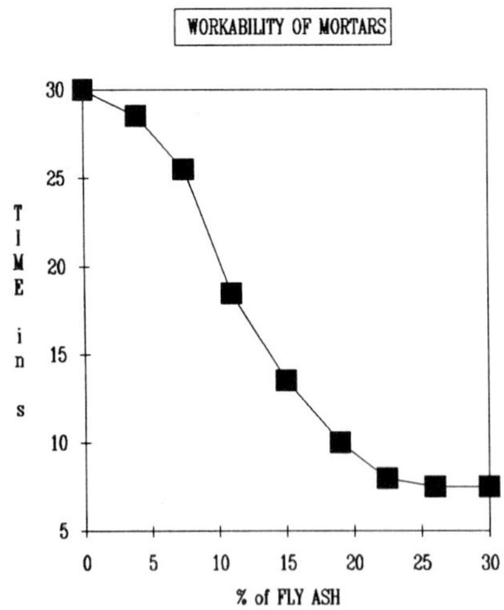
WORKABILITY OF MORTARS
TIME in s
% of FLY ASH

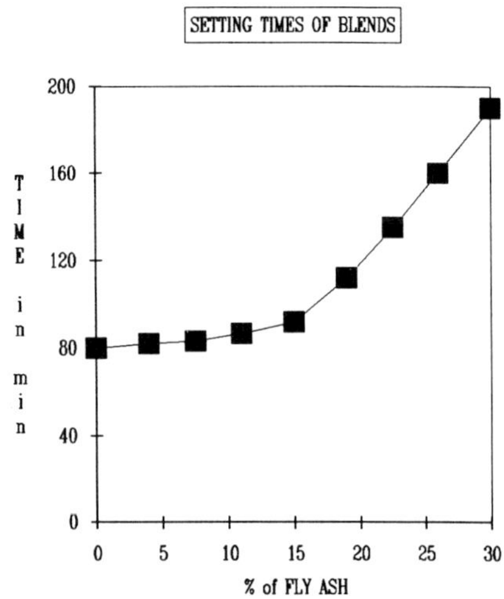
SETTING TIMES OF BLENDS
TIME in min
% of FLY ASH

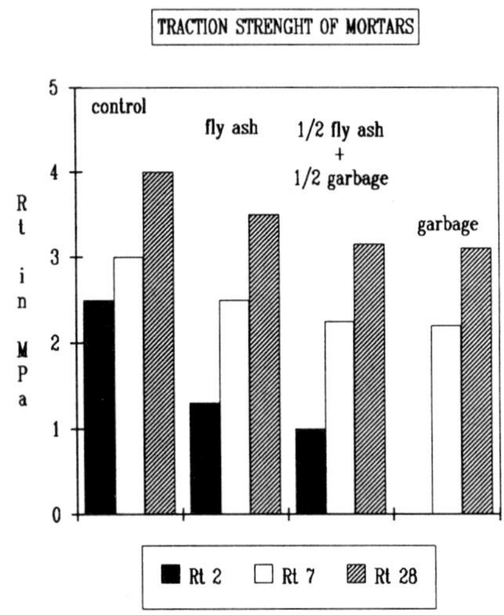
TRACTION STRENGHT OF MORTARS
Rt in MPa
control
fly ash
1/2 fly ash + 1/2 garbage
garbage
Rt 2
Rt 7
Rt 28

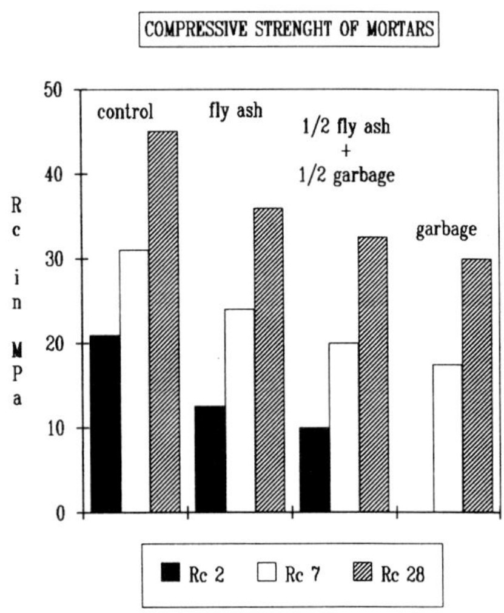
COMPRESSIVE STRENGHT OF MORTARS
Rc in MPa
control
fly ash
1/2 fly ash + 1/2 garbage
garbage
Rc 2
Rc 7
Rc 28

calcium choro- and sulpho-aluminate.

2.3 Mechanical strength and size variations

Figs 3 and 4 show the delaying effect of incinerated ash (negative strength after 2 days) and the positive contribution of fly ash, which can also reduce size variations in mortar.

3. VERIFICATION ON CONCRETE

Were moulded concrete samples 30 MPa on day 28 (Abrams cone slump approx 9 cms). The rate of substitution in the cement was 20 %. Mechanical tests on day 7 gave the following results :

	reference sample	fly ash	mixture ½
R traction (MPa)	2,4	2,1	1,6
R compression	22	18	12

Note the good result for ash mixture under traction.

CONCLUSIONS

Mixing incinerated ash with at least the same amount of silicoaluminous ash enables it to be used in concrete manufacture and thus reduces storage problems at the refuse tip.

Waste Materials in Construction.
J.J.J.R. Goumans, H.A. van der Sloot and Th.G. Aalbers (Editors)

READY-TO-USE MIXTURE BASED ON THE WASTE RAW MATERIALS FOR REPAIR WORKS

S.MILETIĆ[1], M.STEFANOVIĆ[2] and R.DJURIČIĆ[1]

[1]Institute for Testing Materials of Republic of Serbia, Belgrade, Bul.vojv. Mišića 43, (Yugoslavia)

[2]Cement Plant "Novi Popovac", Popovac kod Paraćina, (Yugoslavia)

SUMMARY

This paper presents results of investigations on the ready-to-use mixture for repair works based on the waste raw materials such as fly ash, lignosulphonate slurry and recycling waste sand. Prepared mixture showed characteristics required for repair works, groutings, filling etc.

1. MATERIALS

1.1. Fly ash is from lignite coal, but belongs to class F (ASTM C 618) with the 35.3% particles retained on the 0.045 mm sieve.

1.2. Lignosulphonate slurry has the surface tencion 72.73 10^{-5}N/cm, dry material content 55%, pH value 4.60, without chloride and reducing material 13%.

1.3. Waste sand has the grading berween 0 and 0.71 mm with SiO_2 content 86%, loss on ignition 8.5% (organic matter), Al_2O_3 content 2.5% and Fe_2O_3 content 2%.

2. OPTIMIZATION OF READY-TO-USE MIXTURES

With the mentioned waste raw materials, inorganic binder and chemical admixtures, ready-to-use mixtures were prepared and examined with the water content to produce flow of 150 and 200 mm.

3. CONCLUSION

According to test results and long-term observation on behaviour of ready--to-use mixtures from Cement Plant "Novi Popovac" it could be concluded:

- All mentioned waste raw materials could be used for the ready-to-use mixtures production.
- Fly ash and cyclone ash are from lignite coal, but belong to class F (ASTM C 618).
- Waste sand contains some organic matter in small quantities, and does not effect the mixture quality. It is used only as a partial replacement for fines.
- Test results of most mixtures does not significantly differs from the results of comercialy produced ready-to-use mixture EGE.
- Chosen ready-to-use mixture showed good characteristics for the repair works: low water/mixture ratio, high workability, no segregation and controled expansion.

Table - Test results for the flow 150 mm

Probe No.	Flow mm	Water/ mixture ratio	Water separa- tion,%	Expansi- on or shrinka- ge, %	Entra- ined air,%	Strengths,MPa after 1 day flex.	comp.	28 days flex.	comp.
EGE	151	0.121	0	+0.66	7.5	6.5	33.7	13.4	79.3
M1	150	0.143	0	-0.53	4.2	6.1	28.0	-	-
M2	151	0.147	0	-0.40	3.8	4.5	20.6	-	-
M3	151	0.150	0	+0.91	3.8	4.0	17.0	-	-
M4	150	0.111	0	-0.25	5.2	7.7	40.3	-	-
M5	151	0.109	0	+0.54	5.8	6.2	32.3	-	-
M6	150	0.115	0	+0.24	4.8	7.0	32.7	-	-
M7	152	0.120	0	+0.34	5.0	7.6	32.8	-	-
M8	149	0.115	0	+0.18	5.0	6.1	32.2	-	-
M9	149	0.116	0	+0.74	5.2	7.9	36.1	-	-
EG1	150	0.110	0	-0.14	-	6.2	32.8	13.3	70.8
EG2	150	0.107	0	0.00	-	7.4	34.7	13.2	80.0
ERE	151	0.155	0	+0,68	4.2	4.9	23.4	10.9	71.2
ER1	150	0.128	0	+0.40	-	5.8	27.9	12.8	74.4
ER2	150	0.122	0	+0.42	-	6.9	36.9	14.5	80.6

Table - Test results for the flow 200 mm

Probe No.	Flow mm	Water/ mixture ratio	Water separa- tion,%	Expansi- on or shrinka- ge, %	Entra- ined air,%	Strengths,MPa after 1 day flex.	comp.	28 days flex.	comp.
EGE	202	0.132	0	+0.60	11.5	5.7	26.6	11.4	67.1
M1	199	0.163	0	-0.53	3.4	4.1	17.2	-	-
M2	200	0.170	0	-0.70	3.4	3.3	12.7	-	-
M3	202	0.170	0	+0.79	3.2	3.3	11.7	-	-
M4	199	0.128	0	-0.57	5.0	6.3	30.7	-	-
M5	202	0.126	0	+0.24	7.5	5.6	25.5	-	-
M6	202	0.133	0	+0.46	5.4	5.6	25.8	-	-
M7	202	0.129	0	+0.40	5.8	7.1	28.0	-	-
M8	200	0.129	0	+0.46	5.4	5.6	26.2	-	-
M9	200	0.129	0	+0.90	5.8	7.5	30.0	-	-
EG1	202	0.120	0	-0.14	-	6.4	28.0	12.0	69.1
EG2	202	0.115	0	+0.37	-	6.3	33.1	12.4	74.8
ERE	202	0.173	0	+0.55	4.0	3.8	16.7	10.1	59.6
ER1	202	0.145	0	+0.50	-	5.3	25.8	12.5	74.0
ER2	200	0.139	0	+0.42	-	6.8	32.0	13.4	78.0

Waste Materials in Construction.
J.J.J.R. Goumans, H.A. van der Sloot and Th.G. Aalbers (Editors)

USE OF SCREW-PRESSED PAPER SLUDGE AS LANDFILL COVER

D.L. NUTINI AND R.N. KINMAN

RNK Environmental, Inc., 2643 Crescent Springs Rd, Erlanger, Ky 41018 (U.S.A.)

SUMMARY

Laboratory and field studies were conducted over three years to determine the feasibility of using primary paper sludge as a landfill cover material. This material seemed to be a good candidate as a landfill cover by the nature of its components (particularly clay and calcium carbonate).

1. LABORATORY STUDIES

The primary paper sludge analysis indicated that the sludge was nonhazardous, had a high water content (70%), had a strong odor, and was low in nutrients (nitrogen=1.3g/kg; phosphorus=2%). The solids fraction contained clay-37%, calcium carbonate-30%, cellulose-33%, small quantities of casein binder or soybean protein derivative binder plus the particular dye for whatever color paper was being produced.

It was decided initially that the paper components should be separated to see if they could be used or treated once separated. No separation technique produced a clear separation of sludge components. It was noted that the sludge odor was reduced considerably when the sludge was dried or a chemical additive was used (chlorine).

Since the sludge components couldn't be separated nor treated simply, the study focused on use of the sludge as a cover or liner material. The first step was to determine the sludge's permeability if used as a cover. A good cover material for a landfill requires a permeability of 10^{-7}cm/sec. At the same time, the workability of the material needed to be determined. The primary paper sludge, as obtained from the sewage treatment plant and after passing through a centrifuge, had a permeability of 10^{-4}cm/sec. Various substances were added to the primary paper sludge to attempt to make it less permeable. Some of these additives included clays (natural and koalin), sodium carbonate, and fly ash. These trials decreased the permeability to the 10^{-5} to 10^{-6}cm/sec range depending on the mix used. Through these column studies it was noted that the sludge had a tremendous absorption and filtering capacity. Biological action indicated by areas of anaerobic decomposition and gas formation in the columns was also noted. Some of this biological action may have attributed to the color loss of the sludge. Further attempts were made to get the sludge to a drier condition with various additives. The sludge was dried to 40-55% solids. It was noted that paper sludge plus some natural clay could achieve a permeability in the 10^{-6}cm/sec range very close to 10^{-7}cm/sec range.

At this time a screw press trial was going on at the plant. We obtained samples and determined the sludge's solids content to be around 62%. Column studies were set-up. Results produced a permeability in the 10^{-7}range, a less odorous material, no channeling, a good absorbing material and a filtering aid. At this point, it was recommended to test these results under field conditions.

2. FIELD STUDIES

2.1 Objectives

Three separate field studies were set-up and monitored for a period of one year using the screw-pressed paper sludge. The objectives of the first field cell included: (1) determining the workability of the paper sludge with large landfill equipment; (2) observing for odors and other problems; (3) determining stability of the sludge over time; (4) monitoring any physical changes; (5) determining when leaching would occur, the volume and the characteristics of the leachate once produced.

The second field cell was set-up similarly to the first with one additional objective. This was to determine the ability of the sludge to grow common grasses if the sludge was used as a cover.

The third field study (2 experimental landfills) was set-up to study leachate in more detail. The objectives were to determine (1) when leaching occurs; (2) the amount of leachate from a specified amount of rainfall; (3) the quality of the leachate.

2.2 Materials and Methods

Two field cells were built as shown in Fig. 1 and 2. Construction similarities for both cells were: 15 ft by 15 ft in total area; 3 to 4 ft in depth; built on a 3:1 slope; filled with screw pressed sludge (53-57% solids level); constructed with a bulldozer to test workability; compacted sludge with bulldozer; lined with a 6 ml polyethylene liner with a leachate collection system.

Differences in construction unique to each cell were as follows. Cell 1 (Fig.1) was divided into two subsections. One side had sludge only. The other side had sludge mixed with 10% fly ash. There was a separate leachate collection system for each subcell. Cell 2 (Fig. 2) was filled with paper sludge only. It was quartered on the surface with each section sowed with a different type of grass. The four grasses used included: (1) perenial rye, (2) regal rye, (3) creeping red fescue and (4) KY 31 fescue.

The third field study consisted of two experimental landfills the size of 55 gallon drums (See Fig. 3). They were outfitted with a leachate collection system, loaded with approximately 360 lbs of paper sludge with a density of 1800 lb/cu.yd. Both landfills were artificially rained on each month using a U.S. midwest rainfall pattern (31.5 inches/year). Landfill 2 was left open to receive natural rainfall as well as the artificial rain. We were trying to force the leachate through the landfills. We were also trying to see the effect of evaporation from the sludge mass in the open cell. Leachate was drained monthly and sampled for analysis.

2.3 Results and Discussion

Similar results were noted in both field cells. During construction, the sludge was observed to have various colors. These colors faded to a grayish soil color within a few weeks after construction. At the end of the year, excavations into the sludge indicated all colors were gone. The fresh sludge had an odor, characterized from slight to strong. Within a few days after placement, the odor was not noticeable while standing over the cell. The bulldozer had no handling or workability problems with the sludge. The sludge was compacted easily on a 3:1 slope. The sludge did not stick to the treads of the bulldozer or the bucket of the front-end loader. The equipment did not slide as it worked the sludge on the slope. After

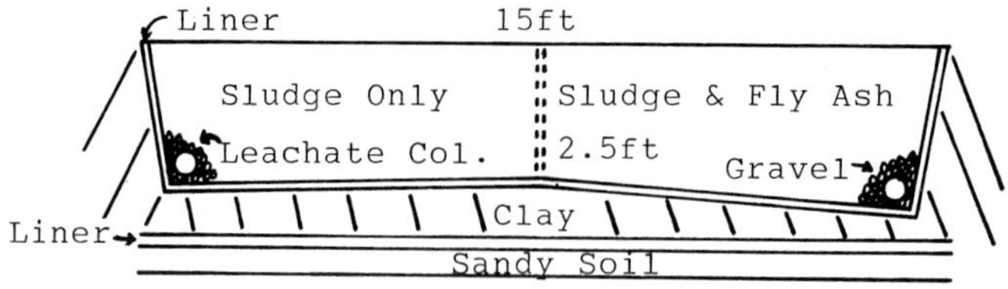

Fig. 1. Front View of Field Cell #1 Construction.

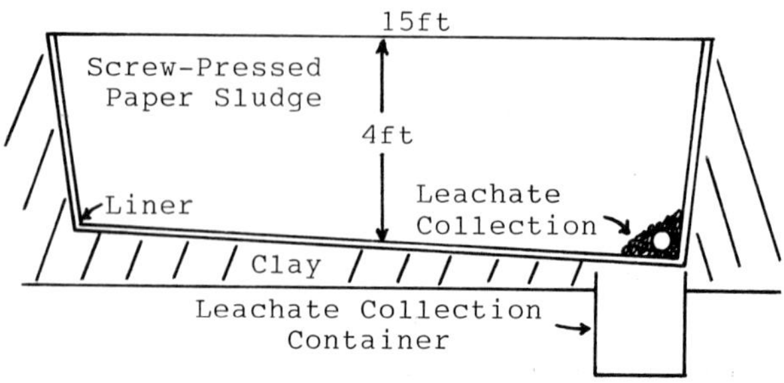

Fig. 2. Front View of Field Cell #2 Construction.

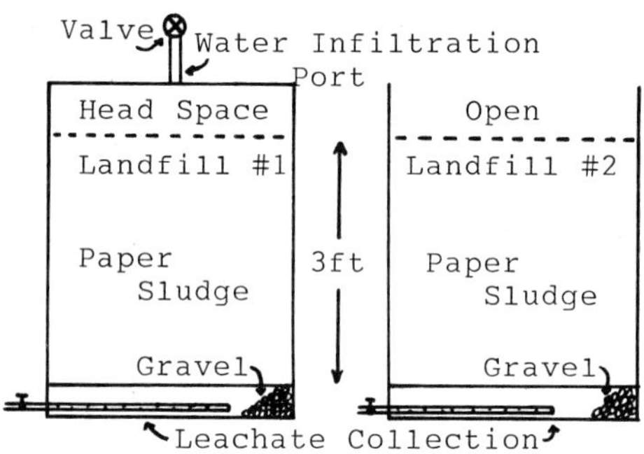

Fig. 3. Diagram of Experimental Landfills.

compaction no channeling in the sludge mass was apparent.

During monitoring, the sludge surface held up well on the 3:1 slope under all weather conditions. Very little material appeared in the runoff collection system. No surface erosion occurred. No channeling occurred. Observations made under light rainy conditions revealed that the sludge mass absorbed the rainfall, creating very little runoff. Although absorbed by the sludge, very little to no measurable leachate was collected.

The sludge acted as a "sponge" during rain-no rain cycles. That is, it soaked up precipitation and then the water evaporated during dry spells. Under heavy rain some minor leaching occurred, but very little in proportion to the amount of rainfall.

Some surface cracking occurred during dry spells but these were shallow cracks and did not interfere with the overall integrity of the liner. The sludge surface was self maintained in this regard. During the freeze-thaw periods of winter, the cracks were eliminated.

The sludge mass settled very little over the one-year period. Settling was uniform. Initial compaction appeared to be sufficient and the reason for very little settling. The sludge supported the weight of an individual walking over the surface. The sludge would support larger equipment as well.

Field cell 1 was not planted originally, yet supported a wide range of natural plants and grasses over time. Field cell 2 was planted with grasses and supported these very well with no additional nutrients added. No additional soil would be required if used as a landfill cover material. The roots of the grasses did not interfere with the integrity of the system. They penetrated only a few inches into the sludge surface. This enhanced the sludge surface support. It kept the other plants out. Some of the natural plants on field cell 1 had roots penetrating through the sludge mass. However, this did not interfere with the integrity of the liner nor create channeling.

As mentioned above the field cells produced very little leachate. The sludge only side and the sludge/fly ash side of field cell 1 had a total of 7113 liters and 7037 liters of rainfall, respectively. The leachate collected over the one year was 1.6 liters (sludge) and 1.1 liters (sludge/fly ash). Leachate was not produced until the ninth month. Field cell 2 had 28000 liters of rain fall on its surface with only 3 liters of leachate collected. Leachate was produced in the second month. Leachate production was probably a function of when these cells were set-up. Whatever the case, this demonstrated that the sludge would make a good cover material and verified our lab results.

This was further substantiated in our experimental landfills in the field. Landfill 1 had 148.5 liters of water added to it over the months. Landfill 2 had 292.6 liters of water (1321) and rain (160.61) added. Leachate collected from cells 1 and 2 were 35.5 liters and 67.8 liters, respectively. This indicated the sludge could hold a tremendous amount of water as noted in the larger field cells.

Also during this study gas was produced in both cells. The biological decomposition of the organic fraction of the paper sludge (cellulose) may also aid in decreasing permeability. With decomposition of the sludge a "plugging" of the cells prevented a quick release of leachate.

Analysis of the leachate indicated that the small amount of water which may permeate the sludge cover will not impact the landfill. Many of the leachate chemical parameters improve in quality with time. For example, in Landfill #1, pH in the first leachate sample was 6.4 In later months, the pH was around 7.3. This was true for total solids (27.8g/l to 12.7g/l) and COD (25805mg/l to 7528mg/l) as well. Our experience and research on landfills also indicates that this water leaching into the garbage layers will further be treated by biological and chemical action.

3. CONCLUSIONS

-Screw-pressed paper sludge can provide a permeability in the 10^{-7}cm/sec range as required for a landfill cover material.

-Paper sludge is workable with large equipment. It compacted well and could be worked under all weather conditions.

-Paper sludge color faded within a short time of placement.

-Paper sludge odor was not problem once placed.

-Paper sludge maintains itself over time under a variety of weather conditions. It was stable on 3:1 slope. No significant erosion nor channeling occurred over 2.5 years.

-Paper sludge supports vegetative growth. Grasses can be grown with no additional nutrients.

-No significant leachate was generated in both field cell studies.

-Leachate generated would not create any additional landfill problems if sludge is used as a cover material.

-The screw-pressed paper sludge will make an excellent landfill cover. It would function as well if not better than most natural covers (clay) used presently.

Waste Materials in Construction.
J.J.J.R. Goumans, H.A. van der Sloot and Th.G. Aalbers (Editors)

USE OF PROCESSED GARBAGE IN CEMENT CONCRETE

Z. ZHANG and F. H. WITTMANN

Institute for Building Materials, Swiss Federal Institute of Technology Zürich (Switzerland)

SUMMARY

The retarding effect of processed garbage on the hydration process and on setting of cement has been investigated using semi-adiabatic calorimetry. Alkaline hydrolysis, namely limewash method is employed to diminish the influence on hardening. Some mechanical and physical properties of cement concrete are determined as a function of concentration of this admixture.

1. INTRODUCTION

After household solid waste is transformed by a new technology in Switzerland into a range of products, the use of the finest fraction in building materials such as cement concrete is studied. It turned out that a major problem is significant retardation of the hydration process of cement. Different organic compounds in the waste are believed to be the cause. The analysis has shown that these compounds can be for instance carbonhydrates, fats and proteins. The mechanism of set retardation involves the adsorption of a thin layer of compound molecules on the surface of cement particles and on the very first hydration products. The ordinary interactions between water and cement particles are thus reduced or more or less blocked, and as a consequence it makes hydration process diffusion controlled.

Sugars, proteins and fats are essential constituents of our food, and the basic material of living organisms. The changes in pH or temperature can lead to loss of biological activity. In this article we will present the results of alkaline hydrolized household wastes as an ingredient in cement concrete in terms of calorimetry analysis.

2. EXPERIMENTS AND RESULTS

Before mixing the fine fractions of processed garbage with cement, they are exposed to $Ca(OH)_2$ water solution, i.e. organic compounds undergo limewash. It has shown that this alkaline hydrolysis process is independent of time if CaO: H_2O up to 1: 10 by weight. Hydration profiles for Portland cement with different content of hydrolized garbage are represented in Fig. 1. Although set retarding is not completely diminished, it can be seen that it is still acceptable in practice when the processed garbage content is not more than 10% of cement in weight. Cement mortar (cement: sand: water=1: 2. 5: 0. 5) was prepared with 0%, 5%, 10%, 15% processed garbage admixture respectively. Specimen size was chosen to be 40x40x160 mm. After 28-days of hydration under water some mechanical and physical properties are tested. The results are listed in Table 1, where ρ=density; λ=thermal conductivity; α_1=water adsorption coefficient as determined during the first hour (water uptake $=\alpha\sqrt{time}$) while α_2=water adsorption coefficient for water suction after the first hour; σ_b=three- point bending strength; σ_c=compreissive strength.

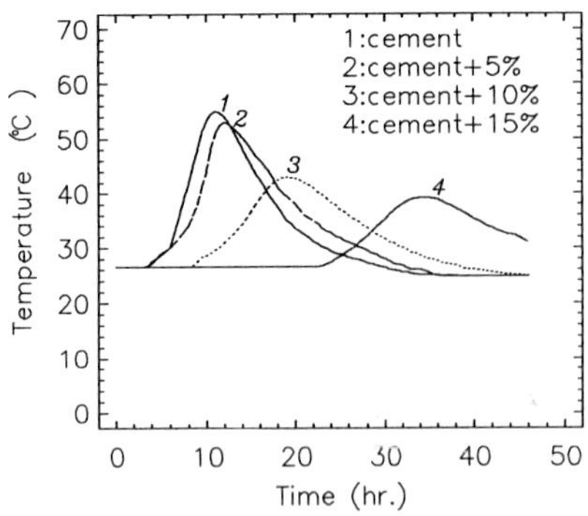

Fig. 1 Heat of Hydration Profiles

Table 1. Mechanical and Physical Properties of Mortars

Admixture Content(%)	ρ g/cm^3	λ $k/mhc°$	α_1 $g/cm^2\sqrt{h}$	α_2 $g/cm^2\sqrt{h}$	σ_b N/mm^2	σ_c N/mm^2
0	2. 2	2. 1	0. 171	0. 146	5. 1	36. 4
5	2. 1	1. 8	0. 114	0. 046	3. 4	19. 9
10	1. 9	1. 5	0. 116	0. 056	2. 0	14. 2
15	1. 8	1. 4	-	-	1. 4	9. 1

3. CONCLUSIONS

$Ca(OH)_2$ alkaline hydrolysis method is a cheap and practical solution for treating fine fraction of processed garbage in order to be used into cement concrete. Compared with normal mortar, the advantages of this mortar are its light weight, its low thermal conductivity and lower capillary suction because of artificially induced pores by this admixture. Possible applications are non load bearing structural elements.

REFERENCES

1 R. Previte, Cement and Concrete Research, Vol. 1(1971) 301-316
2 Jens Akler-Nissen, Enzymic Hydrolysis of Food Proteins, Elsvier Applied Science Publishers 1986, 9-52
3 H. E. Gerthleim, in C. L. Ferrero, M. P. Ferranti, H. Naveau (Ed.), Anaerobic Digestion and Carbohydrate Hydrolysis of Waste, Elsvier Applied Science Publishers 1984, 14-32

Waste Materials in Construction.
J.J.J.R. Goumans, H.A. van der Sloot and Th.G. Aalbers (Editors)

APPLICATION AND REUSE OF LIGHTLY POLLUTED SOIL

J.S. VAN DE GRIENDT and R.G.H. VAN MUILEKOM

TAUW Infra Consult B.V., P.O. Box 479, 7400 AL Deventer (The Netherlands)

1. INTRODUCTION

Quality standards for the chemical composition of soil are laid down in the Guidelines from the Dutch Ministry of Housing, Physical Planning and Environment. When the A-value, or reference value, is surpassed, the soil is no longer suitable for multifunctional soil use. When a value exceeds the C-value, restorative measures should be undertaken and continued until a value is attained which falls below the A-value.

2. LIGHTLY POLLUTED SOIL

Due to the fact that it is often technically impossible to treat the soil until a lower level than the reference value or because the costs of such are often so prohibitive, concentrations of one or more components in the soil originating from soil treatment plants appear higher than the A value, but are simultaneouly lower than the C value. We then talk of lightly polluted soil. In addition to lightly polluted soil originating from remediations, this can also originate in city areas whenever there is talk of increased background concentrations. This soil will often not be considered for remedial purposes.

3. GOVERNMENT POLICY

For lightly polluted soil the governmental statute for restoration comes immediately into force, as this soil is no longer suitable for multifunctional soil purposes (the aim of the ground quality policy). However, due to the threatening shortage of primary building and raw materials, the government's policy is aimed at restricting, whenever possible, incineration or disposal of polluted soil and to encourage, wherever possible, the use and reuse of lightly polluted soil, possibly after being tested, providing certain requirements are met.

The Dutch Service Centre Soil-cleaning manages and coordinates the use and reuse of lightly polluted soil in the Netherlands. In addition to setting standards regarding the quality of the soil, the centre also sets up priorities for soil remediation and offers advice on additional soil protection measures, which can follow as a result of the reuse of lightly polluted soil.

4. BUILDING MATERIAL DECREE

Presently the Provinces are responsible for setting up priorities for soil remediation and the use and reuse of lightly polluted soil. Due to this, the course of action taken can vary in each Province. An attempt is being made, by means of the Building Material Decree, to change this situation and to set up uniform regulations which would be enforced throughout The Netherlands. In future, the Building Material Decree will state when and to what degree lightly polluted soil, derived from soil remediation, (and soil treatment plants) may be used as a secondary building material. The Provisional Building Material Decree concerns itself with "soil like" building material which could be used in construction and ground works. For example sand, gravel, furnace slags, slags from incinerator plants and fly-ash all fall under this decree.

In the Building Material Decree requirements are set, in particular for the concentrations and the mobility of the "contributing" components which could pose a threat to the environment. The mobility is determined by measuring the leaching or the diffusion from the building material. This concerns heavy metals such as cadmium, lead, copper, mercury and zinc as well as organic components such as polycyclic aromatic hydrocarbons (PAHs), polychlorinated biphenyls (PCBs) and mineral oil. A difference is made between form-given and non-form-given building material.

5. ICM CRITERIA

Depending upon the actual properties of the material, the leachability, the form in which it is to be used, and the conditions under which this occurs, additional measures should be taken to prevent soil pollution; the so-called ICM Criteria, Criteria of Isolation, Control and Monitoring laid down in the Soil Protection Act. Depending upon the risk to the public health and the environment, stricter requirements will be imposed.

An example of an isolation provision is a plastic liner underneath a sound dampening wall in which lightly polluted soil is used. Examples of control and monitoring provisions are the regular sampling and analysis of soil and groundwater under or around a sound dampening wall.

Waste Materials in Construction.
J.J.J.R. Goumans, H.A. van der Sloot and Th.G. Aalbers (Editors)

APPLICATIONS OF AAC BY-PRODUCTS

Ingeborg Lang
R & D Centre, YTONG AG, Sandhof 6, 8898 Schrobenhausen, Germany

SUMMARY

During the production of AAC blocs granules and dusts are produced as autoclaved by-products. The versatile properties of autoclaved by-products allow them to be used in many industrial, especially environmental applications for example in flue gas purification, sludge conditioning and as filling material and cat litter.

1. Production of AAC

AAC consists of a mixture of finely ground quartzite sand together with lime and cement binding agents, water and aluminium powder acting as a foaming agent. The initial hardening of the mixture is due to the hydration of the unslaked lime and the cement. Then, after 1 or 2 hours, the curing process in the autoclave takes place with saturated steam at 12 bars and 180 to 200 °C. This procedure lasts for a total of 10 to 12 hours. The cured by-products are produced through product breakage after the autoclaving process.

2. Possible Uses of AAC By-Products

The diagram below shows the main application areas for AAC dusts and granules which are either already being marketed under wellknown brands -such as the case of cat litter and oil binders- or which are part of present or future R & D projects.

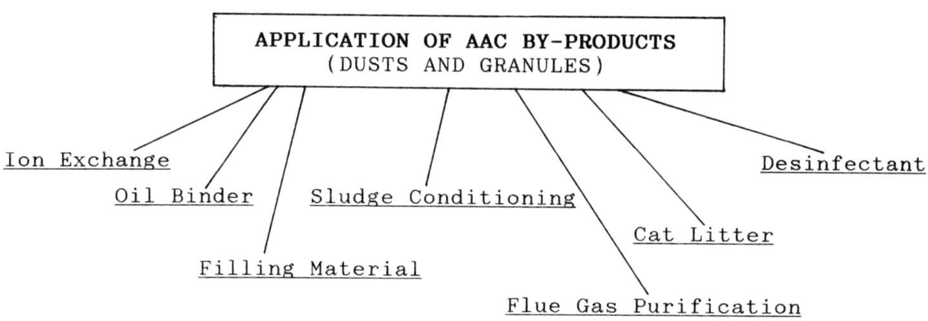

3. Physical and Technical Properties of AAC

The crystalline structure of AAC or CSH products is essentially that of tobermorites with zeolithic properties and quartz. Their specific surface of them range between 20 and 30 m^2/g and the porosity is nearly 80 %. Their pH-value lies between 10 and 11 and they can absorb between 95-100 weight % water.

Owing to these properties AAC-products are very useful for chemisorption of acid gases such as SO_2/SO_3, fluoric and chloric compounds to build $CaSO_4$, CaF_2, $CaCl_2$ etc.

Another very promising property which requires more research is the chemi- and/or physisorption of polar hydrocarbons with high boiling points such as aniline, pyridine, pentachorphenol etc.

4. Flue Gas Purification

Fig.(1) and (2) show the results of measurements of SO_2 - and aniline sorption. The SO_2 sorption reaches a very high efficency over 90 % if the water/SO_2 ratio is between 4 and 5. The gaschromatographically measured retention volume of AAC during aniline sorption demonstrates very well the dependence of the efficency on the specific surface and the temperature, where the first break through will occur.

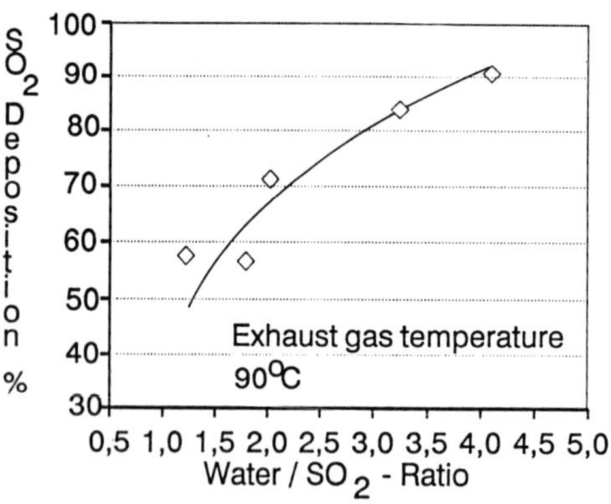

Fig.(1): SO_2 deposition for various water/SO_2 ratios

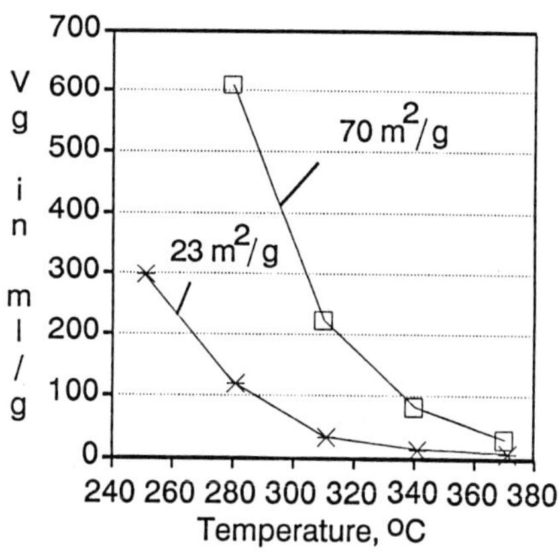

Fig.(2): Retention of aniline as a function of the temperature

It must be emphasized that the current research work and its findings are still at any early stage and that more details on the properties of AAC materials are still to be gathered and investigated.

Waste Materials in Construction.
J.J.J.R. Goumans, H.A. van der Sloot and Th.G. Aalbers (Editors)

PILOT SCALE DISPOSAL OF S/S TREATED SOIL CLEANING RESIDUE

C.W.J. Hooykaas

Pelt & Hooykaas B.V., Bijlstraat 1, P.O. Box 59011, 3008 PA Rotterdam (The Netherlands).

1. INTRODUCTION

In the Netherlands chemically hazardous wastes are disposed on special landfills (C2-/C3-depony) or are exported abroad. In contrast to most industrialized countries it is not allowed to treat chemically hazardous waste prior to disposal. Due to the high costs of disposal and the expected high costs involved in aftercare of these landfills, there is however a growing interest for stabilization/solidification.

Stabilization/solidification (S/S) is used all over the world as a final treatment of chemical waste prior to land disposal. The objectives of S/S are to reduce leachability of hazardous constituents, to improve handling of the waste and to detoxify contaminants. With respect to inorganic wastes, cement-based grouting technology is the most common and accepted S/S technique because of low processing costs and ability to meet stringent performance requirements. Recent studies have demonstrated that advanced S/S-techniques are also applicable to organic wastes [1,2].

One of the types of chemically hazardous waste that present a major concern in the Netherlands, is soil cleaning residu (SCR). SCR is produced at a rate of 40.000 to 80.000 tons per year and contains a mixture of relative mobile anorganic and organic constituents.

The underlying research project consists of a pilot-scale testing and will be performed to support the assertion that S/S-treatment of SCR will produce a waste product that is suitable for a safe and permanent disposal. Other objectives of this project are to evaluate the practical and performance aspects of S/S as a treatment for SCR and to compare field leachate concentrations with results of leaching tests.

2. OUTLINE PILOT SCALE TEST

2.1 Waste treatment

The SCR that will be tested contains large amounts of heavy metals (e.g. As and Hg) and of organics (e.g chlorobenzenes and phenols). The S/S process used here is a cementious process that uses additives that interact with the cement matrix and at the same time adsorbs organics. A mixing plant on the site will mix the SCR with the binders and additives in a weight ratio of about 10/1 to 10/2. The slurry will be pumped into the pilot-scale disposal. A second pilot-scale disposal will be filled with untreated SCR.

2.2 Pilot-scale disposal

The pilot-scale disposal consists of a 10x10x2m construction and is filled with 150 m3 of slurry. Run off water and percolate will be collected separately. Background immission of contaminants and avapotranspiration will be measured using dust and rain collectors respectively.

2.3 Testing program

In order to evaluate leachability and chemical binding, leachingtests will be conducted on core samples of the treated and of the untreated SCR. Test procedures will include extended leaching tests using the Dutch standard leachingtest [3] and the tank leachingtest [4]. Physical integrity of the treated SCR will be evaluate by measuring unconfined compressive strength, weathering and permeability.

2.4 Supervision

For the supervision of the research project a working group has been installed with representation of the Ministry of Housing, Physical Planning and the Environment, Environmental research, the Provence of Zuid-Holland, Rijnmond Central Environmental Control Agency, Service Centre Soil Purification, VBM, IWACO B.V.and Pelt & Hooykaas B.V.

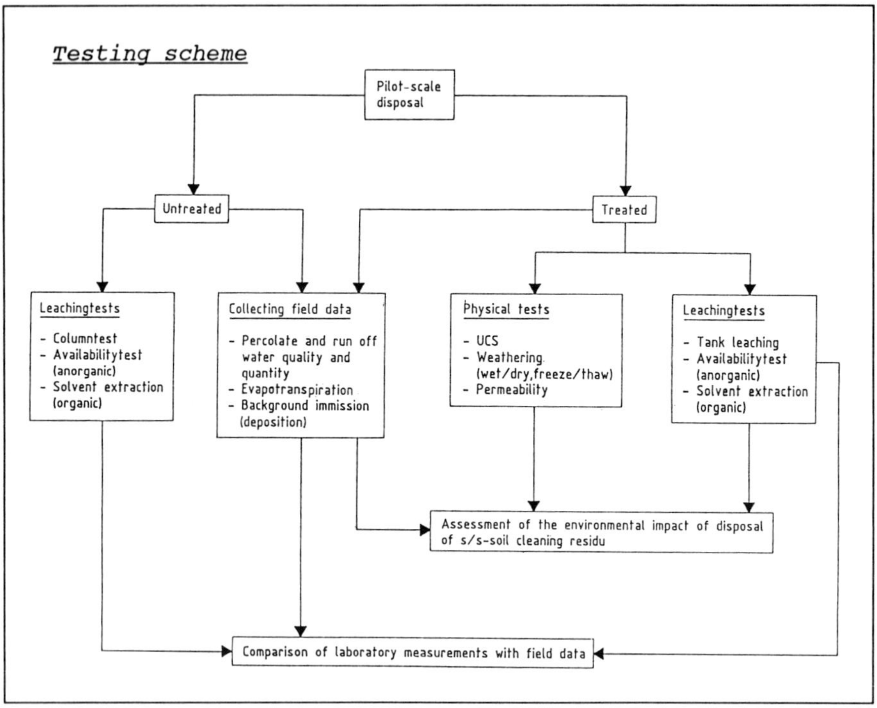

REFERENCES

[1] *R. Soundararajan. Immobilization of polychlorinated biphenyls by oragnophillic binders- a case study. <u>In:</u> Arendt, M. Hinsenveld and W.J. van den Brink (eds.), Contaminated Soil '90, (1990) 1275-1282.*

[2] *T. S. Sheriff, C.J. Sollars, D. Montgomery and R Perry. The use of activated charcoal and tetra-alkylammonium-substituted clays in cement-based stabilisation/solidification of phenols and clorinated phenols. <u>In:</u> P. Cote and M. Gilliam (eds). Environmental aspects os stabilization and solidification of hazardous and radioactive wastes. Atlanta 1987.*

[3] *NVN 2508. Standard leaching test for combustion residues. Netherlands Normalization Institute, Delft, 1988.*

[4] *Draft NVN 5432. Determination of the maximum leachable quantity and the emission of potentially hazardous components from construction materials, monolithic waste materials and stabilized waste products of mainly inorganic character. Netherlands Normalization Institute, Delft, 1989.*

Waste Materials in Construction.
J.J.J.R. Goumans, H.A. van der Sloot and Th.G. Aalbers (Editors)

REACTIVITY OF LOW-CA FLY ASH IN CEMENT

Hans S. Pietersen and Jan M. Bijen*

Delft Technical University, Faculty of Civil Engineering,
Materials Sciences, Stevinweg 1, 2628 CN Delft, The Netherlands
* Also director of INTRON, Institute for Materials and Environment

Introduction

At Delft Technical University a NOVEM supported research program is carried out to investigate the mechanism of fly ash reaction in cement pastes and cementitious binders. The program was started in 1988 and will finish this year. Investigated topics include an investigation of the structure of the fly ash glass, the mechanism of the pozzolanic reaction between fly ash and cement and a study of reaction products formed. Results are related to performance.

Structure of the fly ash glass phase

Fly ash consists of a matrix of alumina-silicate glass, modified by elements such as Ca, K, Na, Fe, Ti. Part of the fly ash consists of crystalline material (common are mullite, quartz and spinel). With Transmission Electron Microscopy (TEM) it has been shown that phase separation occurs in almost all fly ash glasses [1]. In figure 1 variations in Al_2O_3/SiO_2 ratio on respectively micro- and -nanometer scale are illustrated. Some micrographs of fly ash phase separation are shown in figure 2. Compositional gradients at a nanometer scale may create a potential for heterogeneous nucleation at the fly ash cement interface.
Multivariate statistical analysis of chemical analysis of individual fly ash particles with a microprobe (EPMA) indicate that most fly ashes may be divided into groups with differing amounts of glass modifier elements, reflecting the type of precursor (clay-)minerals [2].

Quantification of the glass content in fly ash

Quantitative X-Ray Diffraction (QXRD) has been used to quantify the amount of crystalline matter in a fly ash sample. Upon correction for carbon content and readily soluble surface deposited alkali-salts, the glass content may may be calculated [3]. The glass part of the fly ash is generally believed to be the reactive part. The QXRD method is suitable for routine analysis. Ashes with a relatively high modifier content generally seem to display the highest glass content, due to a lowering of the melting point of the molten clay minerals in the combustion chamber of a coal power plant.

Mechanism of fly ash glass reaction; a chemical view

Size- and density separated fly ash particles of different fly ash batches have been immersed in a sodium-hydroxide solution. The results have shown that fly ashes dissolves congruently at the high pH commonly encountered in cement pastes; fly ash reaction may be described as a "bulk dissolution" mechanism [1]. Fly ash grains originating from different fly ashes, but of similar size- and density, react at an approximately similar rate, when normalized according to glass content- and chemistry. It is well known that fly ashes with a small average particle size react relatively fast with cement; this study indicated that a high glass content also contributes significantly to overall reactivity.

Reaction products formed

Initially, fly ash in portland cement concrete will not react significantly; this period, in which the fly ash serves predominantly as a nucleus for precipitation of CH and CSH (both formed because of the cement hydration reaction), may be called the "induction period". The only reaction product that may form initially in somewhat increased amounts are sulphoaluminates (a.o. AFt or ettringite). At later ages the increased level of bound water in cement-fly ash pastes, as well as the siginicantly higher CH consumption, are indications for the formation of extended amounts of calciumsilicate-hydrates (CSH). This process may be quantified by X-ray mapping of polished sections of fly ash cement paste at different ages [4]. With ^{29}Si NMR it has been found that cement containing 20% m/m fly ash generally develops increased amounts of dense "inner-products CSH", the so-called Q^2 silicon or chain-middle groups, relative to the more open structured "outer-product CSH" (figure 3) [5]. On average, the CSH formed in cement containing fly ash displays a lower Ca/Si ratio, due to the increased availability of Si (from the fly ash) which reacts with CH. Alumina containing reaction products are generally amorphous alumina-silicate hydrates and consequently difficult to detect; minerals like "strätlingite" or "gehlenite hydrate" (both C_2ASH_8) or hydrogarnet (C_4AH_x) are known to occur [6]. However, it is difficult to quantify these reaction products, other than by chemical or thermodynamic modelling of the cement-fly ash paste system.

Performance of cementpaste with fly ash

Initially cement pastes, in which part of the cement is replaced by fly ash, display lower compressive strengths than reference cement pastes. Only after 1 to 3 months equal or higher strengths develop due to the relatively slow, but steady, pozzolanic reaction. The strength increase may go on for as long as 10 years [7]. Durability and permeability tests show that cement pastes containing fly ash are more resistant to chemical attack due to a decrease in the amount of larger pores and to the decrease in free lime [8].

Acknowledgements

This research has been sponsored jointly by the Technical University Delft and by NOVEM, the Dutch Society for Energy and Environment, contract nr. 11238.p117.

References

[1] H.S. Pietersen, A.L.A. Fraay and J.M. Bijen in Materials Research Society Symposia Proceedings, Volume 178, 1990, p.139-157.
[2] H.S. Pietersen, S.P. Vriend, R.E. Poorter and J.M. Bijen in [1] MRS Symposium Proceedings, Volume 178, 1990, p.115-126.
[3] H.S. Pietersen and N.M. van der Pers, Stevinrapport Civiele Techniek, T.U. Delft 25.1-89-19, 1989.
[4] H.S. Pietersen, unpublished research.
[5] H.S. Pietersen, A. Kentgens, W. Veeman and J.M. Bijen, The microstructure of concrete, 19-20 september 1990, University of Oxford.
[6] H.F.W. Taylor, Cement Chemistry, Academic Press, London, 1990.
[7] P.L. Pratt, in [1] MRS Symposium Proceedings, Volume 178, 1990, p.177.
[8] J.A. Larbi, personal communication.

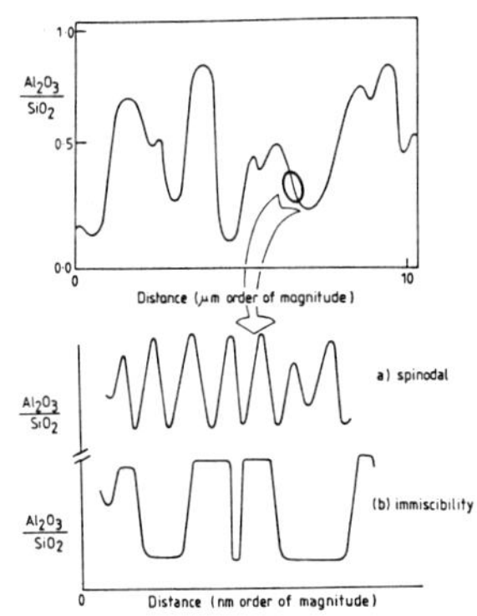

Figure 1.

Upper diagram: Variations in Al_2O_3/SiO_2 ratio in fly ash on a µm scale.
Lower diagram: Variations due to spinodal decomposition (a) and liquid immiscibility (b).

Figure 2.

Two examples of phase separation in fly ash glass.

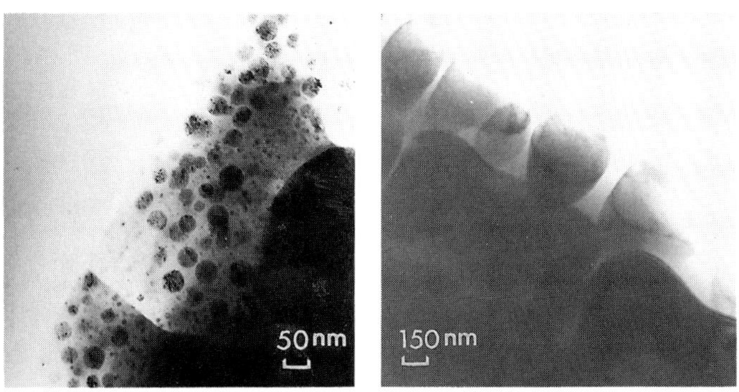

Figure 3.

Differences in distribution of silica in cement containing fly ash. The cement paste with 20% m/m fly ash contains an increased amount of Q^2 type silica polymers or "inner-product" CSH.

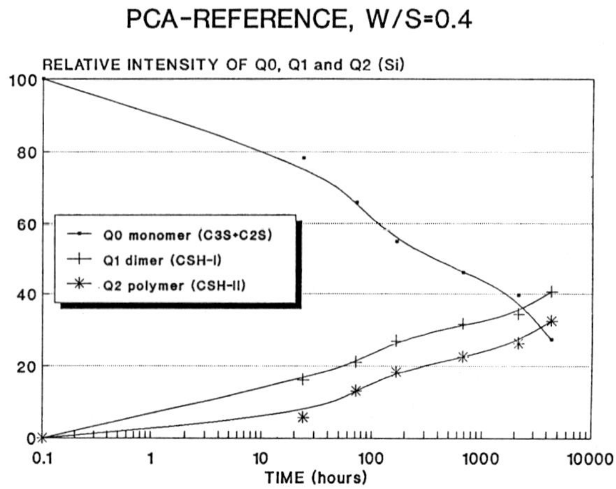

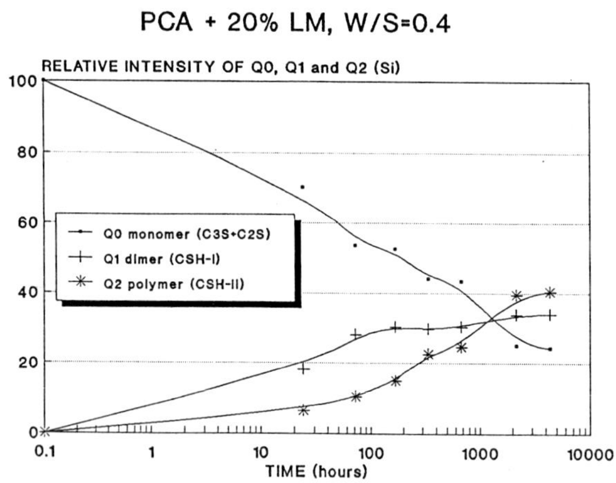

Waste Materials in Construction.
J.J.J.R. Goumans, H.A. van der Sloot and Th.G. Aalbers (Editors)

THE INDAS FOUNDATION, AN INNOVATIVE ROUTE FOR THE UTILIZATION OF INDUSTRIAL ASHES

G.A.O. Teekman, Aardelite Holding B.V., Meesterstraat 5,
3861 RE Nijkerk (The Netherlands)

1. INTRODUCTION

In today's society there is an increasing amount of waste materials and the availability of land for storage of waste diminishes rapidly. The two priorities in solving this problem are: reducing the amount of the waste and separation techniques to make part of the waste into useful products. There will always remain materials that have to be disposed off. Size reduction of these materials is the most obvious solution, which is most effectively done by incineration of the waste. However the combustion residues from this waste have two important negative properties, namely:

1. The ash is a very finely divided powder, which can easily create dust problems.
2. Toxic materials, like heavy metals which were present in the waste, are concentrated in the ash.

Since 1978 Aardelite Holding B.V. invented and developed a process to overcome above mentioned negative properties for pulverized coal fly ash.

2. THE AARDELITE PROCESS

In the middle of the 1970's the Government decided to increase the share of coal in power generation. This inherently increased the amount of fly ash. A number of people were challenged to find solutions to this problem. One such solution is the Aardelite process.

The Aardelite process is based on the puzzolanic properties of fly ash, which are utilized to form rock like materials. The Aardelite process consists of the next main unit operations:

1. Dosing of raw materials
2. Mixing
3. Pelletizing
4. Curing
5. Sizing in fractions

The process flow scheme is developed in such a way that no waste streams are generated. With the use of atmospheric steam,

energy demands are low and the hardening equipment relatively inexpensive.
The final product, a stone hard product, neutralizes the two negative properties. The heavy metals are both chemically and physically immobilized to a very high extend. Furthermore the final product is ideally suited to be used as an aggregate in e.g. concrete or asphalt (1). When doing so, two more advantages emerge, namely:

1. The heavy metals are immobilized to an even higher extend.
2. The saving of natural aggregate.

3. HISTORY

In the region of Rotterdam there are numerous industries concentrated on a relatively small area. After the first initial contact in the early 1980's by separate industries with Aardelite, it was decided to combine the efforts in the foundation INDAS (INDustrial AShes), in which several parties participate.
The goal of the foundation is to build a demonstration plant which should produce INDAS granulate on a semi commercial scale (20.000 tpa). First however, a number of problems had to be tackled in environmental and civil areas.

4. CURRENT STATUS

To overcome the above mentioned problems, a number of studies have been performed up till now, namely:

- Framework for the logistics of the ashes / granules / end products and the recycled end products (2).
- Study which compares all the impacts (environmental, financial, etc.) of INDAS granulate versus dumping of the ashes (3).

A stretch of asphalt road was prepared in Rotterdam with the INDAS granules early 1988. At regular intervals this road is inspected and up till now the results are very promising. Currently a study is performed on the technical and environmental implication when INDAS granulate is used in pavement stones. For this large scale test 50 ton of INDAS was prepared in the pilot plant of Aardelite in Nunspeet, from which Biemans Beton made pavement stones which will be used in Rotterdam.

The bricks from the different batches were tested for total composition. 10 stones per batch were subjected to a diffusiontest to simulate leaching during the life span of the stone, and 10 stones per batch were crushed to smaller 3 mm. and subjected to a column leaching test to simulate the leaching behaviour after the demolition phase.
From the results (4) of the chemical analysis two things were observed, namely:

1. By making a material balance for the heavy metals with the recipe as used, the analysis of the pellets is well within the calculated composition.
2. The analysis from the pellets shows no discontinuity. From which can be concluded that both the raw materials composition as the process control were stable throughout the production.

The results of the column leaching tests confirmed that a good degree of immobilization of heavy metals had taken place during the processing of INDAS aggregate and the subsequent pavement brick manufacturing. When this paper was written, it was not yet under which conditions the utilization of INDAS aggregate will be allowed, complying wiht future regulations. It will probably mean for the INDAS pavement stones that when used the authorities have to be notified and that they are applied in such a manner that they can be reclaimed after their life span. In the demolition phase they have to be kept separate from other demolition waste, but can be used again for new INDAS pavement stones.

5. FUTURE DEVELOPMENTS

At this moment a large scale test is planned with INDAS granulate in an asphalt road. The set-up for this test in view of civil and environmental aspects will be similar to the pavement stone testing.
Furthermore a study done by Aardelite, proved that the economical feasibility heavily depends on the size of this demonstration plant.
The INDAS members therefore agreed that a marketing study should be performed to investigate the market possibilities

and the conditions for acceptance regarding the use of INDAS granulate. This has to determine how large the demonstration plant has to be.
The materials which have been incoperated so far in this INDAS project are: ash from municipal waste incineration (both fly ash and fine fraction of bottom ash), ash from sewage sludge incineration, dewatered potable water sludge, fine fraction of concrete recycling plant and steel grit.
But the technology keeps evolving, enabling it to take other solid wastes in the future too.

6. CONCLUSIONS

Not withstanding the fact that waste should be prevented at the source, it will prove impossible to avoid all waste.
As a last resort this waste can be incinerated. But instead of dumping the incineration-ashes, it is of course much better to use these ashes as a resource. The INDAS foundation with the Aardelite technology is currently proving the environmental, civil and economical feasibility to turn these ashes into a resource.

REFERENCES

1. E. Mulder and A.S.M. Houtepen, "Artificial gravel as a gravel substitute in asphaltic concrete". Proceeding: Ninth International Coal Ash Utilization Symposium, Orlando, Florida, 22-25 January, 1991, ACAA, Washington D.C., 1991, vol 1, pp. 23-1 to 23-11.
2. F.J. ter Heide and T.H. Maas, "Advies over een integraal systeem en een organisatieplan logistiek voor het beheer van de INDAS-keten", Twijnstra Gudde, Deventer, 1990.
3. R. Schelwald, "Het milieu-rendement van INDAS-korrels", Ingenieursbureau Geotechniek en Milieu, Rotterdam, 1991.
4. M.J.M. Stam, "Milieu-hygiënisch onderzoek aan beton-straatklinkers met INDAS-korrels als grindvervanging", KEMA, Arnhem, 1990.

Waste Materials in Construction.
J.J.J.R. Goumans, H.A. van der Sloot and Th.G. Aalbers (Editors)

HYDROTHERMAL SYNTHESIS OF LIGHT-WEIGHT INSULATING MATERIAL USING FLY-ASH

B. BORST and P. KRIJGSMAN

Ceramic Design International Holding BV, P.O.Box 68, 8050 AB Hattem (The Netherlands)

SUMMARY

Fly-ash and lime are used as raw materials for the hydrothermal production of a light-weight insulating material. The process is carried out in an aqueous suspension heated to a temperature of 190-240 °C. Compared to a hydrothermally produced calcium-silicate material made from silica-fume and lime, the limiting temperature is much lower.

Fly-ash can also be used as a free additive for the material made from silica-fume. This reduces the thermal shrinkage of the material.

1. INTRODUCTION

Hydrothermal processes are very promising for making added value products out of fly-ash (1,2). Alumina and Silica are the interesting constituents of the fly-ash that can be used for the hydrothermal synthesis of ceramic materials. The reactivity is high due to its amorphous nature. Because hydrothermal processes are carried out at moderate temperatures, the negative influence of certain impurities is reduced.

An industrial process is known (3) and in progress for the hydrothermal production of Xonotlite ($Ca_6Si_6O_{17}(OH)_2$) made from lime and silica-fume. Other silica raw materials can be used instead as well (4). Silica-fume is a waste material from the production of ferro-silicon or silicon metal. The Xonotlite is used for the production of light-weight insulating material with a high limiting temperature (1100 °C). Adding some extra silica-fume after the hydrothermal reaction reduces the thermal shrinkage of the material (5).

The aim of this project is to investigate the possibilities of using fly-ash as a raw material for the production of light-weight insulating material or as a free additive to the Xonotlite-product.

2. EXPERIMENTAL

Lime together with fly-ash or silica-fume are mixed with water in an stirred autoclave. The lime-silica ratio on a molar basis was 1.04:1 in all cases. The reaction mixture is then heated to a temperature of 190-240 °C (depending on the reactivity of the raw materials) and the pressure corresponds to the saturated steam pressure. After completion of the reaction, the reaction product is transferred from the autoclave to a receiving vessel at a sub-stantially constant flow rate. Just prior to this transfer, the pressure in the receiving vessel is brought up to the pressure in the autoclave by passing gas from the autoclave to the receiving vessel. A small pressure difference is created between the two vessels by letting down some pressure from the receiving vessel and the reaction product flows through the heat-exchanger to this vessel (see Fig. 1). This process is developed to be able to operate a batchwise hydrothermal process on an industrial scale.

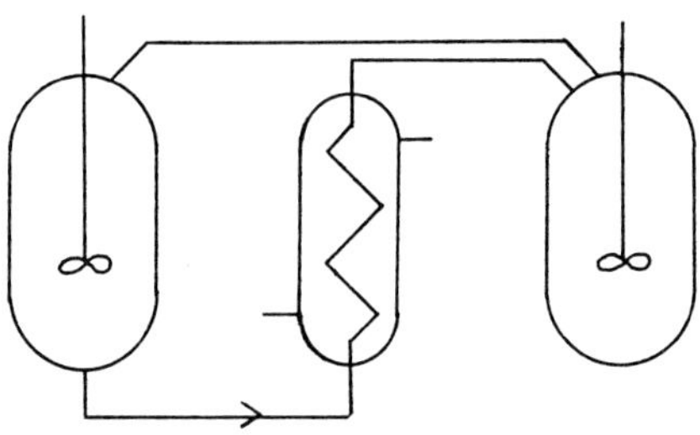

Fig. 1. Lay-out of the process showing reactor, heat-exchanger and receiving vessel.

Free additives like fly-ash or silica-fume are mixed with the thus obtained slurry. A fibrous material like wood pulp is added as well to give strenght to the finished product.

Slabs are pressed using a filterpress. Drying at 120 °C removed the residual water.

2. RESULTS

2.1 Crystal Structure

When silica-fume is used as raw material, the reaction product consists entirely out of the crystal Xonotlite. It is a mixture of Tobermorite and Hydrogrossular when fly-ash is used. The ratio between these two crystal structures depends on the ratio of Al and Si in the fly-ash.

Upon calcination at 1000 °C the material is transformed to Wollastonite when silica-fume is used and to a mixture of Wollastonite and Gehlenite when fly-ash is used.

2.2 Thermal Properties

Slabs are pressed from reaction mixtures made from fly-ash/lime and from silica-fume/lime. The properties of the two different materials can be compared. The results are summarized in tabel 1.

Tabel 1. Some properties of insulating material made from different silica raw materials.

Raw Material	Free Additive	Density [kg/m^3]	Temp. limit [°C]	Thermal shrinkage	L.O.I.
Fly-ash	10 wt% FA + 3 wt% WP	360	650	1.7 % at 650 °C	13.5 %
Silica-fume	10 wt% SF + 3 wt% WP	210	1100	1.0 % at 1000 °C	9.0 %

FA = Fly-ash, SF = Silica-fume, WP = Wood Pulp

One can see that the material made from fly-ash has a much lower temperature limit. This is because it is a much more hydrated material with a higher level of impurities compared to the Xonotlite material. The density of the fly-ash material is higher due to the lower porosity of the particles. This also effects the thermal conductivity as can be seen in figure 2.

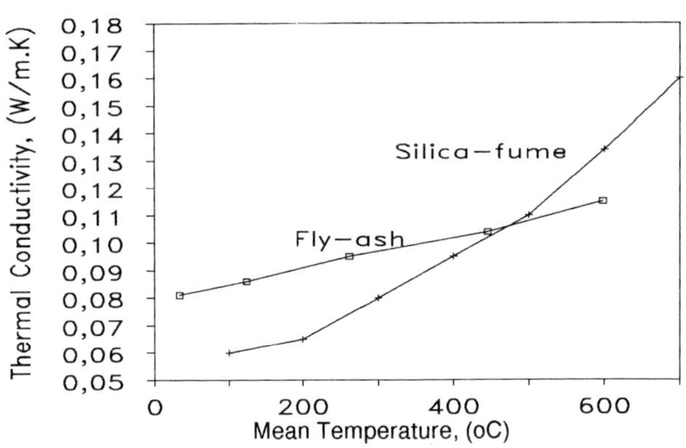

Figure 2. Thermal conductivity of the two materials of tabel 1.

We can conclude that fly-ash can be used as a raw material for the hydrothermal production of light-weight insulating material although it can only be used for lower temperature applications. The temperature limit depends on the ratio between Si and Al in the fly-ash and on the level of impurities.

The effect on some thermal properties is measured when fly-ash is used as a free additive to a Xonotlite product. The results are summarized in tabel 2 together with the results obtained with some other free additives.

Tabel 2. Thermal properties of slabs pressed from Xonotlite slurry together with different free additives.

Additive	Drying shrinkage [%]	Calcination shrink. (24 h at 1000 °C) [%]
--	2.7	3.0
10 wt% FA	2.9	1.6
10 wt% SF	1.4	1.0
3 wt% WP is added in all cases		

Addition of fly-ash effectively reduces the thermal shrinkage although not as good as the addition of silica-fume. Addition of fly-ash has no effect on the drying shrinkage whereas silica-fume is giving a reduction.

3. CONCLUSIONS

A light-weight insulating material with good insulating properties can be made using fly-ash as one of the raw materials in a hydrothermal process. The working temperature of the resulting material is limited to 650 °C.

Fly-ash can also be used as a free additive for a Xonotlite product. This reduces the thermal shrinkage of the insulating material made from this mixture.

REFERENCES

1 E.P. Stambaugh, Materials & Design, 10 (1989) 175-185

2 M. Hirato and Y. Ninomiya, SPEY 16 (Reports of special project research on energy under grant in aid of scientific research, Japan) (1987) 117-122

3 P. Krijgsman, US-Patent 4,753,787 and 4,366,121

4 T. Mitsuda, J. Saito and E. Hattori, in: S. Somiya (Ed.), Proceedings of the 1st International Symposium on Hydrothermal Reactions, Japan, 22-26 March 1982, pp. 823-838

5 P. Krijgsman, US-Patent 4,545,970

Waste Materials in Construction.
J.J.J.R. Goumans, H.A. van der Sloot and Th.G. Aalbers (Editors)

ANALYSIS OF WASTE BUILDING MATERIALS USAGE IN AGRICULTURAL CONSTRUCTION WORKS IN KUBAN REGION.

LISTOPAD IVAN ANDREEVICH
Intertrade Scientific Technical Association "Electronizatsia"
Russian Federative Republic, c. Krasnodar, U.S.S.R.

Kuban, that is the second name of Krasnodar region, is characterized by wide variety of natural and climatic conditions. There are mountains and plains, rivers and seas on its territory. Here are well developed communications. Climate is continential one. Average annual precipitation is 200 - 600 mm, average annual temperature is + 14°C.

More than 50 kinds of natural resourses have been found in the region. Among them are many construction materials such as cement, marl, pebbles, loamy soil, gravel, sand, limestone, chalk, gypsum, schist, graphite, etc.

Reed, growing in boggy places, is also used as a building material.

"Saman" (bricks, made of straw and clay, cement blocks, reed mats, waste of woodworking industry, rice industry waste, chemical industry waste is also used in agricultural construction.

Mentioned above factors help to develop building material industry to obtain varied nomenclature of articles and building constructions. The process is restrained by the use of exclusively traditional kinds of raw material.

The target of the given paper is to analyze the possibilities of rural building materials industry development in Kuban region on the example of one material.

Clay bricks are considered to be traditional agricultural building materials. Their low quality is due to three reasons: bad quality of raw material, usage of out-of-date equipment, low technological discipline. There is enough raw clay materials in the region for essential raise of industrial volume, meanwhile the quality of the one does not meet the requirements.

Some scientists think that reservoirs' silt can also be used for the production of bricks.

Silt is characterized with dispersionness, presence of a great number of highly dispersed organic substances and iron oxides. The plasticity of silt provides porousness of bricks, lower density and good coagulation under lower temperature, as iron oxide plays the role of mineralizer.

Many reservoirs of Kuban river are strongly flooded. That prevents free navigation, fishing, recreational possibilities and worsens ecological situation. So silt usage in agricultural construction is very effective.

What are the main Kuban reservoirs' silt characteristics?

40% constitute slim clay fractions with dimentions of less than 0,005 mm. Its qualities are close to typical kaolin clays, having high ceramic qualities. Silt is usually related to medium- and high-plastic clays .

Chemical composition of silt is close to clays with high content of iron oxide and organics, the fact presupposes its hightened plasticity, porosity, coagulation, hightened content of components stipulating higher plasticity and solidity after baking.

During the conduction of experiments (1) the following composition was used: 40-60% of clay, 30-40% of silt, 8-10% of sawdust, 0-3% of fire clay. The process of drying lasted for 24 hours in natural conditions, and then for 65 hours in tunnel drying rooms under the temperature of 20-25°C in initial stage and 50-54°C during unloading. Residual moisture constituted 6-9%. Kilning of adobes lasted for 50-70 hours under the temperature of 1000°C.

The quality of obtained bricks was: compactness - 1700 kg/m^2, pressing strength - 12 MPa, bending strength - 2,5 MPa, water consumption - 12-14%. Waste of bricks is conditioned by the high silt moisture, which is higher than 45% and therefore by high moulding moisture of ceramic mass.

Done work testifies perspectiveness of the given initial stuff for the production of bricks, meanwhile high plasticity and moulding moisture impede qualitative mixture of silt mass with purifying additions. Besides, due to its high moulding moisture, silt requires softer and longer drying regime.

Reference:

1. Tokarev P.Y., Kasakov A.A., Gryaznov V.K., Lewash M.V., Optimum contents of raw clay masses on the basis of silt for bricks production. - Krasnodar, 1989.

Waste Materials in Construction.
J.J.J.R. Goumans, H.A. van der Sloot and Th.G. Aalbers (Editors)

RE-USE OF WASTE MATERIALS IN CONSTRUCTIONAL WORKS; EXPERIENCES IN THE CITY OF ROTTERDAM, THE NETHERLANDS

W.G. DASSEN, W. PIERSMA, R. SCHELWALD and I.M.J. VRIES

Department of Public Works Rotterdam, P.O. Box 6633, 3002 AP Rotterdam (The Netherlands)

SUMMARY

The city of Rotterdam initiates research programs and is involved in several studies on the re-use or application of waste materials. Three specific studies on environmental aspects of re-use of waste materials in Rotterdam are presented shortly in this poster.
1. the filling in of harbours with minestone;
2. application of incineration slags in road foundations;
3. the environmental yield of re-use of industrial ashes as a substitute for gravel in concrete, asphalt and bricks.

1. INTRODUCTION

The city of Rotterdam is part of the highly industrialized and densily populated area of the Rijnmond in the West of the Netherlands. The city has some 600,000 inhabitants. Since many years the city of Rotterdam is striving after and stimulating the re-use of waste materials.

Several motives can be put forward:
1. in the city larges quantities of waste are produced;
2. due to the limited space available landfilling should be minimized;
3. re-use of waste materials reduce the need for new raw materials and natural resources;
4. due to the increasing tariffs for landfilling re-use of waste materials is preffered to landfilling.

Therefore the city of Rotterdam:
1. requires of contractors that waste material is suitable for re-use (e.g. demolition rubble, asphalt);
2. requires of contractors that waste-derived materials are applied where possible (e.g. application of rubble and fly-

ash from coal fired powerplants in concrete);
3. initiates research on the physical and chemical porperties of waste-derived materials (polluted soil, slags and fly-ash from waste incinerators, industrial ashes);
4. is involved in studies on the consequences of applications of waste-derived materials (mine-stone, polluted dredged material) for the environment.

In this way the city of Rotterdam sets an example to initiate contractors to use waste materials for construction.

This poster presents three specific studies on environmental aspects of re-use of waste-derived materials.

2. FILLING IN HARBOURS WITH MINESTONE

The city of Rotterdam is planning to fill in (parts of) harbours, which are no longer used for shipment activities. In the "Waal-Eemhaven"-area, where more space is needed to create container terminals, the filling in requires a maximum of 27 million tons of material. Instead of sand, waste material (minestone) from the coalmining industry in Germany may be applicable.

The department of Public Works Rotterdam has studied the effects of filling in harbours with minestone on the quality of water, sediment and groundwater. The potential (maximum) concentrations in water, sediment, porewater and groundwater were predicted, using the results of leaching tests and specific dispersionmodels.

Cascade tests were considered to be relevant to predict concentrations in water and sediment, because of the high Liquid/Solid-ratio (L/S=100) and the intensive contact of water with minestone. To study the effects on groundwater, a modified column test (L/S=10) was used to simulate a percolating system. The leaching fluid used in this test was saltish, in order to simulate the local environmental conditions in the harbour-area.

Reviewing the results of the leaching tests, SO_4, Ba, Sr, Ni, naftalene and benzo-a-pyrene were considered to be relevant to describe the possible effects on the environment. For these parameters the flux was estimated and used as input for the dispersion models.

The conclusion of the study is that the use of minestone for filling in harbours is environmentally safe, provided that minestone is applied under water in order to avoid the oxydation of pyrites.

At the moment, the department of Public Works Rotterdam is studying the possibility to use other waste materials (minestone from the Province of Limburg, dredged material and slightly contaminated soil) for filling in harbours.

3. THE ENVIRONMENTAL YIELD OF USING INDUSTRIAL ASHES AS A SUBSTITUTE FOR GRAVEL IN CONCRETE, ASPHALT AND BRICKS

In different waste-incineration processes, at first sight, not usable rest products are formed such as slags and fly-ashes. Land-filling of these products must be minimized.

For the production of concrete, aphalt and bricks, gravel and sand are obtained from former riverbanks in the Netherlands and neighbouring countries. In the near future the winning of these materials will be restricted as a result of nature conservation policy.

The department of Public Works Rotterdam participates in a research program on creating artificial gravel from industrial ashes. Industrial ashes are kit together by a low temperature process into grains. The physical structure and properties are sufficient for substituting gravel and sand in concrete, asphalt or bricks.

The environmental consequences are subject to research at this stage. From different points of view, such as energy, ecology and potential risk analyses, an integrated **"environmental yield"** is calculated. The chemical properties, such as leaching, are also subject to research.

Re-use of industrial ashes in grains as an additive in concrete, asphalt or bricks seems to have a positive environmental yield and provides an excellent solution to the above described problems.

4. APPLICATION OF INCINERATION SLAGS IN CONSTRUCTION PROJECTS

In the Rotterdam region 350.000 tons of incineration slags are produced each year. The policy of the city of Rotterdam is directed to use these slags in construction projects. The following projects are presented in this poster:

1. Slags in road foundation ("Vondelingenweg")

This concerns a pilot study to determine environmental consequences of application of incineration slags as road foundation under concrete block pavement. The study consisted of a field experiment and laboratory tests. During the field experiment the quality of groundwater and subsoil were monitored. The laboratory tests comprised of determining the physical and chemical properties (leaching) of the slags. For measuring the leaching several tests were applied.

From the field experiments it was calculated that in the long term only a limited contamination of the subsoil and groundwater may be expected.

2. Slags in underwater applications

The department of Public Works of Rotterdam has conducted laboratory and small scale pilot tests in cooperation with the Netherlands Energy Research Foundation (ECN).

The quality of leachate and percolate was monitored for several types of application: with and without impermeable liner and using a layer of clay. In this poster the results of the monitoring and leaching tests will be discussed.

Since the results indicate that the effect of application of slags on the environment may be limited a new research project is being developed.

This project concerns the filling of a former construction dock ("Van Brienenoord") and represents a large-scale application of incineration slags under groundwater level, e.g. for the filling in of harbours. The former construction dock will have a lining of clay to minimize the exchange of contaminants between percolate/porewater and surrounding groundwater.

Waste Materials in Construction.
J.J.J.R. Goumans, H.A. van der Sloot and Th.G. Aalbers (Editors)

THE APPLICATION OF METALLURGICAL SLAGS FOR THE BUILDING MATERIALS PRODUCTION IN POLAND.

J. MAŁOLEPSZY, J. DEJA and W. BRYLICKI
Academy of Mining and Metallurgy 30-059 Kraków, al. Mickiewicza 30, Poland

The total annual output of the granulated blast furnace slags in Poland is about 8 000 000 tons. The authors have been working for many years in the field of the special building materials production with the utilization of granulated blast furnace slags as a binder. In "classic" technologies the granulated blast furnace slag is used as a pozzolanic admixture ground with the OPC clinker. The authors propose another way of slag utilization consisting in the activation by alkaline compounds such as Na_2CO_3, NaOH, water glass added to the finely ground slag (specific surface 320 - 360 m^2/kg). The fol lowing materials can be thus obtained :

1 - High strength concretes with higher corrosion resistance.

Depending on the activator used, the concrete with different high compressive strength are produced. Activation by Na_2CO_3 gives the concretes of R_c in the range 20 - 40 MPa . The concretes produced with NaOH or water glass give the compressive strength in the range 50 - 80 MPa. The corrosion resistance of these materials is of particular interest (1). The durability of alkali activated blast furnace slag concretes stored for 1.5 year in the highly concentrated solution of chlorides is illustrated in table 1.

TABLE 1

Corrosion resistance of different concretes after 1.5 year storage

Compressive strength (MPa)	OPC		Copper slag + bfs + + NaOH		bfs + Na_2CO_3	
	Water	chloride solution	Water	chloride solution	Water	chloride solution
	53.2	26.0	41.1	43.7	42.2	41.5

2 - Alkali activated slag pastes for dilling operation.
In 1990 the alkali activated slag paste was used for tightening of some drill-holes and as the screening to filtration (2).

3 - Alkali activated slag plaster "Tynkol".
The new plaster "Tynkol" basing on the slag binder with mineral activator and organic resin has been produced since 1985 on a large scale. This plaster shows very high colour fastness, resistance to atmospheric corrosion and insular properties (3). The elevation about 5 000 000 m^2 has been covered with this plaster until now.

4 - Granulated blast furnace slags as concrete admixture.
The utilization of the ground, granulated bfs reduces the cement consumption in concrete about 10 - 15% thus lowering the concrete production costs (4).

REFERENCES

1 Deja J., Małolepszy J., Third CANMET/ACI International Conference, Vol. 2, 1547-1563, Trondheim, Norway, 1989.

2 Stryczek S., Brylicki W., Małolepszy J., Unpublished report, AGH, Kraków 1990.

3 Małolepszy J., Deja J., 2 th International Seminar "Durability of Concrete", 77 - 84, Gothenburg, Sweden, 1989.

4 Małolepszy J., Deja J., Unpublished report, AGH, Kraków, 1989.